T0324046

SEX DIFFERENCES IN PHYSIOLOGY

SEX DIFFERENCES IN PHYSIOLOGY

Edited by

GRETCHEN N. NEIGH

Department of Anatomy and Neurobiology, Virginia Commonwealth University, Richmond, VA, United States

MEGAN M. MITZELFELT

American Physiological Society, Bethesda, MD, United States

AMSTERDAM • BOSTON • HEIDELBERG • LONDON
NEW YORK • OXFORD • PARIS • SAN DIEGO
SAN FRANCISCO • SINGAPORE • SYDNEY • TOKYO
Academic Press is an imprint of Elsevier

Academic Press is an imprint of Elsevier
125 London Wall, London EC2Y 5AS, United Kingdom
525 B Street, Suite 1650, San Diego, CA 92101, United States
50 Hampshire Street, 5th Floor, Cambridge, MA 02139, United States

Notices

Knowledge and best practice in this field are constantly changing. As new research and experience broaden our understanding, changes in research methods, professional practices, or medical treatment may become necessary.

Practitioners and researchers must always rely on their own experience and knowledge in evaluating and using any information, methods, compounds, or experiments described herein. In using such information or methods they should be mindful of their own safety and the safety of others, including parties for whom they have a professional responsibility.

To the fullest extent of the law, neither the Publisher nor the authors, contributors, or editors, assume any liability for any injury and/or damage to persons or property as a matter of products liability, negligence or otherwise, or from any use or operation of any methods, products, instructions, or ideas contained in the material herein.

British Library Cataloguing-in-Publication Data
A catalogue record for this book is available from the British Library.

Library of Congress Cataloging-in-Publication Data
A catalog record for this book is available from the Library of Congress.

ISBN: 978-0-12-802388-4

Publisher: Mica Haley
Acquisition Editor: Stacy Masucci
Editorial Project Manager: Samuel Young
Production Project Manager: Kirsty Halterman and Karen East
Designer: Matthew Limbert

Typeset by MPS Limited, Chennai, India

Contents

8. Sex Differences in Gastrointestinal Physiology and Diseases: From Endogenous Sex Hormones to Environmental Endocrine Disruptor Agents

ERIC HOUDEAU

9. Sex and Gender Differences in Body Composition, Lipid Metabolism, and Glucose Regulation

KELLY ETHUN

10. Sex Hormone Influenced Differences in Skeletal Muscle Responses to Aging and Exercise

MARYBETH BROWN AND PETER TIIDUS

11. Strategies and Approaches for Studying Sex Differences in Physiology

MARGARET M. McCARTHY

11.1. How to Study Sex Differences Caused by "Sex Chromosome Effects" on Tissues

ARTHUR P. ARNOLD

11.2. Organizational Influences of the Gonadal Steroid Hormones: Lessons Learned Through the Hypothalamic-Pituitary-Adrenal (HPA) Axis

VICTOR VIAU AND LEYLA INNALA

11.3. Strategies and Approaches for the Study of Activational Influences of Gonadal Hormones on Sex Differences in Physiology and Behavior

JILL B. BECKER

11.4. Human Methodologies in the Study of Sex Differences

EMILY J. BARTLEY AND MARGARETE RIBEIRO-DASILVA

List of Contributors

Allan K. Alencar Department of Anesthesiology, Wake Forest School of Medicine, Medical Center Boulevard, Winston-Salem, NC, United States

Arthur P. Arnold Department of Integrative Biology & Physiology, UCLA, Los Angeles, CA, United States

Massimo Bardi Department of Psychology, Randolph-Macon College, Ashland, VA, United States

Emily J. Bartley University of Florida, College of Dentistry, Pain Research and Intervention Center of Excellence, Gainesville, FL, United States

Jill B. Becker Patricia Y. Gurin Collegiate Professor of Psychology, Molecular & Behavioral Neuroscience Institute, University of Michigan, Ann Arbor, MI, United States

Marybeth Brown Department of Physical Therapy, School of Health Professions, University of Missouri, Columbia, MO, United States

Iwona A. Buskiewicz Department of Pathology, University of Vermont, Burlington, VT, United States

Carolyn M. Ecelbarger Department of Medicine, Center for the Study of Sex Differences in Health, Aging, and Disease, Georgetown University, Washington, DC, United States

Kelly Ethun Department of Pathology and Laboratory Medicine, Emory University School of Medicine, Atlanta, GA, United States

DeLisa Fairweather Department of Cardiovascular Diseases, Mayo Clinic, Jacksonville, FL, United States

Leanne Groban Department of Anesthesiology, Wake Forest School of Medicine, Medical Center Boulevard, Winston-Salem, NC, United States

Eric Houdeau Intestinal Development, Xenobiotics & ImmunoToxicology, Research Centre in Food Toxicology (INRA Toxalim UMR 1331), Toulouse, France

Sally A. Huber Department of Pathology, University of Vermont, Burlington, VT, United States

Leyla Innala Laboratory of Neuroendocrine Function, Department of Cellular and Physiological Sciences, University of British Columbia, Vancouver, BC, Canada

Craig Kinsley Department of Psychology, University of Richmond, Richmond, VA, United States

Kelly Lambert Department of Psychology, Randolph-Macon College, Ashland, VA, United States

Sarah H. Lindsey Department of Pharmacology, School of Medicine, Tulane University, New Orleans, LA, United States

Margaret M. McCarthy Department of Pharmacology, University of Maryland School of Medicine, Baltimore, MD, United States

Liana Merrill Department of Physiology, Emory University College of Medicine, Atlanta, GA, United States

Virginia M. Miller Departments of Surgery, Physiology and Biomedical Engineering, Mayo Clinic, Rochester, MN, United States

Gretchen N. Neigh Department of Anatomy and Neurobiology, Virginia Commonwealth University, Richmond, VA, United States

Y.S. Prakash Department of Physiology and Biomedical Engineering, Mayo Clinic, Rochester, MN, United States; Department of Anesthesiology, Mayo Clinic, Rochester, MN, United States

Margarete Ribeiro-Dasilva University of Florida, College of Dentistry, Pain Research and Intervention Center of Excellence, Gainesville, FL, United States

Venkatachalem Sathish Department of Physiology and Biomedical Engineering, Mayo Clinic, Rochester, MN, United States; Department of Anesthesiology, Mayo Clinic, Rochester, MN, United States

Peter Tiidus Department of Kinesiology, Brock University, St. Catharines, ON, Canada

Victor Viau Laboratory of Neuroendocrine Function, Department of Cellular and Physiological Sciences, University of British Columbia, Vancouver, BC, Canada

Hao Wang Department of Anesthesiology, Wake Forest School of Medicine, Medical Center Boulevard, Winston-Salem, NC, United States

CHAPTER

1

Introduction for Sex Differences in Physiology

Virginia M. Miller

Departments of Surgery, Physiology and Biomedical Engineering, Mayo Clinic, Rochester, MN, United States

The study of the human body dates back to ancient times but was not named as the discipline "physiology" until the 16th century by the French physician Jean François Fernel, who introduced the term to describe the study of bodily functions. Since that time, physiologists have contributed fundamental and critical information needed for the evidence-based practice of modern medicine. However, like all scientific disciplines, physiology and physiologists are not immune from political, societal, and cultural trends. In part, because science was historically a male-dominated profession, except for studies related to the physiology of reproduction, most human and animal physiological studies enrolled male volunteers and utilized male animals. Other considerations impacting a male bias in research included concerns about variability in measured parameters resulting from cyclic hormonal variation in females and potential risk for teratogenic effects of interventions and procedures to the fetus in women of child-bearing age. Although the human population can be defined by sex as either male or female, assigned by chromosomal complement and reproductive organs (XX for female and XY for male) [1], sex as a biological variable is rarely considered in the design of basic physiological studies. Thus, physiological principles contained in classical physiological and medical textbooks and graduate and medical curricula have been based on the 70 kg healthy male (usually between 18 and 40 years of age) or on male animals [1].

In 2001, the Institute of Medicine report "Exploring the Biological Contribution of Sex" concluded that sex matters in all aspects of cellular function and physiology from "womb to tomb" [1]. What logically follows, then, is that physiological principles and regulatory mechanisms need to be defined in males and females (animals and humans), so that findings from basic science can be translated to clinical research for the development of evidence-based, individualized medical strategies or practice guidelines.

In the United States, a legislative approach was taken to correct the scientific problem of too few women in clinical trials by the passage of the National Institutes of Health (NIH) Revitalization Act of 1993. This law mandated that women be included in human studies supported by the NIH. Although women have since been included in clinical studies, results of those studies in the clinical setting have rarely reported data separated by sex, thus making it difficult, if not impossible, to understand where the two sexes fell within the distribution of results. In the NIH Revitalization Act, there was no mention of basic human physiological functions or mechanistic studies utilizing isolated cells or tissues. Although the Office of Research on Women's Health of the NIH was founded in 1991, it was not until 2002 that the Office developed and implanted an interdisciplinary targeted funding mechanism (Specialized Centers of Research on Sex Differences) specifically to begin to fill the knowledge gap in information for areas of women's health and sex differences research. Since the inception of the program, 33 awards have been made to 26 academic centers (see http://orwh.od.nih.gov/sexinscience/researchtrainingresources/scor.asp). Advocacy groups such as the Society of Women's Health Research and scientists themselves through original research articles, editorials, and editorial policies of professional societies (eg, Organization for the Study of Sex Differences, that was founded in 2006, the American Physiological Society, and the Endocrine Society) began to draw attention to the lack of experiments and reporting of data on females in basic and translational animal and human studies [2–7] (see http://genderedinnovations.stanford.edu for an up-to-date list of editorial policies for other journals). As a result of these efforts, the

Sex Differences in Physiology
DOI: http://dx.doi.org/10.1016/B978-0-12-802388-4.00001-X

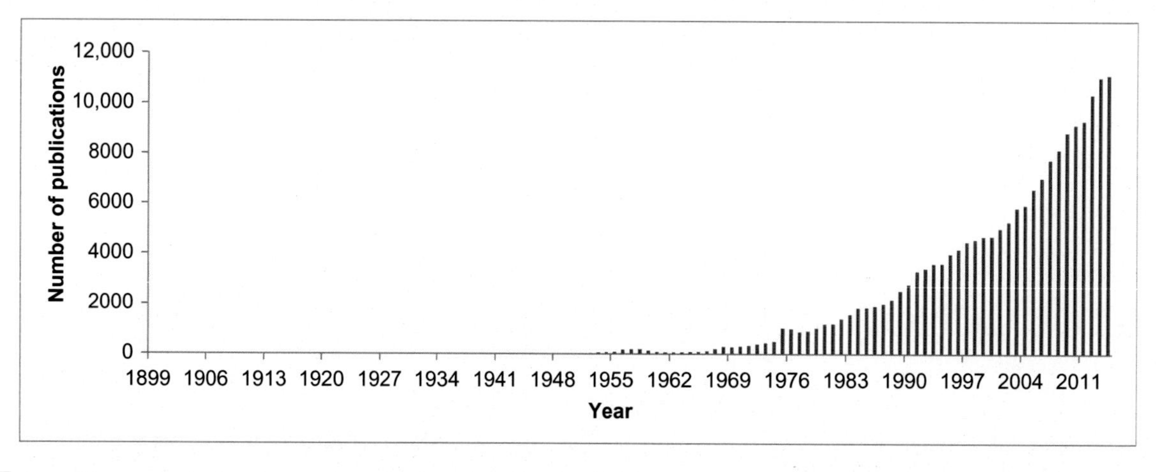

FIGURE 1.1 The graph depicts the number of publications listed in PubMed (www.pubmed.gov), a NIH resource for biomedical literature, that included reference to "sex differences" for each year from 1899–2014.

number of publications addressing biological sex differences has risen dramatically (Fig. 1.1). In addition, problems with reproducibility of basic science experiments, including lack of reporting of the sex of the experimental animals or human participants prompted announcements in 2014 by the NIH that steps would be taken to address these deficiencies [8,9]. These steps were announced in Jun. 2015 (NOT-OD-15-102: Consideration of Sex as a Biological Variable in NIH Funded Research and NOT-OD-15-103: Enhancing Reproducibility through Rigor and Transparency) to be implemented in grant applications funded by the NIH starting in 2016.

So, where do we stand with data upon which to construct a book on "Sex Differences in Physiology"? Epidemiological studies have consistently identified differences in disease incidence, prevalence, morbidity, and mortality between men and women. These statistics reflect differences in the underlying physiological processes that arise from the basic genetic difference between males/females and men/women, coupled with modulation of these processes by hormonal fluctuations and aging. Chromosomal and hormonal differences allow for reproductive competence which brings us full circle: sex differences in reproduction (specific to one sex) drive physiological processes that express as sex differences throughout the lifespan. What is meant by this? Consider the physiological processes that allow successful pregnancy: changes in respiration, metabolism, blood volume, renal function, cardiac output, musculoskeletal locomotion, neurological regulatory processes, and sensory and immune function. These changes are female-specific and suggest differences in underlying mechanisms of adaptability in female physiological processes that may (or may not) be present in males. Although males do not undergo the tremendous physiological changes associated with pregnancy and birth, this does not definitely indicate that there are no hormonally mediated shifts in physiological processes within males. Conversely, if there is an absence of adaptability of physiology within males, could this absence of adaptation render males susceptible to diseases or disorders? Further, although regulatory processes maintain homeostasis in both males and females, for example, blood pressure or glucose or electrolytes, within defined ranges, differing underlying processes may facilitate these regulatory processes and these variations could vary across the lifespan. These processes need to be understood in order to inform adequate diagnosis and treatment of disease and disorders in both males and females.

It is difficult to gather sufficient information about physiological processes in females, as the data may be published either as single sex studies, or as comparative studies labeled as "sex-differences" or "gender differences." Studies utilizing cell cultures rarely are comparative or are based on cell lines of unidentified or single sex. Although gender is related to sex, gender defines behavioral, psychological, and cultural characteristics that are influenced by sociocultural expectations [10]. Curricula in graduate and medical courses do not specifically address differences in physiology between males and females because data and resources are sparse. Both sex and behaviors influenced by gender will affect physiology and pathophysiology. Development of resources and consideration of physiological principles are often segregated into women's specific or men's specific knowledge repositories. Since the first textbooks on *Principles of Sex-Based Differences in Physiology* [11] and *Principles of Gender-Specific Medicine* [12] were published in 2004, updates and additional collective resources are few [13–15]. Thus, *Sex Differences in Physiology* provides an important update and focus on basic physiological control systems and mechanisms in females and males that contribute to health and disease across the lifespan. The approach is systematic considering the first developmental aspects of sexual differentiation including neuroanatomical and neurophysiological aspects of brain function. Each physiological system is then considered separately, including highlighting body composition and metabolism with influenced risk factors for pathophysiology. In the final section, experts in sex-differences research provide guidelines for strategies to study sex differences.

In order to improve the health of women and men, it is essential for scientists and clinicians to consider sex differences as one of the underlying physiological mechanisms of disease. These chapters will lead the way to new discoveries about basic female physiology across the lifespan and about differences in physiology between females and males, thus providing building blocks for evidence-based, individualized medicine.

Acknowledgments

Dr Miller's research program is funded by National Institute of Health AG 44170, HD 65987, HL 83947, HL 90639. Dr Miller is a past president of the Organization for the Study of Sex Differences.

References

[1] Wizemann TM, Pardue ML. Board on Health Sciences Policy. In: Wizemann TM, Pardue M-L, editors. Exploring the biological contributions to human health: does sex matter? Washington, DC: Institute of Medicine; 2001.

[2] Zucker I, Beery AK. Males still dominate animal studies. Nature 2010;465:690.

[3] Taylor KE, Vallejo-Giraldo C, Schaible NS, Zakeri R, Miller VM. Reporting of sex as a variable in cardiovascular studies using cultured cells. Biol Sex Differ 2011;2:11.

[4] Shah K, McCormack CE, Bradbury NA. Do you know the sex of your cells? Am J Physiol Cell Physiol 2014;306:C3–18.

[5] Wadman M. NIH mulls rules for validating key results. Nature 2013;500:14–16.

[6] Miller VM. In pursuit of scientific excellence: sex matters. Am J Physiol Heart Circ Physiol 2012;302:H1771–2.

[7] Blaustein JD. Animals have a sex, and so should titles and methods sections of articles in endocrinology. Endocrinology 2012;153:2539–40.

[8] Clayton JA, Collins FS. Policy: NIH to balance sex in cell and animal studies. Nature 2014;509:282–3.

[9] Collins FS, Tabak LA. Policy: NIH plans to enhance reproducibility. Nature 2014;505:612–13.

[10] Mahalik JR, Locke BD, Ludlow LH, et al. Development of the conformity to masculine norms inventory. Psychol Men Masc 2003;3–25.

[11] Miller VM, Hay M. In: Bittar EE, editor. Principles of sex-based differences in physiology. The Netherlands: Elsevier; 2004.

[12] Legato MJ, editor. Principles of Gender-Specific Medicine. London: Elsevier Academic Press; 2004.

[13] Legato MJ. Principles of gender-specific medicine. 2nd ed. London: Elsevier Academic Press; 2009.

[14] Oertelt-Prigione S, Regitz-Zagrosek V, editors. Sex and gender aspects in clinical medicine. London: Springer-Verlag; 2012.

[15] Spangenburg E, editor. Integrative biology of women's health. New York: Springer; 2013.

CHAPTER

2

Chromosomal and Endocrinological Origins of Sex

Craig Kinsley[1], Massimo Bardi[2], Gretchen N. Neigh[3] and Kelly Lambert[2]

[1]Department of Psychology, University of Richmond, Richmond, VA, United States [2]Department of Psychology, Randolph-Macon College, Ashland, VA, United States [3]Department of Anatomy and Neurobiology, Virginia Commonwealth University, Richmond, VA, United States

INTRODUCTION

As described in chapter "Introduction," this book provides a survey of each of the organ systems of the body that is influenced by the biological variable of sex. An obvious omission is the reproductive system, which we exclude because of the well-developed literature on sex differences in the male and female reproductive systems. Before we progress through each of the systems of the body, we first provide a review in this chapter that covers the process of sexual differentiation and the forces that can influence this process. We focus on mammals throughout this book, although the process of sexual differentiation occurs in other classes and under other strictures, and still involves powerful and long-lasting/permanent effects of steroid hormones. We will describe the genetic and endocrinological origins of sex and sexual differentiation throughout different stages of male and female development. For instance, the amount of steroid hormones available pre- versus postpubertal, is significant [1], which, in turn, reshapes the activity of brain circuits during adolescent development and affects numerous other physiological functions. Furthermore, the sensitivity of the system, as reflected by the presence of receptors for the aforementioned hormones, likewise changes. The net result is a system that is responsive to the specific endocrine milieu characteristic of the stage of development, as well as the specific male and female environments [2].

CHROMOSOMAL SEX

The basics of sex determination and sexual differentiation are well understood [3] (see Fig. 2.1). In the human, males and females produce gametes, ova in females and spermatozoa in males, in which the number of chromosomes, compared to the other cells of the body, are halved. Thus, instead of the 23 pairs (46 total), the production of gametes involves a process (meiosis) that results in one member of the 23 pairs of chromosomes, compared to the diploid parent cells (which contain 23 pairs), plus a sex chromosome. The latter in the male's spermatozoa contains a particular genetic trait, the presence of a Y-chromosome or an X-chromosome. When added to the female's egg-bearing X-chromosome, the resulting genotype will be XX and female. If paired with a Y-bearing sperm, the resulting ovum will be XY or male. Thus, in the mammal, for example, the father determines sex by providing an X or a Y chromosome, but in what manner does the XX versus the XY chromosomal pairing initiate its effects? Nature produces in the XX or XY model the potential to grow and elaborate a single underlying substrate into the male or female phenotype. The Y-chromosome contains the *SRY* gene, which encodes the proteins that will facilitate the development of the male phenotype. Regardless of sex/genotype, the embryo has an internal undifferentiated gonad that will develop—or not—depending on the hormonal milieu to which it is exposed. The version that will be present

Sex Differences in Physiology
DOI: http://dx.doi.org/10.1016/B978-0-12-802388-4.00002-1

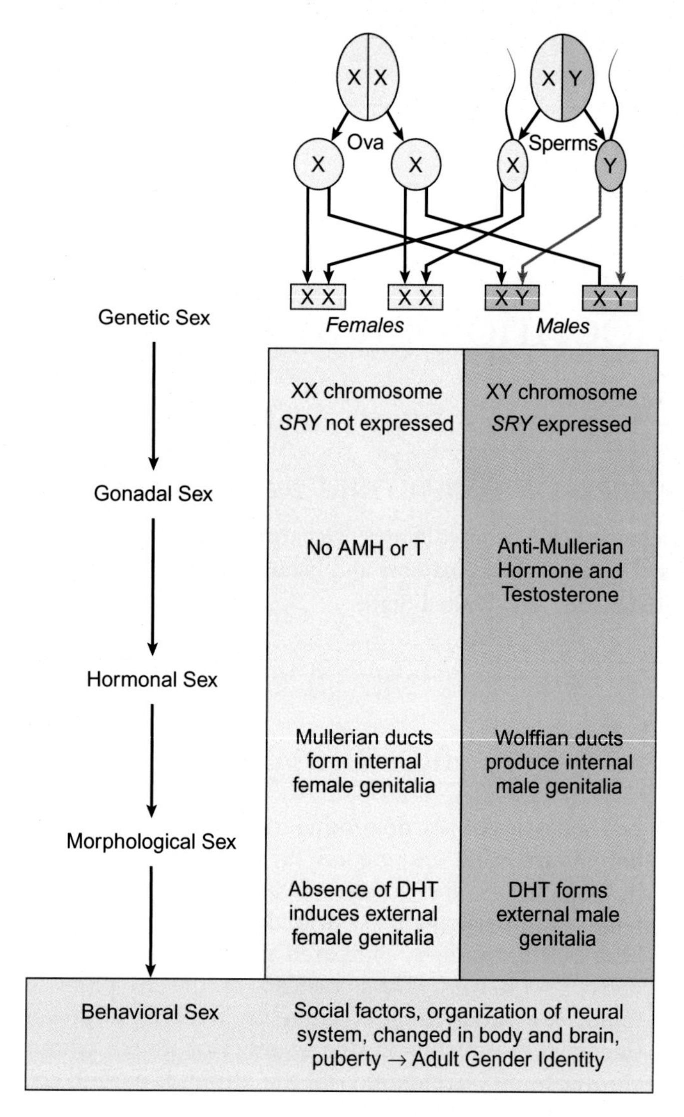

FIGURE 2.1 **Process of sexual differentiation in the human.** This figure displays and summarizes the process of sexual differentiation in the human. As can be seen, the process represents the interaction among a large number of phenomena and events, any point along which variation can occur subtly shifting the manner of development and its outcomes.

in the average female is the Müllerian system; in the male, the Wolffian. The tendency then is to produce the female version, which develops in the absence of any alternative signal. The Y-chromosome codes for the production of Müllerian-inhibiting factor (MIF), which suppresses the development of the Mullerian system and expresses the development of the Wolffian system, and the male-like structures that follow. The bipotential gonad and its associated structures follow the sex-typical program and either develop or wither away.

In the human, during the first 6 weeks of embryonic development, the gonadal ridge, germ cells, internal ducts, and external genitalia are formed—the basic anatomy of the reproductive tract. Unlike most other developing organs in the embryo that have a specific developmental trajectory, the gonads are bipotential in both genetic configurations (46-XX and 46-XY). At this stage, even after the initial genetic sex determination, the embryonic structures determining the individual's sex can develop either into ovaries or testes (in relation to the tissue of the gonadal ridge), oocytes or spermatocytes (regarding the germ cells), the male's or female's internal organs (the trajectory of the internal ducts), and average external masculine or feminine genitalia. The bipotential gonadal ridge is located medially on the urogenital ridge. Germ cells migrate to the gonadal ridge at approximately 5 weeks of development. Finally, these cells undergo rapid mitotic divisions in both the fetal testis and ovary [4,5].

In mammals, the primordial gonad, for both male and female duct systems, is expressed in the embryonic excretory organ known as the mesonephros. The Wolffian duct system will form male-typical structures such as the epididymis, vas deferens, and seminal vesicles; the Müllerian duct system will form the fallopian tubes, uterus, and posterior part of the vagina. Experimental work in animal models has shown that differentiation and subsequent gonad development is dependent on the inductive interaction between the Wolffian ducts and the intermediate mesoderm [6]. At 5 weeks of embryonic development, tissue destined to form the external genitalia is detectable at the cranial region of the cloacal folds, and it is still bipotential, basically identical in both male and female embryos [7].

GONADAL/HORMONAL SEX

On average, in the human, the bipotential gonads and germ cells begin to form either testis or the ovary around 6 weeks of embryonic development—or around 16% into the embryonic development. In general, in mammals, gonadal sex determination is regulated by a DNA-binding protein expressed in the Y chromosome by the gene known as *SRY* (sex-determining region on the Y chromosome), followed by its downstream mediators, including Sf1 (encoding for the steroidogenic factor 1) and SOX9 (encoding for the transcription factor SOX-9), which, interestingly, induces and maintains neural stem cells—further strengthening the ties between the brain and the reproductive system [8]. The evidence suggests that at this early stage the supporting cell precursors can develop into either Sertoli cells, which promote the development of the testis and their powerful chemical products, or ovary follicle cells [9]. Activation of the encoded proteins cited above can cause cells of the bipotential tissue to develop in the core regions at the expense of the outer layer. This, then, initiates a process that will lead to the formation of Sertoli cells, which in turn, organize themselves into the seminiferous tubules. Sertoli cells then secrete Müllerian-inhibiting chemicals, including the anti-Müllerian hormone, which causes regression of the Müllerian or female-typical morphological system. The next step is marked by the appearance of Leydig cells, which are capable of producing and secreting testosterone, the powerful, largely male steroid hormone. The final step in male sex determination is constituted by the development of male germ cells, which are influenced by retinoid signals within the mesonephros [10].

In contrast to the continuous proliferation of male germ cells, female germ cell development occurs only during embryogenesis, very early in the female's life. This process is characterized by the proliferation and morphogenesis of the granulosa cells into their cuboidal state, which induces oocytes to increase in size, the production of the zona pellucida, an extracellular glycoprotein matrix deposit between the oocytes and the granulosa cells, and following production of thecal cells [11]. In other words, the embryonic female produces the eggs she will carry with her for the rest of her life.

After differentiation of the gonads, sex steroid hormones are responsible for the next crucial stage of sexual determination: organizational effects of sex steroids. It should be noted that sex hormones must bind to the proper receptors on or within the cell in order to be physiologically active. Many of these hormone receptors are nuclear receptors that function as transcription factors enabling them to exert widespread effects on cell function. Thus, any environmental or cellular phenomenon interfering with the correct binding can prevent fully developed sexual differentiation. For example, as discussed later in the chapter, certain environmental agents, fertilizers, for example, can wreak havoc with the above cellular events. The hormone–receptor complex activates a specific response of the promoter of steroid-responsive genes and interacts with RNA polymerase II to form a large transcriptional activation complex, which in turn is responsible for the appropriate protein synthesis. If any one of the steps involved in the androgen transcription/translation is defective, the result is lack of masculinization of internal sex ducts and external genitalia [11]. It is like disconnecting some links in a chain: what worked before no longer does.

Müllerian-inhibiting substances (MISs) are important for the regression of the Müllerian duct system, which contributes to the successful proliferation of the male, or Wolffian duct system, and its associated testicular development. MISs are also detected in females, but only later on, after the Müllerian ducts have already begun their development.

At this stage of development (8–13 weeks, approx. 22–36% through prenatal development), testosterone is produced by fetal Leydig cells. This hormone is crucial for the development of both the Wolffian duct system and the masculinization of the external genitalia. The biochemical precursor of testosterone is cholesterol, which is produced using a biochemical pathway involving four recognized enzymes. Moreover, cells of the external masculine genitalia contain a 5α-steroid reductase which potentiates masculinization by transforming testosterone to dihydrotestosterone, the most biologically active androgen [12].

In females, the path to feminization is led by a specific cytochrome P450, aromatase, which converts androgens to estrogens [13] and is detectable in fetal tissue. Once again, feminization constitutes the default state, and masculinization can only be started as an active process involving the pathway detailed above.

Hormonal effects on phenotypical sex include discrete anatomical features, internal ducts, and external genitalia. External genitalia in males typically begins forming early in gestation. If androgens are not present until week 12 or later, full masculinization cannot take place. Whereas testosterone is of critical importance for the development of internal ducts, dihydrotestosterone is crucial in the development of external genitalia [13]. Masculinization in this stage includes increasing anogenital distance, fusion of urethral folds, and growth of scrotal swelling. The penis forms from the genital tubercle and continues to grow throughout gestation. In the absence of androgens, the labia majora and labia minora form from the genital swelling and urethral folds. The clitoris, therefore, forms from the genital tubercle.

Other phenotypical sex differences include qualitative differences (males are generally larger and heavier), different trajectory for puberty (females generally reach puberty sooner than males), and the development of secondary sex characteristics (body and facial hair, change in voice, and so forth) [14]. These differences also affect behavioral characteristics and qualitative as well as quantitative differences in behavior between the sexes. The classic dogma concerning phenotypical activation of sex differences links gonadal hormones as the basic factor controlling the sex differentiation of nongonadal tissue, including the brain. Classic studies showed the importance of the influence of sex steroids on brain sexual differentiation [15]. In these studies the removal of the testes early in neonatal life resulted in feminization of brain-regulated functions and behavior in adulthood; whereas, administration of exogenous testosterone to the neonatal female induced masculinization. Several more recent reports, however, have indicated that sexual differentiation of the embryonic neural tissue occurs before the activation of the gonadal hormones. These studies suggest that early events in sex differentiation, such as cell migration and the activation of sex germ cells, are also dependent on the activation of several genes linked to the Y chromosome [16]. Furthermore, appreciation has grown for direct effects of chromosomal sex on both physiology and behavior, in some cases, separate from the contribution of hormones (see Box 2.1).

Many aspects of the mechanisms of actions of the hormonal sex differentiation of the brain have been demonstrated in the last two decades [14]. In the brain, the involvement of a large variety of intracellular pathways mediated by steroid actions can explain sex behavioral differences. The involvement of neurotransmitters in sexual differentiation of the brain and behavior is now well understood. In most cases, the neurotransmitters act as mediators of steroid action, initiating biological negative feedback loops and acting as modulators of steroid activity, like a volume or gain switch on one's stereo. These are, needless to say, complex, subtle, and multilayered. The influence of sex on brain development is discussed in detail in the chapter "Sex Differences in Neuroanatomy and Neurophysiology: Implications for Brain Function, Behavior, and Neuropsychiatric Disease."

PUBERTY

The influence of sex steroids on development, physiology, and behavior is relatively quiescent from approximately 6 months after birth until the onset of puberty. Puberty specifically consists of the hormonal changes that lead to the sexual maturation of an organism. The increases in sex steroids during puberty also exert activational effects on other organ systems and many of these are discussed in the subsequent chapters of this book. Before we proceed with a discussion of puberty and the governing biological signals, let us first distinguish between puberty and adolescence. Puberty refers specifically to sexual maturation and the related hormonally driven events. Adolescence is a longer period of time and consists of both biological changes and sociocultural influences. We focus here on puberty, but for a discussion of adolescence and related neuronal and behavioral changes, see the work of Blakemore and Robbins [17]. Regarding puberty, this is not a single event but rather a process that occurs over a normal

BOX 2.1

SEPARATING CHROMOSOMAL AND ENDOCRINOLOGICAL SEX: FOUR-CORE GENOTYPES MODEL

Researchers have long sought to understand the physiological effects of sex chromosomes independently from sex hormones and have traditionally manipulated these two systems through gonad removal. This technique results in an immediate cessation of gonadally derived sex hormone production and a precipitous drop in circulating sex steroids. It also allows for the addition of exogenous hormones to determine the impact of sex chromosomes (without hormone replacement) and of varying concentrations of sex hormones (with hormone replacement) in both males and females. Depending on the timing of gonadectomy, partial isolation of organizational and activational effects of hormones is possible. Sex steroids have dramatic organizational influences on the developing fetus, including masculinization or feminization of the genitalia and brain. Recent studies also indicate that sex steroids exert some permanent organizational effects across a broad developmental window from perinatal to the end of puberty. These organizational influences are permanent, remaining into adulthood, and are not affected by gonad removal.

A relatively new animal model, called the Four-Core Genotypes Model, addresses these drawbacks by removing the *Sry* gene from the Y chromosome (Y^-) in mice. Given that the *Sry* gene is necessary for the development of the male phenotype, XY^- are characteristically female. Additionally, the model incorporates mice expressing the *Sry* transgene on an autosome to produce genetically female mice (XX) that express *Sry* and are thus phenotypically male, as well as to produce XY^- mice expressing *Sry* independently of the Y chromosome. This model and others that are targeted at isolation of the origin of observed sex differences are described in greater detail in the chapter "Strategies and Approaches for Studying Sex Differences in Physiology."

range of ages in the human. For girls, puberty is generally initiated between 8 and 12 years of age. For boys, the changes, on average, occur later with a window of 10–14 years of age. When considering common animal models such as rodents, puberty is initiated in the range of 32–38 days postnatally [18].

Major Hormonal Events of Puberty

Gonadotropin-releasing hormone (GnRH) or luteinizing hormone–releasing hormone is well-established as the essential trigger of puberty. GnRH is present during fetal and early postnatal development but then becomes quiescent from about 6 months of age until puberty approaches. As puberty approaches, pulses of GnRH increase in frequency and eventually surges in the hormone trigger puberty. The release of GnRH leads to the release of luteinizing hormone (LH) and follicle-stimulating hormone (FSH). These events are similar in both males and females but the subsequent events diverge between the sexes.

The main site of action for GnRH-induced release of FSH and LH in females is the ovary. Stimulation of the ovaries by LH and FSH leads to secretion of estrogen and progesterone and follicular development. Estrogen mediates the appearance of secondary sex characteristics and the maturation of the genital organs. Following the hormone events of female puberty, the pattern of hormonal secretion transitions to the adult ovarian cycle (Fig. 2.2). In the adult female, there is an alternating cyclic pattern of GnRH release which shifts between tonic and cyclic release, an effect which influences the estrous cycle. The pattern of the cyclic release is essential to appropriate hormone coordination to maintain reproductive viability. Hypothalamic release of GnRH stimulates the anterior pituitary to produce LH and FSH which stimulate estradiol (E2) and progesterone (P4) production in the ovaries. E2 rises during the follicular phase and P4 rises during the luteal phase of the menstrual cycle. This cyclicity in the female will continue until menopause (discussed below).

Puberty is also stimulated by GnRH in the male, but the subsequent events diverge from the female. GnRH-stimulated release of LH leads to the production of testosterone in the Leydig cells and thereby the initiation of spermatogenesis and development of secondary sex characteristics. Testosterone also serves as a modulator of negative feedback on the hypothalamus and anterior pituitary. FSH stimulates the Sertoli cells to release Inhibin which provides negative feedback at the level of the anterior pituitary. The pulsatile release of GnRH and the subsequent negative feedback that is initiated at puberty continues throughout the life of the adult male and no cyclic pattern in activity is present.

Initiation of Puberty

We have established that GnRH is essential to puberty and there is a wealth of literature demonstrating the essential nature of GnRH. So, when one asks "what triggers puberty" what they are really asking is "what triggers the pubertal increase in GnRH?" There are two basic means by which GnRH surges could originate at puberty. One possibility is that the release of GnRH is stimulated de novo at puberty. The second possibility is that tonic inhibition of GnRH is released at puberty. The first of these possibilities is referred to as the Gonadostat Hypothesis and was proposed by Dohrn & Hohlweg in 1931. The Gonandostat Hypothesis proposed that before puberty minute quantities of estrogen can block GnRH function and over time this inhibitory ability decreases releasing GnRH activity. If this were true, high doses of exogenous estrogen should stimulate puberty. This appears to be true for sheep but has not been found to account for puberty in other species. The second hypothesis is termed the GnRH Pulse Hypothesis. The basic principle of this hypothesis is that GnRH is inhibited centrally until puberty, but not by sex steroids. A stimulus triggers an increase in GnRH pulsatility which then leads to puberty. If this hypothesis is true, then a hypothalamic lesion which disinhibits GnRH should lead to puberty. Two lines of evidence for this hypothesis are that GnRH secretion in the absence of gonads remains intact until puberty despite the absence of gonadal steroids and precocious puberty is, in fact, caused by hypothalamic lesions, which is reversible with amelioration of the lesion.

Collectively, it appears that in most mammals GnRH is present but tonically inhibited. The determination of what specifically is inhibiting GnRH until puberty has been an area of intense study [19]. A few important themes from this work are as follows. The control of GnRH pre- and postpubertally is not identical. GABA, leptin, and NPY all contribute to the start of puberty but no one independently accounts for the initiation. Norepinephrine, opioids, and other neurotransmitters and neuropeptides are involved in the coordination of GnRH release—but none of these messengers completely accounts for the initiation of GnRH pulsatile release at puberty. A relatively new neuropeptide first documented in 2000, kisspeptin, appears to be the critical factor in the initiation of pubertal patterns of GnRH release.

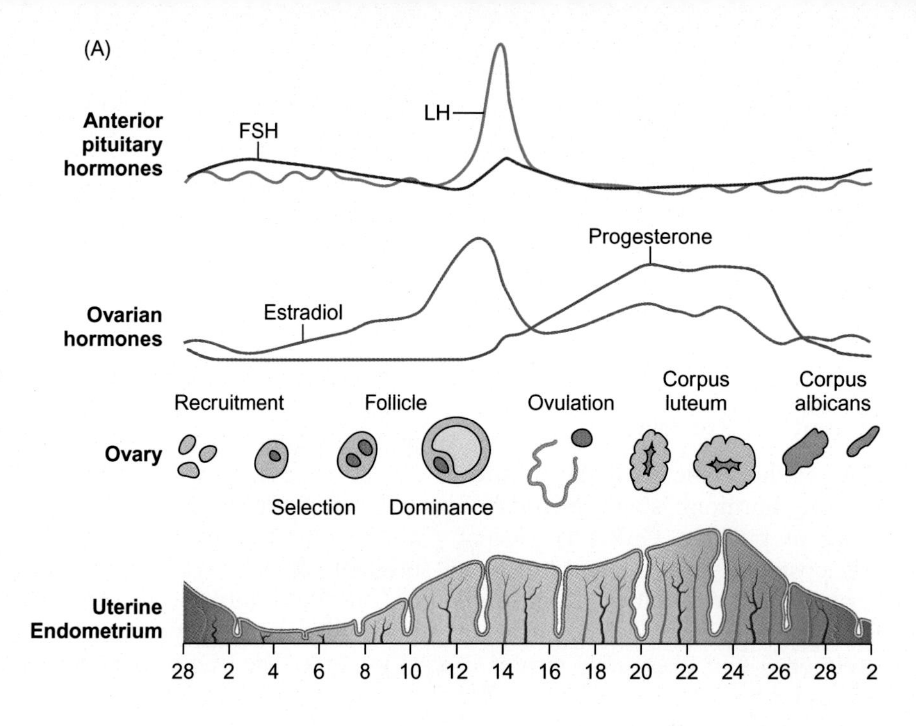

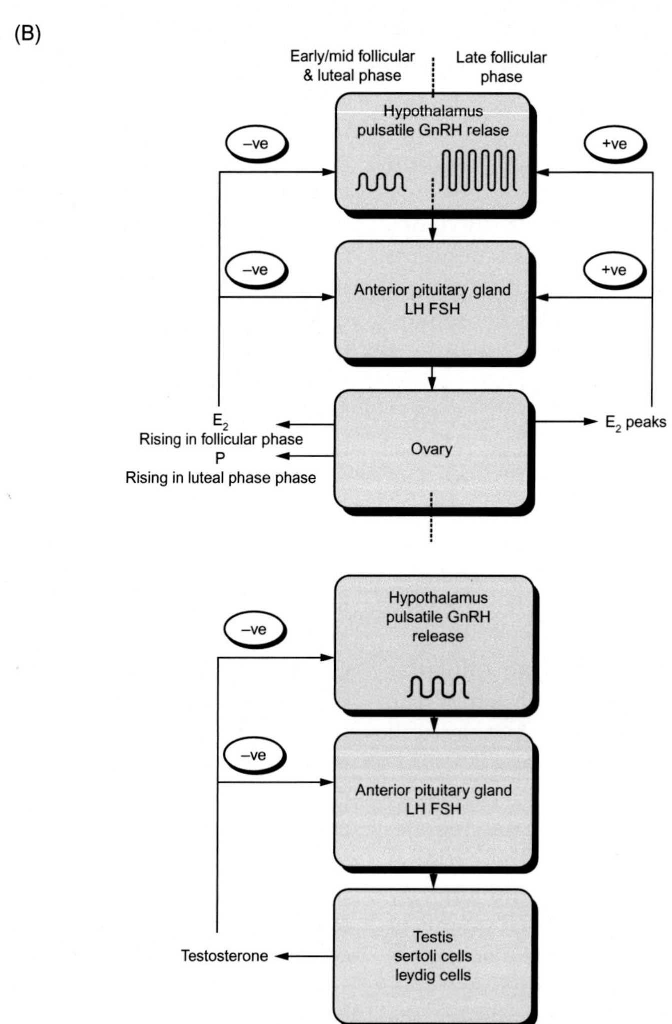

FIGURE 2.2 **Hormonal variations across the menstrual cycle.** (A) The coordinated release of LH and FSH along with changes in estradiol and progesterone coordinate the adult female menstrual cycle. The alternating pattern between tonic pulsatile release of GnRH and coordinated surges is unique to the female once puberty is complete and essential to the continuation of the cycle. (B) In the adult female, there is a cyclic pattern of GnRH release which stimulates the cycle. The pattern of the cyclic release is essential to appropriate hormone coordination to maintain reproductive viability. In the adult male, pulsatile release of GnRH continues, but no surge activity or variable cyclic pattern is present.

The discovery of kisspeptin stems from the documentation of a family with a high incidence of hypogonadotropic hypogonadism. Multiple members of this family had an absence of spontaneous puberty and partial or absent LH pulses. However, the individuals had a normal response to GnRH replacement. This pinpointed the abnormality to GnRH synthesis, secretion, or activity. Genetic analysis of the family determined that there was a leucine to serine substitution at position 148 on the GPR54 gene in the affected individuals [20]. Further study revealed that Gpr54 is the cognate ligand of kisspeptin, GnRH neurons express GPR54, and GPR54 mRNA expression increases in the hypothalamus at puberty. In addition, Gpr54 knockout mice fail to initiate puberty [20]. Adding to the connection between GnRH, kisspeptin, and puberty are demonstrations of increasing kisspeptin neuron apposition to GnRH neurons with progression toward puberty in rodents [21]. Furthermore, kisspeptin antagonists blunt the release of GnRH suggesting a causal relationship [22]. The precise relationship between kisspeptin and GnRH is still being established and the factors that initiate maturation of kisspeptin are still being recognized, but it is clear at this time that kisspeptin is a strong initiating factor in the stimulation of GnRH pulses which initiate puberty. For more detailed information on the initiation of puberty, detailed reviews are available [23–25].

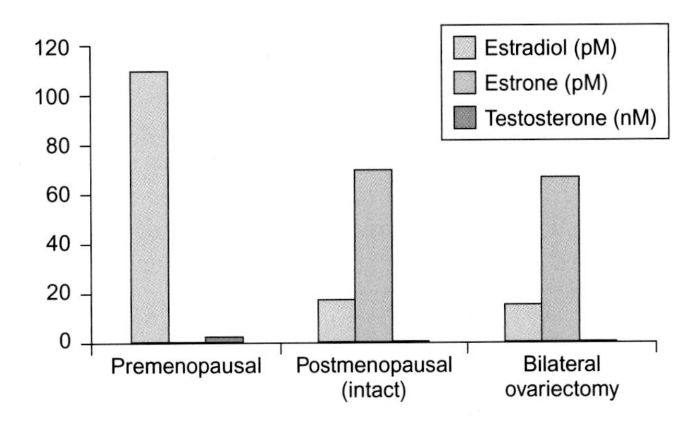

FIGURE 2.3 **Changes in circulating levels of sex steroid hormones across life span in women.** In premenopausal women, circulating estradiol varies from 18 to 110 pM, depending on the phase of the menstrual cycle. After menopause or bilateral ovariectomy, levels drop to 16–18 pM. Estrone is not produced in premenopausal women and increases to ~70 pM following menopause (intact and ovariectomized). In premenopausal women, testosterone varies from 1.3 to 2.6 nM and drops to 0.29–0.4 nM after menopause (intact and ovariectomized). *Source: Figure adapted from Knochenhauer E, Azziz R. Ovarian hormones and adrenal androgens during a woman's life span.* J Am Acad Dermatol *2001; 45:S105–15.*

ENDOCRINE FUNCTIONS ACROSS LIFE SPAN: AN EMPHASIS ON MAMMALIAN MENOPAUSE

Levels of reproductive hormones vary greatly across the life span. During human fetal development, reproductive hormones are high but then enter a phase of quiescence during childhood before a cascade of endocrine changes commence during puberty [26]. In human adults, reproductive hormones in men are generally consistent across the life span although slow age-related declines in hormone and sexual activity exist [27,28]. Further, whereas sperm production diminishes across the male's life span, fertility may exist throughout the aging process [26]. On the contrary, endocrine hormones fluctuate across the menstrual cycles in women and, as they age, women experience drastic reductions in endocrine function (Fig. 2.3) [29]. In women, menopause refers to the final menstrual cycle associated with the natural cessation of ovulation [30]. Around the time of menopause, physiological changes affecting secondary sex characteristics, vasomotor instability in the form of hot flashes, loss of bone density, and increased cardiovascular vulnerabilities are observed. Further, research suggests a decrease in sexual desire in menopausal women, accompanied by diminished frequency of sexual activity [31], an effect that may be mitigated with hormone replacement therapy [32].

Although many reproductive processes have been conserved through evolution, the phenomenon of menopause is rather unique. Specifically, as human females age, intervals between menstrual cycles increase, accompanied by lower estradiol levels and higher FSH concentrations [33], decreased ovary size, and increased androgen secretion [30,34,35]. Although human females' average life span is 80 years, they experience menopause at approximately 50 years of age, resulting in an extended postmenopausal phase at the end of the life span that is unique among mammals [30]. In fact, using the criterion of the cessation of breeding many years prior to the end of the life span, across all mammals, only pilot and killer whales are thought to have a similar menopause to humans [36]. However, other nonhuman primates are also considered to undergo variations of menopause [30].

Although certain mammals have evolved to live a significant portion of their lives in a postmenopausal state, this final reproductive stage in females presents many questions about the adaptive significance of this phenomenon [36]. There are also many additional effects of drastic reductions in reproductive hormones in women, apart from the physiological effects described above. For example, cognition and mood have been reported to be adversely affected, resulting in attempts to mitigate these symptoms with interventions such as hormone replacement therapy [37]. With all the uncertainties surrounding these questions about variations in reproductive endocrine levels in aging females, more research is necessary to provide the most informed responses to questions about how women can live the healthiest postmenopausal lives.

INFLUENCES ON SEXUAL DIFFERENTIATION AND DEVELOPMENT

Thus far in this chapter, the trajectory of sexual differentiation has been presented in a typical manner in which developmental events follow in an expected sequence; however, exceptions to these developmental processes exist that can disrupt the organizational impact of sex hormones. Genetic variations, modified environmental and chemical contexts, and altered endocrinological patterns may contribute to various forms of alterations in sexual differentiation.

Environmental influences—Given the findings that ambient temperature influences gonadal sexual differentiation in amphibians and reptiles [38], it is interesting to consider the potential role of various environmental factors on sexual differentiation in mammals. For example, exposure to environmental chemicals has been investigated with some substances, mostly in the pesticide category, identified as endocrine disruptors [39]. A primary route of action for these endocrine disruptors is their binding to estrogen and androgen receptors and subsequent mimicking of natural hormones.

An unfortunate lesson was learned about artificial manipulations of fetal estrogens via pharmacological manipulations when women were prescribed the drug diethylstilbestrol (DES) to reduce the risk of spontaneous abortion in the 1940s–60s. After about 25 years and millions of fetuses exposed to this drug, causal connections were made between the drug and effects on the reproductive organs in females, including deformed uteri and increased risk for vaginal cancer (as well as immune and behavioral adverse effects). Consequently, DES was banned in 1972 [40]. In fact, evidence from both animal and human studies suggest an epigenetic effect for offspring of individuals exposed to DES during fetal development [41].

Outside of the pharmacological realm, environmental chemicals have also been found to have an impact on sexual development [14]. Bisphenol A, a chemical released from polycarbonate products such as water and baby bottles is an estrogen receptor ligand and, consequently, considered an endocrine-disrupting chemical [40]. In females, fetal exposure to this chemical results in mammary gland and vaginal alterations as well as accelerated puberty and, in males, increased prostate weight [42]. Other chemicals known as alkyl phenols, found in products such as paints, detergents, and herbicides also bind to estrogen receptors as well as exert a weak androgenic effect, have been found to reduce the synthesis of testosterone and size of testes and masculinize females by interrupting steroid feedback sensitivity [43–45]. Certain chemicals found in natural sources such as plants also have effects on sexual development; specifically, isoflavones have been reported to feminize male-typical behavior and masculinize gonadal, endocrine, and behavioral sex-typical responses in females [46–48].

Finally, natural variations in the uterine environment have been found to influence sexual differentiation in rodents. For example, the specific position of fetuses in litter-bearing mammals such as rats and mice may influence sexual development. Research confirms that having a uterine position between male and female fetuses differentially exposes the fetuses to varied levels of gonadal steroids and affects sexual differentiation, especially in females [42]. Considering that during sexual differentiation male fetuses produce high levels of testosterone that can pass into adjacent fetuses, the fetus positioned next to the male fetuses experiences supplemental testosterone levels. Thus, a female positioned between two male fetuses experiences a different fetal hormonal milieu than in the same uterus as a female positioned between two female fetuses [49]. In humans, however, uterine exposure to gonadal steroid levels during fetal development varies according to variables such as maternal age, reproductive history, and diet [50].

Natural genetic and endocrine variations have both short-term and long-term effects on sexual differentiation. In humans, case studies have been monitored to characterize the general effects of these naturally varied conditions. One of the most extensively studied medical conditions affecting sexual differentiation is the genetic condition congenital adrenal hyperplasia (CAH), a spectrum of related conditions resulting from exposure to higher than normal androgen levels during gestation due to enzymatic disruptions in steroid hormone production in the adrenal glands [51]. This condition often results in ambiguous-looking genitalia in females prompting early hormone therapy following birth. Researchers have followed these cases throughout childhood and adulthood to determine long-term effects of high levels of androgens during fetal development. As previously described, results have revealed more masculine play than control female children; further, as adults, these individuals express more interest in male-dominated careers such as airline pilot, construction worker, or engineering-related careers [52,53]. Research also indicates that women with CAH diagnosed at birth have better spatial abilities than matched control subjects [54]. Considering sexual orientation, most of these individuals are heterosexual but, compared to non-CAH women, higher rates of homosexuality and bisexuality are reported in CAH individuals [52,55].

Whereas CAH is more influential in developing female fetuses, another genetic condition, 5-alpha reductase deficiency, has a dramatic influence on developing male fetuses and subsequent pubescent development. Individuals with this condition lack the ability to produce sufficient levels of 5-alpha reductase that is necessary for the development of the external male genitalia. Thus, these XY fetuses fail to develop masculinized genitalia and present a more female-typical genital phenotype at birth. However, since this enzyme is no longer necessary for masculinization in the form of development of male secondary sex characteristics during puberty, the individual starts developing more male-typical genitalia at puberty as described earlier in this chapter [56,57]. Such findings provide support for possible plasticity of the organized effects of sex-specific physiology and behavior. The effects of puberty are varied, however. Some individuals raised as females fail to develop male genitalia during puberty but experience a lack of female-typical secondary characteristics such as menarch and breast development. In these cases surgical reconstruction of the genitalia and hormone replacement therapy have been used to achieve a more female-typical phenotype [58].

Another genetic condition that influences sexual differentiation of XY fetuses is androgen insensitivity syndrome (AIS). In this condition, genetic males lack the appropriate sensitivity to androgens during fetal development and, consequently, the fetus develops female-typical genitalia similar to the outcome of 5-alpha reductase deficiency. These individuals are raised as females and develop some degree of secondary sex characteristics, except for menarche [59]. Although the genetic composition between XX women and AIS women varies, female-typical sexual behavior has been observed to be similar in both groups. For example, a neuroimaging study indicated that when these two groups of women were presented with sexually arousing images, both groups exhibited less amygdala activation than male subjects [60].

Thus factors from both exogenous and endogenous/natural origins have been shown to alter typical sexual differentiation, suggesting that the organizational effects of genetics and reproductive hormones during critical periods are more fluid than originally envisioned. Disruptions of the typical developmental patterns produce a plethora of variations; as we learn more about these conditions and outcomes in appropriate animals models and case studies, more information will be available about outcomes related to specific surgical and endocrine interventions.

CONCLUSIONS

This chapter provided an overview of the chromosomal (genetic) and endocrinological origins of sex and sexual differentiation across the life span. Subsequent chapters in this book explore the role of both sex chromosomes and hormones to influence organ function and, ultimately, disease propensity and outcomes.

References

[1] Sisk CL, Foster DL. The neural basis of puberty and adolescence. Nat Neurosci 2004;7:1040–7.

[2] Blaustein JD, Ismail N, Holder MK. Review: puberty as a time of remodeling the adult response to ovarian hormones. J Steroid Biochem Mol Biol 2015. Available from: http://dx.doi.org/10.1016/j.jsbmb.2015.05.007. [Epub ahead of print].

[3] Rodgers JE. Sex: a natural history. New York, NY: Henry Holt and Company; 2002.

[4] Kim Y, Capel B. Balancing the bipotential gonad between alternative organ fates: a new perspective on an old problem. Dev Dyn 2006;235:2292–300.

[5] Sadler TW. Langman's medical embryology. Baltimore, MD: Lippincott Williams & Wilkins; 2012.

[6] Fernandez-Teran M, Piedra ME, Simandl BK, Fallon JF, Ros MA. Limb initiation and development is normal in the absence of the mesonephros. Dev Biol 1997;189:246–55.

[7] Eicher EM, Washburn LL. Genetic control of primary sex determination in mice. Annu Rev Genet 1986;20:327–60.

[8] Scott CE, Wynn SL, Sesay A, Cruz C, Cheung M, Gomez Gaviro MV, et al. SOX9 induces and maintains neural stem cells. Nat Neurosci 2010;13:1181–9.

[9] Palmer SJ, Burgoyne PS. In situ analysis of fetal, prepuberal and adult XX–XY chimeric mouse testes: sertoli cells are predominantly, but not exclusively, XY. Development 1991;112:265–8.

[10] Bowles J, Knight D, Smith C, Wilhelm D, Richman J, Mamiya S, et al. Retinoid signaling determines germ cell fate in mice. Science 2006;312:596–600.

[11] Wilhelm D, Palmer S, Koopman P. Sex determination and gonadal development in mammals. Physiol Rev 2007;87:1–28.

[12] Maclaughlin DT, Donahoe PK. Sex determination and differentiation. N Engl J Med 2004;350:367–78.

[13] Sharpe RM. Pathways of endocrine disruption during male sexual differentiation and masculinization. Best Pract Res Clin Endocrinol Metab 2006;20:91–110.

[14] Wilson CA, Davies DC. The control of sexual differentiation of the reproductive system and brain. Reproduction 2007;133:331–59.
[15] Jost A, Vigier B, Prepin J, Perchellet JP. Studies on sex differentiation in mammals. Recent Prog Horm Res 1973;29:1–41.
[16] Tobet SA. Genes controlling hypothalamic development and sexual differentiation. Eur J Neurosci 2002;16:373–6.
[17] Blakemore SJ, Robbins TW. Decision-making in the adolescent brain. Nat Neurosci 2012;15:1184–91.
[18] Vetter-O'hagen CS, Spear LP. Hormonal and physical markers of puberty and their relationship to adolescent-typical novelty-directed behavior. Dev Psychobiol 2012;54:523–35.
[19] Terasawa E, Fernandez DL. Neurobiological mechanisms of the onset of puberty in primates. Endocr Rev 2001;22:111–51.
[20] Seminara SB, Messager S, Chatzidaki EE, Thresher RR, Acierno Jr JS, Shagoury JK, et al. The GPR54 gene as a regulator of puberty. N Engl J Med 2003;349:1614–27.
[21] Clarkson J, Herbison AE. Postnatal development of kisspeptin neurons in mouse hypothalamus; sexual dimorphism and projections to gonadotropin-releasing hormone neurons. Endocrinology 2006;147:5817–25.
[22] Roseweir AK, Kauffman AS, Smith JT, Guerriero KA, Morgan K, Pielecka-Fortuna J, et al. Discovery of potent kisspeptin antagonists delineate physiological mechanisms of gonadotropin regulation. J Neurosci 2009;29:3920–9.
[23] Plant TM. Neuroendocrine control of the onset of puberty. Front Neuroendocrinol 2015;38:73–88.
[24] Bhattacharya M, Babwah AV. Kisspeptin: beyond the brain. Endocrinology 2015;156:1218–27.
[25] Teles MG, Silveira LF, Tusset C, Latronico AC. New genetic factors implicated in human GnRH-dependent precocious puberty: the role of kisspeptin system. Mol Cell Endocrinol 2011;346:84–90.
[26] Cameron JL. Hormonal mediation of physiological and behavioral processes that influence fertility. In: Wachter KW, Bulatao RA, editors. Offspring: human fertility behavior in biodemographic perspective, xv. Washington, DC: National Academies Press; 2003. p. 379.
[27] Bain J. Andropause. Testosterone replacement therapy for aging men. Can Fam Physician 2001;47:91–7.
[28] Morales A, Heaton JP. Hormonal erectile dysfunction. Evaluation and management. Urol Clin North Am 2001;28:279–88.
[29] Burger HG, Dudley EC, Robertson DM, Dennerstein L. Hormonal changes in the menopause transition. Recent Prog Horm Res 2002;57:257–75.
[30] Walker ML, Herndon JG. Menopause in nonhuman primates? Biol Reprod 2008;79:398–406.
[31] Mccoy NL, Davidson JM. A longitudinal study of the effects of menopause on sexuality. Maturitas 1985;7:203–10.
[32] Sherwin BB. The impact of different doses of estrogen and progestin on mood and sexual behavior in postmenopausal women. J Clin Endocrinol Metab 1991;72:336–43.
[33] Sherman BM, Korenman SG. Hormonal characteristics of the human menstrual cycle throughout reproductive life. J Clin Invest 1975;55:699–706.
[34] Clement PB. Histology of the ovary. Am J Surg Pathol 1987;11:277–303.
[35] Manieri C, Di Bisceglie C, Fornengo R, Grosso T, Zumpano E, Calvo F, et al. Postmenopausal virilization in a woman with gonadotropin dependent ovarian hyperthecosis. J Endocrinol Invest 1998;21:128–32.
[36] Johnstone RA, Cant MA. The evolution of menopause in cetaceans and humans: the role of demography. Proc Biol Sci 2010;277:3765–71.
[37] Frye CA. Steroids, reproductive endocrine function, and affect. A review. Minerva Ginecol 2009;61:541–62.
[38] Dournon C, Houillon C, Pieau C. Temperature sex-reversal in amphibians and reptiles. Int J Dev Biol 1990;34:81–92.
[39] Mnif W, Hassine AI, Bouaziz A, Bartegi A, Thomas O, Roig B. Effect of endocrine disruptor pesticides: a review. Int J Environ Res Public Health 2011;8:2265–303.
[40] Newbold R. Cellular and molecular effects of developmental exposure to diethylstilbestrol: implications for other environmental estrogens. Environ Health Perspect 1995;103(Suppl. 7):83–7.
[41] Newbold RR, Hanson RB, Jefferson WN, Bullock BC, Haseman J, Mclachlan JA. Increased tumors but uncompromised fertility in the female descendants of mice exposed developmentally to diethylstilbestrol. Carcinogenesis 1998;19:1655–63.
[42] Nagel SC, Vom Saal FS. Endocrine control of sexual differentiation: effects of the maternal-fetal environment and endocrine disrupting chemicals. In: Miller V, Hay M, editors. Principles of sex-based difference in physiology. New York, NY: Elsevier; 2004.
[43] Aydogan M, Barlas N. Effects of maternal 4-*tert*-octylphenol exposure on the reproductive tract of male rats at adulthood. Reprod Toxicol 2006;22:455–60.
[44] Pocock VJ, Sales GD, Wilson CA, Milligan SR. Effects of perinatal octylphenol on ultrasound vocalization, behavior and reproductive physiology in rats. Physiol Behav 2002;76:645–53.
[45] Willoughby KN, Sarkar AJ, Boyadjieva NI, Sarkar DK. Neonatally administered tert-octylphenol affects onset of puberty and reproductive development in female rats. Endocrine 2005;26:161–8.
[46] Kouki T, Kishitake M, Okamoto M, Oosuka I, Takebe M, Yamanouchi K. Effects of neonatal treatment with phytoestrogens, genistein and daidzein, on sex difference in female rat brain function: estrous cycle and lordosis. Horm Behav 2003;44:140–5.
[47] Levy JR, Faber KA, Ayyash L, Hughes Jr CL. The effect of prenatal exposure to the phytoestrogen genistein on sexual differentiation in rats. Proc Soc Exp Biol Med 1995;208:60–6.
[48] Wisniewski AB, Cernetich A, Gearhart JP, Klein SL. Perinatal exposure to genistein alters reproductive development and aggressive behavior in male mice. Physiol Behav 2005;84:327–34.
[49] Vom Saal FS. Variation in phenotype due to random intrauterine positioning of male and female fetuses in rodents. J Reprod Fertil 1981;62:633–50.
[50] Swan SH, Vom Saal FS. Alterations in male reproductive development: the role of endocrine disrupting chemicals. In: Metzler M, editor. Endocrine disruptors, part II. Berlin Heidelberg: Springer-Verlag; 2002. p. 131–70.
[51] Przybylik-MazurekE, KurzynskaA, SkalniakA, Hubalewska-DydejczykA. (2015). Current approaches to diagnosis classical form of congenital adrenal hyperplasia. Recent Pat Endocr Metab Immune Drug Discov.
[52] Meyer-Bahlburg HF, Dolezal C, Baker SW, Ehrhardt AA, New MI. Gender development in women with congenital adrenal hyperplasia as a function of disorder severity. Arch Sex Behav 2006;35:667–84.
[53] Berenbaum SA, Himes M. Early androgens are related to childhood sex-typed toy preferences. Psychol Sci 1992;3:203–6.
[54] Hampson E, Rovet JF, Altmann D. Spatial reasoning in children with congenital adrenal hyperplasia due to 21-hydroxylase deficiency. Dev Neuropsychol 1998;14:299–320.
[55] Kanhere M, Fuqua J, Rink R, Houk C, Mauger D, Lee PA. Psychosexual development and quality of life outcomes in females with congenital adrenal hyperplasia. Int J Pediatr Endocrinol 2015;2015:21.

[56] Berenbaum SA, Beltz AM. Sexual differentiation of human behavior: effects of prenatal and pubertal organizational hormones. Front Neuroendocrinol 2011;32:183–200.

[57] Ruble DN, Martin CL, Berenbaum SA. Gender development. In: Eisenberg N, editor. Social, emotional, and personality development. New York, NY: Wiley; 2006. p. 858–932.

[58] Makeki N, Hormozi MK, Alamdari MI, Tavosi Z. "5-Alpha-reductase 2 deficiency in a woman with primary amenorrhea." In: Case reports in endocrinology; 2013.

[59] Gingu C, Dick A, Patrascoiu S, Domnisor L, Mihai M, Harza M, et al. Testicular feminization: complete androgen insensitivity syndrome. Discussions based on a case report. Rom J Morphol Embryol 2014;55:177–81.

[60] Hamann S, Stevens J, Vick JH, Bryk K, Quigley CA, Berenbaum SA, et al. Brain responses to sexual images in 46,XY women with complete androgen insensitivity syndrome are female-typical. Horm Behav 2014;66:724–30.

CHAPTER

3

Sex Differences in Neuroanatomy and Neurophysiology: Implications for Brain Function, Behavior, and Neurological Disease

Gretchen N. Neigh[1] and Liana Merrill[2]

[1]Department of Anatomy and Neurobiology, Virginia Commonwealth University, Richmond, VA, United States
[2]Department of Physiology, Emory University College of Medicine, Atlanta, GA, United States

INTRODUCTION

There has been a long-standing interest surrounding sex differences in the brain, probably due at least in part to stereotypes regarding cognitive abilities of both men and women. For example, women are thought to have better verbal skills, while men are often praised for their superior spatial abilities. Sex differences in brain size have been apparent for some time; women have smaller brain volumes than men, and this can be partially explained by the smaller stature of women. However, there remain many more sexually dimorphic characteristics within the central nervous system. The purpose of this chapter is to describe these complex sex differences within the brain, focusing on differences in neuronal signaling, body functions and behavior, and how some of these sexually dimorphic characteristics have influenced neurological disease. The majority of the differences discussed are modulated by steroid hormones and our collective knowledge of the mediators and modifiers of sex-dependent differences has been facilitated by preclinical research.

STEROID HORMONES

Steroid hormones are a group of signaling molecules that can travel great distances through the circulation from their site of synthesis to target organs [1]. One such target organ is the brain, where steroid hormones act in a local autocrine or paracrine manner. Sex differences in the brain and in behavior are due in part to the potent actions of three different families of steroid hormones within the nervous system: estrogens, progestins, and androgens. These hormones exert both organizational and activational effects, which were originally described by Phoenix et al. (1959). As discussed in a previous chapter, organizational effects occur in the developing brain during a perinatal sensitive period and ensure that the brain of an animal is set up for the hormonal signals it will be exposed to in adulthood (activational effects). For example, a male-organized brain is less likely to produce female sexual behavior in adulthood, even in the presence of female steroid hormone replacement [2]. An elegant example of the interplay between organizational and activational effects of steroid hormones on behavior is depicted by studies of the Japanese quail (see Box 3.1).

Mechanism of Action of Steroid Hormones

There are several modes of action that steroid hormones use to modulate the nervous system and, in turn, cause sex differences in brain function and behavior. Steroid hormones each bind to their own cognate receptors, which will be discussed in further detail in subsequent sections. These receptors are

Sex Differences in Physiology
DOI: http://dx.doi.org/10.1016/B978-0-12-802388-4.00003-3

BOX 3.1

TRANSPLANTATION OF THE FOREBRAIN IN MALE AND FEMALE JAPANESE QUAIL EMBRYOS—IMPLICATIONS FOR SUBSEQUENT SEXUAL BEHAVIOR

Sexual differentiation of the brain and behavior is thought to occur via the canonical model. The canonical model of sexual differentiation suggests that genetics determine the development of the gonad, which leads to sex-dependent release of steroid hormones that exert epigenetic actions to specify the brain as male or female [3,4]. Work contradicting the canonical model includes study of the vocal system of the zebra finch [5], where sex reversal of the female gonad does not produce a male-typical vocal system. Rodent studies also demonstrate incomplete sex reversals of behavior after hormone treatment during early development [6,7]. To further evaluate the concept of gonadal hormone-dependent determination of brain sex and behavior [8], an elegant study using nonendocrine manipulation was performed in the Japanese quail model. Normal sexual differentiation in the Japanese quail has been extensively studied, with male brain development considered default. Evidence supports the idea of default male brain development; gonadal estrogen production at embryonic day 6 (E6) in females drives ovary formation, estrogen treatment of male quail embryos is demasculinizing, and inhibiting estrogen production before E14 in female embryos masculinizes the brain and behavior [9,10]. The nonendocrine manipulation in the current study involved the transfer of a male or female brain before sexual differentiation of the gonad on E2. Male donor-to-female host (MF), female-to-male (FM), male-to-male (MM), and female-to-female (FF) transplants were performed, with the forebrain including the hypothalamus originating from the donor [11,12]. The hypothesis then, if male brain development is considered default and female development only occurs in the presence of ovarian hormones, was that development should occur in a host-typical manner where female brains in male bodies (FM) would develop into male brains, and male brains in female bodies (MF) would develop into female brains. What happened was that the gonads of the host developed normally, but quails with male gonads but female brains (FM) did not develop normally and MF transplants developed normally [8]. For example, FM transplants did not express male or female sex behavior and only sparse male behaviors like crowing. This study suggests that there may not be a default brain sex, as both male and female brains were able to respond to a female endocrine environment. Thus, brain sex appears to be set in stone at the start of brain development, and steroid hormones may then modulate sexual differentiation during development and into adulthood via epigenetic mechanisms. Studies in the Japanese quail, as well as the zebra finch and other rodent models, highlight the existence of brain-intrinsic and genetically influenced mechanisms of sexual differentiation, which remain to be verified in other vertebrate species including humans.

members of a nuclear receptor superfamily and exert their effects in four specific ways:

1. In the classical mode of steroid hormone action, receptors act as transcription factors. Following dimerization, steroid hormone receptors translocate to the nucleus and bind to appropriate hormone response elements on DNA; [13,14]
2. Steroid hormone receptors can also regulate transcription by interacting with cofactors (eg, c-fos) that bind to specific sequences of DNA on gene promoters; [15,16]
3. Steroid hormones can regulate gene transcription independently of DNA by activating kinases, proteases, and other molecules that initiate signaling cascades; [17,18] and
4. Steroid hormones can rapidly change cell physiology by acting on transmembrane receptors [19].

Estrogen and Receptors

Estrogen is the most potent steroid hormone, as it is active at concentrations in the femtomolar range. It is the end product of six to seven enzymatic conversions of its precursor, cholesterol [1]. The most critical enzyme is aromatase which provides the rate-limiting step in the synthesis of estrogen from androgen precursors. In contrast to the old view of estrogen as a slow mediator of distant cellular processes, it is now known that neural cells synthesize estrogen both locally and quickly, and therefore is now thought to function in a manner similar to neurotransmitters [20,21]. Recent studies have demonstrated the dynamic effects of estrogen aside from its classic role as a nuclear transcription factor. For example, estrogen acting at the cell membrane can lead to permanent organization of dendritic morphology in the hypothalamus [22]. Also, glia can synthesize estrogen independently, which likely has effects on the interaction between neurons and nonneuronal cells [23–25].

Estrogen binds to one of two different estrogen receptor (ER) isoforms, ERα and ERβ, members of the nuclear receptor transcription factor superfamily [1]. Transgenic mouse models utilizing ERα and ERβ knockout mice suggest that ERα modulates masculinization while ERβ is more closely coupled with defeminization [26]. As transcription factors, ERα and ERβ can exert their effects by the classical mode of steroid receptors, acting directly at the genome as part of a transcription complex that recognizes hormone response elements in promoter regions to either induce or repress gene expression [27]. These receptors can also act at neuronal membranes [19], where they directly interact with cellular kinases (eg, MAP kinases and PI3 kinase) and other membrane receptors such as mGluR [28]. Furthermore, estrogen can bind to the G-protein-coupled receptors ER-X and GPR30. ER-X is thought to activate intracellular signaling cascades by acting at the cell membrane [29,30], while GPR30 has been shown to act on intracellular calcium signaling due to its high expression levels on the endoplasmic reticulum [31]. These rapid effects of estrogen at the neuronal membrane contrast with the slow, classical mechanism of action in the nucleus. However, this dual action of estrogen greatly expands the range of target endpoints and becomes important in understanding behavioral outcomes.

Progesterone and Receptors

Similar to estrogen, progesterone also has both genomic and membrane effects. Progesterone can bind to its cognate nuclear receptors, progesterone receptors (PR) A and B, thus activating transcription of target genes. The two receptors are coexpressed in most tissues and their expression is strongly upregulated by estrogen in the anterior pituitary and in many other brain regions. For example, estrogen administration to ovariectomized female rats increases PR expression in the anterior pituitary as well as hypothalamic, preoptic, and septal tissues, but not in the hippocampus and amygdala [32,33]. Progesterone also downregulates its own receptor expression. Effects of progesterone at the cell membrane include the interaction of progesterone with ion channels, neurotransmitter receptors, and peptide hormone receptors. For example, progesterone can bind to and inhibit the function of the oxytocin receptor in the rat uterus [34]. Progesterone also interacts with opioid [35,36], cholinergic [37], and GABAergic signaling [38] in the CNS.

Progesterone and its receptors have received less attention compared to the other steroid hormones, likely because levels do not differ in male and female neonates [39]. However, sex differences in PR expression exist in the developing hypothalamus and males have greater expression of PRs in several sexually dimorphic brain regions (discussed in a subsequent section), including the AVPV, medial POA, and ARC [40–42]. PRs have been highly studied in the context of the facilitation of female sexual behavior, which will be discussed in a subsequent section. Some attention has been given to allopregnalone, an endogenous inhibitory pregnane neurosteroid synthesized from progesterone that is a positive modulator of the $GABA_A$ receptor. Evidence suggests that sex differences in susceptibility to certain neurological disorders like epilepsy, anxiety, depression, and Alzheimer's disease could be due to neurosteroids like allopregnalone (see Box 3.2) [43,44].

Androgen and Receptors

Androgens are involved in many biological processes, including the development and maintenance of male reproductive organs. Most actions of androgens are mediated by binding to its single cognate nuclear receptor, the androgen receptor (AR). Like the other steroid hormone receptors, the AR is a member of the steroid hormone nuclear receptor superfamily and acts as a ligand-inducible transcription factor [55]. The AR has two endogenous ligands, testosterone and the more potent metabolite, 5α-dihydrotestosterone (DHT). Both testosterone and DHT, when bound to the AR, activate target gene expression by binding to specific DNA elements known as androgen responsive elements (ARE) located on promoters [56].

Classic studies in vertebrates, including mice, have demonstrated that masculinization of the brain and behavior requires both estrogen and testosterone signaling and subsequent binding of their cognate receptors [57,58]. While estrogen is practically nonexistent in the male circulation, testosterone is a precursor for estrogen in vivo and thus can act as a source of estrogen in the brain. Therefore, it appears that testosterone serves a dual role in that it activates the AR in neural circuits that control male-typical behaviors, and testosterone can also indirectly activate the ER after the synthesis of estrogen from testosterone in order to influence separate estrogen-mediated neural circuits.

BOX 3.2

ALLOPREGNANOLONE AS A THERAPEUTIC FOR ALZHEIMER'S DISEASE—BUT WHAT ABOUT SEX?

Allopregnanolone (3α-hydroxy-5α-pregnan-20-one), a metabolite of progesterone that is synthesized de novo in the embryonic as well as the adult CNS [45,46], has demonstrated promising therapeutic potential in the treatment of Alzheimer's disease due to its promise in the area of neurogenesis [44]. Whereas most neurogenesis occurs during development, there are two brain regions that continue to generate new neurons in adulthood, an area of the frontal cortex called the subventricular zone (SVZ), and the dentate gyrus (DG) in the hippocampus [47,48]. However, the ability to proliferate declines with both age and neurodegenerative diseases such as Alzheimer's disease [49], as do neurosteroids like allopregnanolone in the brain [50,51]. Using a transgenic mouse model of Alzheimer's disease, researchers have demonstrated the ability of allopregnanolone to reverse both neurogenic and cognitive deficits in these mice [52], thus demonstrating the therapeutic potential of allopregnanolone in neurodegenerative diseases like Alzheimer's disease. The triple transgenic Alzheimer's disease mouse (3xTgAD) carries three human Alzheimer's disease-related gene mutations and displays age-dependent neuropathology of the two hallmarks of the disease, β-amyloid plaque formation and neurofibrillary tangles [53]. Excitingly, it was discovered that allopregnanolone could reverse the deficits in cell proliferation in the SGZ before AD pathology was evident, as well as the cognitive deficits seen in a hippocampal-dependent associative learning and memory task in the 3xTgAD mice [52]. Most importantly, these reversals were such that levels of neurogenesis and learning and memory in 3xTgAD mice treated with allopregnanolone were not significantly different from nontransgenic control mice. Due to these and other supporting results [46,54], allopregnanolone has exuberant clinical promise and has been suggested as a therapeutic that should begin clinical trials [44]. It is important to note however that studies thus far have been carried out only in males. While the clinical promise is evident, and these studies highlight that female-specific steroid hormones have dramatic effects in males, it is still crucial to examine the effects of allopregnanolone in females before continuing on to clinical trials. As just one example, allopregnanolone may have similar effects in males and females, but females may require different dosing regimens. Regardless, the findings of the effects of a lesser-recognized steroid hormone on a devastating disease such as Alzheimer's are exciting and further support of the clinical promise in both sexes is welcomed and encouraged.

SEX AND NEURONAL SIGNALING

Sex differences in the brain are large and diverse, and range from structural to neurochemical to molecular. Here we will discuss several sex differences in neuronal signaling in defined brain regions (ie, structural), as well as in particular neurotransmitter and hormone systems (ie, neurochemical). Steroid hormones, particularly estrogen, play a large role in the regulation of these sex differences. As will become evident, the vast differences in neuronal signaling between the sexes point out the inappropriateness of using only one sex in experimental procedures and the difficulty in drawing conclusions from the studies that have focused on one sex alone.

Brain Regions

Sex differences in the volume of particular brain regions are frequently observed in nature. One of the first reported was the size of the song control nuclei of birds [59]. Volumetric differences in brain nuclei are established by steroid hormone exposure during development, causing sex differences in the size of brain regions, such as some nuclei of the hypothalamus. Two of the best-characterized brain regions are the sexually dimorphic nucleus of the preoptic area (SDN-POA) and the anteroventral periventricular nucleus (AVPV). The differences in the size of these two brain regions are due to the effects of steroid hormones on cell death, either promoting or preventing cell death depending on the region [60]. There are also brain regions within the hypothalamus that display prominent sex differences in cell morphology: the arcuate nucleus (ARC), the preoptic area (POA), and the medial basal hypothalamus (MBH) (Table 3.1).

Sexually Dimorphic Nucleus of the Preoptic Area

The SDN-POA is more than five times larger in males than females due to the fact that it contains many more cells in males. Cells in females undergo apoptosis caused

TABLE 3.1 Brain Regions with Documented Sex Differences

Region	Abbreviation	Primary function	Noted sex difference	Origin of sex difference	Citations
Sexually dimorphic nucleus of the preoptic area	SDN-POA	Control of male copulatory behavior	SDN-POA is markedly larger in the male than the female	Androgen-derived estrogens prevent apoptosis in the male	[60–67]
Anteroventral periventricular nucleus	AVPV	Control of GnRH and LH release necessary for ovulation	AVPV is larger in females	Reduced cell death in the female	[68–72]
Arcuate nucleus	ARC	Regulation of hormone release	Predisposition toward GABAergic synapses in the male and glutamatergic synapses in the female	Organizational effects of estrogen	[23,73–78]
Preoptic area	POA	Expression of male sexual behavior	Sex differences in synaptic connectivity; twice as many dendritic spines in male than female	Estrogen-induced effects on prostaglandin E2 signaling	[79–84]
Medial basal hypothalamus/ ventromedial nucleus of the hypothalamus	MBH/VMN	Critical for the display of female sexual behavior	Sex differences in dendritic architecture and spine distribution	Effects of testosterone and estrogen on morphology and synaptic transmission	[22,75,85,86]
Bed nucleus of the stria terminalis	BNST	Modulation of anxiety and the stress response	Enlarged in males with higher neuron content	Alternations in Bax-mediated apoptosis and histone acetylation	[87–95]
Hippocampus		Learning and memory	Sex differences in anatomical structure, neurochemistry, and cell genesis	Not fully defined by sex steroids play a role	[96–105]
Amygdala		Emotion and fear	Sex difference in morphology, neurochemistry, and remodeling following developmental stress	Endocannabinoid signaling and hormone influences	[96,106,107]

by a lack of estrogen during development, whereas androgen-derived estrogens prevent cell death in this brain region in males [60–62]. Female-specific apoptosis occurs through NMDA receptor activation and downstream expression of the antiapoptotic protein bcl-2 [63,64]. Treatment of females with estrogens or androgens during development will rescue the cells from death, resulting in a permanently masculinized, larger nucleus [65]. Studies utilizing transgenic mice that lack *Bax*, a proapoptotic gene that codes for a protein localized to the cytosol of healthy cells, confirm that the higher rate of cell death in females is a result of apoptosis [66]. Although this is an active area of research, the molecular mechanisms of sexually dimorphic cell death are incompletely defined. In contrast to the increased apoptosis in females, there is also a later onset of cell genesis in males, thus leading to greater retention of SDN-POA volume in males well into adulthood [67].

Anteroventral Periventricular Nucleus

The AVPV is involved in the control of gonadotropin-releasing hormone (GnRH) neurons and the production of the luteinizing hormone (LH) surge that is necessary for ovulation. Not surprisingly, the AVPV is larger in females. Like the SDN-POA, the increased size of the AVPV is due to a larger number of cells in the female, particularly dopaminergic cells that express the enzyme tyrosine hydroxylase [68], estrogen-sensitive GABAergic/glutamatergic cells [69], and kisspeptin-positive cells [70]. The sex differences in cell number in the AVPV likely result from higher rates of estrogen-induced cell death during male development [70], as the sex difference in the total number of cells in the AVPV is eliminated in *Bax* knockout mice [66]. In the male, cell death involves distinct mechanisms depending on the cell type. Estrogen-induced activation of caspase-dependent cell death occurs in dopaminergic cells of the male AVPV [71]. The death of GABAergic cells is due to the suppression of tumor necrosis factor alpha (TNFα), a proinflammatory cytokine that activates nuclear factor kappa-light-chain-enhancer of B cell (NFκB) receptors and promotes cell survival by increased expression of an associated protein called TRAF-interacting protein (TRIP) [72]. Like the SDN-POA, less cell genesis occurs during the peripubertal period in males compared to females, thus maintaining the larger brain region size in females into adulthood [67].

Arcuate Nucleus

The ARC, located around the third ventricle near the median eminence, is involved in many processes including regulating the release of hormones (eg, GnRH and prolactin) from the anterior pituitary, the LH surge [23], lactation [73], appetite and growth hormone release [74]. Robust sex differences in cell morphology in the ARC are mediated by organizational effects of estrogen and include more GABAergic axosomatic synapses in the male and more glutamatergic axodendritic spine synapses in the female [75,76]. The presence of estrogen in the male ARC also leads to more complex astrocytes [77,78].

Preoptic Area

The medial POA, which is critical for the expression of male sexual behavior in adulthood [79], receives input from the amygdala, bed nucleus of the stria terminalis (BNST), and septum, in addition to sensory areas that convey information regarding sexually relevant stimuli [80]. Sex-specific differences in synaptic connectivity are apparent in the POA, once again due to estrogen exposure during development [81]. Given that the POA is critical for male-specific reproductive behavior, it is not surprising that there are twice as many dendritic spines in male POA neurons than females [82], and that males have larger numbers of and longer astrocytic processes in this brain region compared to females [83]. These sex differences in cell morphology in the POA appear to be due to downstream effects of prostaglandin E_2 (PGE_2), a lipid molecule synthesized following upregulation of cyclooxygenase-2 by estrogen, which upregulates the dendritic spine protein spinophilin [84].

Medial Basal Hypothalamus

The MBH lies in close proximity to the POA, and contains the sexually dimorphic ventromedial nucleus of the hypothalamus (VMN). The VMN is critical for the display of female sexual behavior; stimulation of this brain region facilitates female sexual receptivity and lordosis, a female-specific posture characterized by an arching of the spine that allows a female rat to take part in sexual behavior [85]. Testosterone-mediated sex differences in patterns of synaptic transmission and cell morphology also exist in the VMN; males have more synapses located on dendritic shafts and spines [86] and more branches on dendrites [75]. These differences appear to be mediated by glutamate; estrogen increases glutamate release, which binds to postsynaptic receptors and promotes spine formation [22].

Bed Nucleus of the Stria Terminalis

The BNST is a sexually dimorphic brain region involved in the modulation of known sex differences in anxiety and stress responses [87]. This nucleus is larger in volume and possesses more neurons in male rodents compared to females [88–90]. This sex difference is controlled by Bax-mediated apoptosis [91], but histone acetylation has recently been implicated in the BNST as well [92], possibly by direct or indirect epigenetic mechanisms. Disruption of histone deacetylation during development using pharmacological techniques results in feminization of the BNST. In addition, the adult male BNST has more vasopressin (VP)-expressing [93], aromatase positive, and AR positive neurons [94]. The BNST also projects to the AVPV, and these projections are more abundant in males [95].

Hippocampus and Amygdala

In contrast to sex differences in cell death that are highly studied in brain nuclei of the hypothalamus, there are also differences in cell proliferation and neurogenesis that result in small sex differences in volume of other brain regions, including the hippocampus and the amygdala. Hippocampi of males and females exhibit significant differences in anatomical structure and neurochemistry. Imaging studies provide evidence that the hippocampus is larger in women relative to total brain size [96]. Furthermore, male rats have a larger volume of and greater number of cells in the CA1 region of the hippocampus, and have greater neuronal density in the dentate gyrus [97]. There is also emerging evidence of sex differences in cell proliferation in the rat hippocampus. Studies indicate that twice as many new cells are born in the male rat hippocampus during development compared to the female [98], resulting in the modestly larger size of the hippocampus in males [99]. Neurogenesis in the hippocampus is a result of higher endogenous estrogen action in males [100]. In the adult female, estrogen promotes neurogenesis particularly in the olfactory bulb and dentate gyrus of the hippocampus [88]. Given the role of the hippocampus in learning and memory, it is not surprising that structural differences in the hippocampus are coupled with sex differences in performance of hippocampus-dependent behaviors such as spatial learning and memory as demonstrated by performance in the Morris water maze and differences in the formation and retention of fear information indicated through assessments of fear conditioning.

In addition to general sex differences in volume and cell number in the hippocampus, morphological changes occur across the 4–5-day estrous cycle of the female rat. Changes are especially prominent during the 24 h period between the proestrus phase of the estrous cycle, when the estrogen levels are highest, and the estrus phase during which estrogen is low [101,102]. There is an approximately 30% increase in synaptic density in the CA1 region of the hippocampus during proestrus, which correlates with increased plasticity characterized by increased long-term potentiation (LTP) [103]. The relationship between proestrus and increased plasticity may have implications for hippocampus-dependent behaviors like the Morris water maze and fear conditioning. Water maze and fear conditioning performance are both reduced during proestrus, indicating an inverse relationship between synaptic plasticity in the hippocampus and hippocampus-dependent behavioral ability. In other words, higher levels of endogenous estrogen appear to reduce hippocampus-dependent learning [104,105] suggesting that estrogen may impair learning and memory in females.

Whereas more neurogenesis occurs in the male rat hippocampus during development, the opposite is true in the amygdala. More new cells (neurogenic cells) are generated during development in the amygdala of female rats, mediated by sex differences in endogenous endocannabinoid signaling [106]. The medial nucleus of the amygdala is well-known to be a sexually dimorphic brain region, and more recently, almost the entire amygdala. The amygdala is significantly larger in men than women [96], and a recent study in rodents further exemplifies this sex difference during development. Male rodent pups exposed to temporary separation stress exhibit increased numbers of serotonin receptors in the amygdala compared to decreased receptor expression in female pups [107].

Neuronal Structure

Sex differences also exist in certain structural aspects within the brain, like synapses. For example, estrogen has an organizational effect on the ratio of the number and density of dendritic spine to axosomatic synapses, resulting in differences between males and females best characterized in the SDN-POA, VMN, and ARC [65]. For example, in the SDN-POA, there is greater density of dendritic spine synapses in males compared to females [82]. Neuronal dendrites in the male VMN are longer and more highly branched, and males have more overall dendritic spine synapses in the VMN than females [108]. In contrast, female neurons have twice the density of dendritic spine synapses in the ARC compared to males [75]. Given the role of dendritic spine plasticity in learning and memory, sex differences in synaptic density are positioned to contribute to differences in behavioral outcomes between the sexes. Therefore, it becomes apparent that one sex cannot be substituted for the other when looking at functional outcomes.

Neurotransmitter Systems

It is difficult to relate structural sex differences in the brain to sexually dimorphic functions, as the sexually dimorphic brain regions discussed above are implicated in many different functions and the connectivity of these brain regions remains largely unknown. Several of the neurotransmitter systems that innervate these brain regions are sensitive to steroid hormone levels in adulthood, and neurochemical studies have shown that neurotransmitter synthesis, content, and metabolism are sexually differentiated in certain brain areas [109]. Therefore, understanding these systems can help relate sex differences in structure to function.

Norepinephrine

Norepinephrine (NE) was one of the first neurotransmitters shown to play a role in female sexual behavior. NE appears to have both activational and inhibitory actions on lordosis [109]. This relationship remains poorly understood; however, NE appears to regulate the synthesis of estrogen and PR in the VMN, which may play a role in the ability to regulate female sex behavior [109].

Serotonin

One of the sexually dimorphic neurotransmitter systems in the POA is its serotonin innervation, which is thought to play a role in the inhibition of female sexual behavior, as depletion of serotonin in the POA facilitates lordosis in both males and females [110]. In 1960, Kato reported higher levels of serotonin in female than in male brains [111], but it was later reported that males have more serotonin-1 receptors in the POA than females [112]. It is unclear how the higher receptor number in males relates to the sexually dimorphic serotonin innervation in the POA because females have more serotonin fibers in the MBH than males [113], and estrogen facilitates an increase in serotonin-1 receptor levels in the POA of females but a decrease in males [112,114]. It is critical to further understand sex differences in serotonin, as therapeutic treatments targeting serotonin like selective serotonin reuptake inhibitors (SSRIs) may have different outcomes in males and females.

Acetylcholine

Research suggests that cholinergic systems can facilitate lordosis by acting on muscarinic receptors [115]. Intraventricular administration of cholinergic agonists can restore lordosis abolished by lesions within the hypothalamus. These cholinergic actions likely take place in the VMN and POA, but studies also suggest a more widespread action of acetylcholine [116,117]. Sexually dimorphic effects of cholinergic systems appear to be organizational in nature, as estrogen treatment increases the number of muscarinic receptors in the hypothalamus of gonadectomized females but not males [118].

Luteinizing Hormone

LH, secreted by the anterior pituitary gland, surges periodically, under the influence of estrogen, in order to trigger ovulation and development of the corpus luteum. The function of the LH surge appears to be due, at least in part, to organizational effects of steroid hormones. For example, castration of male rodents during development induces the ability to display an estrogen-induced LH surge in adulthood similar to normal adult females. However, male adults that are administered the same hormonal treatment that triggers the LH surge in females do not surge [119,120]. On the other hand, administration of testosterone to female rodents during development blocks the ability to produce the LH surge in adulthood [57,88,121]. The impact of steroid hormones on the LH surge illustrates the ability of organizational and activational hormone effects to collectively control this biological process.

Although poorly defined, there is also evidence that GABA/glutamate neurons in the AVPV may contribute to neuronal signaling between the AVPV and GnRH neurons, thus potentially modulating the LH surge. For example, antagonists of GABA and glutamate receptors block GnRH release and the LH surge in rats [69,122]. Other evidence suggests that synaptic terminals of GABA and glutamate neurons in the AVPV make direct contact with GnRH neurons in rats, and that estrogen inhibits GABA release and increases glutamate release during the LH surge [69]. Thus, steroid hormones can interact with sexually dimorphic neurotransmitter systems like the GABAergic system to modulate the LH surge.

Kisspeptin

Kisspeptin is a neuropeptide encoded by the *Kiss1* gene that binds to the G-protein-coupled receptor Kiss1R [123–125] and has been detected in both the ARC and the AVPV. Female mice have 10 times more kisspeptin neurons in the AVPV than males [126], and male rats express extremely low levels of kisspeptin [70]. A wide array of evidence suggests a role for kisspeptin in the modulation of GnRH secretion and reproductive function [127]. Kisspeptin-containing fibers in the ARC and AVPV make direct contact with Kiss1R-expressing GnRH neurons [126,128], suggesting that the neuropeptide may act directly on GnRH neurons. Furthermore, kisspeptin induces c-Fos expression in GnRH neurons, increases electrical activity of GnRH neurons, and causes the secretion of GnRH in hypothalamic explants [128–131]. Administration of kisspeptin to multiple species (eg, rodents, sheep, monkeys, humans) increases circulating LH and FSH levels [132,133], and experiments examining alterations in *Kiss1* and kisspeptin receptor genes result in infertility and impaired maturation [134–136]. Taken together, these studies suggest an important role for kisspeptin in reproductive function, likely through the modulation of GnRH signaling.

It is no surprise given the role of kisspeptin in reproductive function and GnRH signaling that these neurons are sexually differentiated, although only in the AVPV. *Kiss1*-expressing cells in the ARC do not appear to be sexually dimorphic in adult rodents, but may be so during the prepubertal period in mice, regardless of activational effects of steroid hormones [70,137]. Adult females express greater levels of *Kiss1* mRNA [70] and protein [126,138] compared to males, and *Kiss1* expression is stimulated by estrogen in the AVPV [139]. This increase of *Kiss1* expression in females appears to be organized early in development by steroid hormone signaling, as male and female rats that are gonadectomized in adulthood and treated with exogenous estrogen still demonstrate sexually dimorphic *Kiss1* expression in the AVPV [70,140]. Furthermore, castration of male rats at birth produces a permanent feminization of developing *Kiss1* cells in the AVPV [138].

Evidence also suggests a role for kisspeptin in the sexually differentiated LH surge. For example, *Kiss1* neuronal activity in the AVPV functions in synchrony with the circadian timing of the LH surge [141]. Transgenic mice lacking the *Kiss1* or *Kiss1R* gene cannot produce an LH surge, even in the presence of estrogen [142]. Furthermore, administration of a kisspeptin receptor antagonist in cycling female rats can block the LH surge [143].

Vasopressin and Oxytocin

Vasopressin and oxytocin are hormones produced by the hypothalamus and secreted by the posterior pituitary gland. Sex differences in vasopressin pathways are widespread in nature and are present in mammals including mice [144] and hamsters [145], as well as nonmammalian vertebrates including reptiles [146], birds [147], and amphibians [148]. Specifically, it appears that vasopressin has an inhibitory effect on reproductive behavior, in that vasopressin inhibits lordosis in both females and males [149]. Intracerebroventricular (icv) administration of vasopressin during the dark period (when the lordosis response is high) inhibits lordosis in females, while icv administration of a vasopressin antagonist during the light period (when the lordosis response is low) facilitates lordosis [150], indicating a role for vasopressin in time-dependent sexual reproduction. Vasopressin provides sexually dimorphic innervation to several brain areas that control sexual behavior, including the septum and the POA [151–153]. For example, males have more vasopressin fibers in the lateral septum [154], and lesions of this area increase the display of lordosis in males [155] and eliminate vasopressin fiber staining [156]. Males also have twice as many vasopressin-immunoreactive cells in the BNST than females [157]. As the BNST is indicated in anxiety, this increase in vasopressin in the BNST of males could play a role in the increased anxiety levels observed in males in tests like the elevated plus maze (EPM) [158].

A role for oxytocin in female sexual receptivity has been suggested, as icv injections of oxytocin in estrogen-treated females stimulates receptivity [159], but these effects may be region-specific. For example, binding of oxytocin in the VMN increases after treatment with estrogen [160,161]. In the paraventricular nucleus, levels of oxytocin vary across the estrous cycle [162], and in the POA, the number of oxytocin-expressing cells decreases after mating in females primed with estrogen [163]. Sex differences, not just in vasopressin and oxytocin, but in all of the described areas of neuronal signaling, provide evidence for the necessary consideration of both sexes at all levels of research.

NEUROLOGICAL CONTROL OF BODY FUNCTIONS

In addition to structural and hormonal sex differences within the central nervous system, there are also differences in several bodily functions, including body weight and body fluid balance, controlled by the nervous system. Dysregulation of the neuronal signaling involved in body weight and food intake can cause obesity, the leading cause for the development of adverse metabolic effects, including type II diabetes mellitus, dyslipidemia, and cardiovascular disease [164,165]. Sex differences exist in the prevalence of these metabolic diseases, and estrogen appears to play a protective role. Prior to menopause, women possess fewer obesity-related metabolic disorders than men, and the prevalence in women increases dramatically after menopause [166]. Regulation of body fluid homeostasis also involves sex differences in the cardiovascular and the renin-angiotensin systems, as well as water and salt intake, which are thought to be modulated in part by estrogen.

Body Weight and Food Intake

Many mammalian species display sex differences in body weight in adulthood, with males typically being larger and heavier than females. These sex differences have long been associated with both organizational and activational effects of steroid hormones [167,168]. For instance, guinea pigs, gonadectomized within 2 days of birth, show a small but significant (6–8%) sex difference in body weight in adulthood, suggesting an organizational impact of sex steroids [167]. On the other hand, treatment with estrogen decreases body weight in adult guinea pigs, while androgen treatment increases body weight [169,170]. Furthermore, gonadectomy in adulthood extinguishes the normal 20–25% difference in body weight between male and female guinea pigs. Therefore, evidence suggests both organizational and activational steroid hormone control of body weight that produces distinguishable sex differences in adulthood.

The regulation of body weight occurs through negative feedback mechanisms whereby signals act in the brain to regulate food intake by modulating the amount of calories stored in adipose tissue and maintaining constant overall adiposity levels [171]. Signaling molecules include leptin, insulin, and estrogen, which are all thought to play a role by relaying information to the brain regarding the overall level of adiposity and body fat distribution. Sex differences in these signaling molecules contribute to sex differences in body weight and food intake in adulthood.

Leptin signaling in the brain, which occurs through its secretion from adipose tissue and subsequent penetration of the blood–brain barrier to bind to leptin receptors in the hypothalamus and brainstem [172], inhibits food intake and increases thermogenesis [173]. Transgenic animals that lack leptin, the leptin receptor, or downstream leptin signaling pathways display profound obesity [174]. Females possess higher circulating levels of leptin, even prior to puberty, than males, independent of differences in body composition [175]. Following puberty, estrogen and testosterone control leptin synthesis and secretion. This is thought to occur via steroid hormone receptor-dependent transcriptional mechanisms [176]. Because females have more subcutaneous fat than males, sex differences in leptin signaling to the brain regarding adiposity are apparent [177,178]. Leptin levels are also inversely correlated with testosterone; exposure of human fat cells to testosterone inhibits leptin expression [179,180].

Insulin, another adiposity signaling molecule in the brain, is secreted from pancreatic β cells in response to increases in circulating glucose levels. Insulin, like leptin, crosses the blood-brain barrier and binds to receptors in order to regulate energy balance. Sex differences have been reported in sensitivity to the catabolic actions of insulin. While female rats are more sensitive to the catabolic action of centrally administered leptin, male rats are more sensitive to the catabolic action of centrally administered insulin [177,178]. Illustrating this sex difference, men, but not women, lose body weight, body fat, and waist circumference following intranasal insulin administration [181].

Changes in food intake are difficult to study in relation to the cycling of estrogen in women due to small differences in consumption over the days of the menstrual cycle. However, the role of estrogen in food intake and body weight has been extensively studied in adult females of many other species [182–186]. For example, food intake decreases during the estrus phase in female rats [184], and progressive decreases in eating through the follicular phase in old-world monkeys have been reported [187,188]. Ovariectomy produces increased daily food intake and promotes weight gain in rodents [177], and subsequent treatment with estrogen normalizes meal size, food intake, and body weight gain to levels observed in intact animals [189]. It appears that estrogen regulates body weight and food intake via interactions with ERα. Male and female transgenic mice lacking the ERα subunit exhibit increased adiposity [190], and site-specific deletion of ERα in the VMH results in obesity [191]. Therefore, it is apparent that sex differences exist in body weight, likely due to steroid hormone actions on signaling in the brain through leptin and insulin.

Body Fluid Balance and Regulation

Body fluid regulation is a dynamic process that, like other regulated physiological systems, involves negative feedback mechanisms as well as a narrow range at which the function of the system is optimal. Body fluid homeostasis involves volume, pressure, and osmotic regulation, all of which are influenced by ovarian hormones. Body fluid balance fluctuates during normal menstrual cycling in tandem with fluctuating ovarian hormones [192–195], and drastically changes during pregnancy. Changes in blood pressure, plasma sodium levels, and vascular volume are subtle during normal reproductive cycling. High estrogen concentrations are associated with basal hypervolemia (increased blood volume) and basal hyponatremia (decreased plasma sodium concentration). During gestation, these effects are enhanced; blood pressure decreases, plasma sodium levels decline profoundly, and vascular volume increases by as much as 50% [196,197]. These changes, while dangerous in a normal individual, are essential for normal fetal development during gestation [198,199].

Cardiovascular Effects

Cardiac function is an important reciprocal component of body fluid balance and regulation in which sex differences exist. For example, the risk of cardiovascular disorders including ischemic heart disease, hypertension, arrhythmias, and heart failure are lower in premenopausal women compared to men, and the incidence of disease increases in women after menopause [200,201], suggesting a role for ovarian hormones. In rodent models, estrogen reverses or prevents experimentally induced hypertension [202,203] and decreases resting blood pressure and heart rate in nonhypertensive rats [194,204]. These protective effects of estrogen result from both an indirect [205] and a direct [206,207] effect of the hormone. However, findings regarding the protective effects of estrogen are inconsistent. For example, estrogen replacement therapy has been shown to increase cardiovascular events in postmenopausal women [208,209]. These inconsistencies in a role for estrogen, as well as apparent sex differences, should be taken into consideration when developing therapeutic targets for cardiac dysfunction.

Estrogen also modulates the cardiovascular response to blood pressure challenges. For example, ovariectomy with estrogen replacement exacerbates both bradycardia in response to phenylephrine-induced hypertension and tachycardia in response to hypotension in rodents compared to those not given estrogen replacement [210–212]. It appears that these effects are receptor-dependent. The enhancement of bradycardia but not tachycardia is absent in ERα knockout mice [213], suggesting that estrogen control of reflex bradycardia requires ERα. Other steroid hormones may also play a role in cardiovascular changes. For example, ARs have been identified in the myocardium of many species, including mice [214], and testosterone has been reported to transcriptionally regulate cardiac gene expression [215,216].

Estrogen acts at multiple regions within the hindbrain that are involved in the neural circuitry of body fluid regulation to control cardiovascular function. Administration of estrogen into the nucleus of the solitary tract (NTS), the primary terminal site of baroreceptor afferents, decreases blood pressure and heart rate and increases vagal efferent activity in ovariectomized rats [217]. Administration of estrogen to the rostral ventral lateral medulla (RVLM) decreases blood pressure and renal sympathetic nerve activity. Injection of estrogen into the nucleus ambiguous decreases heart rate and increases vagal efferent activity [217]. Finally, estrogen administration to the pontine parabrachial nucleus, the midbrain relay center, increases vagal efferent activity and decreases blood pressure and heart rate in ovariectomized rats with estrogen replacement compared to those without estrogen replacement [217]. Given the complex role of steroid hormones in cardiac function, it is evident that findings in one sex cannot be generalized across both sexes.

Renin-Angiotensin System

Renin-angiotensin system (RAS) is a key system in the control of blood pressure and body fluid volume. Sex differences in RAS activity are apparent, and may play a role in the development and progression of cardiovascular diseases [218]. Plasma renin activity, angiotensin-converting enzyme (ACE) activity and expression, and angiotensin-1 (AT1) receptor expression levels are greater in kidneys of males compared to females [219]. Females appear to be less sensitive than males to the hypertensive effects of angiotensin II. Angiotensin II levels peak during the luteal phase of the menstrual cycle when estrogen levels are highest [220], and in postmenopausal women, angiotensin II levels increase following estrogen treatment [221].

However, evidence for the role of estrogen in the RAS is mostly indirect. Estrogen replacement therapy in ovariectomized mice has been shown to reverse the hypertension produced by angiotensin II administration [213] and the angiotensin II-mediated attenuation of reflex bradycardia reported in male rats [222] and mice [213]. At least in part, ERα mediates this reversal, as ovariectomized transgenic mice lacking ERα possess the same angiotensin II-mediated attention of reflex bradycardia as males, regardless of estrogen replacement [213]. Estrogen may also attenuate the effects of angiotensin II by interacting with angiotensin II to decrease activation of AT1 receptors prominent in the area postrema (AP) [223,224]. The AP also contains ERs [225,226], and estrogen has been shown to inhibit neuronal responses to angiotensin II both in dissociated AP neurons [227] and in vivo [228]. Both ACE inhibitors and angiotensin receptor blockers (ARBs) have been repeatedly shown to block the development of renal injury in male animal models [229,230], yet less is known regarding the effects on renal injury in female animals. Despite known sex differences in the RAS, ACE inhibitors and ARBs are the most commonly prescribed treatment for renal disease in both sexes. Understanding renal injury more fully in female animal models will shed light on potentially other possible therapeutics for renal disease.

Water and Salt Intake

A role for estrogen has been suggested in certain behaviors associated with body fluid balance, including water and salt intake. Water consumption has been reported to fluctuate across the estrous cycle in intact female rats with free access to food and water [184,231,232], and there is evidence that estrogen has independent effects on food and water intake in a guinea pig model [169]. This free access model is complicated by the fact that food consumption also varies across the estrous cycle [184], so the decrease in water intake could be secondary to decreased food intake. Therefore, studies also examine estrogen's effects on water intake using various other methodologies, including water deprivation, systemic salt load, volume loss, or hypotension. Results of studies using these techniques suggest that estrogen affects water consumption in response to body fluid challenges related to volume [232–234], but not osmotic [233,235] challenges. For example, female rats consume more water following water deprivation compared to male rats [236]. Furthermore, ovariectomized rats with estrogen replacement before water deprivation or isoproterenol (a β-adrenergic antagonist) treatment consumed approximately half as much water as rats without estrogen treatment [233], suggesting that estrogen is sufficient to inhibit water intake stimulated by angiotensin II.

Sex differences in salt intake remain elusive due to conflicting reports regarding the role of ovarian hormones. One study reports that female rats lick for very concentrated salt solutions at higher rates compared to males, suggesting that salt taste may be less aversive or more palatable to females [237]. However, in free access paradigms, increases in salt intake by estrogen in rats have been reported [238,239], but reports of estrogen-induced decreases in salt intake also exist [240,241]. Speculation remains regarding whether sex differences in salt intake are due to enhanced salt intake by estrogen or to attenuation of salt intake by testosterone. For example, researchers have reported decreases in salt intake in ovariectomized female rats [242,243], while others have reported no effect of ovariectomy, but increased salt intake in castrated male rats [239,244]. Therefore, many questions remain regarding sex differences in salt intake, but it is clear that sex differences in water and salt intake discourage drawing conclusions from one sex without studying the other.

Stress

The stress response provides the framework for survival during hostile conditions. The hypothalamic-pituitary-adrenal (HPA) axis mediates a portion of the stress response and closely interacts with the hypothalamic-pituitary-gonadal (HPG) axis. Thus not surprisingly, sex differences manifest in the HPA axis and arise largely during puberty [245]. Sex differences exist at many levels throughout the HPA axis, particularly at the level of the adrenal cortex and the release of glucocorticoids into the bloodstream, as glucocorticoids are the main player of HPA axis signaling. Glucocorticoids interact with steroid hormones and the HPG axis in an inhibitory manner at all levels of the HPA axis. Sex differences in glucocorticoid actions are present at several levels, particularly at the level of the glucocorticoid receptor (GR). For example, GR mRNA in the anterior pituitary gland is increased following castration in rats, and this effect is reversed by exogenous estrogen administration [246]. Furthermore, recruited chaperones facilitate the transport of the GR to the nucleus to influence GR sensitivity [247], and sex differences in HPA axis function are potentially mediated by these cochaperones and coregulators of glucocorticoids and their effects on GR translocation to the nucleus [248]. Sex differences in region-specific GR activation also exist. Female rats, for example, exhibit increased activation of the GR in the hypothalamus following acute and chronic stress compared to males [249]. The complex interaction between the HPA and HPG axes suggests that sex differences in HPA axis responses to stress likely affect sex differences in stress-induced pathophysiology.

SEX AND BEHAVIOR

Steroid hormones play a large role in species-specific sexually dimorphic behaviors displayed early and late in life by all sexually reproducing animals. For example, estrogens act on ERs to both masculinize (enhance male-typical) and defeminize (suppress female-typical) behavior. These differences are most prominent in reproductive behaviors such as displays of courtship, aggression, mating, and parental care. The biological basis for these behavioral differences is centered in the nervous system, which, as previously discussed, is tightly regulated by steroid hormones. Therefore, this section will focus on sexually dimorphic behaviors and their regulation by steroid hormones.

Reproductive Behavior

Mating behavior is a genetically hardwired sexually dimorphic behavior, in that animals exhibit reproductive behavior without any training or previous experience. Activation of the underlying neural circuits is controlled by both sensory cues as well as steroid hormone secretion by the gonads, the latter of which will be the focus of this section. While feminization and masculinization are determined during development independent of gonadally released hormones, sexual maturation and subsequent reproductive behavior are controlled by binding of steroid hormones to their cognate receptors during puberty and into adulthood. Early studies in songbirds correlated the presence of male song with the presence of testicular secretions [250,251]. Work in rodents established that testosterone could restore mating behavior in castrated male rats [252], and ovarian hormones (estrogen and progesterone) could restore the estrous cycle and reproductive behavior in ovariectomized female rats [253]. Taken together, these studies suggest that circulating steroid hormones in adulthood are necessary and sufficient for the display of mating behavior.

While mating behavior patterns vary across different species, they can also be surprisingly similar. For example, male flies of different drosophilid species, as well as male songbirds of different species, sing a species-specific song to attract a female fly or songbird of the corresponding species only [254,255]. Furthermore, while the genes involved in sex determination and differentiation have diverged rapidly over time, the neurotransmitter systems that modulate reproductive behaviors are quite similar across multiple species.

Rodent models have been the most well studied in regards to mating behavior, as they provide a readily accessible and easily observable behavioral pattern. Female reproductive behavior is traditionally discussed using two distinct characteristics: sexual receptivity and sexual proceptivity. Sexual receptivity is characterized by the display of lordosis, and is quantified by the lordosis quotient (LQ = lordosis to mount ratio). Sexually proceptive behaviors include ear wiggling as well as hopping and darting of the female rat [256]. Male rodent mating behavior is characterized by mounts, intromissions, and ejaculations.

Steroid hormones are important for both female- and male-typical reproductive behavior. In females, ovariectomy abolishes lordosis in response to male mounting behavior. Treatment with estrogen partially restores the lordosis response [257,258], and additional treatment with progesterone dramatically increases the display of female sexual receptivity [259,260]. In fact, treatment with progesterone is necessary for the display of proceptive behaviors [261,262].

Males and females have the ability to display both male- and female-typical mating behavior. Two scenarios to explain this phenomenon have been suggested. The first implies that males and females share the neural circuitry underlying sexual behavior, but that specific factors determine whether the circuitry is activated or repressed. For example, fruit flies produce song for courtship by rubbing their wings together rapidly while near a female. Studies suggest that females have the neural circuitry for this behavior, but that it is repressed in the normal condition, and when activated exogenously is different than the song of males [263]. The second scenario is that male and female brains use different neural circuitry to generate both male- and female-typical behaviors, which is implied in functional imaging studies where men and women perform similarly on multiple tasks [264,265]. Classic studies in rodents suggest that males and females possess the neural circuitry needed for all aspects of male- and female-typical mating behavior. For example, castrated male rats given exogenous estrogen and progesterone display lordosis as well as proceptive behaviors [266]. In addition, female mice treated with testosterone or estrogen exhibit mounting behavior [267,268], and most untreated female rats spontaneously display male mating behavior [269].

Studies utilizing transgenic mouse models have confirmed these earlier reports as well as shed light on ER-dependent sexual behaviors. For example, female mice constitutively null for PR or ERα are not sexually receptive [270,271], but those lacking ERβ retain reproductive behavior [271]. Male mice lacking ERα show greatly reduced sex behavior (although retain simple mounting behavior), while ERβ knockout mice have no impairments in sex behavior [272]. However, ERβ is implicated in the suppression of female sex behavior (defeminization) in males [273]. Not surprisingly, male mice lacking both ERα and ERβ receptors exhibit complete sexual dysfunction [274]. In addition to the ER, studies using manipulation of the gene coding for aromatase reiterate the importance of estrogen for masculinizing sex behavior; [275] male mice lacking the AR have profound impairments in sexual behavior [58,276]. One must understand these behavioral sex differences when examining mating behavior endpoints.

Aggressive Behavior

Aggression, another sexually dimorphic behavior, is closely linked with mating behavior. Aggressive behavior is typically characterized in rodents using the resident/intruder paradigm, where a stimulus (intruder) rodent is placed into the home cage of a resident rodent. Each test for aggression lasts for a certain amount of time (~5 min) or until the first aggressive attack, whichever occurs first. An aggressive attack is characterized usually by the combination of at least two aggressive behaviors, including biting, chasing, wrestling, or lunging [277]. Latency to aggressive attack can also be recorded, typically by direct observation of a single individual. Just as transgenic male mice that lack aromatase or the AR exhibit deficits in mating, they also display profound deficits in aggressive behavior [278,279]. Furthermore, ERα but not ERβ is essential for aggressive behavior in adult males [280]. Male rodents, especially, are more inherently aggressive than their female counterparts, which should be taken into consideration when studies utilize aggressive behavior outcomes.

Addictive Behavior

It is clear that sex differences exist for compulsive, addictive behaviors, particularly in humans. For example, women proceed from initial, casual drug use to compulsive, habitual drug-related behavior more rapidly than men, despite adverse consequences [281]. In addiction, the key player appears to be the neurotransmitter, dopamine. There are clear sex differences in the levels of dopamine in several brain regions, as well as differences in the responsiveness of dopamine to stimulation by amphetamine and steroid hormones [282]. In the context of alcohol dependence and abuse, rates are higher in men compared to women, and being male is a prominent risk factor for alcohol abuse [283]. However, in rodent models, females consume more alcohol than males when exposed to free-drinking paradigms [284,285]. Addiction, therefore, may be a more complicated translational model.

In regards to opiate addiction, sex differences are apparent not only in the analgesic/antinociceptive properties of opioids, but also in opioid-induced side effects including morphine-induced sedation [286], respiratory depression [287], locomotor activity [288], urinary retention [289], seizure susceptibility [290], and learning and memory [291]. Men and women have different sensitivities to noxious stimuli and analgesic drugs; women exhibit lower sensitivity thresholds, greater discrimination ability, and higher ratings of pain when exposed to noxious somatic stimuli compared to men. In the clinic, women show more sensitivity than men to certain opioid analgesics in postoperative surgery [292,293].

Sex differences in responses to opioids also exist in animal models, which have shown differences in sensitivity to the effects of opioids in males and females. Male rats, for example, are more sensitive to the antinociceptive properties of morphine than females in several different pain models [294–296], but females are generally more sensitive to noxious somatic stimuli than males [297,298]. However, studies examining the role of steroid hormones in these sex differences remain inconsistent. For example, intact female rats are more sensitive to the analgesic effects of systemic administration of morphine during diestrus and proestrus [299], which is in line with the interaction of progesterone and opioid receptors mentioned in an earlier section. However, opioid sensitivity does not vary throughout the estrous cycle in mice [300]. Also, conflicting results have been observed using the same nociceptive assay. For example, ovariectomy in adult female rats has been shown to both reduce [301] and increase [302] the magnitude of morphine analgesia on the tail-flick test, as well as produce no change [294]. Further research examining sex differences in opioid sensitivity may shed light on opioid addiction in humans and suggest new rehabilitation or therapeutic practices.

Social Behavior

Social play, which generally takes the form of rough and tumble play, has been studied in several species including dogs, horses, monkeys, rodents, and humans [303]. In rodents, play behavior begins around the time of weaning, peaks at about postnatal day 35, and slowly fades following puberty [304,305]. Social play exhibits a prominent sex difference in that males, regardless of species, engage in more frequent and more intense physical interactions as a part of social play than females [306]. Like mating behavior, this sex difference in social behavior is established early in development by steroid hormones. Both androgens and estrogens have been suggested as primary modulators of this difference. However, the emergence of social play occurs prior to puberty, at a time when there is little to no activational steroid hormone exposure. Therefore, social play is an example of a sexually dimorphic behavior that is not dependent on steroid hormone activation, and the mechanisms underlying sex differences in social play remain largely unknown [303]. The amygdala has been considered the primary brain region involved in juvenile social behavior, but recent evidence supports a role for the prefrontal cortex (PFC) and striatum as well [307].

One theory of the purpose of social play is that it functions to facilitate social development, which is carried over into adulthood. For example, rats that have been deprived of social play during peak age by social isolation, regrouped in adolescence and tested in adulthood show impaired behavioral and neurological responses when confronted with a social stressor (ie, resident-intruder paradigm) [305]. Rats that had been isolated took longer to assume a submissive posture upon attack, displayed decreased immobility and increased exploration and grooming, and had lower adrenaline and corticosterone levels compared to rats that had not been isolated and thus engaged in social play during development. This study and others suggest that social play during development is crucial for the progression of survival skills in adulthood [305,308].

Influence of Genetics on Behavior

The long-standing idea behind sex differences in vertebrate behavior has been a critical role for both organizational and activational effects of gonadal hormones. Recently, a clear influence of genetics on these sex differences has become apparent, not only for behaviors associated with reproduction, but also other behaviors like habit formation and alcohol preference. For example, studies in the zebra finch suggest an interaction between sex chromosome genes and steroid hormones in the development of the sexually dimorphic song nuclei [309]. Furthermore, a relatively new transgenic mouse model called the four core genotypes (FCG) model has shed light on the potential genetic influences of certain sexually dimorphic behaviors. The FCG mouse model dissociates sex chromosome complement from gonadal phenotypes [310] in order to independently study the functional involvement of sex chromosome genes and gonadal sex [311]. This transgenic mouse model has demonstrated a clear genetic role in sexually dimorphic behaviors including aggression [311], habit formation [312], pain perception [313], and alcohol preference [314]. Therefore, it is important to take into consideration both genetic as well as hormonal influences when examining sex differences in the brain and behavior.

SEX AND NEUROLOGICAL DISEASE

It was not until recently that studies began to examine sex differences in the disease progression, symptoms, and treatment of several neurological disorders. Susceptibility to disease can be as great as two- to fivefold more likely in one sex. For example, higher rates of neuropsychiatric and learning disorders are observed in males. A role for steroid hormones, particularly in terms of a protective effect of estrogen, in several conditions, including Alzheimer's disease, PTSD, schizophrenia, stroke, multiple sclerosis, autism, addiction, fibromyalgia, attention deficit disorder, irritable bowel syndrome, Tourette's syndrome, and eating disorders, has been suggested [315,316]. Based on the sex differences that will be discussed in this section, it is imperative to study potential therapeutic targets in preclinical models of both sexes (see Box 3.2).

Alzheimer's Disease

Alzheimer's disease, the leading cause of dementia in elderly populations, is characterized by the presence of plaque deposition in the brain, as well as neurofibrillary pathology. Some studies suggest that Alzheimer's disease is more prevalent in postmenopausal women, indicating that decreasing estrogen levels during and after menopause may influence the pathogenesis of Alzheimer's disease [317,318]. While it remains controversial that estrogen or ER modulators have beneficial effects on disease outcome, both clinical and animal studies have reported a protective role for estrogen in disease onset, progression, and outcome [319]. Several clinical studies have found that estrogen therapy may delay the onset, prevent, and even attenuate Alzheimer's disease [320–322]. For example, the incidence of the disease in women with estrogen treatment is significantly reduced compared to those without estrogen treatment, and meta-analysis studies suggest an overall incidence reduction of 29–44% following estrogen treatment [323,324]. An elegant transgenic mouse model was produced to study the effect of estrogen depletion on a key characteristic of the disease, β-amyloid plaque formation. These mice were produced by crossing aromatase knockout mice with the well-characterized mouse model of Alzheimer's disease, APP23 transgenic mice [325]. This new mouse line exhibits early-onset and increased β-amyloid plaque deposition compared to the APP23 control mice. Furthermore, ovariectomized APP23 control mice display similar β-amyloid plaque pathology to the crossed aromatase/APP23 mice [325]. These studies support the idea that estrogen depletion in the brain is a risk factor for the development of Alzheimer's disease neuropathology.

The mechanism by which estrogen may play a protective role in Alzheimer's disease remains unclear. Furthermore, a protective role does not necessarily imply a cure, as studies examining estrogen treatment after disease diagnosis found generally no change in the decline in cognitive function [326,327], even though there are conflicting reports regarding this matter [320,328]. What is known in terms of mechanism is that a role for estrogen has been shown in several parameters involved in the pathology of Alzheimer's disease. For example, estrogen can protect against β-amyloid toxicity [329–331], oxidative stress [329,332], and excitotoxicity [333,334]. Estrogen can also induce dephosphorylation of the tau protein and prevent its hyperphosphorylation in neurons [335], enhance the activation of Akt, a survival factor, and induce the phosphorylation and deactivation of known death signals, including GSK-3β and BAD in neurons [336,337]. More recently, a role for ERα-mediated inhibition of Death-domain associated protein (Daxx) has been suggested in protection against Alzheimer's disease [318], as well as the *seladin-1* gene, which is downregulated in brain regions associated with Alzheimer's disease and is also upregulated by estrogen. Taken together, these influences of estrogen could play a role in the protective effects observed in Alzheimer's disease, although further research is needed.

A major symptom of individuals with Alzheimer's disease is cognitive decline, and evidence suggests a role for estrogen in cognition and memory. Most observational and longitudinal studies demonstrate that estrogen users perform better on cognitive tests [338]. Particularly, women tend to perform better on tasks of verbal skills, memory, speed, accuracy, and fine motor skills [339,340]. Furthermore, periods of high circulating estrogen levels during the menstrual cycle have been correlated with improved working memory [341]. Postmenopausal women who received estrogen replacement therapy performed better on memory tasks than those without replacement [342], but it appears that the protective effects of estrogen occur at a critical period after menopause. For example, estrogen treatment too long after menopause may have less of a beneficial effect [343]. Animal studies have also supported a role for estrogen in cognition and memory. Estrogen replacement can significantly improve cognitive function in ovariectomized monkeys [344], as well as performance on memory tasks (eg, radial and water mazes, operant alternation task, and active avoidance task) in ovariectomized rats [345–349]. Unfortunately, less is known regarding estrogen's effects on memory in men, but studies suggest similar benefits. Estrogen appears to enhance short-term memory in men taking estrogen to change their gender [350], and estrogen improved memory performance in men with prostate cancer on androgen deprivation [351], suggesting the possible therapeutic potential of estrogen treatment for Alzheimer's disease in men as well as women.

The hippocampus is thought to be the primary site of action of estrogen in regards to cognition [352]. Treatment of ovariectomized rats with estrogen increases hippocampal neuronal excitability [353], as well as hippocampal CA1 dendritic spines and synapse density in rats and monkeys [354–357]. Estrogen also increases spine density and neurite outgrowth in hippocampal cell cultures [358,359]. Furthermore, estrogen enhances LTP in the hippocampus [360,361], and induction of LTP in the CA1 region is enhanced during proestrus and produces enhanced spine density [103,362]. LTP is one of many models for synaptic plasticity, which may underlie learning and memory. As the hippocampus is a sexually dimorphic brain region, it is useful to examine changes in the hippocampus following estrogen treatment and the implications on cognition in both sexes to further establish potential treatments for memory-related symptoms of Alzheimer's disease.

Parkinson's Disease

Parkinson's disease is a chronic neurodegenerative disorder characterized by the destruction of dopaminergic neurons in the substantia nigra. This neuronal loss leads to the well-known symptoms of the disease: tremor, bradykinesia (slowness of movement), rigidity, and balance problems. Sex differences in the risk of Parkinson's disease, symptom severity, and treatment outcome have been reported, including reports of a higher incidence [363–365] and a 1.5 times greater risk [366] in men. Some studies have also found less frequent [367] and less severe symptoms [368,369] in postmenopausal women. In regards to treatment, one study found greater significant improvement in motor function in women treated with levodopa compared to men, but that women were more likely to develop drug-related dyskinesias (124).

Growing evidence in animal models suggests a neuroprotective role of estrogen in the disease state. A commonly used animal model of Parkinson's disease is 1-methyl-4-phenyl-1,2,3,6-tetrahydropyridine (MPTP)-induced nigrostriatal lesions, which exemplifies this neuroprotective role of estrogen [370]. Estrogen has been shown to prevent MPTP-induced depletion of striatal dopamine, as well as dopamine transporter binding and expression. Estrogen also decreases the MPTP-induced tyrosine hydroxylase-immunoreactive neuronal loss and attenuates the glial activation induced by MPTP [371–375]. Furthermore, estrogen has been reported to reduce the dyskinetic effects produced by MPTP in a monkey model [376]. In human studies, postmenopausal women given estrogen replacement therapy are less likely to develop Parkinson's disease than those without [377], and estrogen replacement therapy increases dopamine transporter availability in the caudate and putamen [378], suggesting that estrogen may have a protective role in disease progression.

Stroke

Reports suggest that women have a lesser incidence of stroke compared to men; however, stroke increases in women following menopause [379,380], suggesting a potential role for ovarian hormones. Studies also suggest that stroke outcome is worse in postmenopausal women, with a higher disability and fatality rate compared to men [381,382], and sex differences exist in risk factors for stroke. For example, there are risk factors that are unique to women like reproductive factors, and those that are more common in women, including migraine with aura, obesity, metabolic syndrome, and atrial fibrillation [383]. Animal models also shed light on a potential role of estrogen in stroke outcome. For example, female gerbils with cerebral ischemia-induced brain damage have a lower incidence and less severe brain damage compared to males [384]. A similar sex difference was found in studies using mice and rats, where females possess smaller infarct volume than males following middle cerebral artery occlusion, and this protective effect was abolished by ovariectomy [385,386]. In a traumatic brain injury model, female rats have greater survival rates compared to males [387].

A protective role for estrogen in cerebral ischemia has been supported by numerous studies. For example, exogenous estrogen administration has been shown to reduce infarct volume following focal or global cerebral ischemia in several models, including ovariectomized female mice [386,388], rats [389,390], and gerbils [391,392], as well as male rats [393] and aged female rats [390]. In a systematic review of more than 160 publications, estrogen was able to dose-dependently reduce lesion volume after both transient and permanent ischemia in the majority of studies [394]. Furthermore, serum estrogen levels have been shown to be inversely correlated with ischemic stroke damage in intact animals, and treatment with the anti-ER compound ICI182,780 significantly enhances stroke infarct size in intact female mice [395,396]. Studies using aromatase knockout mice also demonstrate increased infarct size, and aromatase inhibitors cause increases in cortical and striatal damage following cerebral ischemia [397]. The complexities surrounding sex differences in stroke outcome suggest that further research is crucial in order to treat and ideally prevent the incidence of stroke.

Depression and Anxiety

Depression and anxiety, two types of affective disorders, are often discussed together because they are frequently comorbid disorders that result in psychological, physiological, and behavioral symptoms that have a significant impact on quality of life. Affective or mood disorders are significantly more prevalent in women, and women are twice as likely to suffer from depression and anxiety than men [398,399]. Not surprisingly, estrogen is thought to play an important role in the sex difference in incidence of mood disorders. Women are more likely to suffer from depression and anxiety during time periods of hormonal flux (eg, premenstrual, postpartum, and perimenopausal periods) [400,401]. Like other disorders described in this section, it has been suggested that estrogen may have a protective role when it comes to mood disorders, although some studies find no changes in mood following estrogen treatment [402–404]. In addition, the incidence of mood disorders increases closer toward menopause in parallel with declining estrogen levels [405], further suggesting a role for estrogen in the incidence of depression and anxiety.

Studies using animal models of anxiety and depression have further examined sex differences in mood disorders and the potential role for steroid hormones. One of the many rodent models commonly used to test depression-like behavior is the forced swim test (FST) [406]. Rodents exhibit either active behaviors (eg, swimming, diving, headshakes, and climbing) or passive behaviors (eg, immobility and floating) in the FST, and the amount of time spent exhibiting passive behaviors is indicative of depression-like behavior. Rodent behavior in the FST varies during the estrous cycle; passive behaviors are reduced during the proestrus (high circulating estrogen) phase compared to females in the metestrus or diestrus phases, which are associated with lower estrogen levels [407,408]. Immobility is also reduced during pregnancy [409,410] and increased following ovariectomy [411,412]. Importantly, this last effect can be reversed by estrogen replacement, suggestive of a protective role of estrogen.

The EPM and open field tests provide two common rodent models of anxiety-like behavior. Rodents possess an innate aversion to open or brightly lit spaces and therefore express a cessation in exploratory behaviors in these tests, for example the open arms of the EPM and the center of the open field. Several studies demonstrate significant increases in open arm time in the EPM and exploratory behavior in the open field in cycling females during the proestrus phase [408,413], which are suggestive of decreased anxiety-like behavior. However, there is conflicting evidence in the literature suggestive of the opposite effect of estrogen [414,415], and another group has reported no differences [416]. Contradictions are also present in studies using ovariectomized females, with reported decreases in anxiety-like behavior in the EPM and open field [417,418] as well as reported increases [415,419]. Thus, it is important to consider multiple factors when examining anxiety-like behavior in rodents, including species differences, time of day of testing, light-dark cycles, hormonal dosage, route of administration, and especially previous exposure to other behavior tasks, as multiple behavioral exposures may have an effect on anxiety-like behavior [420].

Important considerations regarding mechanism include binding of ERs and in what brain areas estrogen may be having an effect in regards to depression- and anxiety-like behavior. For example, both depression- and anxiety-like behaviors are significantly increased in the FST, EPM, and open field in female ERβ knockout mice compared to wild-type [421,422], but no differences are observed in ERα knockout mice [422], suggesting an implication for ERβ receptor binding in these behaviors. In addition, estrogen administered directly to the hippocampus or medial amygdala decreases passive behavior in the FST as well as anxiety-like behaviors in the EPM and open field tests [417,423]. Therefore, outcomes in these tests may be dependent on receptor binding in specific brain regions.

CONCLUSIONS

Sex differences in neurophysiology and behavior are extremely important to consider, especially as more research is uncovered regarding neurological disease. While necessary to understand, the traditional theory that sex differences are due solely to steroid hormone actions including estrogen, progesterone, and testosterone is now encompassing other aspects including environmental and genetic factors. Long-standing differences exist in brain structure and function in sexually dimorphic regions like the SDN-POA and AVPV, neurological signaling including neurotransmitter systems, bodily functions, and reproductively associated behaviors like mating and aggressive behavior. These sexually dimorphic characteristics, including possible environmental and genetic influences, must be further evaluated as the search continues for therapeutic targets for neurological disease, especially those that exhibit clear sex differences like Alzheimer's disease and stroke.

References

[1] McCarthy MM. The two faces of estradiol: effects on the developing brain. Neuroscientist 2009;15(6):599–610.

[2] Lenz KM, McCarthy MM. Organized for sex—steroid hormones and the developing hypothalamus. Eur J Neurosci 2010;32(12):2096–104.

[3] Jost A. Genetic and hormonal factors in sex differentiation of the brain. Psychoneuroendocrinology 1983;8(2):183–93.

[4] Phoenix CH, Goy RW, Gerall AA, Young WC. Organizing action of prenatally administered testosterone propionate on the tissues mediating mating behavior in the female guinea pig. Endocrinology 1959;65:369–82.

[5] Gahr M, Metzdorf R. The sexually dimorphic expression of androgen receptors in the song nucleus hyperstriatalis ventrale pars caudale of the zebra finch develops independently of gonadal steroids. J Neurosci 1999;19(7):2628–36.

[6] Sayag N, Robinzon B, Snapir N, Arnon E, Grimm VE. The effects of embryonic treatments with gonadal hormones on sexually dimorphic behavior of chicks. Horm Behav 1991; 25(2):137–53.

[7] Wilson JA, Glick B. Ontogeny of mating behavior in the chicken. Am J Physiol 1970;218(4):951–5.

[8] Gahr M. Male Japanese quails with female brains do not show male sexual behaviors. Proc Natl Acad Sci USA 2003;100(13):7959–64.

[9] Bruggeman V, Van As P, Decuypere E. Developmental endocrinology of the reproductive axis in the chicken embryo. Comp Biochem Physiol A Mol Integr Physiol 2002;131(4):839–46.

[10] Smith CA, Sinclair AH. Sex determination in the chicken embryo. J Exp Zool 2001;290(7):691–9.

[11] Balaban E, Teillet MA, Le Douarin N. Application of the quail-chick chimera system to the study of brain development and behavior. Science 1988;241(4871):1339–42.

[12] Gahr M, Balaban E. The development of a species difference in the local distribution of brain estrogen receptive cells. Brain Res Dev Brain Res 1996;92(2):182–9.

[13] King WJ, Greene GL. Monoclonal antibodies localize oestrogen receptor in the nuclei of target cells. Nature 1984;307 (5953):745–7.

[14] O'Malley BW, Tsai MJ. Molecular pathways of steroid receptor action. Biol Reprod 1992;46(2):163–7.

[15] Paech K, Webb P, Kuiper GG, Nilsson S, Gustafsson J, Kushner PJ, et al. Differential ligand activation of estrogen receptors ERalpha and ERbeta at AP1 sites. Science 1997;277(5331):1508–10.

[16] Uht RM, Anderson CM, Webb P, Kushner PJ. Transcriptional activities of estrogen and glucocorticoid receptors are functionally integrated at the AP-1 response element. Endocrinology 1997;138(7):2900–8.

[17] Abraham IM, Todman MG, Korach KS, Herbison AE. Critical in vivo roles for classical estrogen receptors in rapid estrogen actions on intracellular signaling in mouse brain. Endocrinology 2004;145(7):3055–61.

[18] Zadran S, Qin Q, Bi X, Zadran H, Kim Y, Foy MR, et al. 17-Beta-estradiol increases neuronal excitability through MAP kinase-induced calpain activation. Proc Natl Acad Sci USA 2009;106(51):21936–41.

[19] Mermelstein PG, Micevych PE. Nervous system physiology regulated by membrane estrogen receptors. Rev Neurosci 2008;19(6):413–24.

[20] Balthazart J, Ball GF. Is brain estradiol a hormone or a neurotransmitter? Trends Neurosci 2006;29(5):241–9.

[21] Remage-Healey L, Maidment NT, Schlinger BA. Forebrain steroid levels fluctuate rapidly during social interactions. Nat Neurosci 2008;11(11):1327–34.

[22] Schwarz JM, Liang SL, Thompson SM, McCarthy MM. Estradiol induces hypothalamic dendritic spines by enhancing glutamate release: a mechanism for organizational sex differences. Neuron 2008;58(4):584–98.

[23] Micevych P, Kuo J, Christensen A. Physiology of membrane oestrogen receptor signalling in reproduction. J Neuroendocrinol 2009;21(4):249–56.

[24] London SE, Remage-Healey L, Schlinger BA. Neurosteroid production in the songbird brain: a re-evaluation of core principles. Front Neuroendocrinol 2009;30(3):302–14.

[25] Garcia-Segura LM, Lorenz B, DonCarlos LL. The role of glia in the hypothalamus: implications for gonadal steroid feedback and reproductive neuroendocrine output. Reproduction 2008;135(4):419–29.

[26] Kudwa AE, Michopoulos V, Gatewood JD, Rissman EF. Roles of estrogen receptors alpha and beta in differentiation of mouse sexual behavior. Neuroscience 2006;138(3):921–8.

[27] Beato M, Klug J. Steroid hormone receptors: an update. Hum Reprod Update 2000;6(3):225–36.

[28] Kuo J, Hariri OR, Bondar G, Ogi J, Micevych P. Membrane estrogen receptor-alpha interacts with metabotropic glutamate receptor type 1a to mobilize intracellular calcium in hypothalamic astrocytes. Endocrinology 2009;150(3):1369–76.

[29] Qiu J, Bosch MA, Tobias SC, Grandy DK, Scanlan TS, Ronnekleiv OK, et al. Rapid signaling of estrogen in hypothalamic neurons involves a novel G-protein-coupled estrogen receptor that activates protein kinase C. J Neurosci 2003; 23(29):9529–40.

[30] Toran-Allerand CD, Guan X, MacLusky NJ, Horvath TL, Diano S, Singh M, et al. ER-X: a novel, plasma membrane-associated, putative estrogen receptor that is regulated during development and after ischemic brain injury. J Neurosci 2002;22(19):8391–401.

[31] Revankar CM, Cimino DF, Sklar LA, Arterburn JB, Prossnitz ER. A transmembrane intracellular estrogen receptor mediates rapid cell signaling. Science 2005;307(5715):1625–30.

[32] Parsons B, Rainbow TC, MacLusky NJ, McEwen BS. Progestin receptor levels in rat hypothalamic and limbic nuclei. J Neurosci 1982;2(10):1446–52.

[33] Romano GJ, Krust A, Pfaff DW. Expression and estrogen regulation of progesterone receptor mRNA in neurons of the mediobasal hypothalamus: an in situ hybridization study. Mol Endocrinol 1989;3(8):1295–300.

[34] Grazzini E, Guillon G, Mouillac B, Zingg HH. Inhibition of oxytocin receptor function by direct binding of progesterone. Nature 1998;392(6675):509–12.

[35] Ferin M, Wehrenberg WB, Lam NY, Alston EJ, Vande Wiele RL. Effects and site of action of morphine on gonadotropin secretion in the female rhesus monkey. Endocrinology 1982;111 (5):1652–6.

[36] Casper RF, Alapin-Rubillovitz S. Progestins increase endogenous opioid peptide activity in postmenopausal women. J Clin Endocrinol Metab 1985;60(1):34–6.

[37] Valera S, Ballivet M, Bertrand D. Progesterone modulates a neuronal nicotinic acetylcholine receptor. Proc Natl Acad Sci USA 1992;89(20):9949–53.

[38] Rick CE, Ye Q, Finn SE, Harrison NL. Neurosteroids act on the GABA(A) receptor at sites on the N-terminal side of the middle of TM2. Neuroreport 1998;9(3):379–83.

[39] Weisz J, Ward IL. Plasma testosterone and progesterone titers of pregnant rats, their male and female fetuses, and neonatal offspring. Endocrinology 1980;106(1):306–16.

[40] Quadros PS, Goldstein AY, De Vries GJ, Wagner CK. Regulation of sex differences in progesterone receptor expression in the medial preoptic nucleus of postnatal rats. J Neuroendocrinol 2002;14(10):761–7.

[41] Quadros PS, Pfau JL, Goldstein AY, De Vries GJ, Wagner CK. Sex differences in progesterone receptor expression: a potential mechanism for estradiol-mediated sexual differentiation. Endocrinology 2002;143(10):3727–39.
[42] Quadros PS, Pfau JL, Wagner CK. Distribution of progesterone receptor immunoreactivity in the fetal and neonatal rat forebrain. J Comp Neurol 2007;504(1):42–56.
[43] Reddy DS. Neurosteroids: endogenous role in the human brain and therapeutic potentials. Prog Brain Res 2010;186:113–37.
[44] Irwin RW, Brinton RD. Allopregnanolone as regenerative therapeutic for Alzheimer's disease: translational development and clinical promise. Prog Neurobiol 2014;113:40–55.
[45] Brinton RD. The neurosteroid 3 alpha-hydroxy-5 alpha-pregnan-20-one induces cytoarchitectural regression in cultured fetal hippocampal neurons. J Neurosci 1994;14(5 Pt 1):2763–74.
[46] Wang JM, Johnston PB, Ball BG, Brinton RD. The neurosteroid allopregnanolone promotes proliferation of rodent and human neural progenitor cells and regulates cell-cycle gene and protein expression. J Neurosci 2005;25(19):4706–18.
[47] Jessberger S, Gage FH. Fate plasticity of adult hippocampal progenitors: biological relevance and therapeutic use. Trends Pharmacol Sci 2009;30(2):61–5.
[48] Toni N, Laplagne DA, Zhao C, Lombardi G, Ribak CE, Gage FH, et al. Neurons born in the adult dentate gyrus form functional synapses with target cells. Nat Neurosci 2008;11 (8):901–7.
[49] Kuhn HG, Dickinson-Anson H, Gage FH. Neurogenesis in the dentate gyrus of the adult rat: age-related decrease of neuronal progenitor proliferation. J Neurosci 1996; 16(6):2027–33.
[50] Marx CE, Trost WT, Shampine LJ, Stevens RD, Hulette CM, Steffens DC, et al. The neurosteroid allopregnanolone is reduced in prefrontal cortex in Alzheimer's disease. Biol Psychiatry 2006;60(12):1287–94.
[51] Weill-Engerer S, David JP, Sazdovitch V, Liere P, Eychenne B, Pianos A, et al. Neurosteroid quantification in human brain regions: comparison between Alzheimer's and nondemented patients. J Clin Endocrinol Metab 2002;87(11):5138–43.
[52] Wang JM, Singh C, Liu L, Irwin RW, Chen S, Chung EJ, et al. Allopregnanolone reverses neurogenic and cognitive deficits in mouse model of Alzheimer's disease. Proc Natl Acad Sci USA 2010;107(14):6498–503.
[53] Oddo S, Caccamo A, Shepherd JD, Murphy MP, Golde TE, Kayed R, et al. Triple-transgenic model of Alzheimer's disease with plaques and tangles: intracellular Abeta and synaptic dysfunction. Neuron 2003;39(3):409–21.
[54] Chen S, Wang JM, Irwin RW, Yao J, Liu L, Brinton RD. Allopregnanolone promotes regeneration and reduces beta-amyloid burden in a preclinical model of Alzheimer's disease. PLoS One 2011;6(8):e24293.
[55] Mangelsdorf DJ, Thummel C, Beato M, Herrlich P, Schutz G, Umesono K, et al. The nuclear receptor superfamily: the second decade. Cell 1995;83(6):835–9.
[56] Matsumoto T, Shiina H, Kawano H, Sato T, Kato S. Androgen receptor functions in male and female physiology. J Steroid Biochem Mol Biol 2008;109(3–5):236–41.
[57] Morris JA, Jordan CL, Breedlove SM. Sexual differentiation of the vertebrate nervous system. Nat Neurosci 2004;7(10):1034–9.
[58] Juntti SA, Coats JK, Shah NM. A genetic approach to dissect sexually dimorphic behaviors. Horm Behav 2008;53(5):627–37.
[59] Nottebohm F, Arnold AP. Sexual dimorphism in vocal control areas of the songbird brain. Science 1976;194(4261):211–13.
[60] Arai Y, Sekine Y, Murakami S. Estrogen and apoptosis in the developing sexually dimorphic preoptic area in female rats. Neurosci Res 1996;25(4):403–7.
[61] Gorski RA, Gordon JH, Shryne JE, Southam AM. Evidence for a morphological sex difference within the medial preoptic area of the rat brain. Brain Res 1978;148(2):333–46.
[62] Rhees RW, Shryne JE, Gorski RA. Onset of the hormone-sensitive perinatal period for sexual differentiation of the sexually dimorphic nucleus of the preoptic area in female rats. J Neurobiol 1990;21(5):781–6.
[63] Hsu HK, Shao PL, Tsai KL, Shih HC, Lee TY, Hsu C. Gene regulation by NMDA receptor activation in the SDN-POA neurons of male rats during sexual development. J Mol Endocrinol 2005;34(2):433–45.
[64] Hsu HK, Yang RC, Shih HC, Hsieh YL, Chen UY, Hsu C. Prenatal exposure of testosterone prevents SDN-POA neurons of postnatal male rats from apoptosis through NMDA receptor. J Neurophysiol 2001;86(5):2374–80.
[65] McCarthy MM, Arnold AP. Reframing sexual differentiation of the brain. Nat Neurosci 2011;14(6):677–83.
[66] Forger NG, Rosen GJ, Waters EM, Jacob D, Simerly RB, de Vries GJ. Deletion of Bax eliminates sex differences in the mouse forebrain. Proc Natl Acad Sci USA 2004; 101(37):13666–71.
[67] Ahmed EI, Zehr JL, Schulz KM, Lorenz BH, DonCarlos LL, Sisk CL. Pubertal hormones modulate the addition of new cells to sexually dimorphic brain regions. Nat Neurosci 2008; 11(9):995–7.
[68] Simerly RB, Swanson LW, Gorski RA. The distribution of monoaminergic cells and fibers in a periventricular preoptic nucleus involved in the control of gonadotropin release: immunohistochemical evidence for a dopaminergic sexual dimorphism. Brain Res 1985;330(1):55–64.
[69] Ottem EN, Godwin JG, Krishnan S, Petersen SL. Dual-phenotype GABA/glutamate neurons in adult preoptic area: sexual dimorphism and function. J Neurosci 2004;24 (37):8097–105.
[70] Kauffman AS, Gottsch ML, Roa J, Byquist AC, Crown A, Clifton DK, et al. Sexual differentiation of Kiss1 gene expression in the brain of the rat. Endocrinology 2007;148(4):1774–83.
[71] Waters EM, Simerly RB. Estrogen induces caspase-dependent cell death during hypothalamic development. J Neurosci 2009;29(31):9714–18.
[72] Krishnan S, Intlekofer KA, Aggison LK, Petersen SL. Central role of TRAF-interacting protein in a new model of brain sexual differentiation. Proc Natl Acad Sci USA 2009;106(39):16692–7.
[73] Smith MS, Grove KL. Integration of the regulation of reproductive function and energy balance: lactation as a model. Front Neuroendocrinol 2002;23(3):225–56.
[74] Bouret SG, Simerly RB. Developmental programming of hypothalamic feeding circuits. Clin Genet 2006;70(1):295–301.
[75] Mong JA, Glaser E, McCarthy MM. Gonadal steroids promote glial differentiation and alter neuronal morphology in the developing hypothalamus in a regionally specific manner. J Neurosci 1999;19(4):1464–72.
[76] Matsumoto A, Arai Y. Sexual dimorphism in "wiring pattern" in the hypothalamic arcuate nucleus and its modification by neonatal hormonal environment. Brain Res 1980; 190(1):238–42.
[77] Mong JA, Kurzweil RL, Davis AM, Rocca MS, McCarthy MM. Evidence for sexual differentiation of glia in rat brain. Horm Behav 1996;30(4):553–62.
[78] Mong JA, Roberts RC, Kelly JJ, McCarthy MM. Gonadal steroids reduce the density of axospinous synapses in the developing rat arcuate nucleus: an electron microscopy analysis. J Comp Neurol 2001;432(2):259–67.
[79] Larsson K, Heimer L. Mating behaviour of male rats after lesions in the preoptic area. Nature 1964;202:413–14.

[80] Simerly RB, Swanson LW. The organization of neural inputs to the medial preoptic nucleus of the rat. J Comp Neurol 1986;246(3):312–42.
[81] Raisman G, Field PM. Sexual dimorphism in the neuropil of the preoptic area of the rat and its dependence on neonatal androgen. Brain Res 1973;54:1–29.
[82] Amateau SK, McCarthy MM. Induction of PGE2 by estradiol mediates developmental masculinization of sex behavior. Nat Neurosci 2004;7(6):643–50.
[83] Amateau SK, McCarthy MM. Sexual differentiation of astrocyte morphology in the developing rat preoptic area. J Neuroendocrinol 2002;14(11):904–10.
[84] Amateau SK, McCarthy MM. A novel mechanism of dendritic spine plasticity involving estradiol induction of prostaglandin-E2. J Neurosci 2002;22(19):8586–96.
[85] Pfaff DW, Sakuma Y. Facilitation of the lordosis reflex of female rats from the ventromedial nucleus of the hypothalamus. J Physiol 1979;288:189–202.
[86] Matsumoto A, Arai Y. Male-female difference in synaptic organization of the ventromedial nucleus of the hypothalamus in the rat. Neuroendocrinology 1986;42(3):232–6.
[87] Walker DL, Toufexis DJ, Davis M. Role of the bed nucleus of the stria terminalis versus the amygdala in fear, stress, and anxiety. Eur J Pharmacol 2003;463(1–3):199–216.
[88] Simerly RB. Wired for reproduction: organization and development of sexually dimorphic circuits in the mammalian forebrain. Annu Rev Neurosci 2002;25:507–36.
[89] Cooke B, Hegstrom CD, Villeneuve LS, Breedlove SM. Sexual differentiation of the vertebrate brain: principles and mechanisms. Front Neuroendocrinol 1998;19(4):323–62.
[90] Hines M, Allen LS, Gorski RA. Sex differences in subregions of the medial nucleus of the amygdala and the bed nucleus of the stria terminalis of the rat. Brain Res 1992;579(2):321–6.
[91] Gotsiridze T, Kang N, Jacob D, Forger NG. Development of sex differences in the principal nucleus of the bed nucleus of the stria terminalis of mice: role of Bax-dependent cell death. Dev Neurobiol 2007;67(3):355–62.
[92] Murray EK, Hien A, de Vries GJ, Forger NG. Epigenetic control of sexual differentiation of the bed nucleus of the stria terminalis. Endocrinology 2009;150(9):4241–7.
[93] De Vries GJ, Panzica GC. Sexual differentiation of central vasopressin and vasotocin systems in vertebrates: different mechanisms, similar endpoints. Neuroscience 2006;138(3):947–55.
[94] Shah NM, Pisapia DJ, Maniatis S, Mendelsohn MM, Nemes A, Axel R. Visualizing sexual dimorphism in the brain. Neuron 2004;43(3):313–19.
[95] Hutton LA, Gu G, Simerly RB. Development of a sexually dimorphic projection from the bed nuclei of the stria terminalis to the anteroventral periventricular nucleus in the rat. J Neurosci 1998;18(8):3003–13.
[96] Goldstein JM, Seidman LJ, Horton NJ, Makris N, Kennedy DN, Caviness VS, et al. Normal sexual dimorphism of the adult human brain assessed by in vivo magnetic resonance imaging. Cereb Cortex 2001;11(6):490–7.
[97] Madeira MD, Lieberman AR. Sexual dimorphism in the mammalian limbic system. Prog Neurobiol 1995;45(4):275–333.
[98] Zhang JM, Konkle AT, Zup SL, McCarthy MM. Impact of sex and hormones on new cells in the developing rat hippocampus: a novel source of sex dimorphism? Eur J Neurosci 2008;27(4):791–800.
[99] Isgor C, Sengelaub DR. Prenatal gonadal steroids affect adult spatial behavior, CA1 and CA3 pyramidal cell morphology in rats. Horm Behav 1998;34(2):183–98.
[100] Bowers JM, Waddell J, McCarthy MM. A developmental sex difference in hippocampal neurogenesis is mediated by endogenous oestradiol. Biol Sex Differ 2010;1(1):8.
[101] Woolley CS, Gould E, Frankfurt M, McEwen BS. Naturally occurring fluctuation in dendritic spine density on adult hippocampal pyramidal neurons. J Neurosci 1990;10(12):4035–9.
[102] Woolley CS, McEwen BS. Estradiol mediates fluctuation in hippocampal synapse density during the estrous cycle in the adult rat. J Neurosci 1992;12(7):2549–54.
[103] Warren SG, Humphreys AG, Juraska JM, Greenough WT. LTP varies across the estrous cycle: enhanced synaptic plasticity in proestrus rats. Brain Res 1995;703(1–2):26–30.
[104] Markus E, Zecevic M. Sex differences and estrous cycle changes in hippocampus-dependent fear conditioning. Psychobiology 1997;25(3):246–52.
[105] Frye CA. Estrus-associated decrements in a water maze task are limited to acquisition. Physiol Behav 1995;57(1):5–14.
[106] Krebs-Kraft DL, Hill MN, Hillard CJ, McCarthy MM. Sex difference in cell proliferation in developing rat amygdala mediated by endocannabinoids has implications for social behavior. Proc Natl Acad Sci USA 2010;107(47):20535–40.
[107] Ziabreva I, Poeggel G, Schnabel R, Braun K. Separation-induced receptor changes in the hippocampus and amygdala of Octodon degus: influence of maternal vocalizations. J Neurosci 2003;23(12):5329–36.
[108] Todd BJ, Schwarz JM, Mong JA, McCarthy MM. Glutamate AMPA/kainate receptors, not GABA(A) receptors, mediate estradiol-induced sex differences in the hypothalamus. Dev Neurobiol 2007;67(3):304–15.
[109] Vries GJ. Sex differences in neurotransmitter systems. J Neuroendocrinol 1990;2(1):1–13.
[110] Luine VN, Thornton JE, Frankfurt M, MacLusky NJ. Effects of hypothalamic serotonin depletion on lordosis behavior and gonadal hormone receptors. Brain Res 1987;426(1):47–54.
[111] Kato R. Serotonin content of rat brain in relation to sex and age. J Neurochem 1960;5:202.
[112] Fischette CT, Biegon A, McEwen BS. Sex differences in serotonin 1 receptor binding in rat brain. Science 1983; 222(4621):333–5.
[113] Simerly RB, Swanson LW, Gorski RA. Demonstration of a sexual dimorphism in the distribution of serotonin-immunoreactive fibers in the medial preoptic nucleus of the rat. J Comp Neurol 1984;225(2):151–66.
[114] Biegon A, Fischette CT, Rainbow TC, McEwen BS. Serotonin receptor modulation by estrogen in discrete brain nuclei. Neuroendocrinology 1982;35(4):287–91.
[115] Clemens LG, Barr P, Dohanich GP. Cholinergic regulation of female sexual behavior in rats demonstrated by manipulation of endogenous acetylcholine. Physiol Behav 1989;45(2):437–42.
[116] Dohanich GP, Barr PJ, Witcher JA, Clemens LG. Pharmacological and anatomical aspects of cholinergic activation of female sexual behavior. Physiol Behav 1984; 32(6):1021–6.
[117] Dohanich GP, Clemens LG. Brain areas implicated in cholinergic regulation of sexual behavior. Horm Behav 1981;15 (2):157–67.
[118] Olsen KL, Edwards E, Schechter N, Whalen RE. Muscarinic receptors in preoptic area and hypothalamus: effects of cyclicity, sex and estrogen treatment. Brain Res 1988;448(2):223–9.
[119] Kalra SP, Kalra PS. Neural regulation of luteinizing hormone secretion in the rat. Endocr Rev 1983;4(4):311–51.
[120] Dyer RG. Sexual differentiation of the forebrain—relationship to gonadotrophin secretion. Prog Brain Res 1984;61:223–36.
[121] Goldman BD, Gorski RA. Effects of gonadal steroids on the secretion of LH and FSH in neonatal rats. Endocrinology 1971;89(1):112–15.
[122] Brann DW, Mahesh VB. Glutamate: a major neuroendocrine excitatory signal mediating steroid effects on gonadotropin secretion. J Steroid Biochem Mol Biol 1995;53(1–6):325–9.

[123] Kotani M, Detheux M, Vandenbogaerde A, Communi D, Vanderwinden JM, Le Poul E, et al. The metastasis suppressor gene KiSS-1 encodes kisspeptins, the natural ligands of the orphan G protein-coupled receptor GPR54. J Biol Chem 2001;276(37):34631–6.
[124] Muir AI, Chamberlain L, Elshourbagy NA, Michalovich D, Moore DJ, Calamari A, et al. AXOR12, a novel human G protein-coupled receptor, activated by the peptide KiSS-1. J Biol Chem 2001;276(31):28969–75.
[125] Ohtaki T, Shintani Y, Honda S, Matsumoto H, Hori A, Kanehashi K, et al. Metastasis suppressor gene KiSS-1 encodes peptide ligand of a G-protein-coupled receptor. Nature 2001;411(6837):613–17.
[126] Clarkson J, Herbison AE. Postnatal development of kisspeptin neurons in mouse hypothalamus; sexual dimorphism and projections to gonadotropin-releasing hormone neurons. Endocrinology 2006;147(12):5817–25.
[127] Kauffman AS, Clifton DK, Steiner RA. Emerging ideas about kisspeptin- GPR54 signaling in the neuroendocrine regulation of reproduction. Trends Neurosci 2007;30(10): 504–11.
[128] Irwig MS, Fraley GS, Smith JT, Acohido BV, Popa SM, Cunningham MJ, et al. Kisspeptin activation of gonadotropin releasing hormone neurons and regulation of KiSS-1 mRNA in the male rat. Neuroendocrinology 2004;80(4):264–72.
[129] Kauffman AS, Park JH, McPhie-Lalmansingh AA, Gottsch ML, Bodo C, Hohmann JG, et al. The kisspeptin receptor GPR54 is required for sexual differentiation of the brain and behavior. J Neurosci 2007;27(33):8826–35.
[130] Han SK, Gottsch ML, Lee KJ, Popa SM, Smith JT, Jakawich SK, et al. Activation of gonadotropin-releasing hormone neurons by kisspeptin as a neuroendocrine switch for the onset of puberty. J Neurosci 2005;25(49):11349–56.
[131] Messager S, Chatzidaki EE, Ma D, Hendrick AG, Zahn D, Dixon J, et al. Kisspeptin directly stimulates gonadotropin-releasing hormone release via G protein-coupled receptor 54. Proc Natl Acad Sci USA 2005;102(5):1761–6.
[132] Navarro VM, Castellano JM, Fernandez-Fernandez R, Tovar S, Roa J, Mayen A, et al. Effects of KiSS-1 peptide, the natural ligand of GPR54, on follicle-stimulating hormone secretion in the rat. Endocrinology 2005;146(4):1689–97.
[133] Navarro VM, Castellano JM, Fernandez-Fernandez R, Tovar S, Roa J, Mayen A, et al. Characterization of the potent luteinizing hormone-releasing activity of KiSS-1 peptide, the natural ligand of GPR54. Endocrinology 2005;146(1):156–63.
[134] Seminara SB, Messager S, Chatzidaki EE, Thresher RR, Acierno Jr JS, Shagoury JK, et al. The GPR54 gene as a regulator of puberty. N Engl J Med 2003;349(17):1614–27.
[135] Funes S, Hedrick JA, Vassileva G, Markowitz L, Abbondanzo S, Golovko A, et al. The KiSS-1 receptor GPR54 is essential for the development of the murine reproductive system. Biochem Biophys Res Commun 2003;312(4):1357–63.
[136] de Roux N, Genin E, Carel JC, Matsuda F, Chaussain JL, Milgrom E. Hypogonadotropic hypogonadism due to loss of function of the KiSS1-derived peptide receptor GPR54. Proc Natl Acad Sci USA 2003;100(19):10972–6.
[137] Kauffman AS, Navarro VM, Kim J, Clifton DK, Steiner RA. Sex differences in the regulation of Kiss1/NKB neurons in juvenile mice: implications for the timing of puberty. Am J Physiol Endocrinol Metab 2009;297(5):E1212–21.
[138] Homma T, Sakakibara M, Yamada S, Kinoshita M, Iwata K, Tomikawa J, et al. Significance of neonatal testicular sex steroids to defeminize anteroventral periventricular kisspeptin neurons and the GnRH/LH surge system in male rats. Biol Reprod 2009;81(6):1216–25.
[139] Smith JT, Popa SM, Clifton DK, Hoffman GE, Steiner RA. Kiss1 neurons in the forebrain as central processors for generating the preovulatory luteinizing hormone surge. J Neurosci 2006;26(25):6687–94.
[140] Adachi S, Yamada S, Takatsu Y, Matsui H, Kinoshita M, Takase K, et al. Involvement of anteroventral periventricular metastin/kisspeptin neurons in estrogen positive feedback action on luteinizing hormone release in female rats. J Reprod Dev 2007;53(2):367–78.
[141] Robertson JL, Clifton DK, de la Iglesia HO, Steiner RA, Kauffman AS. Circadian regulation of Kiss1 neurons: implications for timing the preovulatory gonadotropin-releasing hormone/luteinizing hormone surge. Endocrinology 2009;150(8):3664–71.
[142] Clarkson J, d'Anglemont de Tassigny X, Moreno AS, Colledge WH, Herbison AE. Kisspeptin-GPR54 signaling is essential for preovulatory gonadotropin-releasing hormone neuron activation and the luteinizing hormone surge. J Neurosci 2008;28(35):8691–7.
[143] Pineda R, Garcia-Galiano D, Roseweir A, Romero M, Sanchez-Garrido MA, Ruiz-Pino F, et al. Critical roles of kisspeptins in female puberty and preovulatory gonadotropin surges as revealed by a novel antagonist. Endocrinology 2010; 151(2):722–30.
[144] Mayes CR, Watts AG, McQueen JK, Fink G, Charlton HM. Gonadal steroids influence neurophysin II distribution in the forebrain of normal and mutant mice. Neuroscience 1988;25(3):1013–22.
[145] Buijs RM, Pevet P, Masson-Pevet M, Pool CW, de Vries GJ, Canguilhem B, et al. Seasonal variation in vasopressin innervation in the brain of the European hamster (Cricetus cricetus). Brain Res 1986;371(1):193–6.
[146] Stoll CJ, Voorn P. The distribution of hypothalamic and extrahypothalamic vasotocinergic cells and fibers in the brain of a lizard, Gekko gecko: presence of a sex difference. J Comp Neurol 1985;239(2):193–204.
[147] Voorhuis TA, Kiss JZ, de Kloet ER, de Wied D. Testosterone-sensitive vasotocin-immunoreactive cells and fibers in the canary brain. Brain Res 1988;442(1):139–46.
[148] Moore FL, Miller LJ. Arginine vasotocin induces sexual behavior of newts by acting on cells in the brain. Peptides 1983;4(1):97–102.
[149] Sodersten P, De Vries GJ, Buijs RM, Melin P. A daily rhythm in behavioral vasopressin sensitivity and brain vasopressin concentrations. Neurosci Lett 1985;58(1):37–41.
[150] Sodersten P. Sexual differentiation: do males differ from females in behavioral sensitivity to gonadal hormones? Prog Brain Res 1984;61:257–70.
[151] Zasorin NL. Suppression of lordosis in the hormone-primed female hamster by electrical stimulation of the septal area. Physiol Behav 1975;14(5): 595–9.
[152] Nance DM, Shryne J, Gorski RA. Septal lesions: effects on lordosis behavior and pattern of gonadotropin release. Horm Behav 1974;5(1):73–81.
[153] Powers B, Valenstein ES. Sexual receptivity: facilitation by medial preoptic lesions in female rats. Science 1972;175(4025):1003–5.
[154] de Vries GJ, Buijs RM, Swaab DF. Ontogeny of the vasopressinergic neurons of the suprachiasmatic nucleus and their extrahypothalamic projections in the rat brain—presence of a sex difference in the lateral septum. Brain Res 1981; 218(1–2):67–78.
[155] Yamanouchi K, Arai Y. Female lordosis pattern in the male rat induced by estrogen and progesterone: effect of interruption of the dorsal inputs to the preoptic area and hypothalamus. Endocrinol Jpn 1975;22(3):243–6.

[156] De Vries GJ, Buijs RM. The origin of the vasopressinergic and oxytocinergic innervation of the rat brain with special reference to the lateral septum. Brain Res 1983;273(2):307–17.

[157] van Leeuwen FW, Caffe AR, De Vries GJ. Vasopressin cells in the bed nucleus of the stria terminalis of the rat: sex differences and the influence of androgens. Brain Res 1985;325 (1–2):391–4.

[158] Johnston AL, File SE. Sex differences in animal tests of anxiety. Physiol Behav 1991;49(2):245–50.

[159] Gorzalka BB, Lester GL. Oxytocin-induced facilitation of lordosis behaviour in rats is progesterone-dependent. Neuropeptides 1987;10(1):55–65.

[160] de Kloet ER, Voorhuis DA, Boschma Y, Elands J. Estradiol modulates density of putative "oxytocin receptors" in discrete rat brain regions. Neuroendocrinology 1986;44(4):415–21.

[161] Johnson AE, Coirini H, Ball GF, McEwen BS. Anatomical localization of the effects of 17 beta-estradiol on oxytocin receptor binding in the ventromedial hypothalamic nucleus. Endocrinology 1989;124(1):207–11.

[162] Greer ER, Caldwell JD, Johnson MF, Prange Jr AJ, Pedersen CA. Variations in concentration of oxytocin and vasopressin in the paraventricular nucleus of the hypothalamus during the estrous cycle in rats. Life Sci 1986;38(25):2311–18.

[163] Caldwell JD, Jirikowski GF, Greer ER, Stumpf WE, Pedersen CA. Ovarian steroids and sexual interaction alter oxytocinergic content and distribution in the basal forebrain. Brain Res 1988;446(2):236–44.

[164] Cummings DE, Schwartz MW. Genetics and pathophysiology of human obesity. Annu Rev Med 2003;54:453–71.

[165] Eckel RH, Barouch WW, Ershow AG. Report of the National Heart, Lung, and Blood Institute-National Institute of Diabetes and Digestive and Kidney Diseases Working Group on the pathophysiology of obesity-associated cardiovascular disease. Circulation 2002;105(24):2923–8.

[166] Ford ES. Risks for all-cause mortality, cardiovascular disease, and diabetes associated with the metabolic syndrome: a summary of the evidence. Diabetes Care 2005;28(7):1769–78.

[167] Slob AK, Goy RW, Van der Werff ten B. Sex differences in growth of guinea-pigs and their modification by neonatal gonadectomy and prenatally administered androgen. J Endocrinol 1973;58(1):11–19.

[168] Slob AK, Van der Werff Ten Bosch JJ. Sex differences in body growth in the rat. Physiol Behav 1975;14(3):353–61.

[169] Czaja JA, Butera PC, McCaffrey TA. Independent effects of estradiol on water and food intake. Behav Neurosci 1983;97 (2):210–20.

[170] Czaja JA, McCaffrey TA, Butera PC. Effects of female hormonal condition on body weight of male partners: dependence on testicular factors. Behav Neurosci 1983;97(6):984–93.

[171] Woods SC, Seeley RJ, Porte Jr D, Schwartz MW. Signals that regulate food intake and energy homeostasis. Science 1998;280 (5368):1378–83.

[172] Schwartz MW, Woods SC, Porte Jr D, Seeley RJ, Baskin DG. Central nervous system control of food intake. Nature 2000;404(6778):661–71.

[173] Seeley RJ, Woods SC. Monitoring of stored and available fuel by the CNS: implications for obesity. Nat Rev Neurosci 2003;4 (11):901–9.

[174] Munzberg H, Bjornholm M, Bates SH, Myers Jr MG. Leptin receptor action and mechanisms of leptin resistance. Cell Mol Life Sci 2005;62(6):642–52.

[175] Demerath EW, Towne B, Wisemandle W, Blangero J, Chumlea WC, Siervogel RM. Serum leptin concentration, body composition, and gonadal hormones during puberty. Int J Obes Relat Metab Disord 1999;23(7):678–85.

[176] Machinal F, Dieudonne MN, Leneveu MC, Pecquery R, Giudicelli Y. In vivo and in vitro ob gene expression and leptin secretion in rat adipocytes: evidence for a regional specific regulation by sex steroid hormones. Endocrinology 1999;140(4):1567–74.

[177] Clegg DJ, Brown LM, Woods SC, Benoit SC. Gonadal hormones determine sensitivity to central leptin and insulin. Diabetes 2006;55(4):978–87.

[178] Clegg DJ, Riedy CA, Smith KA, Benoit SC, Woods SC. Differential sensitivity to central leptin and insulin in male and female rats. Diabetes 2003;52(3):682–7.

[179] Kristensen K, Pedersen SB, Richelsen B. Regulation of leptin by steroid hormones in rat adipose tissue. Biochem Biophys Res Commun 1999;259(3):624–30.

[180] Wabitsch M, Blum WF, Muche R, Braun M, Hube F, Rascher W, et al. Contribution of androgens to the gender difference in leptin production in obese children and adolescents. J Clin Invest 1997;100(4):808–13.

[181] Hallschmid M, Benedict C, Schultes B, Fehm HL, Born J, Kern W. Intranasal insulin reduces body fat in men but not in women. Diabetes 2004;53(11):3024–9.

[182] Gong EJ, Garrel D, Calloway DH. Menstrual cycle and voluntary food intake. Am J Clin Nutr 1989;49(2):252–8.

[183] Czaja JA, Goy RW. Ovarian hormones and food intake in female guinea pigs and rhesus monkeys. Horm Behav 1975;6((4):329–49.

[184] Eckel LA, Houpt TA, Geary N. Spontaneous meal patterns in female rats with and without access to running wheels. Physiol Behav 2000;70(3–4):397–405.

[185] Friend DW. Self-selection of feeds and water by swine during pregnancy and lactation. J Anim Sci 1971;32(4):658–66.

[186] Houpt KA, Coren B, Hintz HF, Hilderbrant JE. Effect of sex and reproductive status on sucrose preference, food intake, and body weight of dogs. J Am Vet Med Assoc 1979;174(10):1083–5.

[187] Czaja JA. Ovarian influences on primate food intake: assessment of progesterone actions. Physiol Behav 1978; 21(6):923–8.

[188] Rosenblatt H, Dyrenfurth I, Ferin M, vande Wiele RL. Food intake and the menstrual cycle in rhesus monkeys. Physiol Behav 1980;24(3):447–9.

[189] Asarian L, Geary N. Cyclic estradiol treatment normalizes body weight and restores physiological patterns of spontaneous feeding and sexual receptivity in ovariectomized rats. Horm Behav 2002;42(4):461–71.

[190] Heine PA, Taylor JA, Iwamoto GA, Lubahn DB, Cooke PS. Increased adipose tissue in male and female estrogen receptor-alpha knockout mice. Proc Natl Acad Sci USA 2000;97(23):12729–34.

[191] Musatov S, Chen W, Pfaff DW, Mobbs CV, Yang XJ, Clegg DJ, et al. Silencing of estrogen receptor alpha in the ventromedial nucleus of hypothalamus leads to metabolic syndrome. Proc Natl Acad Sci USA 2007;104(7):2501–6.

[192] Summy-Long JY, Kadekaro M. Role of circumventricular organs (CVO) in neuroendocrine responses: interactions of CVO and the magnocellular neuroendocrine system in different reproductive states. Clin Exp Pharmacol Physiol 2001;28(7):590–601.

[193] Barron WM, Schreiber J, Lindheimer MD. Effect of ovarian sex steroids on osmoregulation and vasopressin secretion in the rat. Am J Physiol 1986;250(4 Pt 1):E352–61.

[194] Takezawa H, Hayashi H, Sano H, Saito H, Ebihara S. Circadian and estrous cycle-dependent variations in blood pressure and heart rate in female rats. Am J Physiol 1994;267(5 Pt 2):R1250–6.

[195] Stachenfeld NS, DiPietro L, Kokoszka CA, Silva C, Keefe DL, Nadel ER. Physiological variability of fluid-regulation hormones in young women. J Appl Physiol 1999;86(3):1092–6.
[196] Ganzevoort W, Rep A, Bonsel GJ, de Vries JI, Wolf H. Plasma volume and blood pressure regulation in hypertensive pregnancy. J Hypertens 2004;22(7):1235–42.
[197] Chapman AB, Abraham WT, Zamudio S, Coffin C, Merouani A, Young D, et al. Temporal relationships between hormonal and hemodynamic changes in early human pregnancy. Kidney Int 1998;54(6):2056–63.
[198] Duvekot JJ, Cheriex EC, Pieters FA, Menheere PP, Schouten HJ, Peeters LL. Maternal volume homeostasis in early pregnancy in relation to fetal growth restriction. Obstet Gynecol 1995;85(3):361–7.
[199] Duvekot JJ, Cheriex EC, Pieters FA, Peeters LL. Severely impaired fetal growth is preceded by maternal hemodynamic maladaptation in very early pregnancy. Acta Obstet Gynecol Scand 1995;74(9):693–7.
[200] Safar ME, Smulyan H. Hypertension in women. Am J Hypertens 2004;17(1):82–7.
[201] Rossouw JE. Hormones, genetic factors, and gender differences in cardiovascular disease. Cardiovasc Res 2002;53(3):550–7.
[202] Haywood JR, Hinojosa-Laborde C. Sexual dimorphism of sodium-sensitive renal-wrap hypertension. Hypertension 1997;30(3 Pt 2):667–71.
[203] Xue B, Pamidimukkala J, Hay M. Sex differences in the development of angiotensin II-induced hypertension in conscious mice. Am J Physiol Heart Circ Physiol 2005;288(5):H2177–84.
[204] Hernandez I, Delgado JL, Diaz J, Quesada T, Teruel MJ, Llanos MC, et al. 17beta-estradiol prevents oxidative stress and decreases blood pressure in ovariectomized rats. Am J Physiol Regul Integr Comp Physiol 2000;279(5):R1599–605.
[205] Lange DL, Haywood JR, Hinojosa-Laborde C. Role of the adrenal medullae in male and female DOCA-salt hypertensive rats. Hypertension 1998;31(1 Pt 2):403–8.
[206] Hinojosa-Laborde C, Lange DL, Haywood JR. Role of female sex hormones in the development and reversal of dahl hypertension. Hypertension 2000;35(1 Pt 2):484–9.
[207] Crofton JT, Share L. Gonadal hormones modulate deoxycorticosterone-salt hypertension in male and female rats. Hypertension 1997;29(1 Pt 2):494–9.
[208] Rossouw JE, Anderson GL, Prentice RL, LaCroix AZ, Kooperberg C, Stefanick ML, Writing Group for the Women's Health Initiative I, et al. Risks and benefits of estrogen plus progestin in healthy postmenopausal women: principal results from the Women's Health Initiative randomized controlled trial. Jama 2002;288(3):321–33.
[209] Grady D, Wenger NK, Herrington D, Khan S, Furberg C, Hunninghake D, et al. Postmenopausal hormone therapy increases risk for venous thromboembolic disease. The Heart and Estrogen/progestin Replacement Study. Ann Intern Med 2000;132(9):689–96.
[210] Saleh TM, Saleh MC, Connell BJ. Estrogen blocks the cardiovascular and autonomic changes following vagal stimulation in ovariectomized rats. Auton Neurosci 2001;88(1–2):25–35.
[211] Mohamed MK, El-Mas MM, Abdel-Rahman AA. Estrogen enhancement of baroreflex sensitivity is centrally mediated. Am J Physiol 1999;276(4 Pt 2):R1030–7.
[212] He XR, Wang W, Crofton JT, Share L. Effects of 17beta-estradiol on sympathetic activity and pressor response to phenylephrine in ovariectomized rats. Am J Physiol 1998;275 (4 Pt 2):R1202–8.
[213] Pamidimukkala J, Xue B, Newton LG, Lubahn DB, Hay M. Estrogen receptor-alpha mediates estrogen facilitation of baroreflex heart rate responses in conscious mice. Am J Physiol Heart Circ Physiol 2005;288(3):H1063–70.
[214] Marsh JD, Lehmann MH, Ritchie RH, Gwathmey JK, Green GE, Schiebinger RJ. Androgen receptors mediate hypertrophy in cardiac myocytes. Circulation 1998;98(3):256–61.
[215] Koenig H, Goldstone A, Lu CY. Testosterone-mediated sexual dimorphism of the rodent heart. Ventricular lysosomes, mitochondria, and cell growth are modulated by androgens. Circ Res 1982;50(6):782–7.
[216] Morano I, Gerstner J, Ruegg JC, Ganten U, Ganten D, Vosberg HP. Regulation of myosin heavy chain expression in the hearts of hypertensive rats by testosterone. Circ Res 1990;66(6):1585–90.
[217] Saleh MC, Connell BJ, Saleh TM. Autonomic and cardiovascular reflex responses to central estrogen injection in ovariectomized female rats. Brain Res 2000;879(1–2):105–14.
[218] Sullivan JC. Sex and the renin-angiotensin system: inequality between the sexes in response to RAS stimulation and inhibition. Am J Physiol Regul Integr Comp Physiol 2008;294(4): R1220–6.
[219] Fischer M, Baessler A, Schunkert H. Renin angiotensin system and gender differences in the cardiovascular system. Cardiovasc Res 2002;53(3):672–7.
[220] Chidambaram M, Duncan JA, Lai VS, Cattran DC, Floras JS, Scholey JW, et al. Variation in the renin angiotensin system throughout the normal menstrual cycle. J Am Soc Nephrol 2002;13(2):446–52.
[221] Harvey PJ, Morris BL, Miller JA, Floras JS. Estradiol induces discordant angiotensin and blood pressure responses to orthostasis in healthy postmenopausal women. Hypertension 2005;45(3):399–405.
[222] Cox BF, Bishop VS. Neural and humoral mechanisms of angiotensin-dependent hypertension. Am J Physiol 1991;261 (4 Pt 2):H1284–91.
[223] Mendelsohn FA, Quirion R, Saavedra JM, Aguilera G, Catt KJ. Autoradiographic localization of angiotensin II receptors in rat brain. Proc Natl Acad Sci USA 1984;81(5):1575–9.
[224] Sirett NE, McLean AS, Bray JJ, Hubbard JI. Distribution of angiotensin II receptors in rat brain. Brain Res 1977;122 (2):299–312.
[225] Simerly RB, Chang C, Muramatsu M, Swanson LW. Distribution of androgen and estrogen receptor mRNA-containing cells in the rat brain: an in situ hybridization study. J Comp Neurol 1990;294(1):76–95.
[226] Simonian SX, Herbison AE. Differential expression of estrogen receptor and neuropeptide Y by brainstem A1 and A2 noradrenaline neurons. Neuroscience 1997;76(2):517–29.
[227] Pamidimukkala J, Hay M. 17 beta-Estradiol inhibits angiotensin II activation of area postrema neurons. Am J Physiol Heart Circ Physiol 2003;285(4):H1515–20.
[228] Krause EG, Curtis KS, Markle JP, Contreras RJ. Oestrogen affects the cardiovascular and central responses to isoproterenol of female rats. J Physiol 2007;582(Pt 1):435–47.
[229] Teng J, Fukuda N, Suzuki R, Takagi H, Ikeda Y, Tahira Y, et al. Inhibitory effect of a novel angiotensin II type 1 receptor antagonist RNH-6270 on growth of vascular smooth muscle cells from spontaneously hypertensive rats: different antiproliferative effect to angiotensin-converting enzyme inhibitor. J Cardiovasc Pharmacol 2002;39(2):161–71.
[230] Kobori H, Ozawa Y, Suzaki Y, Nishiyama A. Enhanced intrarenal angiotensinogen contributes to early renal injury in spontaneously hypertensive rats. J Am Soc Nephrol 2005;16 (7):2073–80.
[231] Tarttelin MF, Gorski RA. Variations in food and water intake in the normal and acyclic female rat. Physiol Behav 1971;7 (6):847–52.
[232] Kucharczyk J. Localization of central nervous system structures mediating extracellular thirst in the female rat. J Endocrinol 1984;100(2):183–8.

[233] Krause EG, Curtis KS, Davis LM, Stowe JR, Contreras RJ. Estrogen influences stimulated water intake by ovariectomized female rats. Physiol Behav 2003;79(2):267–74.
[234] Fregly MJ, Thrasher TN. Attenuation of angiotensin-induced water intake in estrogen-treated rats. Pharmacol Biochem Behav 1978;9(4):509–14.
[235] Findlay AL, Fitzsimons JT, Kucharczyk J. Dependence of spontaneous and angiotensin-induced drinking in the rat upon the oestrous cycle and ovarian hormones. J Endocrinol 1979;82(2):215–25.
[236] Kaufman S. A comparison of the dipsogenic responses of male and female rats to a variety of stimuli. Can J Physiol Pharmacol 1980;58(10):1180–3.
[237] Curtis KS, Davis LM, Johnson AL, Therrien KL, Contreras RJ. Sex differences in behavioral taste responses to and ingestion of sucrose and NaCl solutions by rats. Physiol Behav 2004; 80(5):657–64.
[238] Wolf G. Refined salt appetite methodology for rats demonstrated by assessing sex differences. J Comp Physiol Psychol 1982;96(6):1016–21.
[239] Chow SY, Sakai RR, Witcher JA, Adler NT, Epstein AN. Sex and sodium intake in the rat. Behav Neurosci 1992; 106(1):172–80.
[240] Stricker EM, Thiels E, Verbalis JG. Sodium appetite in rats after prolonged dietary sodium deprivation: a sexually dimorphic phenomenon. Am J Physiol 1991;260(6 Pt 2):R1082–8.
[241] Scheidler MG, Verbalis JG, Stricker EM. Inhibitory effects of estrogen on stimulated salt appetite in rats. Behav Neurosci 1994;108(1):141–50.
[242] Krecek J, Novakova V, Stibral K. Sex differences in the taste preference for a salt solution in the rat. Physiol Behav 1972; 8(2):183–8.
[243] Flynn FW, Schulkin J, Havens M. Sex differences in salt preference and taste reactivity in rats. Brain Res Bull 1993;32(2):91–5.
[244] Krecek J. Sex differences in salt taste: the effect of testosterone. Physiol Behav 1973;10(4):683–8.
[245] Panagiotakopoulos L, Neigh GN. Development of the HPA axis: where and when do sex differences manifest? Front Neuroendocrinol 2014;35(3):285–302.
[246] Peiffer A, Barden N. Estrogen-induced decrease of glucocorticoid receptor messenger ribonucleic acid concentration in rat anterior pituitary gland. Mol Endocrinol 1987;1(6):435–40.
[247] Binder EB. The role of FKBP5, a co-chaperone of the glucocorticoid receptor in the pathogenesis and therapy of affective and anxiety disorders. Psychoneuroendocrinology 2009;34 (Suppl. 1):S186–95.
[248] Bourke CH, Raees MQ, Malviya S, Bradburn CA, Binder EB, Neigh GN. Glucocorticoid sensitizers Bag1 and Ppid are regulated by adolescent stress in a sex-dependent manner. Psychoneuroendocrinology 2013;38(1):84–93.
[249] Zavala JK, Fernandez AA, Gosselink KL. Female responses to acute and repeated restraint stress differ from those in males. Physiol Behav 2011;104(2):215–21.
[250] Bottjer SW, Hewer SJ. Castration and antisteroid treatment impair vocal learning in male zebra finches. J Neurobiol 1992;23(4):337–53.
[251] Marler P, Peters S, Ball GF, Dufty Jr AM, Wingfield JC. The role of sex steroids in the acquisition and production of birdsong. Nature 1988;336(6201):770–2.
[252] Beeman EA. The effect of male hormone on aggressive behavior in mice. Physiol Zool 1947;20(4):373–405.
[253] Wiesner BP. The hormones controlling reproduction. Eugen Rev 1930;22(1):19–26.
[254] Wheeler DA, Kyriacou CP, Greenacre ML, Yu Q, Rutila JE, Rosbash M, et al. Molecular transfer of a species-specific behavior from Drosophila simulans to Drosophila melanogaster. Science 1991;251(4997):1082–5.
[255] Konishi M. Birdsong: from behavior to neuron. Annu Rev Neurosci 1985;8:125–70.
[256] Beach FA. Sexual attractivity, proceptivity, and receptivity in female mammals. Horm Behav 1976;7(1):105–38.
[257] Pfaff D. Nature of sex hormone effects on rat sex behavior: specificity of effects and individual patterns of response. J Comp Physiol Psychol 1970;73(3):349–58.
[258] Sodersten P, Eneroth P. Serum levels of oestradiol-17 beta and progesterone in relation to sexual receptivity in intact and ovariectomized rats. J Endocrinol 1981;89(1):45–54.
[259] Glaser JH, Rubin BS, Barfield RJ. Onset of the receptive and proceptive components of feminine sexual behavior in rats following the intravenous administration of progesterone. Horm Behav 1983;17(1):18–27.
[260] McGinnis MY, Parsons B, Rainbow TC, Krey LC, McEwen BS. Temporal relationship between cell nuclear progestin receptor levels and sexual receptivity following intravenous progesterone administration. Brain Res 1981;218(1–2):365–71.
[261] Tennent BJ, Smith ER, Davidson JM. The effects of estrogen and progesterone on female rat proceptive behavior. Horm Behav 1980;14(1):65–75.
[262] Fadem BH, Barfield RJ, Whalen RE. Dose-response and time-response relationships between progesterone and the display of patterns of receptive and proceptive behavior in the female rat. Horm Behav 1979;13(1):40–8.
[263] Rideout EJ, Billeter JC, Goodwin SF. The sex-determination genes fruitless and doublesex specify a neural substrate required for courtship song. Curr Biol 2007;17(17):1473–8.
[264] Grabowski TJ, Damasio H, Eichhorn GR, Tranel D. Effects of gender on blood flow correlates of naming concrete entities. NeuroImage 2003;20(2):940–54.
[265] Piefke M, Weiss PH, Markowitsch HJ, Fink GR. Gender differences in the functional neuroanatomy of emotional episodic autobiographical memory. Hum Brain Mapp 2005;24(4):313–24.
[266] Sodersten P. Lordosis behaviour in male, female and androgenized female rats. J Endocrinol 1976;70(3):409–20.
[267] Wersinger SR, Sannen K, Villalba C, Lubahn DB, Rissman EF, De Vries GJ. Masculine sexual behavior is disrupted in male and female mice lacking a functional estrogen receptor alpha gene. Horm Behav 1997;32(3):176–83.
[268] Jyotika J, McCutcheon J, Laroche J, Blaustein JD, Forger NG. Deletion of the Bax gene disrupts sexual behavior and modestly impairs motor function in mice. Dev Neurobiol 2007;67(11):1511–19.
[269] Sodersten P, Larsson K, Ahlenius S, Engel J. Stimulation of mounting behavior but not lordosis behavior in ovariectomized female rats by p-chlorophenylalanine. Pharmacol Biochem Behav 1976;5(3):329–33.
[270] Blaustein JD. Neuroendocrine regulation of feminine sexual behavior: lessons from rodent models and thoughts about humans. Annu Rev Psychol 2008;59:93–118.
[271] Kudwa AE, Rissman EF. Double oestrogen receptor alpha and beta knockout mice reveal differences in neural oestrogen-mediated progestin receptor induction and female sexual behaviour. J Neuroendocrinol 2003;15(10):978–83.
[272] Ogawa S, Eng V, Taylor J, Lubahn DB, Korach KS, Pfaff DW. Roles of estrogen receptor-alpha gene expression in reproduction-related behaviors in female mice. Endocrinology 1998;139(12):5070–81.

[273] Kudwa AE, Bodo C, Gustafsson JA, Rissman EF. A previously uncharacterized role for estrogen receptor beta: defeminization of male brain and behavior. Proc Natl Acad Sci USA 2005;102(12):4608–12.
[274] Ogawa S, Chester AE, Hewitt SC, Walker VR, Gustafsson JA, Smithies O, et al. Abolition of male sexual behaviors in mice lacking estrogen receptors alpha and beta (alpha beta ERKO). Proc Natl Acad Sci USA 2000;97(26):14737–41.
[275] Honda S, Harada N, Ito S, Takagi Y, Maeda S. Disruption of sexual behavior in male aromatase-deficient mice lacking exons 1 and 2 of the cyp19 gene. Biochem Biophys Res Commun 1998;252(2):445–9.
[276] Juntti SA, Tollkuhn J, Wu MV, Fraser EJ, Soderborg T, Tan S, et al. The androgen receptor governs the execution, but not programming, of male sexual and territorial behaviors. Neuron 2010;66(2):260–72.
[277] Selmanoff MK, Maxson SC, Ginsburg BE. Chromosomal determinants of intermale aggressive behavior in inbred mice. Behav Genet 1976;6(1):53–69.
[278] Sato T, Matsumoto T, Kawano H, Watanabe T, Uematsu Y, Sekine K, et al. Brain masculinization requires androgen receptor function. Proc Natl Acad Sci USA 2004;101(6):1673–8.
[279] Matsumoto T, Honda S, Harada N. Alteration in sex-specific behaviors in male mice lacking the aromatase gene. Neuroendocrinology 2003;77(6):416–24.
[280] Scordalakes EM, Rissman EF. Aggression in male mice lacking functional estrogen receptor alpha. Behav Neurosci 2003;117 (1):38–45.
[281] Lynch WJ, Roth ME, Carroll ME. Biological basis of sex differences in drug abuse: preclinical and clinical studies. Psychopharmacology (Berl) 2002;164(2):121–37.
[282] Becker JB. Gender differences in dopaminergic function in striatum and nucleus accumbens. Pharmacol Biochem Behav 1999;64(4):803–12.
[283] Kalaydjian A, Swendsen J, Chiu WT, Dierker L, Degenhardt L, Glantz M, et al. Sociodemographic predictors of transitions across stages of alcohol use, disorders, and remission in the National Comorbidity Survey Replication. Compr Psychiatry 2009;50(4):299–306.
[284] Lancaster FE, Brown TD, Coker KL, Elliott JA, Wren SB. Sex differences in alcohol preference and drinking patterns emerge during the early postpubertal period. Alcohol Clin Exp Res 1996;20(6):1043–9.
[285] Middaugh LD, Kelley BM, Bandy AL, McGroarty KK. Ethanol consumption by C57BL/6 mice: influence of gender and procedural variables. Alcohol 1999;17(3):175–83.
[286] Craft RM, Heideman LM, Bartok RE. Effect of gonadectomy on discriminative stimulus effects of morphine in female versus male rats. Drug Alcohol Depend 1999;53(2):95–109.
[287] Dahan A, Kest B. Recent advances in opioid pharmacology. Curr Opin Anaesthesiol 2001;14(4):405–10.
[288] Craft RM, Clark JL, Hart SP, Pinckney MK. Sex differences in locomotor effects of morphine in the rat. Pharmacol Biochem Behav 2006;85(4):850–8.
[289] Craft RM, Ulibarri CM, Raub DJ. Kappa opioid-induced diuresis in female vs. male rats. Pharmacol Biochem Behav 2000;65 (1):53–9.
[290] Riazi K, Honar H, Homayoun H, Rashidi N, Dehghani M, Sadeghipour H, et al. Sex and estrus cycle differences in the modulatory effects of morphine on seizure susceptibility in mice. Epilepsia 2004;45(9):1035–42.
[291] Dahan A, Kest B, Waxman AR, Sarton E. Sex-specific responses to opiates: animal and human studies. Anesth Analg 2008;107(1):83–95.
[292] Berkley KJ. Sex differences in pain. Behav Brain Sci 1997;20 (3):371–80: discussion 435–513.
[293] Mayer EA, Berman S, Chang L, Naliboff BD. Sex-based differences in gastrointestinal pain. Eur J Pain 2004;8(5):451–63.
[294] Cicero TJ, Nock B, Meyer ER. Gender-related differences in the antinociceptive properties of morphine. J Pharmacol Exp Ther 1996;279(2):767–73.
[295] Ali BH, Sharif SI, Elkadi A. Sex differences and the effect of gonadectomy on morphine-induced antinociception and dependence in rats and mice. Clin Exp Pharmacol Physiol 1995;22(5):342–4.
[296] Baamonde AI, Hidalgo A, Andres-Trelles F. Sex-related differences in the effects of morphine and stress on visceral pain. Neuropharmacology 1989;28(9):967–70.
[297] Kim SJ, Calejesan AA, Li P, Wei F, Zhuo M. Sex differences in late behavioral response to subcutaneous formalin injection in mice. Brain Res 1999;829(1–2):185–9.
[298] Chanda ML, Mogil JS. Sex differences in the effects of amiloride on formalin test nociception in mice. Am J Physiol Regul Integr Comp Physiol 2006;291(2):R335–42.
[299] Stoffel EC, Ulibarri CM, Craft RM. Gonadal steroid hormone modulation of nociception, morphine antinociception and reproductive indices in male and female rats. Pain 2003;103(3):285–302.
[300] Moskowitz AS, Terman GW, Carter KR, Morgan MJ, Liebeskind JC. Analgesic, locomotor and lethal effects of morphine in the mouse: strain comparisons. Brain Res 1985;361 (1–2):46–51.
[301] Kepler KL, Kest B, Kiefel JM, Cooper ML, Bodnar RJ. Roles of gender, gonadectomy and estrous phase in the analgesic effects of intracerebroventricular morphine in rats. Pharmacol Biochem Behav 1989;34(1):119–27.
[302] Kasson BG, George R. Endocrine influences on the actions of morphine: IV. Effects of sex and strain. Life Sci 1984;34 (17):1627–34.
[303] Auger AP, Olesen KM. Brain sex differences and the organisation of juvenile social play behaviour. J Neuroendocrinol 2009;21(6):519–25.
[304] Panksepp J, Siviy S, Normansell L. The psychobiology of play: theoretical and methodological perspectives. Neurosci Biobehav Rev 1984;8(4):465–92.
[305] Vanderschuren LJ, Niesink RJ, Van Ree JM. The neurobiology of social play behavior in rats. Neurosci Biobehav Rev 1997;21 (3):309–26.
[306] Olioff M, Stewart J. Sex differences in the play behavior of prepubescent rats. Physiol Behav 1978;20(2):113–15.
[307] van Kerkhof LW, Damsteegt R, Trezza V, Voorn P, Vanderschuren LJ. Social play behavior in adolescent rats is mediated by functional activity in medial prefrontal cortex and striatum. Neuropsychopharmacology 2013;38(10):1899–909.
[308] Siviy SM. Effects of pre-pubertal social experiences on the responsiveness of juvenile rats to predator odors. Neurosci Biobehav Rev 2008;32(7):1249–58.
[309] Agate RJ, Grisham W, Wade J, Mann S, Wingfield J, Schanen C, et al. Neural, not gonadal, origin of brain sex differences in a gynandromorphic finch. Proc Natl Acad Sci USA 2003;100 (8):4873–8.
[310] De Vries GJ, Rissman EF, Simerly RB, Yang LY, Scordalakes EM, Auger CJ, et al. A model system for study of sex chromosome effects on sexually dimorphic neural and behavioral traits. J Neurosci 2002;22(20):9005–14.
[311] Gatewood JD, Wills A, Shetty S, Xu J, Arnold AP, Burgoyne PS, et al. Sex chromosome complement and gonadal sex influence aggressive and parental behaviors in mice. J Neurosci 2006;26(8):2335–42.

[312] Quinn JJ, Hitchcott PK, Umeda EA, Arnold AP, Taylor JR. Sex chromosome complement regulates habit formation. Nat Neurosci 2007;10(11):1398–400.
[313] Gioiosa L, Chen X, Watkins R, Klanfer N, Bryant CD, Evans CJ, et al. Sex chromosome complement affects nociception in tests of acute and chronic exposure to morphine in mice. Horm Behav 2008;53(1):124–30.
[314] Barker JM, Torregrossa MM, Arnold AP, Taylor JR. Dissociation of genetic and hormonal influences on sex differences in alcoholism-related behaviors. J Neurosci 2010;30(27):9140–4.
[315] Klein LC, Corwin EJ. Seeing the unexpected: how sex differences in stress responses may provide a new perspective on the manifestation of psychiatric disorders. Curr Psychiatry Rep 2002;4(6):441–8.
[316] Shors TJ. Opposite effects of stressful experience on memory formation in males versus females. Dialogues Clin Neurosci 2002;4(2):139–47.
[317] Barnes LL, Wilson RS, Bienias JL, Schneider JA, Evans DA, Bennett DA. Sex differences in the clinical manifestations of Alzheimer disease pathology. Arch Gen Psychiatry 2005;62 (6):685–91.
[318] Lan YL, Zhao J, Li S. Update on the neuroprotective effect of estrogen receptor alpha against Alzheimer's disease. J Alzheimer's Dis 2015;43(4):1137–48.
[319] Peri A, Serio M. Estrogen receptor-mediated neuroprotection: the role of the Alzheimer's disease-related gene seladin-1. Neuropsychiatr Dis Treat 2008;4(4):817–24.
[320] Fillit H, Weinreb H, Cholst I, Luine V, McEwen B, Amador R, et al. Observations in a preliminary open trial of estradiol therapy for senile dementia—Alzheimer's type. Psychoneuroendocrinology 1986;11(3):337–45.
[321] Ohkura T, Isse K, Akazawa K, Hamamoto M, Yaoi Y, Hagino N. Evaluation of estrogen treatment in female patients with dementia of the Alzheimer type. Endocr J 1994;41(4):361–71.
[322] Paganini-Hill A, Henderson VW. Estrogen deficiency and risk of Alzheimer's disease in women. Am J Epidemiol 1994;140 (3):256–61.
[323] Yaffe K, Sawaya G, Lieberburg I, Grady D. Estrogen therapy in postmenopausal women: effects on cognitive function and dementia. Jama 1998;279(9):688–95.
[324] Hogervorst E, Williams J, Budge M, Riedel W, Jolles J. The nature of the effect of female gonadal hormone replacement therapy on cognitive function in post-menopausal women: a meta-analysis. Neuroscience 2000;101(3):485–512.
[325] Yue X, Lu M, Lancaster T, Cao P, Honda S, Staufenbiel M, et al. Brain estrogen deficiency accelerates Abeta plaque formation in an Alzheimer's disease animal model. Proc Natl Acad Sci USA 2005;102(52):19198–203.
[326] Mulnard RA, Cotman CW, Kawas C, van Dyck CH, Sano M, Doody R, et al. Estrogen replacement therapy for treatment of mild to moderate Alzheimer disease: a randomized controlled trial. Alzheimer's Disease Cooperative Study. Jama 2000;283 (8):1007–15.
[327] Henderson VW, Paganini-Hill A, Miller BL, Elble RJ, Reyes PF, Shoupe D, et al. Estrogen for Alzheimer's disease in women: randomized, double-blind, placebo-controlled trial. Neurology 2000;54(2):295–301.
[328] Asthana S, Craft S, Baker LD, Raskind MA, Birnbaum RS, Lofgreen CP, et al. Cognitive and neuroendocrine response to transdermal estrogen in postmenopausal women with Alzheimer's disease: results of a placebo-controlled, double-blind, pilot study. Psychoneuroendocrinology 1999;24(6):657–77.
[329] Goodman Y, Bruce AJ, Cheng B, Mattson MP. Estrogens attenuate and corticosterone exacerbates excitotoxicity, oxidative injury, and amyloid beta-peptide toxicity in hippocampal neurons. J Neurochem 1996;66(5):1836–44.
[330] Green PS, Gridley KE, Simpkins JW. Estradiol protects against beta-amyloid (25–35)-induced toxicity in SK-N-SH human neuroblastoma cells. Neurosci Lett 1996;218(3):165–8.
[331] Pike CJ. Estrogen modulates neuronal Bcl-xL expression and beta-amyloid-induced apoptosis: relevance to Alzheimer's disease. J Neurochem 1999;72(4):1552–63.
[332] Biewenga E, Cabell L, Audesirk T. Estradiol and raloxifene protect cultured SN4741 neurons against oxidative stress. Neurosci Lett 2005;373(3):179–83.
[333] Singer CA, Rogers KL, Strickland TM, Dorsa DM. Estrogen protects primary cortical neurons from glutamate toxicity. Neurosci Lett 1996;212(1):13–16.
[334] Weaver Jr CE, Park-Chung M, Gibbs TT, Farb DH. 17beta-Estradiol protects against NMDA-induced excitotoxicity by direct inhibition of NMDA receptors. Brain Res 1997;761(2):338–41.
[335] Alvarez-de-la-Rosa M, Silva I, Nilsen J, Perez MM, Garcia-Segura LM, Avila J, et al. Estradiol prevents neural tau hyperphosphorylation characteristic of Alzheimer's disease. Ann N Y Acad Sci 2005;1052:210–24.
[336] Znamensky V, Akama KT, McEwen BS, Milner TA. Estrogen levels regulate the subcellular distribution of phosphorylated Akt in hippocampal CA1 dendrites. J Neurosci 2003;23(6):2340–7.
[337] Zhang L, Rubinow DR, Xaing G, Li BS, Chang YH, Maric D, et al. Estrogen protects against beta-amyloid-induced neurotoxicity in rat hippocampal neurons by activation of Akt. Neuroreport 2001;12(9):1919–23.
[338] Sherwin BB. Estrogen and cognitive functioning in women. Endocr Rev 2003;24(2):133–51.
[339] Kimura D, Clarke PG. Women's advantage on verbal memory is not restricted to concrete words. Psychol Rep 2002;91(3 Pt 2):1137–42.
[340] Speck O, Ernst T, Braun J, Koch C, Miller E, Chang L. Gender differences in the functional organization of the brain for working memory. Neuroreport 2000;11(11):2581–5.
[341] Rosenberg L, Park S. Verbal and spatial functions across the menstrual cycle in healthy young women. Psychoneuroendocrinology 2002;27(7):835–41.
[342] Zec RF, Trivedi MA. The effects of estrogen replacement therapy on neuropsychological functioning in postmenopausal women with and without dementia: a critical and theoretical review. Neuropsychol Rev 2002;12(2):65–109.
[343] Matthews K, Cauley J, Yaffe K, Zmuda JM. Estrogen replacement therapy and cognitive decline in older community women. J Am Geriatr Soc 1999;47(5):518–23.
[344] Rapp PR, Morrison JH, Roberts JA. Cyclic estrogen replacement improves cognitive function in aged ovariectomized rhesus monkeys. J Neurosci 2003;23(13):5708–14.
[345] Daniel JM, Fader AJ, Spencer AL, Dohanich GP. Estrogen enhances performance of female rats during acquisition of a radial arm maze. Horm Behav 1997;32(3):217–25.
[346] O'Neal MF, Means LW, Poole MC, Hamm RJ. Estrogen affects performance of ovariectomized rats in a two-choice water-escape working memory task. Psychoneuroendocrinology 1996;21(1):51–65.
[347] Rhodes ME, Frye CA. ERbeta-selective SERMs produce mnemonic-enhancing effects in the inhibitory avoidance and water maze tasks. Neurobiol Learn Mem 2006;85(2):183–91.
[348] Dohanich GP, Fader AJ, Javorsky DJ. Estrogen and estrogen-progesterone treatments counteract the effect of scopolamine on reinforced T-maze alternation in female rats. Behav Neurosci 1994;108(5):988–92.
[349] Singh M, Meyer EM, Millard WJ, Simpkins JW. Ovarian steroid deprivation results in a reversible learning impairment and compromised cholinergic function in female Sprague-Dawley rats. Brain Res 1994;644(2):305–12.

[350] Friedman G. The effects of estrogen on short-term memory in genetic men. J Am Med Dir Assoc 2000;1(1):4–7.
[351] Beer TM, Bland LB, Bussiere JR, Neiss MB, Wersinger EM, Garzotto M, et al. Testosterone loss and estradiol administration modify memory in men. J Urol 2006;175(1):130–5.
[352] Sandstrom NJ, Williams CL. Memory retention is modulated by acute estradiol and progesterone replacement. Behav Neurosci 2001;115(2):384–93.
[353] Bimonte-Nelson HA, Francis KR, Umphlet CD, Granholm AC. Progesterone reverses the spatial memory enhancements initiated by tonic and cyclic oestrogen therapy in middle-aged ovariectomized female rats. Eur J Neurosci 2006;24(1):229–42.
[354] Gould E, Woolley CS, Frankfurt M, McEwen BS. Gonadal steroids regulate dendritic spine density in hippocampal pyramidal cells in adulthood. J Neurosci 1990;10(4):1286–91.
[355] Adams MM, Shah RA, Janssen WG, Morrison JH. Different modes of hippocampal plasticity in response to estrogen in young and aged female rats. Proc Natl Acad Sci USA 2001;98(14):8071–6.
[356] Hao J, Janssen WG, Tang Y, Roberts JA, McKay H, Lasley B, et al. Estrogen increases the number of spinophilin-immunoreactive spines in the hippocampus of young and aged female rhesus monkeys. J Comp Neurol 2003;465(4):540–50.
[357] Choi JM, Romeo RD, Brake WG, Bethea CL, Rosenwaks Z, McEwen BS. Estradiol increases pre- and post-synaptic proteins in the CA1 region of the hippocampus in female rhesus macaques (Macaca mulatta). Endocrinology 2003;144(11):4734–8.
[358] Murphy DD, Segal M. Regulation of dendritic spine density in cultured rat hippocampal neurons by steroid hormones. J Neurosci 1996;16(13):4059–68.
[359] Zhao L, Chen S, Ming Wang J, Brinton RD. 17beta-estradiol induces Ca2+ influx, dendritic and nuclear Ca2+ rise and subsequent cyclic AMP response element-binding protein activation in hippocampal neurons: a potential initiation mechanism for estrogen neurotrophism. Neuroscience 2005;132(2):299–311.
[360] Cordoba Montoya DA, Carrer HF. Estrogen facilitates induction of long term potentiation in the hippocampus of awake rats. Brain Res 1997;778(2):430–8.
[361] Foy MR, Xu J, Xie X, Brinton RD, Thompson RF, Berger TW. 17beta-estradiol enhances NMDA receptor-mediated EPSPs and long-term potentiation. J Neurophysiol 1999;81(2):925–9.
[362] Good M, Day M, Muir JL. Cyclical changes in endogenous levels of oestrogen modulate the induction of LTD and LTP in the hippocampal CA1 region. Eur J Neurosci 1999;11(12):4476–80.
[363] Shulman LM, Bhat V. Gender disparities in Parkinson's disease. Expert Rev Neurother 2006;6(3):407–16.
[364] Bower JH, Maraganore DM, McDonnell SK, Rocca WA. Incidence and distribution of parkinsonism in Olmsted County, Minnesota, 1976–1990. Neurology 1999;52(6):1214–20.
[365] Baldereschi M, Di Carlo A, Rocca WA, Vanni P, Maggi S, Perissinotto E, et al. Parkinson's disease and parkinsonism in a longitudinal study: two-fold higher incidence in men. ILSA Working Group. Italian Longitudinal Study on Aging. Neurology 2000;55(9):1358–63.
[366] Wooten GF, Currie LJ, Bovbjerg VE, Lee JK, Patrie J. Are men at greater risk for Parkinson's disease than women? J Neurol Neurosurg Psychiatry 2004;75(4):637–9.
[367] Scott B, Borgman A, Engler H, Johnels B, Aquilonius SM. Gender differences in Parkinson's disease symptom profile. Acta Neurol Scand 2000;102(1):37–43.
[368] Lyons KE, Hubble JP, Troster AI, Pahwa R, Koller WC. Gender differences in Parkinson's disease. Clin Neuropharmacol 1998;21(2):118–21.
[369] Baba Y, Putzke JD, Whaley NR, Wszolek ZK, Uitti RJ. Gender and the Parkinson's disease phenotype. J Neurol 2005;252(10):1201–5.
[370] Callier S, Morissette M, Grandbois M, Pelaprat D, Di Paolo T. Neuroprotective properties of 17beta-estradiol, progesterone, and raloxifene in MPTP C57Bl/6 mice. Synapse 2001;41(2):131–8.
[371] Grandbois M, Morissette M, Callier S, Di Paolo T. Ovarian steroids and raloxifene prevent MPTP-induced dopamine depletion in mice. Neuroreport 2000;11(2):343–6.
[372] Callier S, Morissette M, Grandbois M, Di Paolo T. Stereospecific prevention by 17beta-estradiol of MPTP-induced dopamine depletion in mice. Synapse 2000;37(4):245–51.
[373] Ramirez AD, Liu X, Menniti FS. Repeated estradiol treatment prevents MPTP-induced dopamine depletion in male mice. Neuroendocrinology 2003;77(4):223–31.
[374] Shughrue PJ. Estrogen attenuates the MPTP-induced loss of dopamine neurons from the mouse SNc despite a lack of estrogen receptors (ERalpha and ERbeta). Exp Neurol 2004;190(2):468–77.
[375] Tripanichkul W, Sripanichkulchai K, Finkelstein DI. Estrogen down-regulates glial activation in male mice following 1-methyl-4-phenyl-1,2,3,6-tetrahydropyridine intoxication. Brain Res 2006;1084(1):28–37.
[376] Gomez-Mancilla B, Bedard PJ. Effect of estrogen and progesterone on L-dopa induced dyskinesia in MPTP-treated monkeys. Neurosci Lett 1992;135(1):129–32.
[377] Currie LJ, Harrison MB, Trugman JM, Bennett JP, Wooten GF. Postmenopausal estrogen use affects risk for Parkinson disease. Arch Neurol 2004;61(6):886–8.
[378] Gardiner SA, Morrison MF, Mozley PD, Mozley LH, Brensinger C, Bilker W, et al. Pilot study on the effect of estrogen replacement therapy on brain dopamine transporter availability in healthy, postmenopausal women. Am J Geriatr Psychiatry 2004;12(6):621–30.
[379] Murphy SJ, McCullough LD, Smith JM. Stroke in the female: role of biological sex and estrogen. ILAR J 2004;45(2):147–59.
[380] Roquer J, Campello AR, Gomis M. Sex differences in first-ever acute stroke. Stroke 2003;34(7):1581–5.
[381] Niewada M, Kobayashi A, Sandercock PA, Kaminski B, Czlonkowska A, International Stroke Trial Collaborative G. Influence of gender on baseline features and clinical outcomes among 17,370 patients with confirmed ischaemic stroke in the international stroke trial. Neuroepidemiology 2005;24(3):123–8.
[382] Hochner-Celnikier D, Manor O, Garbi B, Chajek-Shaul T. Gender gap in cerebrovascular accidents: comparison of the extent, severity, and risk factors in men and women aged 45–65. Int J Fertil Womens Med 2005;50(3):122–8.
[383] Bushnell C, McCullough LD, Awad IA, Chireau MV, Fedder WN, Furie KL, American Heart Association Stroke C, Council on C, Stroke N, Council on Clinical C, Council on E, Prevention, Council for High Blood Pressure R, et al. Guidelines for the prevention of stroke in women: a statement for healthcare professionals from the American Heart Association/American Stroke Association. Stroke 2014;45 (5):1545–88.
[384] Hall ED, Pazara KE, Linseman KL. Sex differences in postischemic neuronal necrosis in gerbils. J Cereb Blood Flow Metab 1991;11(2):292–8.
[385] Alkayed NJ, Harukuni I, Kimes AS, London ED, Traystman RJ, Hurn PD. Gender-linked brain injury in experimental stroke. Stroke 1998;29(1):159–65: discussion 166.
[386] Park EM, Cho S, Frys KA, Glickstein SB, Zhou P, Anrather J, et al. Inducible nitric oxide synthase contributes to gender differences in ischemic brain injury. J Cereb Blood Flow Metab 2006;26(3):392–401.
[387] Roof RL, Hall ED. Estrogen-related gender difference in survival rate and cortical blood flow after impact-acceleration head injury in rats. J Neurotrauma 2000;17(12):1155–69.

[388] McCullough LD, Zeng Z, Blizzard KK, Debchoudhury I, Hurn PD. Ischemic nitric oxide and poly (ADP-ribose) polymerase-1 in cerebral ischemia: male toxicity, female protection. J Cereb Blood Flow Metab 2005;25(4):502–12.
[389] Simpkins JW, Rajakumar G, Zhang YQ, Simpkins CE, Greenwald D, Yu CJ, et al. Estrogens may reduce mortality and ischemic damage caused by middle cerebral artery occlusion in the female rat. J Neurosurg 1997;87(5):724–30.
[390] Dubal DB, Kashon ML, Pettigrew LC, Ren JM, Finklestein SP, Rau SW, et al. Estradiol protects against ischemic injury. J Cereb Blood Flow Metab 1998;18(11):1253–8.
[391] Shughrue PJ, Merchenthaler I. Estrogen prevents the loss of CA1 hippocampal neurons in gerbils after ischemic injury. Neuroscience 2003;116(3):851–61.
[392] Plahta WC, Clark DL, Colbourne F. 17beta-estradiol pretreatment reduces CA1 sector cell death and the spontaneous hyperthermia that follows forebrain ischemia in the gerbil. Neuroscience 2004;129(1):187–93.
[393] Toung TJ, Traystman RJ, Hurn PD. Estrogen-mediated neuroprotection after experimental stroke in male rats. Stroke 1998;29(8):1666–70.
[394] Gibson CL, Gray LJ, Murphy SP, Bath PM. Estrogens and experimental ischemic stroke: a systematic review. J Cereb Blood Flow Metab 2006;26(9):1103–13.
[395] Liao S, Chen W, Kuo J, Chen C. Association of serum estrogen level and ischemic neuroprotection in female rats. Neurosci Lett 2001;297(3):159–62.
[396] Sawada M, Alkayed NJ, Goto S, Crain BJ, Traystman RJ, Shaivitz A, et al. Estrogen receptor antagonist ICI182,780 exacerbates ischemic injury in female mouse. J Cereb Blood Flow Metab 2000;20(1):112–18.
[397] McCullough LD, Blizzard K, Simpson ER, Oz OK, Hurn PD. Aromatase cytochrome P450 and extragonadal estrogen play a role in ischemic neuroprotection. J Neurosci 2003;23(25):8701–5.
[398] Bekker MH, van Mens-Verhulst J. Anxiety disorders: sex differences in prevalence, degree, and background, but gender-neutral treatment. Gender Med 2007;4(Suppl B):S178–93.
[399] Breslau N, Chilcoat H, Schultz LR. Anxiety disorders and the emergence of sex differences in major depression. J Gender Specific Med 1998;1(3):33–9.
[400] Joffe H, Cohen LS. Estrogen, serotonin, and mood disturbance: where is the therapeutic bridge? Biol Psychiatry 1998;44(9):798–811.
[401] Douma SL, Husband C, O'Donnell ME, Barwin BN, Woodend AK. Estrogen-related mood disorders: reproductive life cycle factors. ANS Adv Nurs Sci 2005;28(4):364–75.
[402] Iatrakis G, Haronis N, Sakellaropoulos G, Kourkoubas A, Gallos M. Psychosomatic symptoms of postmenopausal women with or without hormonal treatment. Psychother Psychosom 1986;46(3):116–21.
[403] Hays J, Ockene JK, Brunner RL, Kotchen JM, Manson JE, Patterson RE, Women's Health Initiative I, et al. Effects of estrogen plus progestin on health-related quality of life. N Engl J Med 2003;348(19):1839–54.
[404] Morrison MF, Kallan MJ, Ten Have T, Katz I, Tweedy K, Battistini M. Lack of efficacy of estradiol for depression in postmenopausal women: a randomized, controlled trial. Biol Psychiatry 2004;55(4):406–12.
[405] Tangen T, Mykletun A. Depression and anxiety through the climacteric period: an epidemiological study (HUNT-II). J Psychosom Obstet Gynaecol 2008;29(2):125–31.
[406] Lucki I. The forced swimming test as a model for core and component behavioral effects of antidepressant drugs. Behav Pharm 1997;8(6–7):523–32.
[407] Contreras CM, Molina M, Saavedra M, Martinez-Mota L. Lateral septal neuronal firing rate increases during proestrus-estrus in the rat. Physiol Behav 2000;68(3):279–84.
[408] Frye CA, Walf AA. Changes in progesterone metabolites in the hippocampus can modulate open field and forced swim test behavior of proestrous rats. Horm Behav 2002; 41(3):306–15.
[409] Molina-Hernandez M, Tellez-Alcantara NP. Antidepressant-like actions of pregnancy, and progesterone in Wistar rats forced to swim. Psychoneuroendocrinology 2001; 26(5):479–91.
[410] Frye CA, Walf AA. Hippocampal 3alpha,5alpha-THP may alter depressive behavior of pregnant and lactating rats. Pharmacol Biochem Behav 2004;78(3):531–40.
[411] Rachman IM, Unnerstall JR, Pfaff DW, Cohen RS. Estrogen alters behavior and forebrain c-fos expression in ovariectomized rats subjected to the forced swim test. Proc Natl Acad Sci USA 1998;95(23):13941–6.
[412] Estrada-Camarena E, Fernandez-Guasti A, Lopez-Rubalcava C. Antidepressant-like effect of different estrogenic compounds in the forced swimming test. Neuropsychopharmacology 2003;28(5):830–8.
[413] Marcondes FK, Miguel KJ, Melo LL, Spadari-Bratfisch RC. Estrous cycle influences the response of female rats in the elevated plus-maze test. Physiol Behav 2001;74(4–5):435–40.
[414] Mora S, Dussaubat N, Diaz-Veliz G. Effects of the estrous cycle and ovarian hormones on behavioral indices of anxiety in female rats. Psychoneuroendocrinology 1996;21(7):609–20.
[415] Morgan MA, Pfaff DW. Estrogen's effects on activity, anxiety, and fear in two mouse strains. Behav Brain Res 2002;132 (1):85–93.
[416] Solomon MB, Karom MC, Huhman KL. Sex and estrous cycle differences in the display of conditioned defeat in Syrian hamsters. Horm Behav 2007;52(2):211–19.
[417] Frye CA, Walf AA. Estrogen and/or progesterone administered systemically or to the amygdala can have anxiety-, fear-, and pain-reducing effects in ovariectomized rats. Behav Neurosci 2004;118(2):306–13.
[418] Bowman RE, Ferguson D, Luine VN. Effects of chronic restraint stress and estradiol on open field activity, spatial memory, and monoaminergic neurotransmitters in ovariectomized rats. Neuroscience 2002;113(2):401–10.
[419] Morgan MA, Pfaff DW. Effects of estrogen on activity and fear-related behaviors in mice. Horm Behav 2001;40(4):472–82.
[420] Solomon MB, Herman JP. Sex differences in psychopathology: of gonads, adrenals and mental illness. Physiol Behav 2009;97 (2):250–8.
[421] Imwalle DB, Gustafsson JA, Rissman EF. Lack of functional estrogen receptor beta influences anxiety behavior and serotonin content in female mice. Physiol Behav 2005;84(1):157–63.
[422] Krezel W, Dupont S, Krust A, Chambon P, Chapman PF. Increased anxiety and synaptic plasticity in estrogen receptor beta-deficient mice. Proc Natl Acad Sci USA 2001; 98(21):12278–82.
[423] Walf AA, Frye CA. Administration of estrogen receptor beta-specific selective estrogen receptor modulators to the hippocampus decrease anxiety and depressive behavior of ovariectomized rats. Pharmacol Biochem Behav 2007;86(2):407–14.

CHAPTER

4

Sex Hormone Receptor Expression in the Immune System

Iwona A. Buskiewicz[1], *Sally A. Huber*[1] *and DeLisa Fairweather*[2]

[1]Department of Pathology, University of Vermont, Burlington, VT, United States [2]Department of Cardiovascular Diseases, Mayo Clinic, Jacksonville, FL, United States

INTRODUCTION

The efficiency of the immune system depends on its coordinated response, which is affected by various exogenous and endogenous factors. Environmental, hormonal, and genetic effects have been proposed to explain differences of activity and strength in the immune reactions of females compared to males in response to pathogens [1]. The concept of "Sex" and "Gender" is commonly confounded, even though each term has a distinct connotation. Whereas "sex" defines the biological state of being male versus female, "gender" refers to the differences between men and women in a social and cultural context, that is, social factors that determine "masculine" and "feminine" [2]. Although gender is to a certain degree ambiguous and can change over time within a given culture, it influences all aspects of human life; hence it is plausible that gender plays a role as an immune modulator. Examples of exogenous factors that contribute to various immunological responses to antigens in men and women include aging and access to health care [3]. In contrast, endogenous factors contributing to biological sex differences are more readily quantifiable. Importantly, sex-based differences in immune responses are known to contribute directly to the pathogenesis of infectious and chronic inflammatory diseases in males and females [4]. Generally, females have a more robust immune response and correspondingly lower burden of bacterial, viral, and parasitic infections [5]. Possible causes for these sex differences include endocrine–immune interactions, regulation of the immune response through steroid hormones including estrogens, androgens, and progesterone, and genetics.

The male-defining Y chromosome encodes fewer than a 100 genes, whereas the X chromosome encodes approximately 1100 genes with many X-linked genes involved in immunity [6,7]. One of the X chromosomes is (randomly) transcriptionally silenced during female development through CpG methylation [8]. As a result, only the maternal (Xm) or paternal (Xp) X chromosome is transcriptionally active. Since males do not undergo X chromosomal inactivation, 100% of cells in males will express an X-linked gene mutation, whereas any X-linked gene mutation will potentially only be expressed in 50% of female cells. Therefore, X-linked diseases invoke more severe phenotypes in males. For example, a mutation in X-linked inhibitor of apoptosis leads to immunodeficiency through a lymphoproliferative syndrome [9], X-linked severe combined immunodeficiency is caused by a defect in the interleukin (IL)-2 receptor gamma chain, mutations in forkhead box P3 (FoxP3) lead to immune dysregulation, polyendocrinopathy, and enteropathy [10]. Sex-related hormones, also known as sex steroids, are the other significant contributors to the increased male susceptibility to infection. The effects of sex hormones on the immune response are the primary focus of this review.

TRADITIONAL SEX HORMONE RECEPTORS IN THE IMMUNE SYSTEM

The major sex hormones that have been studied in immunity are estrogens, testosterone, and

Sex Differences in Physiology
DOI: http://dx.doi.org/10.1016/B978-0-12-802388-4.00004-5

TABLE 4.1 Putative Hormone Receptor Expression by Cells of the Immune System

	ERα		ERβ		AR		PR		GPER
Cell type	**Nuclear**	**Membrane**	**Nuclear**	**Membrane**	**Nuclear**	**Membrane**	**Nuclear**	**Membrane**	**ER**
T lymphocytes	+				+	+			
CD4	+								+
CD8								+	
Treg			+				+		+
NKT									
γδ								+	
B lymphocytes	+		+		+	+	+		+
Monocyte	+	+	+	+	+	+	+		+
Macrophage	+	+	+	+	+	+	+		+
DCs									
Myeloid	+	+	+				+		
Plasmacytoid	+								
NK cells	+		+				+		
Granulocytes									
MCs	+	+	+	+	+	+	+	+	
Neutrophils									
Eosinophils									
Thymus	+			+					
BM	+		+		+				

progesterone; and all of these hormones and their associated receptors are present in both males and females on many immune cells (Table 4.1). Sex hormones bind to specific nuclear and membrane-associated protein receptors resulting in diverse and often contradictory affects in innate and adaptive immunity. The two classic nuclear receptors estrogen receptor alpha (ERα) and estrogen receptor beta (ERβ) are sequestered in the cell cytoplasm bound to heat shock proteins (HSPs) which act as chaperones [11]. Estrogens diffusing through the cell membrane bind to the ERs displacing the chaperones and cause translocation of the receptor–ligand complexes to the nucleus as either homodimers (ERα/ERα or ERβ/ERβ) or heterodimers (ERα/ERβ) [12]. Here, the ER–estrogen complex may bind directly to estrogen response elements (EREs) in the promoter region of specific genes, such as the interferon (IFN)-γ gene [13], or indirectly activate gene transcription by binding to specific transcription factors (TFs) such as SP-1, AP-1, NFκB, or c-jun which bind to gene promoters without ERE sites [12]. Approximately 35% of estrogen-responsive human genes may be activated through this indirect ER–TF–DNA interaction. Examples include low density lipoprotein receptor and endothelial nitric oxide synthase (eNOS) [12]. Furthermore, when both ERs are expressed in the same cell they may have opposing activities on gene expression. ERα–AP-1 complexes increase cyclin D expression whereas ERβ–AP-1 complexes suppress expression in HeLa cells [14]. Additionally, ligand (estrogen)-independent nuclear signal transduction can occur as a result of cellular kinases such as protein kinase C (PKC), protein kinase A (PKA), mitogen-activated protein kinase (MAPK), and phosphoinositide 3-kinase (PI3K) phosphorylating the ER leading to dimerization and TF interaction in the absence of estrogen. This process has not been specifically studied in immunity but may be of significance since signaling of the T cell receptor (TCR) and accessory molecule CD28 during T cell activation proceeds through PI3K activation [15].

An alternative, nonnuclear pathway for sex steroid signaling has gained increased attention over the last two decades that involves membrane associated hormone receptors. Evidence that membrane ERs (mER) are derived from the same genes as nuclear ERs comes from studies where deletion of the ERα or ERβ gene eliminates the function of both types of receptors [16]. Signaling through mER is more rapid than through nuclear receptors and often occurs in seconds to minutes. These mERs are splice variants of the full length nuclear ERs lacking the A/B domains, are palmitolated in the Golgi, bind to caveolins in the cytoplasm, and are transported and anchored in the plasma membrane by binding to c-Src [17–19]. In the presence of estrogen (ie, estradiol, E2), mERs form either homo- or heterodimers. Dimerization is required for mER function since if dimerization is prevented, downstream signaling does not occur. The mER itself lacks intrinsic kinase activity but can directly bind to the PI3K regulatory subunit (p85α) resulting in activation of the PI3K or PI3K/Akt pathway, which may include a complex of other molecules such as c-Src, Gαq, Gαs, RAS, IGFR1, EGFR, and eNOS. Importantly, the specific constituents of the mER signaling complex or signalosome may differ between cells and this can lead to cell—and tissue—specific hormone signal signatures [20]. For example, in endothelial cells mER interacts with c-Src, PI3K, Akt kinase, and HSP90 to activate eNOS and release nitric oxide [21]. In cardiac myocytes, mER signaling through PI3K occurs without activation of Akt resulting in increased modulatory calcineurin-interacting protein-1 expression, which is an inhibitor of calcineurin and prevents cardiac hypertrophy [22]. Signalosomes containing G proteins that dissociate into Gα and Gβγ subunits act as a bridge between the mER and the MAPK cascade pathway [19]. The type of caveolin can dictate the nature of the mER signal produced. Caveolin-1 causes mERα to bind glutamate receptor 1 (mGuR1) leading to MAPK-dependent phosphorylation of CREB in neurons, whereas caveolin-3 results in mER binding to mGuR2/3 and a decrease in L-type calcium channel-dependent CREB phosphorylation [23].

Another ER that is less well characterized is termed the G protein-coupled ER (GPER), which is a 7-transmembrane protein that is genetically unrelated to ERα or ERβ. There is controversy about whether GPER is located in the endoplasmic reticulum or Golgi of the cell cytoplasm, in the nucleus, or on the cell membrane [19]. Another major question is whether GPER interacts with the classical ERs, ERα, or ERβ. Studies indicate that GPER can be activated with physiological E2 concentrations (10 nM) [24]. Cell expression of the various ERs differs so that some cells express the classical ERs but not GPER and vice versa [19]. This likely leads to diverse estrogen responses in various cells. Even when multiple ERs are simultaneously expressed in the same cell, GPER may contribute unique signal cascades to the total estrogen response separate from that provided by the classical ERα or ERβ responses [25].

Testosterone, which is converted to dihydrotestosterone through the 5-α reductase enzyme, binds to androgen receptors (ARs) in the cytoplasm causing receptor dimerization, nuclear translocation, cofactor recruitment, and gene transcription [26]. Similar to estrogens, testosterone induces rapid, membrane-associated calcium flux. Testosterone binding sites have been demonstrated on the cell membranes of macrophages [27] and T cells [28] resulting in rapid calcium flux. However, activation could not be inhibited by cyproterone, an antagonist of the classical AR. This suggests the existence of a nonclassical testosterone receptor homologous to GPER, but this type of receptor has not been isolated yet. Other studies indicate that a functional AR exists in lymphoid cells because conditional knockout (KO) mice where the AR gene has been selectively deleted in specific cell types including T cells, B cells, and monocytes resulted in distinct effects on immune function [29]. Whether this effect was mediated through nuclear receptor translocation or membrane-associated receptor activation remains unclear. Nongenomic actions of the AR can be mediated either by AR binding to steroid hormone-binding globulin (SHBG), which activates the SHBG receptor leading to cAMP and PKA activation, or by activation of c-Src and MAPK [30]. As with ERs, the AR contains a 9 amino acid sequence in the ligand binding region that is conducive to membrane translocation after palmitoylation and mutations in this region reduce membrane localization and MAPK activation [17]. As mER activate c-Src and MAPK, a mAR would likely mediate a similar response.

As with estrogen and testosterone, progesterone uses both nuclear and membrane progesterone receptors (mPRs). The PR, which in humans is located on chromosome 11 and consists of 8 exons [31], can be expressed as either a PR-A or PR-B isoform under the control of two distinct promoters [32]. Both isoforms contain identical DNA-binding domains and hormone (ligand) binding domains but differ in the length of the amino-terminal domain with truncation of PR-A. This result potentially in distinct tissue tropism, regulation of different subsets of genes, or cross-regulation of PR-B by PR-A when both isoforms are co-expressed [33]. Other splice variants or isoforms of PR have been described, especially in female reproductive tissues, but often the physiological roles of these isoforms are not well delineated [34]. Since both PR-A and PR-B are derived from the same gene, a major question is what factors determine which isoform is produced. Evidence suggests that the concentration of progesterone (P4) is crucial with high P4 levels resulting in higher PR-A expression compared to PR-B while low P4 concentrations have the opposite effect [34]. As with the other hormone receptors, PRs are sequestered in the cytosol to chaperones including HSP70 and HSP90, but are released when P4 diffuses through the plasma membrane resulting in receptor dimerization, nuclear translocation, and binding to hormone response elements in gene promoter regions, recruitment of TFs and coactivation factors, and gene activation [34]. P4 also causes rapid nonnuclear cell signaling through membrane associated PR-B via Src/MAPK (PR-A is putatively not involved in nonnuclear signal cascades) [35]. However, as P4 has been shown to cause nonnuclear signaling in classical PR-deficient cells, other membrane associated mechanisms for signal transduction have also been implicated including progesterone membrane component 1 and progesterone membrane component 2, which interact with plasminogen activator inhibitor 1 RNA-binding protein (SERBP1) and activate protein kinase G, PR, and adipoQ receptor (PAQR). The mPRs belonging to the PAQR family are represented by three 7-transmembrane protein isoforms (mPRα, mPRβ, and mPRγ) that are encoded by separate genes, and when MAPK, phospholipase C (PLC), PKC are activated can result in increased Ca^{2+} release from internal stores [34].

VITAMIN D RECEPTOR: A SEX STEROID RECEPTOR

Vitamin D is a fat-soluble pro-hormone that regulates serum calcium levels and bone homeostasis, but also functions as a sex steroid that influences immune function. Vitamin D is synthesized by keratinocytes in the skin after exposure to the sun (UVB) or from dietary sources and is hydroxylated in the liver by 25-hydroxylase (Cyp2R1) and carried in the bloodstream by vitamin D binding protein (DBP) to the kidney where it is converted by 25-OH-D-1α-hydroxylase (Cyp27B1) to the active form of vitamin D (ie, 1α,25-dihydroxyvitamin D_3), which binds the vitamin D receptor (VDR) [36–38]. VDR is a member of the steroid nuclear receptor superfamily and expressed on the surface of and/or intracellularly in a wide variety of cells including keratinocytes, fibroblasts, and most immune cells [39]. Many tissues not only express the VDR but also express the key enzyme to convert 25 $(OH)_2D$, the form of vitamin D in the serum, to its active form [40]. Macrophages possess all of the components necessary to import/synthesize cholesterol and convert it to active vitamin D including Cyp2R1 and Cyp27B1, as well as expressing the VDR [41,42]. Vitamin D bound to the VDR forms a complex with the retinoid X receptor and this complex has been estimated to regulate around 3% of the human genome via activation of vitamin D response elements (VDREs) [43]. Many genes important for immune function have a VDRE in their promoter, like tumor necrosis factor (TNF), for example [41]. Vitamin D has been found to be important in mediating protective immunity against infections, particularly during innate immunity, while low levels of vitamin D have been associated with an increased risk of developing inflammatory diseases like autoimmune diseases, for example [44]. Although vitamin D is a sex steroid that significantly alters immune cell function, few studies exist comparing sex differences in the immune response to vitamin D [45].

SEX DIFFERENCES IN INNATE IMMUNITY

The immune response is a complex interaction of cells and is designed to identify and eliminate anything considered by the system to be "nonself." The immune system can be generally divided into innate and adaptive immunity with the former designed to respond rapidly (minutes/hours) to "foreign" agents while the adaptive immune response requires 1–2 weeks to develop optimally when first exposed to the initiating agent (primary immune response). The adaptive response also depends primarily on the innate immune response to provide "direction" (Table 4.2). This means that the innate immune response reacts broadly to molecules shared by many microbes whereas the adaptive immune response responds specifically to molecules expressed uniquely to a single or limited number of microbes. Cells comprising the innate immune response are generally considered to be granulocytes, mast cells (MCs), macrophages, natural killer (NK) cells, NKT cells, and γδ T cells, while cells associated with adaptive immunity include T and B lymphocytes.

TABLE 4.2 Differences between Innate and Adaptive Immunity

Innate immunity	Adaptive immunity
• Cells: macrophage, DCs, MCs, NK cells, NKT cells, γδT cells	• Cells: T and B lymphocytes including regulatory T cells
• Rapid: minutes to hours	• Slow: primary response to antigen requires 7–10 days; memory responses are faster but still require 2–4 days
• Broad specificity: recognize molecules shared among many microbes or tumor cells such as PAMPs	• Highly specific to eliciting antigen
• Antigen processing and presentation not required (except for NKT cells)	• Antigen processing and presentation required for T cell activation; presentation must be on same MHC as T cell (MHC restriction)
• No memory response (except for NKT cells)	• Memory response to antigen

The innate immune system is designed to respond to pathogen-associated molecular patterns (PAMPs) using pattern recognition receptors (PRRs). PAMPs are conserved microbial peptides present in infectious agents that are critical for microbial infection so that mutating or eliminating them would drastically inhibit microbial infectivity [46]. The host immune response to PAMPs allows very rapid cytokine release and initiation of an inflammatory response. The most recognized of the PRRs are the Toll-like receptors (TLRs), which are type I integral membrane glycoproteins that have extracellular domains with variable numbers of leucine-rich repeat motifs and conserved intracellular domains with a homologous TLR/IL-1R response domain [46,47]. There are 12 TLRs in humans and mice that distinguish different molecular patterns on microbes [46,48–52]. Lymphoid cell populations express distinct sets of TLRs. For example, $CD8^+$ T cells express high levels of TLR3 but low levels of TLR2 and TLR4. B cells express high levels of TLR9 and TLR10. Thus, some immune effector cells may respond more robustly to certain microbes over others (reviewed in [53] and [54]). Generally, TLR 3, 5, and 9 are considered to preferentially activate T helper (Th)1 responses while TLR2 and TLR4 have been shown to induce Th2 immunity [54,55]. In addition to the type of TLR, the intensity of signaling through a TLR can determine the type of Th response. Low dose lipopolysaccharide (LPS) stimulation of TLR4 promotes Th2 responses, while higher doses promote Th1 responses. Interestingly, naïve $CD4^+$ T cells may express TLR2 and TLR4 mRNA and intracellular protein but not cell-surface protein, but after activation $CD4^+$ T cells rapidly express cell-surface TLR2 and TLR4 [56]. Only TLR2 may be functional in these cells since TLR2 stimulation of activated $CD4^+$ T cells increases IFNγ, IL-2, and TNFα expression, but LPS stimulation of TLR4 has little effect on these cytokines.

Recently, several studies have evaluated the differences in PRR expression between males and females. The *TLR7* gene is encoded on the X chromosome and one possibility is that this gene undergoes duplication resulting in more copies of *TLR7* gene in some individuals [57]. Possibly, posttranscriptional differences in gene expression such as mRNA stability vary between males and females, since exposure of plasmacytoid dendritic cells (DCs) from women to TLR7 ligands in vitro caused higher production of IFNα than cells from men, even though *TLR7* mRNA levels were similar between the sexes [58]. Other studies have found that female mice express significantly greater amounts of TLR2 while males express higher levels of TLR4 following coxsackievirus B3 infection [59,60]. In contrast, treatment of cells with CpG, a TLR9 ligand, found no sex difference in IFNα production [58].

Sex hormones also impact effector cells of the innate immune system such as NK cells, MCs, and macrophages. Testosterone inhibits NK cell activity by inducing the production of anti-inflammatory cytokines such as IL-10 and by reducing the production of pro-inflammatory cytokines like TNFα through inhibition of NFκB [61]. This leads to increased susceptibility to certain infections, particularly viruses, and an increased mortality in men [62–64]. In some cases, the limited pro-inflammatory response to viral infection due to reduced NK cell activity can lead to a latent infection that results in prolonged, recurrent, or chronic infection. Progesterone can also inhibit NK cell function by inducing the production of IL-4, IL-5, and IL-10 and increasing the expression of SOCS1 while inhibiting the production of IFNγ and TNFα, which are needed to prevent viral and bacterial infections, bacteremia, and sepsis [65]. However, when progesterone occurs at high levels, as during pregnancy, it can promote Th1-type immune responses and protect against infection [65]. In contrast, low doses of estrogen may enhance NK cell activity through the activation of NFκB inducing the production of TNFα, IL-1, IL-6, IL-17, and IL-23, while inhibiting the production of IL-4, IL-10, IL-12, and transforming growth factor (TGFβ) thereby allowing bacterial clearance and recovery from infection.

ERα, ERβ, and the AR are expressed on and within MCs and macrophages, which act as antigen-presenting cells (APCs) particularly at certain sites like the skin, gut, vasculature, and peritoneum [44,55]. Human and mouse monocytes, macrophages, and MCs express ERα, ERβ, and the AR [66]. AR expression on human monocytes is higher in men compared to women [67–69]. In general, estrogen has been found to have anti-inflammatory effects on macrophages. Estrogen inhibits LPS-induced gene products like TNF, IL-1, and IL-6 by downregulating NFκB signaling in macrophages [70–74]. Estrogen has also been found to reduce oxidative stress in healthy murine peritoneal macrophages [74], and to skew macrophages to an alternatively activated M2 phenotype [75]. In contrast, Rettew et al. found that testosterone decreased TLR4 expression in the RAW tumor macrophage cell line (which are male cells) [76], while ovariectomy and estrogen replacement in female C57BL/6 mice increased TLR4 expression on macrophages indicating that estrogen increases TLR4 expression [77]. Complicating this issue is the fact that TLR4 promotes both M1/Th1- and M2/Th2-type immune responses [55]. Additionally, it is likely that the effect of sex hormones on normal healthy immune cells is not always the same as their effect during infection and disease (ie, bacterial LPS, viral infection, myocarditis). In support of this idea, we found that TLR4 expression was higher on male than female macrophages and MCs during innate coxsackievirus B3 infection and acute viral myocarditis [60]. These findings highlight some of the difficulties inherent in studying the effect of sex hormones on immune cells.

SEX DIFFERENCES IN THE ADAPTIVE IMMUNE SYSTEM

The complexity of hormonal signaling is partially explained by the mosaic of hormone receptor expression in different tissues and cells within the body. While extensive investigations have been conducted in certain tissues/organs such as reproductive organs, delineation of hormone receptor expression in the various cells comprising the immune system is currently incomplete. The most information is available for ERs. However, information on ARs and PRs has been increasing in recent years. One unresolved problem is that a hormonal effect can be documented by addition of a specific hormone to lymphoid cells or immune cells from gonadectomized animals with or without hormone replacement, while neither mRNA nor protein for the hormone receptor can be detected in the cells [78]. This suggests that, either hormones are acting through cells not directly associated with the immune response, or that techniques for detecting hormone receptors are in some way deficient, and/or that unknown hormone receptors exist in immune cells allowing hormonal influence by potentially unique receptor pathways. We know that hormone receptors are expressed in tissue cells where they regulate tissue/organ physiology essentially as growth factors.

Sex Hormones and T Cell Ontogeny

T lymphocytes develop in the thymus where prothymocytes derived from fetal liver and bone marrow (BM) interact with thymic epithelium to undergo a defined gene expression cascade (double (CD4/CD8) negative, double (CD4/CD8) positive, single (CD4/CD8) positive) and rearrangement of the TCR genes to produce naïve $CD4^+$ and $CD8^+$ T cells that migrate to peripheral lymphoid organs. Nascent T cells present in the thymus with TCR rearrangements capable of recognizing self-molecules (autoimmune) should be deleted (clonal deletion) in a process termed "central tolerance." While clonal deletion should prevent development of autoimmunity, it has several weaknesses. First, only self-molecules (antigens) present in the thymus during T cell development cause clonal deletion so any TCR recognizing self-antigens sequestered in distant organs, such as the central nervous system, will not be deleted. Secondly, clonal deletion occurs through cellular apoptosis and if environmental factors suppress apoptosis, this might allow autoreactive T cell clones to escape deletion. Both estrogens and androgens impact the thymus and T cell development. Estrogen signaling through ERα is crucial for T cell generation in both male and female mice as thymic weight postnatally is reduced 30–55% in ERα KO animals accompanied by significant reductions in total $CD4^+$ and $CD4^+CD8^+$ (double positive) T cells [79]. This effect at least partially reflects estrogen's ability to protect cells from apoptosis, which should allow more cells to survive the T cell ontogeny process. If some of the increased survival includes autoreactive T cell clones, this could promote susceptibility to autoimmunity and help explain why most autoimmune diseases occur in women [80]. While estrogen promotes T cell ontogeny, androgens have the opposite effect. Castration of males results in increased thymic involution and decreased T cell production associated with increases in apoptosis and TGFβ [29,81]. While both immature T cells and thymic epithelial cells express ARs, AR expression by epithelial cells appears to be most important in restricting T cell clonal development [29,81].

Sex Hormones Effect on DCs

Initiation of T cell activation in the periphery requires antigen presentation. DCs are often called "professional APCs" because of their key role in activating T cells. DCs are derived from BM $CD34^+$ precursor stem cells. The two major subsets of DCs are myeloid DCs (mDCs) and plasmacytoid DCs (pDCs), which express different cell markers and have important functional differences. Generally, pDCs are highly effective producers of type 1 IFNα and are closely associated with protective immunity to viral infections and activation of cytolytic NK and $CD8^+$ T cells. pDCs also express direct cytotoxic activity toward target cells. In contrast, subtypes of mDCs produce high levels of IL-12 and/or IFNγ, which polarize naïve $CD4^+$ T cells to a Th1 phenotype [82]. Activated Th1 cells interacting with DCs through CD40/CD40L induce highly potent cytolytic $CD8^+$ T cells in a process called "DC licensing" [82]. The various DC subtypes reside in tissues in an immature state characterized as highly phagocytic but expressing low levels of major histocompatibility complex (MHC) and accessory molecules. If these immature DCs encounter antigen, the antigen is phagocytized which initiates DC maturation and migration of DCs from peripheral tissues to the spleen and draining lymph nodes. The maturation results in increased cell surface expression of antigen-loaded MHC and accessory molecules while decreasing phagocytic activity of the cell [83]. In the lymph node, DCs randomly interact with T cell clones until a clone with a TCR matching the antigen–MH complex is found. Intimate T cell-DC contact lasting at least 6 h is required for T cell proliferation in a primary immune response [84].

ERα has an important role in DC differentiation and activation. Estrogen stimulates differentiation of GM-CSF-induced mDC and augments their capacity to stimulate T cell responses through upregulation of IRF-4 [85–88]. Estrogen also amplifies TLR7/9 signaling in pDCs increasing their functional activity [88]. Estrogen impacts expression levels of accessory molecules such as CD40, which has important immunomodulatory effects on T cell activation by DC [89]. In contrast to estrogen, testosterone and progesterone have largely inhibitory effects on DCs, although this effect is complex and depends on hormone concentration. The pDC subset produces increased levels of IFNα at low concentrations of progesterone but decreased levels at high concentrations. Generally, mDCs show reduced expression of TNFα, IL-6, IL-12, CD80/86, and MHC II following progesterone exposure.

Sex Hormones Effect on T Cell Activation and Polarization

Naïve T cells have little functional activity or cytokine expression except for low levels of IL-2. Once stimulated by APCs, naïve T cells undergo rapid proliferation over the next 7–10 days. Most of these T cells differentiate into effector cells having substantial cytokine-producing or cytolytic activity, while a small percentage will differentiate into memory T cells. T helper cells are usually delineated by the major type of cytokine produced. Initially, Th1 (IFNγ$^+$) and Th2 (IL-4$^+$) subtypes were defined [90], with the former important in promoting CD8$^+$ cytolytic T cell responses and protection from intracellular infections such as viruses, and the latter producing cytokines including IL-4, IL-5, IL-9, and IL-13 that promote class switching of immunoglobulins in B cells. Many more T cell subtypes have now been defined including Th17 cells (associated with TGFβ, IL-1β, IL-6, and IL-23), which are important in promoting neutrophil and macrophage responses to bacteria [91]. Naïve T cells are able to differentiate into any of these various subtypes based on the cytokine environment present during their activation, but also through antigen concentration and avidity of the TCR–MHC interaction.

T cell polarization is strongly affected by sex hormones, but the effect is complex and contradictory between published studies. Hormone concentration is important as high estrogen concentrations drive a Th2 phenotype while low concentrations polarize to a Th1/Th17 phenotype [4]. Multiple types of sex hormone receptors have been found in/on T cells and APCs (Table 4.1), which could have different hormone binding affinities. In studies comparing ER number and hormone binding affinity between BALB/c and MRL inbred strains of mice, MRL splenocytes showed significantly greater estrogen binding affinity than immune cells from BALB/c mice. Furthermore, the number of ERs decreased by half between 3 and 10 weeks of age in BALB/c mice, but increased approximately threefold for the same age range in MRL mice [92]. Differences were also observed between BALB/c and MRL thymus PR expression in response to E2 treatment, with the thymus of MRL mice showing a 2.5-fold increase in PR expression compared to BALB/c mice. This study did not report the specific ERs involved in the response, but indicates that there can be dramatic differences in hormonal receptor expression, binding affinity, and alterations in hormone receptor expression with age in different mouse strains. Changing the number of sex hormone receptors or the specificity of ligand binding can impact the types or intensity of signal they produce. For example, since estrogen promotes DC maturation and antigen presentation including increases in MHC and accessory molecule expression, this could lead to greater amounts of antigen–MH complexes at the DC cell surface and stronger T cell–APC interactions which would impact the polarization process [93]. Similarly, as discussed above, sex hormones affect TLR expression and signaling. This may alter the cytokine environment in which the adaptive immune response develops [94,95]. For example, in one model of experimental coxsackievirus B3-induced myocarditis, which is a male dominant disease in mice and humans, increased acute inflammation is associated with a Th1 cell response in males while lower inflammation in females correlates to a Th2-type immune response [60,96,97]. Testosterone increased viral replication resulting in increased myocarditis [98], a finding that is consistent with the known susceptibility of males to infections. However, it is often stated in the literature that testosterone promotes a Th2 response while estrogens promote a Th1 response [4,29,65], but recall that whether estrogen drives a Th1 or Th2 response depends on its dose. Additionally, few mechanistic studies have examined the effect of testosterone on immune cells or during inflammatory diseases compared to how many have been performed on the effect of estrogen.

Subsequent to the primary T cell response, many other cells can act as APCs and the time required for T cell–APC interaction is dramatically reduced from over 6 h to 30 min or less. Macrophage/monocytes at tissue sites are major APCs for T cells after initial activation and provide important co-stimulatory factors for humoral immunity. Evidence suggests that the activation state of the monocyte affects their relative ER expression. Thus, monocytes express higher levels of ERβ, but once activated expression of ERα dominates [99]. The activation state of the cell not only determines the type of hormone receptor expressed, but in the case of the monocyte/macrophage also controls cell survival. Estrogen signaling through ERβ in monocytes upregulates expression of FasL leading to increased monocyte apoptosis while this does not occur in activated macrophages expressing ERα. Survival of APC should prolong the immune response as the antigen and antigen presentation will be maintained. Progesterone suppresses nitric oxide and TNFα from macrophages [78], but testosterone may increase TNFα production [29].

Once an infection has been cleared, the antigen-specific T cell clones, which have been dramatically increased through proliferation during the immune response, will contract. This allows conservation of resources in the body so that unnecessary cells are not maintained. However, the memory T cells remain and can rapidly respond should the antigen/pathogen return. This contraction phase is crucial to normal health as animals with defective apoptosis (Fas or FasL deficiency) can develop lymphadenopathy and in some cases a "lupus-like" autoimmune disease [100]. While ERα and ERβ bind ligand with similar affinity, there exist some differences in the ligand binding pockets of these receptors and use of selective ER agonists indicates that while ERα stimulates cellular proliferation, ERβ promotes apoptosis of T cells [101]. Similarly, androgens promote apoptosis and would therefore accelerate clonal contraction [102–104].

Sex Hormones Effect on T Regulatory Cells

Regulatory T cells (Tregs) comprise up to 10% of peripheral $CD4^+$ T cells in naïve mice and humans, express CD25 (IL-2 receptor α chain), and are important negative modulators of the immune response [105–107]. Treg represent a heterogeneous population that can be divided into natural (nTreg) and inducible (iTreg) cells. nTreg cells develop in the thymus during normal T cell ontogeny as $CD4^+CD25^+$ T cells and depend on expression of the FoxP3 transcription factor, which is crucial for their immunosuppression activity. Expression of exogenous FoxP3 into $CD4^+CD25^-$ T cells converts them into functional $CD4^+CD25^+$ Treg [106]. During T cell ontogeny, cells with higher affinity TCR for self-antigens most likely become nTreg. Unlike T cells of the adaptive immune system, nTreg cells are functionally mature when leaving the thymus and do not require peripheral antigen exposure for immunosuppressive activity. In contrast to nTreg cells, iTreg are converted from effector T cell populations in the periphery subsequent to antigen challenge. These iTreg cells are $CD4^+CD25^+$ but can either be $FoxP3^+$ or $FoxP3^-$ [108]. Tregs secrete immunosuppressive cytokines [108] and inhibit inflammation through direct cell-to-cell contact. Tregs express similar chemokine receptor patterns as effector T cells and can migrate to peripheral lymphoid tissues and inflammatory sites like effector populations [106].

Several mechanisms for suppression of inflammation by Treg have been hypothesized [106]. First, Treg may out-compete effector T cells for MH–antigen complexes on DCs thereby effectively blocking antigen presentation to effector T cells. Secondly, direct binding of Tregs to DCs through CTLA4 downregulates accessory molecule expression (CD80/CD86) on DCs making them less effective at antigen presentation. Third, Tregs may kill DCs preventing interaction with effector T cells. Additionally, TGFβ produced by Treg can activate NOTCH and its downstream target gene Hes1, which suppresses gene expression in T cells [108]. IL-10 released from Treg blocks CD2, CD28, and inducible T-cell co-stimulator (ICOS) signaling in T cells and SOCS3 signaling in monocytes resulting in reduced T cell proliferation and cytokine responses [109]. Both estrogen and progesterone can enhance Treg cell activation [110]. However, Tregs predominantly express ERβ rather than ERα and high levels of PR. Additionally, the GPER agonist, G-1, has been shown to upregulate expression of FoxP3 and increase Treg activation [111], while G-1 treatment of mice provides protection in an experimental model of the autoimmune disease, multiple sclerosis [112].

Sex Hormones and B Cells

As with T cell ontogeny, androgens negatively regulate B cell ontogeny as castration of male mice results in increased B cell numbers in both the spleen and BM [113]. The impact appears to be primarily mediated through de novo B cell generation in BM. Both B cells and BM stromal cells express AR but it appears that AR expression by the stromal cells is primarily necessary for B-cell lymphopoiesis inhibition [29,113]. Estrogens also decrease B cell progenitor numbers in the BM and this effect can be mediated by either ERα or ERβ, both of which are expressed in B-cell precursors and stromal cells [114].

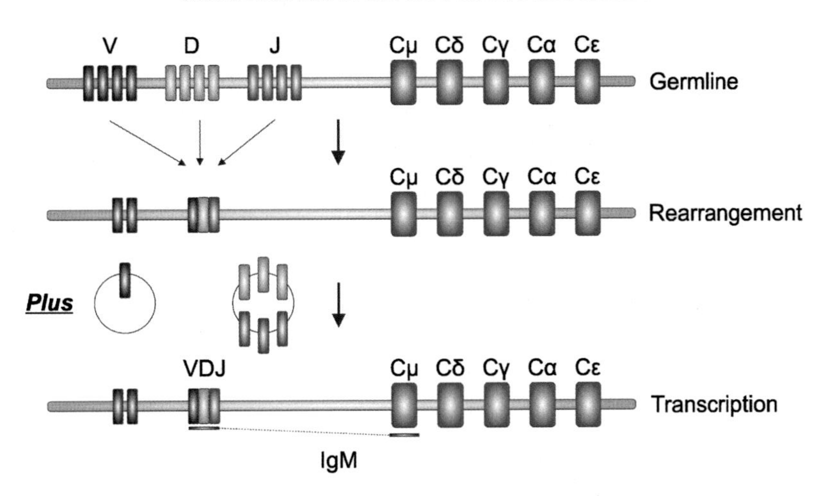

FIGURE 4.1 **Immunoglobulin variable gene rearrangement.** Pre-B cells contain multiple variable (V), diversity (D), and joining (J) gene segments each with its own adjacent recombination signal sequence. Recombination is initiated when recombination activating genes (RAG) 1 and 2, which are primarily expressed in lymphoid cells, induce a double-stranded DNA break near the recombination signal sequences and repair will be mediated through nonhomologous end-joining. Excised V, D, and J segments are eliminated. There are 39 V, 27 D, and 6 J gene sequences for the heavy immunoglobulin chain resulting in the potential for greater than 104 different combinations. Additional variability arises from somatic mutations in the VDJ segments during B cell proliferation. Light chains rearrange V and J gene segments adding to total diversity. The rearranged VDJ segment is transcribed with the germline constant (Cμ) segment giving rise to IgM. For class switching, downstream constant segments corresponding to the other heavy chains (γ, α, ε) contain a cytokine-inducible promoter and a switch intron adjacent to the constant region exon. Under induction by the appropriate T cell dependent cytokine or other factor, IgM positive B cells undergo class switch recombination excising the intervening constant gene segments so that the appropriate Cγ, Cα, or Cε segment is contiguous to the VDJ segment and thus giving rise to IgG, IgA, or IgE, respectively.

Key to B cell development is the rearrangement of immunoglobulin genes to produce a functional antibody. Specificity of the antibody depends on recombination of a single variable (V), diversity (D), and joining (J) segment from each of these segments to form a single VDJ recombinant (Fig. 4.1). While the VDJ combination provides much of the antigen specificity of the antibody, activation and proliferation of mature B cells in response to antigenic stimulation results in further hypermutation (somatic mutation) in the immunoglobulin variable region. This allows the avidity and specificity of the antibody response to evolve during a humoral response. Thus, initially after infection with a virus such as influenza virus, the antibody response will be of relatively low avidity to the pathogen. However, with somatic mutation, some responding B cell clones will mutate their immunoglobulin genes to produce higher avidity antibodies for the virus. As the immune response clears the virus, the lower antigen concentrations available will be selectively captured by B cell clones having the higher avidity antibodies and these will undergo further cellular division and potentially additional somatic mutation. Additionally, humoral responses undergo immunoglobulin class switching. There are five immunoglobulin classes: IgD, IgM, IgG, IgA, and IgE. The VDJ rearrangement in the B cell results from selective deletions of intervening DNA segments between various "V," "D," and "J" segments so that only one of each remains conjoined. Class switching occurs at the level of mRNA transcription by transcribing the VDJ with one of a number of constant regions coding for the delta, mu, gamma, alpha, or epsilon heavy chain. The complete mRNA for the heavy chain will then be translated to the heavy immunoglobulin molecules and combine with either kappa or lamda light chains to make the whole immunoglobulin molecule. Initially, immature B cells express IgD and IgM; with antigen exposure they make IgM in the primary immune response and subsequently, with further antigen stimulation and T cell help in the context of specific cytokines, make either IgG, IgA, or IgE. Both class switching and somatic mutation require a specific enzyme called activation-induced cytidine deaminase (AICDA) [115].

Hormones have multiple effects on mature B cells and humoral immunity. Estrogens increase and progesterone decreases AICDA expression, which enhances (estrogen) and suppresses (progesterone) somatic mutation and class switching. Estrogens also suppress apoptosis promoting B cell survival while androgens increase apoptosis promoting B cell elimination through differential effects on Bcl2 expression, an antiapoptosis gene [114]. Furthermore, estrogen may increase the amount of immunoglobulin produced by each B cell since treating B cell hybridoma clones with E2 resulted in a 255% increase in immunoglobulin secretion, an effect mediated through ERα [116].

EFFECT OF VITAMIN D ON THE INNATE AND ADAPTIVE IMMUNE SYSTEM

Vitamin D via the VDR has important effects on immune cells. In general, vitamin D has been shown in culture studies and animal models to increase Th2 immune responses, IL-10, Treg, alternatively activate M2 macrophages and TGFβ [41,117–120]. Additionally, vitamin D is known to mediate protection against infections where it acts on DCs, MCs, and macrophages to activate the innate immune response. In response to infection vitamin D via the VDR upregulates TLR2, TLR4, CD14 (part of TLR4 signaling complex), the inflammasome (ie, IL-1β, caspase-1), TNF and IFNγ, for example [121–127]. VDR response element activation increases many antimicrobial mediators like β-defensin, cathelicidin, and reactive oxygen species. However, the innate responses induced by VDR signaling described here drive Th1 and M1 macrophage responses. This appears contradictory to the ability of vitamin D to upregulate Th2 and regulatory immune responses. Confusion on this topic exists, at least in part, because most studies of vitamin D do not consider their findings in the context of biologic sex differences. Usually there is no identification of the sex of the cells or animals used to conduct experiments, and no interpretation of data taking sex into consideration. Because vitamin D is a sex steroid, sex differences in its effect on immune cells should exist similar to the sex differences observed for estrogens and androgens. Although almost no studies have been published that examine the issue of sex differences in vitamin D, one study reported that vitamin D administration was only protective in female mice in an animal model of multiple sclerosis [45], an autoimmune disease that occurs more frequently in women than men. To improve our understanding of the role of vitamin D-mediated VDR activation on inflammation and chronic inflammatory diseases it is vital that researchers design experiments, analyze data, and report results according to sex. It must also be kept in mind that although many inflammatory diseases are inhibited by Th2 responses, others like allergy, asthma, and rheumatic autoimmune diseases are promoted by Th2 responses [128,129]. Thus, vitamin D can stimulate protective innate immune responses to infections and may either promote or inhibit chronic inflammatory diseases depending on whether Th1/Th17 or Th2 skewing helps or harms the disease. Research to understand sex differences in the effect of vitamin D and VDR on immune cells in health and disease needs to be conducted.

EFFECT OF MENOPAUSE ON INFLAMMATION

Menopause is defined as the final menstrual period without another menstrual period for 12 months [130–132]. The mean age when menopause occurs in Western cultures is between 49 and 52 years [130,131]. Decreases in estrogen (ie, E2) occur only in the last 6 months before menopause and thereafter [132]. In contrast, testosterone gradually decreases with age in both sexes. There appears to be no period of symptomatic change or "andropause" in men that would be analogous to menopause in women, but instead a steady decline in testosterone levels [133]. After menopause, the primary endogenous source of estrogen is estrone, which is synthesized in adipocytes via aromatase [134]. As in men, testosterone levels decline gradually with age in women [135].

Understanding changes in immune function following menopause are complicated by changes that also occur because of aging [136,137]. Age-related changes in immune function include an impaired ability to respond to new antigens, unsustained memory responses, chronic low-grade inflammation, oligoclonal expansion of B cells, and an increasing propensity for autoimmune responses [138]. Antibody and autoantibody levels continually increase with age. Additionally, low doses of estrogen are likely to be all that is required to promote B cell proliferation and autoantibody production past menopause in women. Support for this idea is the finding that hormone replacement therapy (lifting estrogen levels to a relatively higher dose) decreases antibody levels in menopausal women [139].

Menopause is associated with decreases in $CD4^+$ and $CD8^+$ T cells and NK cell activity [134,140,141]. Importantly, menopause has been found to increase IL-1, IL-6, and TNF levels—cytokines that promote chronic inflammatory diseases. That lower estrogen levels during menopause decrease these cytokines is consistent with the known ability of higher doses of estrogen to downregulate NF-κB, Th1 responses, IL-1, and IL-6 via ERα in various human and murine cell types [71,73,142–144]. Thus, after menopause the protective role of estrogen in decreasing innate immune cell activation is lost allowing increased pro-inflammatory cytokine levels while at the same time antibody and autoantibody levels continue to rise. This may promote the development of certain autoimmune diseases like rheumatoid arthritis and Sjögren's syndrome, which peak after 50 years of age in women [145]. In contrast, in most men androgen levels that promote innate immune cell activation and pro-inflammatory cytokines are declining past age 50 without the increase in antibodies found in women because of estrogen. Interestingly, men die of pro-inflammatory diseases like cardiovascular disease earlier than women.

Epidemiological data indicate that more than 60% of postmenopausal women have low vitamin D levels [146,147]. With aging dermal and epidermal skin

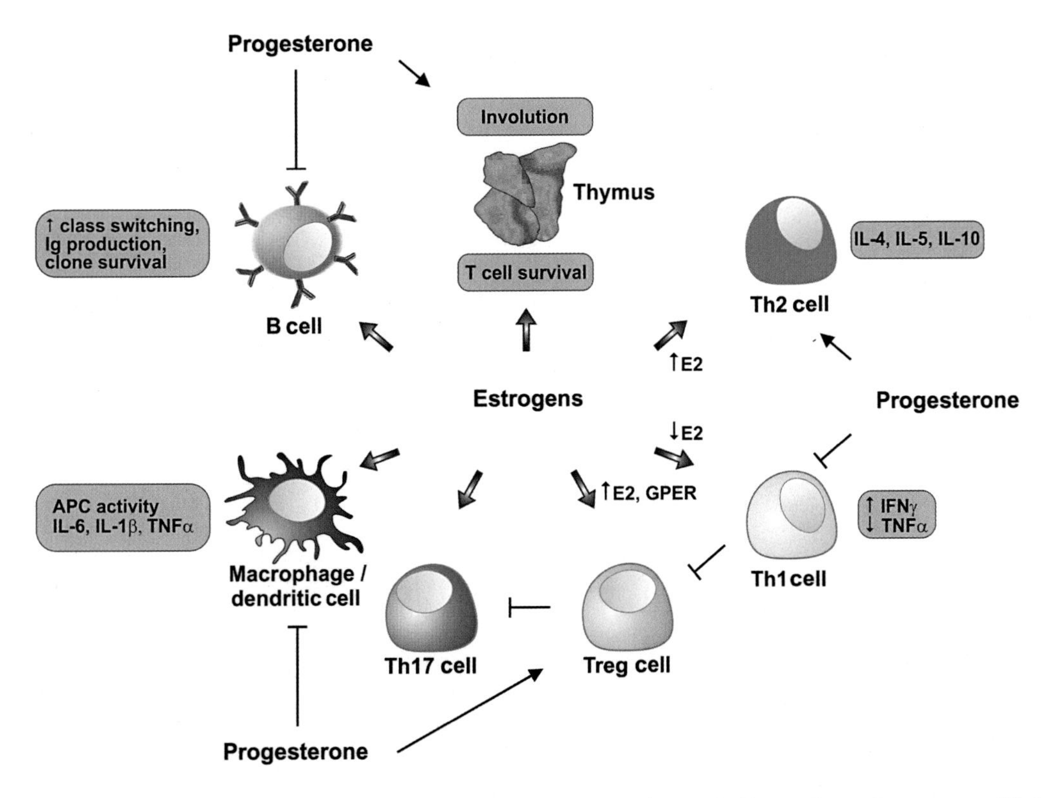

FIGURE 4.2 **Generally accepted hormone effects on the immune system.** Arrows indicate enhanced responses while bars show inhibitory effects.

thickness reduces [148]. Consequently, cutaneous vitamin D synthesis decreases with age and menopause because of smaller stores of the precursor 7-dehydroxycholerstorol in the skin [149–154]. The elderly may be at additional risk of vitamin D deficiency because of decreased mobility and thereby less sun exposure, and because of more kidney or liver disease [155].

CONCLUSIONS AND FUTURE PERSPECTIVES

Although a wealth of information is available about traditional sex hormone signaling, there are many gaps that still need to be filled. Additionally, often classical sex hormones (estrogen, testosterone, progesterone) are reported to have different immunomodulatory affects. For example, estrogen can both promote immunosuppressive Treg activation and pro-inflammatory $CD4^+$ Th1 cell responses. The hormone concentration, the types, and concentrations of hormone receptors in immune cells (including receptor splice variants) and the activation environment during the immune response (which cytokines are present, strength of TLR activation) can have a major impact on the ultimate hormone immunomodulatory effect. Fig. 4.2 provides a summary of the generally accepted hormone effects on the immune system.

Few people realize that vitamin D is a sex steroid that regulates both innate and adaptive immune responses. Similar to traditional sex hormones like estrogen and testosterone, vitamin D levels vary by sex. Environment (ie, sunshine exposure and diet) greatly influences vitamin D levels and may contribute to some of the environmental risk for developing chronic inflammatory diseases. The effect of aging and menopause on immune cell function is also just beginning to be understood. Another emerging area of research is the effect of epigenetic regulation on expression of sex differences in the immune response. Even though researchers are beginning to appreciate the importance of sex differences in their studies, most manuscripts do not list the sex of animals or cells that are used in their experiments nor interpret their data in the context of sex. Additionally, many clinical studies now include both men and women, but few studies analyze their data by sex or interpret the data according to sex. These are just a few of the areas that need more research in the future. One thing that is clear is that sex hormones direct the immune response in both health and disease.

FUNDING

This work was supported by NHLBI awards R01 HL111938 (DF) and R01 HL108371 (SAH); by NIEHS award R21 ES024414 (DF); by COBRE award P20 GM103496 (RC Budd PI; SAH, IB) and by American Heart Association Grant-in-Aid 12GRNT12050000 (DF).

References

[1] Oertelt-Prigione S. The influence of sex and gender on the immune response. Autoimmun Rev 2012;11(6–7):A479–85.

[2] Udry JR. The nature of gender. Demography 1994;31(4):561–73.

[3] Bachmann GA, Mussman B. The aging population: Imperative to uncouple sex and gender to establish "gender equal" health care. Maturitas 2015;80(4):421–5.

[4] Pennell LM, Galligan CL, Fish EN. Sex affects immunity. J Autoimmun 2012;38(2–3):J282–91.

[5] Klein SL. The effects of hormones on sex differences in infection: from genes to behavior. Neurosci Biobehav Rev 2000;24 (6):627–38.

[6] Fish EN. The X-files in immunity: sex-based differences predispose immune responses. Nat Rev Immunol 2008;8(9):737–44.

[7] Libert C, Dejager L, Pinheiro I. The X chromosome in immune functions: when a chromosome makes the difference. Nat Rev Immunol 2010;10(8):594–604.

[8] Cotton AM, Price EM, Jones MJ, Balaton BP, Kobor MS, Brown CJ. Landscape of DNA methylation on the X chromosome reflects CpG density, functional chromatin state and X-chromosome inactivation. Hum Mol Genet 2015;24(6):1528–39.

[9] Latour S, Aguilar C. XIAP deficiency syndrome in humans. Semin Cell Dev Biol 2015;39:115–23.

[10] Torgerson TR, Ochs HD. Immune dysregulation, polyendocrinopathy, enteropathy, X-linked: forkhead box protein 3 mutations and lack of regulatory T cells. J Allergy Clin Immunol 2007;120(4):744–50: quiz 51–2.

[11] Beato M, Klug J. Steroid hormone receptors: an update. Hum Reprod Update 2000;6(3):225–36.

[12] Cui J, Shen Y, Li R. Estrogen synthesis and signaling pathways during aging: from periphery to brain. Trends Mol Med 2013;19 (3):197–209.

[13] Fox HS, Bond BL, Parslow TG. Estrogen regulates the IFN-gamma promoter. J Immunol 1991;146(12):4362–7.

[14] Liu MM, Albanese C, Anderson CM, Hilty K, Webb P, Uht RM, et al. Opposing action of estrogen receptors alpha and beta on cyclin D1 gene expression. J Biol Chem 2002;277 (27):24353–60.

[15] Kurebayashi Y, Nagai S, Ikejiri A, Koyasu S. Recent advances in understanding the molecular mechanisms of the development and function of Th17 cells. Genes Cells 2013;18 (4):247–65.

[16] Razandi M, Pedram A, Greene GL, Levin ER. Cell membrane and nuclear estrogen receptors (ERs) originate from a single transcript: studies of ERalpha and ERbeta expressed in Chinese hamster ovary cells. Mol Endocrinol 1999;13(2):307–19.

[17] Pedram A, Razandi M, Sainson RC, Kim JK, Hughes CC, Levin ER. A conserved mechanism for steroid receptor translocation to the plasma membrane. J Biol Chem 2007;282(31):22278–88.

[18] Schlegel A, Wang C, Katzenellenbogen BS, Pestell RG, Lisanti MP. Caveolin-1 potentiates estrogen receptor alpha (ERalpha) signaling. Caveolin-1 drives ligand-independent nuclear translocation and activation of ERalpha. J Biol Chem 1999;274 (47):33551–6.

[19] Soltysik K, Czekaj P. Membrane estrogen receptors - is it an alternative way of estrogen action? J Physiol Pharmacol 2013;64 (2):129–42.

[20] Moriarty K, Kim KH, Bender JR. Minireview: estrogen receptor-mediated rapid signaling. Endocrinology 2006;147(12):5557–63.

[21] Li L, Haynes MP, Bender JR. Plasma membrane localization and function of the estrogen receptor alpha variant (ER46) in human endothelial cells. Proc Natl Acad Sci USA 2003;100 (8):4807–12.

[22] Pedram A, Razandi M, Aitkenhead M, Levin ER. Estrogen inhibits cardiomyocyte hypertrophy in vitro. Antagonism of calcineurin-related hypertrophy through induction of MCIP1. J Biol Chem 2005;280(28):26339–48.

[23] Luoma JI, Boulware MI, Mermelstein PG. Caveolin proteins and estrogen signaling in the brain. Mol Cell Endocrinol 2008;290(1–2):8–13.

[24] Rettew JA, McCall SH, Marriott I. GPR30/GPER-1 mediates rapid decreases in TLR4 expression on murine macrophages. Mol Cell Endocrinol 2010;328(1–2):87–92.

[25] Filice E, Recchia AG, Pellegrino D, Angelone T, Maggiolini M, Cerra MC. A new membrane G protein-coupled receptor (GPR30) is involved in the cardiac effects of 17beta-estradiol in the male rat. J Physiol Pharmacol 2009;60 (4):3–10.

[26] Centenera MM, Harris JM, Tilley WD, Butler LM. The contribution of different androgen receptor domains to receptor dimerization and signaling. Mol Endocrinol 2008;22(11):2373–82.

[27] Benten WP, Lieberherr M, Stamm O, Wrehlke C, Guo Z, Wunderlich F. Testosterone signaling through internalizable surface receptors in androgen receptor-free macrophages. Mol Biol Cell 1999;10(10):3113–23.

[28] Benten WP, Lieberherr M, Giese G, Wrehlke C, Stamm O, Sekeris CE, et al. Functional testosterone receptors in plasma membranes of T cells. FASEB J 1999;13(1):123–33.

[29] Lai JJ, Lai KP, Zeng W, Chuang KH, Altuwaijri S, Chang C. Androgen receptor influences on body defense system via modulation of innate and adaptive immune systems: lessons from conditional AR knockout mice. Am J Pathol 2012;181 (5):1504–12.

[30] Foradori CD, Weiser MJ, Handa RJ. Non-genomic actions of androgens. Front Neuroendocrinol 2008;29(2):169–81.

[31] Misrahi M, Venencie PY, Saugier-Veber P, Sar S, Dessen P, Milgrom E. Structure of the human progesterone receptor gene. Biochim Biophys Acta 1993;1216(2):289–92.

[32] Mulac-Jericevic B, Conneely OM. Reproductive tissue selective actions of progesterone receptors. Reproduction 2004;128 (2):139–46.

[33] Obr AE, Edwards DP. The biology of progesterone receptor in the normal mammary gland and in breast cancer. Mol Cell Endocrinol 2012;357(1–2):4–17.

[34] Kowalik MK, Rekawiecki R, Kotwica J. The putative roles of nuclear and membrane-bound progesterone receptors in the female reproductive tract. Reprod Biol 2013;13(4):279–89.

[35] Boonyaratanakornkit V, McGowan E, Sherman L, Mancini MA, Cheskis BJ, Edwards DP. The role of extranuclear signaling actions of progesterone receptor in mediating progesterone regulation of gene expression and the cell cycle. Mol Endocrinol 2007;21(2):359–75.

[36] Lavie CJ, Lee JH, Milani RV. Vitamin D and cardiovascular disease will it live up to its hype? J Am Coll Cardiol 2011;58 (15):1547–56.

[37] Plum LA, DeLuca HF. Vitamin D, disease and therapeutic opportunities. Nat Rev Drug Discov 2010;9(12):941–55.
[38] Zittermann A, Gummert JF. Sun, vitamin D, and cardiovascular disease. J Photochem Photobiol B 2010;101(2):124–9.
[39] Norman AW. Minireview: vitamin D receptor: new assignments for an already busy receptor. Endocrinology 2006;147 (12):5542–8.
[40] Mason RS, Reichrath J. Sunlight vitamin D and skin cancer. Anticancer Agents Med Chem 2013;13(1):83–97.
[41] Bikle DD. Vitamin D regulation of immune function. Vitam Horm 2011;86:1–21.
[42] Onyimba JA, Coronado MJ, Garton AE, Kim JB, Bucek A, Bedja D, et al. The innate immune response to coxsackievirus B3 predicts progression to cardiovascular disease and heart failure in male mice. Biol Sex Differ 2011;2:2.
[43] Lisse TS, Hewison M, Adams JS. Hormone response element binding proteins: novel regulators of vitamin D and estrogen signaling. Steroids 2011;76(4):331–9.
[44] Fairweather D. Autoimmune skin diseases: role of sex hormones, Vitamin D and menopause. In: Farage MA, Miller KW, Fugate-Woods N, Maibach HI, editors. Skin, mucosa and menopause: management of clinical issues. Heidelberg: Springer; 2015. p. 359–81.
[45] Spach KM, Hayes CE. Vitamin D3 confers protection from autoimmune encephalomyelitis only in female mice. J Immunol 2005;175(6):4119–26.
[46] Abreu MT, Arditi M. Innate immunity and toll-like receptors: clinical implications of basic science research. J Pediatr 2004;144 (4):421–9.
[47] Means TK, Golenbock DT, Fenton MJ. The biology of Toll-like receptors. Cytokine Growth Factor Rev 2000;11(3):219–32.
[48] Netea MG, Van der Meer JW, Kullberg BJ. Toll-like receptors as an escape mechanism from the host defense. Trends Microbiol 2004;12(11):484–8.
[49] O'Neill LA. TLRs: professor mechnikov, sit on your hat. Trends Immunol 2004;25(12):687–93.
[50] Lauw FN, Caffrey DR, Golenbock DT. Of mice and man: TLR11 (finally) finds profilin. Trends Immunol 2005;26(10):509–11.
[51] Hasan U, Chaffois C, Gaillard C, Saulnier V, Merck E, Tancredi S, et al. Human TLR10 is a functional receptor, expressed by B cells and plasmacytoid dendritic cells, which activates gene transcription through MyD88. J Immunol 2005;174(5):2942–50.
[52] Wang RF, Miyahara Y, Wang HY. Toll-like receptors and immune regulation: implications for cancer therapy. Oncogene 2008;27(2):181–9.
[53] Pulendran B. Variegation of the immune response with dendritic cells and pathogen recognition receptors. J Immunol 2005;174(5):2457–65.
[54] Netea MG, Van der Meer JW, Sutmuller RP, Adema GJ, Kullberg BJ. From the Th1/Th2 paradigm towards a Toll-like receptor/T-helper bias. Antimicrob Agents Chemother 2005;49 (10):3991–6.
[55] Fairweather D. Sex differences in inflammation during atherosclerosis. Clin Med Insights Cardiol 2014;8:49–59.
[56] Liew FY, Komai-Koma M, Xu D. A toll for T cell costimulation. Ann Rheumatic Diseases 2004;63(Suppl. 2):ii76–8.
[57] Pisitkun P, Deane JA, Difilippantonio MJ, Tarasenko T, Satterthwaite AB, Bolland S. Autoreactive B cell responses to RNA-related antigens due to TLR7 gene duplication. Science 2006;312(5780):1669–72.
[58] Berghofer B, Frommer T, Haley G, Fink L, Bein G, Hackstein H. TLR7 ligands induce higher IFN-alpha production in females. J Immunol 2006;177(4):2088–96.
[59] Roberts BJ, Dragon JA, Moussawi M, Huber SA. Sex-specific signaling through toll-like receptors 2 and 4 contributes to survival outcome of coxsackievirus B3 infection in C57Bl/6 mice. Biol Sex Differ 2012;3(1):25.
[60] Frisancho-Kiss S, Davis SE, Nyland JF, Frisancho JA, Cihakova D, Barrett MA, et al. Cutting edge: cross-regulation by TLR4 and T cell Ig mucin-3 determines sex differences in inflammatory heart disease. J Immunol 2007;178(11):6710–14.
[61] Hou J, Zheng WF. Effect of sex hormones on NK and ADCC activity of mice. Int J Immunopharmacol 1988;10(1):15–22.
[62] Flanagan KL, Klein SL, Skakkebaek NE, Marriott I, Marchant A, Selin L, et al. Sex differences in the vaccine-specific and non-targeted effects of vaccines. Vaccine 2011;29(13):2349–54.
[63] Lang TJ. Estrogen as an immunomodulator. Clin Immunol 2004;113(3):224–30.
[64] Styrt B, Sugarman B. Estrogens and infection. Rev Infect Dis 1991;13(6):1139–50.
[65] Tan IJ, Peeva E, Zandman-Goddard G. Hormonal modulation of the immune system - a spotlight on the role of progestogens. Autoimmun Rev 2015;14:536–42.
[66] Gilliver SC. Sex steroids as inflammatory regulators. J Steroid Biochem Mol Biol 2010;120(2–3):105–15.
[67] McCrohon JA, Death AK, Nakhla S, Jessup W, Handelsman DJ, Stanley KK, et al. Androgen receptor expression is greater in macrophages from male than from female donors. A sex difference with implications for atherogenesis. Circulation 2000;101(3):224–6.
[68] Sader MA, McGrath KC, Hill MD, Bradstock KF, Jimenez M, Handelsman DJ, et al. Androgen receptor gene expression in leucocytes is hormonally regulated: implications for gender differences in disease pathogenesis. Clin Endocrinol (Oxf) 2005;62(1):56–63.
[69] Brechenmacher SA, Bruns CJ, Van den Engel NK, Angele P, Loehe F, Jauch KW, et al. Influence of surgical trauma on the mRNA expression of sex hormone receptors in PBMCs in male and female patients. Langenbeck Arch Surg 2008;393(6):871–6.
[70] Deshpande R, Khalili H, Pergolizzi RG, Michael SD, Chang MD. Estradiol down-regulates LPS-induced cytokine production and NFkB activation in murine macrophages. Am J Reprod Immunol 1997;38(1):46–54.
[71] Evans MJ, Eckert A, Lai K, Adelman SJ, Harnish DC. Reciprocal antagonism between estrogen receptor and NF-kappaB activity in vivo. Circ Res 2001;89(9):823–30.
[72] Chadwick CC, Chippari S, Matelan E, Borges-Marcucci L, Eckert AM, Keith Jr. JC, et al. Identification of pathway-selective estrogen receptor ligands that inhibit NF-kappaB transcriptional activity. Proc Natl Acad Sci USA 2005;102 (7):2543–8.
[73] Demyanets S, Pfaffenberger S, Kaun C, Rega G, Speidl WS, Kastl SP, et al. The estrogen metabolite 17beta-dihydroequilenin counteracts interleukin-1alpha induced expression of inflammatory mediators in human endothelial cells in vitro via NF-kappaB pathway. Thromb Haemost 2006;95(1):107–16.
[74] Huang H, He J, Yuan Y, Aoyagi E, Takenaka H, Itagaki T, et al. Opposing effects of estradiol and progesterone on the oxidative stress-induced production of chemokine and proinflammatory cytokines in murine peritoneal macrophages. J Med Invest 2008;55(1–2):133–41.
[75] Okwan-Duodu D, Umpierrez GE, Brawley OW, Diaz R. Obesity-driven inflammation and cancer risk: role of myeloid derived suppressor cells and alternately activated macrophages. Am J Cancer Res 2013;3(1):21–33.
[76] Rettew JA, Huet-Hudson YM, Marriott I. Testosterone reduces macrophage expression in the mouse of toll-like receptor 4, a trigger for inflammation and innate immunity. Biol Reprod 2008;78(3):432–7.

[77] Rettew JA, Huet YM, Marriott I. Estrogens augment cell surface TLR4 expression on murine macrophages and regulate sepsis susceptibility in vivo. Endocrinology 2009;150(8):3877–84.
[78] Hughes GC. Progesterone and autoimmune disease. Autoimmun Rev 2012;11(6–7):A502–14.
[79] Yellayi S, Teuscher C, Woods JA, Welsh Jr. TH, Tung KS, Nakai M, et al. Normal development of thymus in male and female mice requires estrogen/estrogen receptor-alpha signaling pathway. Endocrine 2000;12(3):207–13.
[80] Fairweather D, Root-Bernstein R. Autoimmune disease: mechanisms (version 2). Encyclopedia of life sciences. Chichester: John Wiley & Sons Ltd; 2015.
[81] Olsen NJ, Kovacs WJ. Effects of androgens on T and B lymphocyte development. Immunologic research 2001;23(2–3):281–8.
[82] Wimmers F, Schreibelt G, Skold AE, Figdor CG, De Vries IJ. Paradigm shift in dendritic cell-based immunotherapy: from in vitro generated monocyte-derived DCs to naturally circulating DC subsets. Front Immunol 2014;5:165.
[83] Kiama SG, Cochand L, Karlsson L, Nicod LP, Gehr P. Evaluation of phagocytic activity in human monocyte-derived dendritic cells. J Aerosol Med 2001;14(3):289–99.
[84] Celli S, Lemaitre F, Bousso P. Real-time manipulation of T cell-dendritic cell interactions in vivo reveals the importance of prolonged contacts for CD4+ T cell activation. Immunity 2007;27(4):625–34.
[85] Douin-Echinard V, Laffont S, Seillet C, Delpy L, Krust A, Chambon P, et al. Estrogen receptor alpha, but not beta, is required for optimal dendritic cell differentiation and [corrected] CD40-induced cytokine production. J Immunol 2008;180(6):3661–9.
[86] Carreras E, Turner S, Frank MB, Knowlton N, Osban J, Centola M, et al. Estrogen receptor signaling promotes dendritic cell differentiation by increasing expression of the transcription factor IRF4. Blood 2010;115(2):238–46.
[87] Kovats S. Estrogen receptors regulate an inflammatory pathway of dendritic cell differentiation: mechanisms and implications for immunity. Horm Behav 2012;62(3):254–62.
[88] Seillet C, Rouquie N, Foulon E, Douin-Echinard V, Krust A, Chambon P, et al. Estradiol promotes functional responses in inflammatory and steady-state dendritic cells through differential requirement for activation function-1 of estrogen receptor alpha. J Immunol 2013;190(11):5459–70.
[89] Xie H, Hua C, Sun L, Zhao X, Fan H, Dou H, et al. 17beta-estradiol induces CD40 expression in dendritic cells via MAPK signaling pathways in a minichromosome maintenance protein 6-dependent manner. Arthritis Rheum 2011;63 (8):2425–35.
[90] Mosmann T, Coffman R. Th1 and Th2 cells: different patterns of lymphokine secretion lead to different functional properties. Annu Rev Immunol 1989;7:145–73.
[91] Sun B, Zhang Y. Overview of orchestration of CD4+ T cell subsets in immune responses. Adv Exp Med Biol 2014;841:1–13.
[92] Greenstein B, Roa R, Dhaher Y, Nunn E, Greenstein A, Khamashta M, et al. Estrogen and progesterone receptors in murine models of systemic lupus erythematosus. Int Immunopharmacol 2001;1(6):1025–35.
[93] Coquet JM, Rausch L, Borst J. The importance of co-stimulation in the orchestration of T helper cell differentiation. Immunol Cell Biol 2015;93:780–8.
[94] Zhu J, Yamane H, Paul WE. Differentiation of effector CD4 T cell populations (*). Annu Rev Immunol 2010;28:445–89.
[95] Kanno Y, Vahedi G, Hirahara K, Singleton K, O'Shea JJ. Transcriptional and epigenetic control of T helper cell specification: molecular mechanisms underlying commitment and plasticity. Annu Rev Immunol 2012;30:707–31.
[96] Huber S, Pfaeffle B. Differential Th1 and Th2 cell responses in male and female BALB/c mice infected with coxsackievirus group B type 3. J Virol 1994;68:5126–32.
[97] Fairweather D, Cooper Jr. LT, Blauwet LA. Sex and gender differences in myocarditis and dilated cardiomyopathy. Curr Probl Cardiol 2013;38(1):7–46.
[98] Lyden DC, Olszewski J, Feran M, Job LP, Huber SA. Coxsackievirus B-3-induced myocarditis. Effect of sex steroids on viremia and infectivity of cardiocytes. Am J Pathol 1987;126 (3):432–8.
[99] Mor G, Sapi E, Abrahams VM, Rutherford T, Song J, Hao XY, et al. Interaction of the estrogen receptors with the Fas ligand promoter in human monocytes. J Immunol 2003;170 (1):114–22.
[100] Su HC, Lenardo MJ. Genetic defects of apoptosis and primary immunodeficiency. Immunol Allergy Clin North Am 2008;28 (2):329–51 ix
[101] Yakimchuk K, Jondal M, Okret S. Estrogen receptor alpha and beta in the normal immune system and in lymphoid malignancies. Mol Cell Endocrinol 2013;375(1–2):121–9.
[102] Cutolo M, Sulli A, Capellino S, Villaggio B, Montagna P, Seriolo B, et al. Sex hormones influence on the immune system: basic and clinical aspects in autoimmunity. Lupus 2004;13(9):635–8.
[103] Cutolo M, Sulli A, Craviotto C, Felli L, Pizzorni C, Seriolo B, et al. Modulation of cell growth and apoptosis by sex hormones in cultured monocytic THP-1 cells. Ann N Y Acad Sci 2002;966:204–10.
[104] Cutolo M, Villaggio B, Craviotto C, Pizzorni C, Seriolo B, Sulli A. Sex hormones and rheumatoid arthritis. Autoimmun Rev 2002;1(5):284–9.
[105] Torgerson TR. Regulatory T cells in human autoimmune diseases. Springer Semin Immunopathol 2006;28(1):63–76.
[106] Sakaguchi S, Yamaguchi T, Nomura T, Ono M. Regulatory T cells and immune tolerance. Cell 2008;133(5):775–87.
[107] Sakaguchi S, Setoguchi R, Yagi H, Nomura T. Naturally arising Foxp3-expressing CD25+ CD4+ regulatory T cells in self-tolerance and autoimmune disease. Curr Top Microbiol Immunol 2006;305:51–66.
[108] Ray A, Khare A, Krishnamoorthy N, Qi Z, Ray P. Regulatory T cells in many flavors control asthma. Mucosal Immunol 2010;3(3):216–29.
[109] Ozdemir C, Akdis M, Akdis CA. T regulatory cells and their counterparts: masters of immune regulation. Clin Exp Allergy 2009;39(5):626–39.
[110] Aristimuno C, Teijeiro R, Valor L, Alonso B, Tejera-Alhambra M, de Andres C, et al. Sex-hormone receptors pattern on regulatory T-cells: clinical implications for multiple sclerosis. Clin Exp Med 2012;12(4):247–55.
[111] Brunsing RL, Owens KS, Prossnitz ER. The G protein-coupled estrogen receptor (GPER) agonist G-1 expands the regulatory T-cell population under TH17-polarizing conditions. J Immunother 2013;36(3):190–6.
[112] Blasko E, Haskell CA, Leung S, Gualtieri G, Halks-Miller M, Mahmoudi M, et al. Beneficial role of the GPR30 agonist G-1 in an animal model of multiple sclerosis. J Neuroimmunol 2009;214(1–2):67–77.
[113] Olsen NJ, Gu X, Kovacs WJ. Bone marrow stromal cells mediate androgenic suppression of B lymphocyte development. J Clin Invest 2001;108(11):1697–704.
[114] Sakiani S, Olsen NJ, Kovacs WJ. Gonadal steroids and humoral immunity. Nat Rev Endocrinol 2013;9(1):56–62.
[115] Muramatsu M, Kinoshita K, Fagarasan S, Yamada S, Shinkai Y, Honjo T. Class switch recombination and hypermutation require activation-induced cytidine deaminase (AID), a potential RNA editing enzyme. Cell 2000;102(5):553–63.

[116] Molina J, Masso F, Paez A, Mendez C, Rodriguez E, Mandoki JJ, et al. Differential effect of estradiol on antibody secretion of murine hybridomas. Hybridoma 1999;18(4):377–83.
[117] Litwack G. Vitamins and the immune system [Preface] Vitam Horm 2011;86:xvii–xviii.
[118] Daniel C, Sartory NA, Zahn N, Radeke HH, Stein JM. Immune modulatory treatment of trinitrobenzene sulfonic acid colitis with calcitriol is associated with a change of a T helper (Th) 1/Th17 to a Th2 and regulatory T cell profile. J Pharmacol Exp Ther 2008;324(1):23–33.
[119] Topilski I, Flaishon L, Naveh Y, Harmelin A, Levo Y, Shachar I. The anti-inflammatory effects of 1,25-dihydroxyvitamin D3 on Th2 cells in vivo are due in part to the control of integrin-mediated T lymphocyte homing. Eur J Immunol 2004;34(4):1068–76.
[120] Gregori S, Casorati M, Amuchastegui S, Smiroldo S, Davalli AM, Adorini L. Regulatory T cells induced by 1 alpha,25-dihydroxyvitamin D3 and mycophenolate mofetil treatment mediate transplantation tolerance. J Immunol 2001;167(4):1945–53.
[121] Liu PT, Stenger S, Tang DH, Modlin RL. Cutting edge: vitamin D-mediated human antimicrobial activity against Mycobacterium tuberculosis is dependent on the induction of cathelicidin. J Immunol 2007;179(4):2060–3.
[122] Liu PT, Stenger S, Li H, Wenzel L, Tan BH, Krutzik SR, et al. Toll-like receptor triggering of a vitamin D-mediated human antimicrobial response. Science 2006;311(5768):1770–3.
[123] Schauber J, Dorschner RA, Coda AB, Buchau AS, Liu PT, Kiken D, et al. Injury enhances TLR2 function and antimicrobial peptide expression through a vitamin D-dependent mechanism. J Clin Invest 2007;117(3):803–11.
[124] Feldmeyer L, Keller M, Niklaus G, Hohl D, Werner S, Beer HD. The inflammasome mediates UVB-induced activation and secretion of interleukin-1beta by keratinocytes. Curr Biol 2007;17(13):1140–5.
[125] Baroni E, Biffi M, Benigni F, Monno A, Carlucci D, Carmeliet G, et al. VDR-dependent regulation of mast cell maturation mediated by 1,25-dihydroxyvitamin D3. J Leukoc Biol 2007;81(1):250–62.
[126] Wang TT, Nestel FP, Bourdeau V, Nagai Y, Wang Q, Liao J, et al. Cutting edge: 1,25-dihydroxyvitamin D3 is a direct inducer of antimicrobial peptide gene expression. J Immunol 2004;173(5):2909–12.
[127] Rook GA, Steele J, Fraher L, Barker S, Karmali R, O'Riordan J, et al. Vitamin D3, gamma interferon, and control of proliferation of Mycobacterium tuberculosis by human monocytes. Immunology 1986;57(1):159–63.
[128] Brehm JM, Schuemann B, Fuhlbrigge AL, Hollis BW, Strunk RC, Zeiger RS, et al. Serum vitamin D levels and severe asthma exacerbations in the Childhood Asthma Management Program study. J Allergy Clin Immunol 2010;126(1):52–8.
[129] Wittke A, Weaver V, Mahon BD, August A, Cantorna MT. Vitamin D receptor-deficient mice fail to develop experimental allergic asthma. J Immunol 2004;173(5):3432–6.
[130] Kato I, Toniolo P, Akhmedkhanov A, Koenig KL, Shore R, Zeleniuch-Jacquotte A. Prospective study of factors influencing the onset of natural menopause. J Clin Epidemiol 1998;51(12):1271–6.
[131] Jacobsen BK, Heuch I, Kvale G. Age at natural menopause and all-cause mortality: a 37-year follow-up of 19,731 Norwegian women. Am J Epidemiol 2003;157(10):923–9.
[132] Harlow SD, Gass M, Hall JE, Lobo R, Maki P, Rebar RW, et al. Executive summary of the Stages of Reproductive Aging Workshop + 10: addressing the unfinished agenda of staging reproductive aging. J Clin Endocrinol Metab 2012;97(4):1159–68.
[133] Pines A. Male menopause: is it a real clinical syndrome? Climacteric 2011;14(1):15–17.
[134] Bove R. Autoimmune diseases and reproductive aging. Clin Immunol 2013;149(2):251–64.
[135] Davison SL, Bell R, Donath S, Montalto JG, Davis SR. Androgen levels in adult females: changes with age, menopause, and oophorectomy. J Clin Endocrinol Metab 2005;90(7):3847–53.
[136] Gameiro CM, Romao F, Castelo-Branco C. Menopause and aging: changes in the immune system—a review. Maturitas 2010;67(4):316–20.
[137] Weiskopf D, Weinberger B, Grubeck-Loebenstein B. The aging of the immune system. Transpl Int 2009;22(11):1041–50.
[138] Goronzy JJ, Weyand CM. Understanding immunosenescence to improve responses to vaccines. Nat Immunol 2013;14(5):428–36.
[139] Blum M, Zacharovich D, Pery J, Kitai E. [Lowering effect of estrogen replacement treatment on immunoglobulins in menopausal women]. Revue francaise de gynecologie et d'obstetrique 1990;85(4):207–9.
[140] Shakhar K, Shakhar G, Rosenne E, Ben-Eliyahu S. Timing within the menstrual cycle, sex, and the use of oral contraceptives determine adrenergic suppression of NK cell activity. Br J Cancer 2000;83(12):1630–6.
[141] White HD, Crassi KM, Givan AL, Stern JE, Gonzalez JL, Memoli VA, et al. CD3+ CD8+ CTL activity within the human female reproductive tract: influence of stage of the menstrual cycle and menopause. J Immunol 1997;158(6):3017–27.
[142] Feldman I, Feldman GM, Mobarak C, Dunkelberg JC, Leslie KK. Identification of proteins within the nuclear factor-kappa B transcriptional complex including estrogen receptor-alpha. Am J Obstetr Gynecol 2007;196(4) 394 e1–11; discussion e11–3.
[143] Liu HB, Loo KK, Palaszynski K, Ashouri J, Lubahn DB, Voskuhl RR. Estrogen receptor alpha mediates estrogen's immune protection in autoimmune disease. J Immunol 2003;171(12):6936–40.
[144] Paimela T, Ryhanen T, Mannermaa E, Ojala J, Kalesnykas G, Salminen A, et al. The effect of 17beta-estradiol on IL-6 secretion and NF-kappaB DNA-binding activity in human retinal pigment epithelial cells. Immunol Lett 2007;110(2):139–44.
[145] Farage MA, Millwe DW, Maibach HL. The effects of menopause on autoimmune diseases. In: Farage MA, Miller KW, Fugate-Woods N, Maibach HI, editors. Skin, mucosa and menopause: management of clinical issues. Heidelberg: Springer; 2015. p. 299–318.
[146] Aguado P, del Campo MT, Garces MV, Gonzalez-Casaus ML, Bernad M, Gijon-Banos J, et al. Low vitamin D levels in outpatient postmenopausal women from a rheumatology clinic in Madrid, Spain: their relationship with bone mineral density. Osteoporosis Int 2000;11(9):739–44.
[147] Zold E, Barta Z, Bodolay E. Vitamin D deficiency and connective tissue disease. In: Latwick G, editor. Vitamins and hormones: vitamins and the immune system. Amsterdam: Elsevier; 2011. p. 261–86.
[148] Tang JY, Fu T, Lau C, Oh DH, Bikle DD, Asgari MM. Vitamin D in cutaneous carcinogenesis: part I. J Am Acad Dermatol 2012;67(5):803 e1–12, quiz 15–6.
[149] Lagunova Z, Porojnicu AC, Lindberg F, Hexeberg S, Moan J. The dependency of vitamin D status on body mass index, gender, age and season. Anticancer Res 2009;29(9):3713–20.

[150] Need AG, Morris HA, Horowitz M, Nordin C. Effects of skin thickness, age, body fat, and sunlight on serum 25-hydroxyvitamin D. Am J Clin Nutr 1993;58(6):882–5.
[151] Nordin BE, Polley KJ. Metabolic consequences of the menopause. A cross-sectional, longitudinal, and intervention study on 557 normal postmenopausal women. Calcif Tissue Int 1987;41(Suppl. 1):S1–59.
[152] Webb AR, Pilbeam C, Hanafin N, Holick MF. An evaluation of the relative contributions of exposure to sunlight and of diet to the circulating concentrations of 25-hydroxy vitamin D in an elderly nursing home population in Boston. Am J Clin Nutr 1990;51(6):1075–81.
[153] Vieth R, Ladak Y, Walfish PG. Age-related changes in the 25-hydroxy vitamin D versus parathyroid hormone relationship suggest a different reason why older adults require more vitamin D. J Clin Endocrinol Metab 2003;88(1):185–91.
[154] Gonzalez G, Alvarado JN, Rojas A, Navarrete C, Velasquez CG, Arteaga E. High prevalence of vitamin D deficiency in Chilean healthy postmenopausal women with normal sun exposure: additional evidence for a worldwide concern. Menopause 2007;14(3 Pt 1):455–61.
[155] Hocher B, Reichetzeder C. Vitamin D and cardiovascular risk in postmenopausal women: how to translate preclinical evidence into benefit for patients. Kidney Int 2013;84(1):9–11.

CHAPTER

5

Sex and Gender Differences in Cardiovascular Disease

Leanne Groban[1], Sarah H. Lindsey[2], Hao Wang[1] and Allan K. Alencar[1]

[1]Department of Anesthesiology, Wake Forest School of Medicine, Medical Center Boulevard, Winston-Salem, NC, United States [2]Department of Pharmacology, School of Medicine, Tulane University, New Orleans, LA, United States

INTRODUCTION

Cardiovascular disease (CVD) is the leading cause of death and disability in Western societies. The epidemiology, presentation, pathophysiology, and outcomes of CVD and its various manifestations are significantly different between men and women. An understanding of the reasons underlying these differences is important to make efficient diagnoses and achieve effective treatment. Inherent differences in cardiac and vascular structure and function found among healthy humans and animal model systems may be potential risk factors for sex-associated susceptibility to CVD and cardiopulmonary disease (Box 5.1) [1,2]. While the benefits of hormone therapy in postmenopausal women remain controversial, it is clear that endogenous estrogens exert protective effects in the cardiovascular system. Premenopausal women have a lower cardiovascular risk compared to men, and early menopause or bilateral oophorectomy increases lifetime cardiovascular risk. This chapter examines the sex and gender differences in CVD—specifically in coronary heart disease (CHD), heart failure (HF), and systemic and pulmonary hypertension—in both humans and animal models. The roles that hormonal, genetic, and other factors might have in generating sex and gender differences in these forms of CVD and in cardiopulmonary disease will also be discussed.

CORONARY HEART DISEASE

Epidemiology

CHD is the most common type of CVD, accounting for approximately one-third to one-half of the total cases of CVD. It is the leading cause of death in both men and women, outweighing all cancer-related deaths. Among individuals 35–84 years of age, the total incidence of CHD-related morbidity and mortality is approximately twofold higher in men. The sex gap in morbidity tends to diminish in older age ranges, owing to the increased development of morbidity in women following the menopause, and the tapering of morbidity among men around the same age [3].

Epidemiologic studies reveal age and sex differences in incident rates of CHD and its associated manifestations such as angina pectoris, myocardial infarction (MI), and sudden cardiac death. For persons 40 years of age, the lifetime risk of developing CHD is 49% in men and 32% in women. At age 70, the lifetime risk is 35% in men and 24% in women [4]. The incidence for total coronary events rises steeply with age, with women lagging behind men by 10 years. Beyond the menopause, the incidence and severity of CHD and its serious manifestations such as MI and sudden cardiac death increase abruptly, with rates three times those of women the same age who remain premenopausal [3]. While fewer women than men younger

Sex Differences in Physiology
DOI: http://dx.doi.org/10.1016/B978-0-12-802388-4.00005-7

BOX 5.1

SEX DIFFERENCES IN CV PHYSIOLOGY

Compared to men, women have:

- Lower LV mass
- Greater contractility
- Preserved mass with aging
- Lower rate of apoptosis
- Small coronary vessels
- Lower blood pressure
- Faster resting heart rate
- Less catecholamine mediated vasoconstriction

than 45 years develop CHD [5], women have a worse prognosis than men [6]. A greater proportion of women than men experiencing an MI die before reaching the hospital [7].

Risk Factors

While the modifiable risk factors for CHD, including smoking, dyslipidemia, diabetes, obesity, and hypertension, are virtually the same in men and women, the impact of these risk factors between sexes appears to be different, particularly for smoking and diabetes. Smoking is more deleterious in women than in men (eg, the relative risk of a cardiovascular event is 3.6 vs 2.4 in women vs men, respectively), with an even larger negative impact in women as the total number of cigarettes smoked per day increases [8]. Although the precise mechanisms for the sex differences in CVD risk with smoking are not clear, an antiestrogenic effect in female smokers on the maintenance of normal endothelial and/or platelet function may be profound [9].

Women with diabetes have more than a 40% higher risk of incident CHD than men with diabetes [10,11]. Diabetic women are also at greater risk for cardiovascular complications than their male counterparts. The risk of a fatal CHD event is 50% higher in diabetic women compared with diabetic men [11]. Diabetes is an independent risk factor for poorer outcomes following percutaneous interventions (PCI) and coronary artery bypass graft surgery (CABG) in women but not men [12]. However, given the advancements in PCI techniques in recent years, this may be changing. One study showed that female diabetic patients had greater improvements in outcomes after PCI when compared with nondiabetic patients [13]. The sex differences in CHD mortality among individuals with diabetes may be due to different pathophysiologic processes involving endothelial function, dyslipidemia, and thrombosis [14]. Diabetes abolishes the vascular protection afforded by estrogen in premenopausal women. Steinberg et al. [15] showed that premenopausal women have an enhanced vasodilatory response to endogenously produced nitric oxide (NO) compared with men, and the loss of this effect may be more profound in women with diabetes than their male counterparts. Using an experimental swine model of type 1 diabetes, White and colleagues [16] proposed that diabetes transforms estrogen from a "helpful" into a "harmful" hormone by promoting an estrogen-stimulated oxidative stress milieu within the vasculature of women. Thus, diabetes may be accelerating the normal aging process, particularly in women, promoting cardiovascular dysfunction by estrogen-stimulated oxidative stress.

In addition to smoking and diabetes, triglycerides, high-density lipoprotein (HDL)-cholesterol, metabolic syndrome, and systolic hypertension have a greater influence on CHD risk in women, particularly after the menopause, while total cholesterol, low-density (LDL)-cholesterol, and essential hypertension are more prominent risk factors in men.

Women also have a unique set of risk factors that may be additive in defining CHD risk [17]. Women who have a history of irregular menstrual cycles, estrogen deficiency, and polycystic ovary syndrome have a higher risk of developing heart disease as they age [18,19]. Indeed, early menopause is positively associated with CHD, independent of traditional risk factors. In contrast to these protective estrogenic effects, the use of hormonal contraceptives by women who smoke compounds their risk [20]. Certain pregnancy complications, such as pre-eclampsia, gestational diabetes, and pregnancy-induced hypertension, heighten the potential for CHD, particularly when combined with traditional risk factors and are included in the guidelines for CVD prevention among women [21]. Interestingly, women treated with fertility therapy have fewer cardiovascular events than untreated controls, despite having a higher likelihood of pregnancy-related hypertension and diabetes [22]. Other female sex-specific risk factors for CVD onset and prognosis include systemic autoimmune collagen vascular diseases, such as lupus erythematosus and rheumatoid arthritis, and depression [21].

Clinical Presentation and Diagnosis

The clinical presentation of CHD differs between men and women, which can complicate or delay the diagnosis. The classic symptoms and signs—oppressive or constrictive chest pain and/or pressure, dyspnea, diaphoresis—is the dominant male pattern; most men with an acute coronary event present in this way. While 50% of women present with these classic symptoms, the other 50% may present with "atypical" angina symptoms such as indigestion, nausea, shortness of breath, dizziness, abdominal pain, a vague sense of fullness, pain in the jaw, arm, or shoulder, and overwhelming fatigue. One consideration when interpreting symptoms of CHD in women is the greater likelihood that the symptoms are induced by emotional stress; in men, heavy lifting or physical exertion is the more common symptom-initiating event [23]. The initial presentation of CHD in women is most often angina pectoris (69% women compared to 30% men) [24], whereas infarction predominates in men, and is often preceded by angina. Yet, the incidence of recognized MI is higher in men than in women [25], and almost two-thirds (64%) of women who die suddenly of CHD have no previous symptoms [26,27]. Thus, acknowledging sex differences in the presentation of myocardial ischemia is essential for ensuring appropriate diagnostic and treatment strategies in both sexes.

CHD biomarkers are expressed in sexually dimorphic patterns. In a study of 1865 patients with unstable angina and non-ST elevation MI (34% women), men were more likely to have elevated creatine kinase MB and troponins, whereas women displayed elevated levels of high-sensitivity C-reactive protein (hs-CRP) and B-type natriuretic peptides (BNP) [28,29]. Differences also exist with regard to electrocardiographic findings, as women are more likely than men to present with non-ST elevation. Furthermore, exercise or treadmill stress testing without imaging appears to be insufficient for diagnosis of coronary artery disease (CAD) in women, whereas functional capacity tests, myocardial perfusion imaging, and cardiac magnetic resonance imaging (CMR) may identify female patients whose chest pain is due to myocardial ischemia without obstructive disease [30]. Coronary microvascular reactivity to adenosine (endothelium-independent microvascular coronary reactivity) can improve prediction of major adverse outcomes over angiographic CAD severity and CAD risk factors alone. Women who undergo coronary angiography show a significantly lower prevalence of obstructive epicardial CAD [31,32] when compared to their male counterparts. The presence of apparently "normal" coronaries on angiography often leads to a missed diagnosis of CAD in women who present with symptoms of ischemia.

Sex differences in the signs and symptoms of acute coronary syndromes (ACS) may be related to differences in coronary anatomy and physiology between men and women. Women have smaller coronaries and fewer coronary collateral vessels than men, which may predispose them to increased ischemia with lower levels of exertion. Moreover, in women with angiographically normal coronaries, microvascular dysfunction is marked by reductions in coronary/myocardial blood flow, often related to endothelial and nonendothelial increases in vasoconstriction [7,33]. Studies evaluating risk factors associated with adenosine-related microvascular coronary reactivity in symptomatic women evaluated for CAD found that traditional atherosclerotic risk factors were not associated with defective coronary microvascular reactivity to adenosine in most cases, suggesting that novel gender-specific risk factors need to be identified [34].

Pathophysiology

Men have a greater atheromatous burden in coronary arteries compared to women and surgical/pathological evidence has revealed discrete platelet-rich thrombi and microembolization within their vessels. Plaque rupture associated with sudden cardiac death occurs more frequently in men, whereas superficial plaque erosion with thrombus formation occurs more often in women. The influence of the menopause on CAD pathophysiology in women is unique and important, as the incidence of plaque rupture is higher in older women compared to younger women. Women exhibit a more generalized wall thickening of the coronary arteries [35], and structurally their coronary vessels contain more diffuse atherosclerosis with involvement of the entire circumference of the artery and without discrete plaques [36]. Unlike men, who develop obstructive plaques in medium-sized arteries, coronary pathology is more often present in the small distal arteries in women. "Remodeling" of the entire artery occurs in atherosclerosis in women, thus making the plaques uniform along the wall of the artery. As mentioned previously, this female pattern of coronary disease is not easily detected by angiography, and diagnosis may require intravascular ultrasound (IVUS) and the evaluation of coronary flow dynamics. Microvascular and endothelial dysfunction contribute to the pattern of ischemia that commonly manifests in women. The lack of large obstructive lesions causing blockage and ischemia in women may contribute to the differences in clinical symptoms between women and men.

Models and Mechanisms

The development of CAD in women lags behind men by 10–15 years. This delay is thought to be caused by the protective effects of estrogens against coronary atherosclerosis and a decreased prevalence of coronary risk factors in young women. Hormone changes and fluctuations impact both the larger vasculature as well as the microvasculature. The three major naturally occurring estrogens are 17-β-estradiol, estrone, and estriol, of which estradiol (E2) is the predominant and most active. Since testosterone can be converted to E2 by the enzyme aromatase, which is expressed in the fat tissue and prostate tissue, its antiatherogenic actions are presumed to be mediated, in part, via E2. The effects of E2 on CHD are linked to its diverse actions on the vasculature via both genomic and nongenomic mechanisms, as extensively reviewed by Knowlton and Lee [37], and are mediated by three different estrogen receptors (ERs): the classical steroid receptors ERα and ERβ, and the G protein-coupled estrogen receptor (GPER or GPR30).

Similar to the systemic vasculature (see Hypertension section), multiple mechanisms are involved in estrogen's coronary vasodilatory action, including increased NO, adenylyl cyclase, adenosine, and prostacyclin and opening of calcium-activated $K+$ channels. Moreover, E2 reduces the synthesis of vasoconstrictors such as angiotensin II (Ang II), endothelin-1, and catecholamines. E2 also inhibits the mitogenic effects of various factors generated at the site of vascular endothelial injury (eg, bifurcation areas of sheer stress), which trigger stimulation of hypertrophic growth of vascular smooth muscle cells (SMCs). E2 acts directly on vascular endothelium and smooth muscle to downregulate molecules involved in monocyte adhesion to endothelial cells and to decrease cytokines involved in the migration of monocytes to the subendothelial spaces [38]. E2 also blocks key elements of vascular remodeling, leading to attenuation of inflammation and modulation of signaling factors associated with reactive oxygen species generation.

E2 also has favorable effects on lipid metabolism, which, in turn, modulates the progression of coronary atherosclerosis. In general, E2 increases HDL and reduces LDL cholesterol levels. E2 is also effective at chemically modifying lipoproteins. Free E2 can act either as a peroxyl radical scavenger, due to its hydrogen donating ability, or can act synergistically with other antioxidants [39]. E2 is highly effective at preventing the oxidation of LDL and very low-density lipoprotein (VLDL)-cholesterol, and reducing the formation of minimally modified LDL and its subsequent vascular accumulation. This is important because oxidized LDL enhances monocyte adhesion to vascular endothelial cells, one of the first steps of fatty streak formation, which allows monocytes to enter the subintimal space and become macrophages. Clinical and reverse translational evidence of these antiatherogenic effects of E2 has been discussed in detail elsewhere [40,41]. The loss of estrogen that occurs during aging impacts levels of circulating endothelial progenitor cells, which in turn compromise vascular endothelial repair and increase CAD susceptibility [42].

In addition to the antiatherogenic effects of E2, reduced ischemia-reperfusion (I/R) injury has been observed in female, but not male, animal models, suggesting that endogenous estrogens provide cardioprotection against ischemic insult [43,44]. Data from isolated adult rat hearts subjected to global ischemia show that infarct sizes are smaller (37% vs 48%) [45], and all indices of postischemic myocardial performance are improved in female hearts when compared to age-matched male hearts of the same species [46]. Some of the proposed mechanisms underlying a decrease in susceptibility to I/R injury in female hearts include: (1) a heightened myocardial cardioprotective signaling pathway that involves p-Akt and p-PKC [45]; (2) a diminished inflammatory response that involves decreased myocardial TNF-α, IL-1, and IL-6 expression and decreased activation of the p38 MAPK pathway [46]; and (3) a heightened preconditioning response, as indicated by increased expression or functional response of cardiac K_{ATP} channels [47,48].

The sympathetic nervous system is activated post-MI, and the sexual dimorphism of cardiac injury in the context of adrenergic stimulation has been examined under experimental conditions. Cross et al. [49] showed that isoproterenol or Ca^{2+} pretreatment increased I/R injury to a greater extent in isolated male hearts than female hearts. Endothelial NO synthase (eNOS) expression and NO production were higher in female than male hearts, and L-NAME blocked female sex-related protection, suggesting that female hearts are protected from the detrimental effects of adrenergic stimulation and Ca^{2+} loading via a NOS-mediated mechanism. Moreover, studies using $eNOS^{-/-}$ and neuronal NOS $(nNOS)^{-/-}$ mice suggest that both enzymes play a role in the sex differences observed in I/R injury under adrenergic stimulation [50]. Furthermore, increased S-nitrosylation of L-type Ca^{2+} channels mediates the cardioprotective effects of estrogen in female cardiomyocytes [50].

Various mouse models of coronary occlusion with or without reperfusion, as relevant to the clinical scenario, have been used to examine the sex-specific progression of LV remodeling after an ischemic insult. Following induction of MI, hearts from male mice display greater LV remodeling evidenced by increased LV internal diameter, reduced muscle tensile strength, loss of soluble collagen, and a higher incidence of LV rupture when compared to hearts from females exposed to the same injury [51–55]. A larger local inflammatory response and greater matrix metalloproteinase (MMP)–9 activation in male mice contribute to the sex differences in cardiac rupture after MI, and male mice treated with an MMP-inhibitor prior to MI display reduced MMP activity and a halving of the rupture incidence [52]. These studies further suggest that estrogen and testosterone may have opposing roles in long-term cardiac remodeling post-MI. Cavasin et al. showed in mice [56,57] that estrogens (either endogenous or supplemental) prevent maladaptive chronic remodeling and further deterioration of cardiac performance, whereas testosterone (either endogenous or supplemental) adversely affects myocardial healing (as indicated by a higher rate of cardiac rupture), promotes cardiac dysfunction and remodeling, and exerts pronounced effects when estrogen levels are reduced.

To examine the effects of sex, aging, and diabetes on the susceptibility to myocardial ischemic injury, Desrois et al. [58] used the aging Goto-Kakizaki (GK) rat, a model of type 2 diabetes. Male and female GK rats had heart/body weight ratios 29% and 53% higher, respectively, than their sex-matched wild-type controls, with the female GK rat hearts significantly more hypertrophied than the male. In isolated, perfused hearts, insulin-stimulated ^{3}H-glucose uptake rates were decreased by 23% and 40% in male and female GK rat hearts, respectively. During low-flow ischemia, glucose uptake was 59% lower in female GK hearts, but the same as controls in male GK rat hearts. The recovery of contractile function during reperfusion was 30% lower in female, but the same as controls in male GK rat hearts. These findings suggest that the aging female type 2 diabetic rat heart has increased insulin resistance and greater susceptibility to ischemic injury compared with either nondiabetic female or male type 2 diabetic rat hearts.

ER-specific knockout models and ER-specific agonists, as reviewed by Knowlton and Lee [37], have been used to understand the mechanisms for estrogen-mediated cardioprotection. Chronic genetic deletion of ERβ in the ERβKO mouse, for example, increases I/R myocardial injury compared to wild-type females [59–61]. Moreover, a specific ERβ agonist was cardioprotective against I/R injury in ovariectomized (OVX) female mice [62,63]. Possibilities for ERβ-mediated cardioprotection include upregulation of protective genes involved in NO biosynthesis, fatty acid metabolism, PI3K/Akt activation, and antiapoptotic proteins [44]. There is also evidence for a protective effect of ERα during I/R injury. Wang et al. [64] showed more severe injury in ERαKO female hearts compared to wild-type hearts following global ischemia, and Jeanes [65] demonstrated that chronic pretreatment of OVX rats with ERA-45 (an ERα-specific agonist) mimicked the protective effects of estrogen. Furthermore, E2 replacement in ERαKO OVX females was not protective [66]. The new plasma membrane estrogen receptor, GPER, also has cardioprotective potential in reducing I/R injury. Acute activation by the specific agonist G-1 in Langendorff perfused hearts reduces infarct size and improves functional recovery compared to control hearts [67,68]. GPER-mediated cardioprotection appears to be mediated via an activated Akt and ERK1/2 pathway.

Interestingly, GPER activation also induces coronary artery relaxation and attenuates the proliferation and migration of coronary SMCs. Thus, selective GPER activation has the potential to also increase coronary blood flow and limit the debilitating consequences of coronary atherosclerotic disease [69]. Taken together, male and female differences in response to an ischemic insult may be due to activation of ERs on cardiac cells. As reviewed extensively by Murphy et al. [43], stimulation of the three ER isoforms can mediate acute and chronic protection from I/R injury by altering cardiac gene expression of the NO system and PI3K/Akt pathway. The interaction between signaling by the ERs and differences in CHD between pre- and postmenopausal women remains an area of continued study.

Treatment and Outcomes

The management of CHD is generally similar for both men and women; however, the efficacy of some invasive strategies for ACS may be different between sexes. A meta-analysis of randomized controlled trials of ACS shows that an invasive strategy was more beneficial in women with positive biomarkers compared to women with negative biomarkers; such a difference was not seen in men [70]. Also, women without raised biomarkers may not derive an equal benefit from Glycoprotein IIb/IIIa inhibitors as men or women with positive biomarkers [71]. Women tend to bleed more often with antiplatelet therapy than men, owing to differences in body size and renal function [72]. In both men and women with ACS, early PCI (within the first 48 h) is equally beneficial among sexes. However, because women often present later than men, they may not receive maximal benefit. Women also have a slightly higher periprocedural risk than men, owing to more advanced age at the time of the procedure and a greater burden of risk factors than men. Even so, long-term survival after PCI is better in women than in men. Recent trials show that placement of drug-eluting stents in men and women with ACS has similar outcomes overall [73,74]. Women undergoing CABG with conventional cardiopulmonary bypass have a greater likelihood of early adverse outcomes (eg, neurologic event, HF, perioperative MI, and hemorrhage) than men. This is presumed to be due to patient-related factors in women, including advanced age at presentation, more risk factors, and small body habitus with small coronaries that impose technical challenges and increase the potential for graft failure. Interestingly, off-pump CABG is associated with fewer gender disparities in outcome [75]. Historically, mortality rates post-PCI and CABG have been higher among women than men; however, after correcting for age and comorbidities, recent registry analyses show that women have mortality rates after PCI and CABG that are similar to those in men [76].

The criteria for the use of secondary prevention therapies after MI, including statins, aspirin, nitrates, ACE inhibitors (ACEi), beta-blockers, and aldosterone antagonists are the same for both sexes. For women with nonobstructive CAD or microvascular dysfunction, beta-blockers improve symptoms. Statins and ACEi may be used to improve endothelial function and may also improve symptoms. Exercise training limits symptoms and increases exercise capacity [72]. Future randomized control studies in women with nonobstructive CAD will provide important clinical information regarding the optimal treatment of this female sex-specific form of ischemic heart disease.

HEART FAILURE

Chronic HF continues to increase in incidence and prevalence, both in the United States and worldwide. HF presents as a complex constellation of signs and symptoms, which include dyspnea and exercise intolerance. While the precipitating events and etiologies of HF are diverse, the primary underlying defect is the inability of the LV to maintain an adequate forward stroke volume to meet the metabolic demands of the body, either because of impaired LV contractile function or insufficient LV filling. HF affects either side of the heart and can be systolic or diastolic in nature. Both systolic HF, now termed HF with reduced ejection fraction (HFrEF; EF $\leq$40%) and diastolic HF, termed HF with preserved ejection fraction (HFpEF; EF $\geq$50% and HFpEF, borderline: EF 41–49%), result in a backup of blood in the pulmonary system, leading to pulmonary edema. Failure of the right side of the heart is usually a consequence of left side failure. When the LV begins to fail, pressure backs up through the lungs and into the right side of the heart. As the right side loses its ability to pump against the increasing pressure gradient caused by the failing LV, fluid backs up in the lower part of the body, leading to peripheral edema. HF is considered to be the end of the CVD spectrum, and is often preceded by MI, hypertension, valvular heart disease, and congenital heart disease. In this section, the sex and gender differences in HF epidemiology, etiology, pathophysiology, response to treatment, and impact on quality of life are discussed. While the mechanisms that underlie sex- and gender-related differences in HF remain unresolved, we present the findings of several preclinical studies that provide some clues as to their root causes.

Epidemiology

HF affects nearly 5.7 million people in the United States, with 870,000 new cases reported annually. At the age of 40 years, the lifetime risk of developing HF is one in five [77]. Approximately 20 per 1000 individuals have a diagnosis of HF [77], with the prevalence increasing with age: 1% of individuals 45–54 years of age and more than 8% of individuals older than 74 years have HF [78]. The overall prevalence of HF is equal between genders, and though the prevalence of HF appears to be higher in men before the age of 70 years, the increase in HF prevalence per decade of life is accelerated for women (27% per decade in men and 61% per decade in women) [79]. The higher prevalence of HF in older women is presumed to be related to: (1) a greater life expectancy, (2) a higher prevalence of hypertension with aging when compared to men, and (3) a greater likelihood of developing LV hypertrophy in response to cardiovascular stresses as hypertension, diabetes, and obesity, which may be linked to the loss of estrogenic protection following the menopause. The reported prevalence and incidence of HF in men compared to women varies depending on whether or not HFpEF cases were included, and on the age of patients studied. Patients with HFpEF make up 50% of the HF population, are 1–5 years older than those with HFrEF, and are predominantly female (51–84%) [80]. The incidence of HF has increased by nearly 10% in women over the past 30 years while it has remained relatively stable in men [81]. Among postmenopausal women with established CAD, those with diabetes and an elevated body mass index (BMI) or creatinine have the highest risk for developing HF, with annual incidence rates of 7–13% [82]. Unfortunately, HF trials have been particularly biased against inclusion of women, minorities, and elderly, which does not accurately reflect the general HF patient population [83].

Mortality following a diagnosis of HF remains significant; the 5-year mortality rate is approximately 50% for both men and women. HF prognostic markers differ between genders; for example, atrial fibrillation is a predictor of HF prognosis in women, while QRS duration and BMI are prognostic indicators in men [84]. Underlying CAD increases the risk of HF mortality 2.5 times in women, but only 1.5 times in men. While hospitalization rates and readmission rates for HF are similar between sexes [85], women with nonischemic HF have better survival rates than men, independent of age and disease severity [79,86]. Even so, women with HF tend to have a worse quality of life than men, more physical limitations, longer hospital lengths of stay, and higher rates of depression [85].

Etiology and Precipitating Factors

Similar to CHD, sexually dimorphic patterns of HF have emerged. Most men with HF have reduced systolic function or HFrEF, with ischemic heart disease as the underlying etiology. Older women with HF usually have preserved systolic function or HFpEF, with diastolic dysfunction as the presumed cause. Diastolic dysfunction is a mechanical abnormality whereby the LV does not fill adequately under normal filling pressure conditions. Reduced diastolic filling can be caused by impairment in myocardial relaxation and/or reduction in LV compliance. In aging women, the key factors that impinge on relaxation and LV compliance include systolic hypertension, LV hypertrophy, obesity, and diabetes. Older men can also present with HFpEF; however, the reason for the impaired relaxation is usually ischemia (eg, reductions in ATP energy stores adversely impact Ca^{2+} uptake into the sarcoplasmic reticulum, impairing ventricular lusitropy), which resolves following coronary revascularization. Recent reviews report that HFpEF may not be solely due to diastolic dysfunction. HFpEF is typified by a broad range of cardiac and noncardiac abnormalities, comorbidities, and reduced reserve capacity in multiple organ systems [80,87].

Another form of HF that shows gender differences is stress-induced cardiomyopathy, also called apical ballooning syndrome, broken heart syndrome, and takotsubo cardiomyopathy (TTC). In most studies reported thus far, 90% of those affected with TTC are women, with a mean age of 62 to 76 years. This form of HF is generally characterized by transient systolic dysfunction of the apical and/or mid-segments of the LV in the absence of overt CAD. Commonly, the contractile function of the lower half of the LV is depressed, and there is hyperkinesis of the upper or basal segments, producing a balloon-like appearance of the distal ventricle with systole that is thought to resemble the shape of a "tako-tsubo" octopus trap. Between 1% and 3% of patients presenting with a suspected ACS eventually are diagnosed as having TTC. Its prevalence is slanted toward the female gender (9% in women vs <0.5% in men). Stress cardiomyopathy is usually triggered by an acute medical illness or by intense emotional stress (eg, death of a loved one, an assault, natural disasters) or physical stress (eg, surgical procedures such as colonoscopy or pacemaker implantation). Physical stress as a triggering event is more frequent in male patients with TTC, whereas emotional stress or no identifiable trigger is more prevalent in women [88]. Men with TTC are more likely to have an out-of-hospital arrest, elevated cardiac biomarkers, and evidence of prolonged QT interval on their electrocardiogram. In contrast, women are more likely to present with chest pain. Postulated pathogenic mechanisms include catecholamine excess contributing to multivessel coronary artery spasm and/or and microvascular dysfunction [89]. The initial management of TTC is mainly supportive, including hydration and an attempt to alleviate the triggering physical or emotional stress. While there are no controlled data to define optimal medical regimens, short-term treatment includes standard drugs for the treatment of HF due to systolic dysfunction [see "Management" section below].

Pathophysiology

For both genders, HF is characterized by symptomatic or asymptomatic LV dysfunction. The neurohormonal responses to impaired cardiac performance (salt and water retention, vasoconstriction, sympathetic stimulation) are initially adaptive but, if sustained, become maladaptive, resulting in pulmonary congestion and excessive afterload. This, in turn, leads to a vicious cycle of increases in cardiac energy expenditure and worsening of pump function and tissue perfusion. The cardiorenal and cardiocirculatory branches of the neurohormonal response to HF provide rationale for the use of diuretics, vasodilators, and inotropes in HF. However, results from large, randomized clinical trials in the early 1990s, showed that ACEi [90,91] and angiotensin receptor blockers (ARBs) [92], but not other vasodilators [93], prolonged survival in patients with HF. Beta-blockers, despite their negative inotropic effects, also improve morbidity and mortality in randomized controlled trials in HF [94].

The finding that low-dose aldosterone antagonists, when added to conventional therapy, reduce mortality in patients with severe HF suggests that there is more to the neurohormonal hypothesis of HF than cardiorenal and hemodynamic effects alone. Taken with evidence showing that Ang II is a growth factor and a vasoconstrictor [95], the clinical focus shifted from cardiorenal and cardiocirculatory processes toward cardiac remodeling as the central component in the progression of HF [96]. The renin–angiotensin–aldosterone system [RAAS], excess sympathetic activity, endothelin, and various cytokines induce proliferative signaling that contributes to maladaptive cardiac growth. Ventricular remodeling, or the structural alteration of the heart leading to dilatation and hypertrophy, occurs in addition to counterregulatory hemodynamic responses and exacerbates ventricular dysfunction. Accordingly, current therapeutic interventions for HFrEF target both the maladaptive neurohormonal response and ventricular remodeling.

While activation of the RAAS and sympathetic nervous system plays some role in the pathogenesis of HFpEF, the neutral results from large, randomized control trials involving neurohumoral blockade suggest that the HFpEF syndrome is not "cardio-centric." HFpEF may involve multiple cardiovascular and noncardiovascular mechanisms, including LV systolic, diastolic, and chronotropic reserve; stiffness of ventricles and vasculature; low NO bioavailability; altered myocardial energetics; autonomic imbalance; arterial vasodilatory dysfunction; and abnormal skeletal muscle composition and function [97]. HFpEF has been challenging to understand due to its interaction with various associated comorbidities including diabetes, renal dysfunction, atherosclerosis, chronic obstructive lung disease, anemia, sarcopenia, and frailty [80], all of which can impact cardiac structure and function.

Despite the paucity of data regarding the pathophysiology of HFpEF, and even less data regarding women specifically, clinical and experimental evidence indicates that the female mammalian heart has distinct adaptive mechanisms when subjected to stress. For example, the male LV has a low tolerance for pressure load and becomes dilated with thin walls and a depressed EF in the presence of chronic hypertension or aortic stenosis. By contrast, the female LV tolerates pressure load by developing concentric hypertrophy, allowing it to maintain normal LV size and EF. Unfortunately, the long-term cost of this adaptation is diastolic dysfunction. In a comprehensive review by Piro et al. [98], the factors influencing the differential myocardial response to stress in women and men are discussed, as summarized in Table 5.1.

TABLE 5.1 Factors Influencing Cardiovascular Prognosis in Women and Men

Women	Men
AGING CARDIOMYOPATHY	
Preservation of cardiac weight	Reduction in cardiac weight (1 g/year)
Preservation of myocyte number	Reduction in myocyte number (64 million/year)
Preservation of myocyte volume	Increase in myocyte cell volume
Constant mononucleate/binucleate myocyte ratio	Decreased mononucleate/binucleate myocyte ratio
Low apoptotic index	Apoptotic index threefold higher than women
Increased apoptotic rate	*Decreased apoptotic rate*
MYOCARDIAL RESPONSE TO PRESSURE OVERLOAD	
Earlier improvement in EF after aortic valve replacement	Later improvement in EF after aortic valve replacement
Greater degree of LVH	Lower degree of LVH
Increased LV mass	
Increased relative wall thickness	
Smaller end-diastolic and -systolic dimensions	
Preserved LV function	Impaired LV function
Later onset of impaired systolic pump performance	*Earlier onset of impaired systolic pump performance*
Greater EF	
Greater cardiac index	
Smaller end-diastolic and -systolic volumes	
Higher expression of β-myosin heavy chain	*Lower expression of β-myosin heavy chain*
Higher expression of ANF mRNA	*Higher expression of ANF mRNA*
MYOCARDIAL RESPONSE TO VOLUME OVERLOAD	
Smaller end-diastolic and -systolic volumes	Larger end-diastolic and -systolic volumes
Greater LV mass/volume ratio	Lower LV mass/volume ratio
Concentric hypertrophy	*No concentric hypertrophy*
No impairment of cardiac function	*Impairment of cardiac function*
Minimal ventricular dilation	*Significant ventricular dilation*
No changes in myocardial compliance	*Decreased ventricular compliance*
MYOCARDIAL RESPONSE TO ACUTE MYOCARDIAL ISCHEMIA	
Lower apoptotic rate in peri-infarct region	10-fold higher apoptotic rate in peri-infarct region

(Continued)

TABLE 5.1 (Continued)

Women	Men
Lower *bax* expression in peri-infarct region	Greater *bax* expression in peri-infarct region
Longer duration of the cardiomyopathy	Shorter duration of the cardiomyopathy
Later onset of cardiac decompensation	Earlier onset of cardiac decompensation
Longer interval between heart failure and transplantation	Shorter interval between heart failure and transplantation
Earlier myocardial healing	*Delayed myocardial healing*
Lower infarct expansion index	*Higher infarct expansion index*
Three times lower mortality	*Greater incidence in cardiac rupture*
Better cardiac function	*Worse cardiac function*
Better remodeling	*Maladaptive remodeling*
	Significantly greater dilatation
	Myocyte hypertrophy
	Premature extracellular matrix degradation
	Higher number of neutrophils
	Increased activity of metalloproteinases
CARDIOGENIC SHOCK	
Significantly lower cardiac index	Higher cardiac index
More frequent adverse clinical events	Less frequent adverse clinical events
More frequent mechanical complications	Less frequent mechanical complications
More common low cardiac output syndrome	Less common low cardiac output syndrome
HEART FAILURE	
Preserved LV EF	Impaired LV EF
Smaller LV end-diastolic volume	Greater LV end-diastolic volume
Smaller stroke volumes	Greater stroke volumes
Higher LV end-diastolic pressure	Lower LV end-diastolic pressure
More frequent congestive symptoms	Less frequent congestive symptoms
Greater impairment in diastolic filling	Lower impairment in diastolic filling

Data from animal studies are reported in *italics*.
ANF, atrial natriuretic factor; EF, ejection fraction; LV, left ventricle/ventricular; LVH, left ventricular hypertrophy; mRNA, messenger ribonucleic acid.
Reprinted with permission from Piro, M., Della Bona, R., Abbate, A., Biasucci, L.M., Crea, F. (2010). Sex-related differences in myocardial remodeling. J Am Coll Cardiol 55, 1057–1065.

Models and Mechanisms

Clinical studies document sex differences in the pattern of adaptive LV hypertrophy and transition to HF. The mechanisms involved in this process have been studied in various rat and mouse models.

The pressure overload (PO) model is a well-established animal model in which aortic banding is used to impede aortic outflow. It first causes cardiac hypertrophy and then gradually HF. In the rat PO model, sex significantly influences the evolution of the early response to PO and the transition to HF. Although the magnitude of LV hypertrophy was similar between male and female rats [99,100] early after aortic banding (within 6 weeks of experimentally induced PO), male LVs showed a depressed contractile reserve compared with female LVs [99]. At 20 weeks after aortic banding, male but not female rats showed an early transition to HF, characterized by cavity dilatation and loss of concentric remodeling and contractile ability [99,100].

Estrogen appears to be responsible for the sex differences to PO. Bhuiyan et al. [101] reported that OVX PO rats were unable to compensate for hypertrophy, with deterioration in heart function followed by higher mortality rates compared to estrogen-intact PO rats. Other investigators also demonstrated that in OVX versus intact females, aortic constriction led to significant cardiac remodeling and HF, which was largely reversed by chronic treatment with E2 [102,103]. Estrogen might protect the heart from PO by: (1) restoring the functional activity of Akt and eNOS [101,103], (2) downregulating ERK1/2 phosphorylation and upregulating caveolin-3 expression [102], (3) modulating cardiac calcium dynamics via preservation of sarcoplasmic reticulum Ca^{2+}-ATPase [99], and/or (4) by affecting cardiac mast cell number, composition, and/or chymase release [104]. Indeed, chymase might promote cardiac remodeling directly by activating MMP-9 [105] or indirectly through its enzymatic conversion of Ang-1, or Ang-1–12, to Ang-II [106–108]. Chymase is mainly released from cardiac mast cells. Increased numbers of mast cells have been reported in explanted human hearts with dilated cardiomyopathy and in animal models of experimentally induced hypertension, MI, and volume overload-induced cardiac hypertrophy [105].

Sex differences in the response to PO have also been observed in mouse models of transverse aortic constriction (TAC). Compared to their male counterparts, wild-type female mice subjected to TAC showed less hypertrophy and HF signs [109,110]. Notably, wild-type females developed concentric hypertrophy, while males developed eccentric hypertrophy [110]. Moreover, OVX females subjected to

TAC had exacerbated increases in LV mass when compared to estrogen-intact littermates, while chronic E2 treatment reduced TAC-induced hypertrophy without affecting the degree of PO [111,112].

The molecular mechanisms for the sex-specific differences in LV remodeling in response to PO may be different between mice and rats. In contrast to the rat [102], in the mouse, no sex differences were observed in ERK1/2 and JNK1/2 gene expression [111]. Using microarray analysis, Witt et al. [113] found that classical marker genes of hypertrophy (α−actin, ANP, BNP, CTGF) were similarly upregulated in both sexes in the TAC mouse model. However, 35 genes controlling mitochondrial function (eg, PGC−1, cytochrome oxidase, carnitine palmityl transferase, acyl−CoA dehydrogenase, pyruvate dehydrogenase kinase) had lower expression in males compared to females after TAC. Genes encoding ribosomal proteins and genes associated with extracellular matrix remodeling (collagen 3, MMP 2, TIMP2, and TGFβ2) exhibited relatively higher expression (about twofold) in males after TAC.

DOCA-salt (deoxycorticosterone acetate (DOCA) + salt hypertension (uninephrectomy and 1% saline in drinking water) is a commonly used animal model that mimics most of the changes seen in chronic cardiovascular remodeling in humans including hypertension, cardiac hypertrophy and fibrosis, electrical conduction abnormalities, and HF. Sex differences have also been observed in this model. DOCA-salt male mice develop more LV hypertrophy, stretch-related inflammation, and profibrotic responses compared with female mice, and these differences are independent of blood pressure [114]. The finding that E2 increases serum NO concentrations in DOCA-salt rats may, in part, be responsible for the cardioprotection in this model [115]. In a similar model of hypertrophic remodeling, induced by aldosterone treatment, E2 replacement in OVX female rats attenuated the adverse effects of estrogen loss on cardiac mass, cardiomyocyte cross-sectional area, perivascular collagen accumulation, and vascular osteopontin expression [116]. Cardiac transcripts for calcineurin A and for myocyte-enriched calcineurin interacting protein 1 are upregulated and activated in male but not in female mice, which might also account for the sex-specific differences in remodeling.

Introduction of an arteriovenous (AV) shunt is used for studying volume-overload (VO) induced-remodeling and HF in small animal models. Findings from different groups [117,118] showed that at 8, 16, and 21 weeks after AV shunt creation, both male and female rats developed cardiac hypertrophy; however, the extent of hypertrophy was greater in females. Also, since HF—characterized by increases in ventricular dimensions and LV end-diastolic pressure (LVEDP), and decreases in fractional shortening—occurred only in males [117], these findings suggest that the adaptive response to VO in females may be estrogenic. Moreover, the mortality in female rats was significantly less than in male rats after AV shunting [118]. This sex-specific cardioprotection was lost following OVX and was restored by E2 replacement. Interestingly, E2 supplementation also attenuated ventricular remodeling and disease progression in male rats subjected to chronic VO [119].

The molecular mechanisms involved in the cardiac protective effects of E2 after VO have been examined in a number of studies, with the following findings: (1) E2 inhibits cardiac oxidative stress and circulating endothelin-1 levels, as well as prevents MMP-2 and MMP-9 activation and breakdown of ventricular collagen in the early stages of remodeling [119]; (2) E2 upregulates phospho-Bcl-2 and downregulates the proapoptotic proteins BAX and caspases 3 and 9, thereby preventing cardiomyocyte apoptosis and HF [120]; and (3) E2 modulates catecholamine production and β-adrenoceptor pathways during VO stress, thereby maintaining cardiac function [121]. Moreover, Chancey and colleagues [122] found that mast cell degranulation resulted in reduced collagen volume fraction and ventricular dilatation in hearts of normal males and OVX female rats compared to hearts of intact and estrogen-supplemented OVX females. Estrogen-related cardioprotection of the volume-stressed myocardium might be the result of an altered mast cell phenotype and/or the prevention of mast cell activation.

ERβ appears to mediate the cardioprotective effects of estrogen in response to PO. Cardiac fibrosis and remodeling were more pronounced in wild-type male versus female mice subjected to TAC, while this sex difference was abolished in ERβKO mice [109,110,114]. Moreover, ERβ deletion augments the TAC-induced increase in cardiomyocyte diameter in both male and female mice [109,110]. A set of ERβ-regulated proteins in response to PO has been identified recently that associates with mitochondrial bioenergetics and energy supply [114]. Interestingly, using ERβ-deficient mice with TAC, as well as in vitro cell culture studies, Queirós et al. [123] reported that ERβ regulated the sex-specific expression of a network of miRNAs including fibrosis-related miR-21, -24, -27a, -27b, -106a, and -106b. The activation of ERβ also suppresses Ang II-induced histone deacetylase proteins (HDAC2) production, HDAC-activating phosphorylation, and the prohypertrophic gene expression in the heart [124].

The cardioprotective effects of ERβ have also been confirmed in DOCA-salt rats. In ovariectomized female DOCA-salt rats, the ERβ agonist 8β-VE2 attenuated the effects of estrogen loss on increases in blood pressure, cardiac mass, myocyte cross-sectional area, perivascular collagen accumulation, and vascular osteopontin expression [115]. Accordingly, in female ERβKO mice, the cardioprotective effect of E2 was abolished [125]. The dilated LV response to DOCA-salt in ERβKO female mice was accompanied by increases in collagen deposition and maladaptive remodeling. ERβ-mediated protection from DOCA-salt exposure might be through its induction of adaptive p38, ERK signaling, and/or reductions in maladaptive calcineurin signaling [125]. Recent findings suggest that ERβ-dependent regulation of mammalian target of rapamycin (mTOR) signaling might also be responsible for the adaptive cardiac remodeling in females [126].

Similar to ERβ, activation of ERα by its agonist attenuates cardiac hypertrophy and improves LV function in TAC and DOCA-salt models. However, a study using ERαKO female mice showed that these mice developed cardiac hypertrophy nearly identical to that seen in wild-type littermates within 2 weeks post-TAC [109]. The potential role of ERα activation in response to cardiac stress remains questionable.

Critical work from our lab shows that E2 deprivation and replacement in a congenic model of hypertension due to increased tissue renin gene expression (the mRen2.Lewis female rat) modulates cardiomyocyte and interstitial remodeling, thereby altering lusitropic function and ventricular compliance [127–131]. We also found that chronic activation of the G protein-coupled estrogen receptor, or GPER, by its selective agonist, G-1, mitigated the adverse cardiac effects of estrogen loss in the OVX mRen2.Lewis rat, a model of estrogen-sensitive left ventricular diastolic dysfunction (LVDD). G-1 also inhibited Ang II-induced hypertrophy in cultured cardiomyocytes, and the GPER antagonist, G15, reversed the effects of both E2 and G-1 [128]. G-1 limited the adverse effects of estrogen loss on cardiac mast cell proliferation, chymase expression, and Ang II production without affecting ACE expression or activity in cardiomyocytes, providing the first evidence that GPER interacts with the local renin-angiotensin system (RAS) through regulation of chymase [132]. Moreover, G-1 also inhibited the proliferation of cardiac fibroblasts derived from adult Sprague-Dawley rats. Taken together, these studies reveal the importance of GPER in the maintenance of female sex-specific cardiac structure and function, which likely involves effects on both cardiomyocytes and cardiac fibroblasts, as well as intracellular chymase/RAS. However, a recent study using GPER KO mice showed that male, but not female, mice exhibited impairments in cardiac function without an overt stress. LVs from male GPER KO mice were enlarged, contractile and lusitropic functions were reduced, and LVEDPs were elevated when compared to LVs from GPER KO females [133]. Further studies are needed to understand the cardioprotective potential and mechanisms underlying GPER activation in the context of sex, estrogen status, and cardiac stress.

Management

The clinical implications for sex differences in HF are significant, affecting risk factor screening and targeting of sex-specific interventions. Current guideline-directed medical management of HFrEF is the same for both men and women, and includes ACEi or ARBs and beta-blockers. There are no overt differences in the indications for adding loop diuretics, aldosterone antagonists, and hydral-nitrates for VO, persistent HF symptoms, and/or if the patient is African American. However, some studies suggest that several of these commonly used medications may not be as effective in women. For instance, even though ACEi has been extensively studied in patients with both forms of HF, strong evidence of the effects of these agents on reducing morbidity and improving survival in patients with HFpEF, which effects women twice as often as men, is still lacking. Large trials suggest that the effects of ACEi may be less pronounced in women than in men receiving treatment for hypertension and HF [90,134]. Also, in a subgroup analysis of the main DIG trial, women treated with digoxin had a significantly increased risk for all-cause mortality and a trend toward more hospital admissions than men [135]; which may have been due to their higher serum digoxin concentrations. Beta-blockers and aldosterone receptor antagonists, on the other hand, appear to be equally efficacious in men and women [84,136,137]. As female representation in HF trials expands, sex-specific differences in pharmacotherapy may emerge. This concept also applies to device therapy and resynchronization, both of which are underutilized in women [85]. Interestingly, female gender was recently found to be among the seven patient characteristics predictive of a positive response to resynchronization, defined by reverse remodeling and better outcomes [138].

In contrast to the large randomized trials that have led to the treatment guidelines for HFrEF, large multicenter trials performed in HFpEF patients have produced neutral results on primary outcomes. Consequently, the treatment of diastolic HF remains empiric, irrespective of gender. The general approach to treating HFpEF has three main components. First, treatment should reduce symptoms, primarily by lowering pulmonary venous pressure during rest and exercise by carefully reducing LV volume (diuretics), and maintaining atrial-ventricular synchrony or tachycardia control (beta-blockers). Second, treatment should target the underlying diseases that cause HFpEF. Specifically, ventricular remodeling should be reversed by controlling hypertension (beta-blockers, ACEi, or ARBs), treating ischemia (revascularization), and controlling glycemia in diabetic patients. Third, treatment should attempt to target the underlying mechanisms that are altered by the disease process. Due to our lack of understanding of the pathogenesis of HFpEF [87,88], this third goal and how to achieve it remains in flux. Finally, the cardioprotective potential of hormone replacement therapy in women with HF is not well established and therefore remains an area that requires further investigation.

HYPERTENSION

Epidemiology

Hypertension is a primary and treatable risk factor for cardiovascular mortality, and is the number one cause of death in the United States [139]. An increase in antihypertensive treatment by 10% could save approximately 14,000 lives per year [140]. Hypertension affects 1 in 3 (80 million) Americans and costs more than $100 billion per year [77,141]. The Eighth Joint National Committee (JNC 8) defines hypertension as systolic pressure greater than 140 mm Hg or a diastolic pressure greater than 90 mm Hg [142]. Despite the fact that race is now considered when selecting antihypertensive treatment, sex is not a factor in this clinical algorithm.

There is no sex difference in the overall prevalence of hypertension in men and women in the United States. Approximately 38.3 million (33.5%) males and 41.7 million (31.7%) females are affected [77]. However, sex differences emerge when prevalence is stratified by age. Premenopausal females have a lower incidence of hypertension compared to men, while females >65 years of age are more likely than men to be hypertensive [77]. This clinical data, in addition to results from animal and cell studies, supports a role for endogenous sex hormones in female protection from hypertension and the associated cardiovascular damage. Despite this transient protection during the reproductive years, however, CVD remains the number one killer of both men and women [143].

Pathophysiology

In 90–95% of cases, the etiology for hypertension is unknown, and the clinical diagnosis is essential hypertension. Contributing factors include age, race, family history, salt intake, diabetes, and lifestyle factors such as stress, obesity, lack of exercise, alcohol intake, and cigarette smoking [144]. Although genetic heritability accounts for approximately 30% of cases, hypertension is a complex disease and numerous genes that play a role have been identified [145,146]. Regardless of the origin, an increase in blood pressure is related to either an increase in systemic vascular resistance (SVR) or an increase in cardiac output. Cardiac output is the product of stroke volume and heart rate, and is primarily a function of blood volume and sympathetic nervous system activation. SVR is the ratio of mean arterial pressure to cardiac output and is influenced by myogenic tone, vasoactive metabolites, endothelial factors, autonomic innervation, and circulating hormones. Since cardiac function is reviewed in detail earlier in this chapter, here we will focus on the mechanisms that influence SVR.

Myogenic tone, sometimes called autoregulation, is an important contributor to SVR, especially in small arteries and arterioles. When a blood vessel expands in response to increases in pressure, SMCs contract to maintain proper blood flow to downstream organs [147]. The mechanism for myogenic tone includes stretch-activated ion channels that initiate depolarization and contraction of SMCs. Whether improper myogenic responses contribute to hypertension is unclear; human coronary arterioles from hypertensive patients show greater myogenic tone [148], while animal models show enhancement during the development of hypertension but not after the disease is established [149,150]. While not all animal models show sex differences in myogenic tone [151], some studies report a reduced myogenic response in females versus males [152,153]. Additional studies indicate that estrogen and NO contribute to reduced myogenic tone in females [152,154].

In 1980, Furchgott and colleagues discovered that the presence of endothelial cells switched the acetylcholine response from vasoconstrictor to vasodilator [155]. Endothelial factors are now considered an important component of vascular tone and include both vasodilators such as NO and prostacyclin and vasoconstrictors such as endothelin, thromboxane, and prostaglandins. The potent cardiovascular effects of nitrates were recognized in the mid-19th century [156], and deficits in NO release and bioavailability are extensively studied in hypertension research [157]. Females consistently perform better in flow-mediated dilation, a measurement of endothelial function [158,159]. This sexual dimorphism is attributed to greater production of NO [160]. Many studies demonstrate sex differences in other endothelium-derived factors, including cyclooxygenase products [161,162], endothelin [163], and endothelium-derived hyperpolarizing factor [164].

Autonomic innervation of blood vessels promotes vasoconstriction via α1 and α2 adrenergic receptors and vasodilation via β2 adrenergic receptors. Increased vasoconstrictor responses to α-adrenergic receptor stimulation are frequently found in males. For example, the α1 agonist phenylephrine causes greater contraction in male versus female rat tail arteries [165]. We also find sex differences in phenylephrine-induced contraction of mesenteric arteries in the mRen2.Lewis rat, a rodent model of Ang II-dependent hypertension, as described in the previous section, that displays marked sex differences [166]. Interestingly, this effect was absent in normotensive control Lewis rats, indicating that an enhanced α-adrenergic response may be one mechanism for the sex differences in this hypertensive model. In addition to reduced α-adrenergic vasoconstrictor responses, females also display greater β2 vasodilatory responses. The renal vasculature of a female rat is more sensitive to β stimulation [167], and women have greater forearm vasodilation to a β receptor agonist in comparison to men [168]. Furthermore, sympathetic nerve activity does not correlate with SVR in young females unless a beta-blocker is present [169,170].

Other circulating vasoactive peptides are also implicated in sex differences in hypertension. One of the most obvious players is sex hormones, which have direct effects on vascular tone. Estrogen has acute vasodilatory properties in nonreproductive tissues [171]. Reports of sex differences in estradiol-induced relaxation are varied, with either no difference [172], enhanced responses in female arteries [166], or greater relaxation in male arteries [173]. Interestingly, androgens also induce dilation in isolated vascular preparations, although the limited research on this subject has not yet identified sex differences in this response [174–176].

The most commonly studied mechanism for sex differences in hypertension is the RAS. Differences in expression of various components of this system, including its receptors (AT1R, AT2R, mas), enzymes (renin, ACE, ACE2, neprilysin), and peptides (Ang I, Ang II, Ang-(1–7)) have been characterized in numerous animals models and tissues [177–180]. When considering specific changes in the vasculature, the most striking differences seem to favor an increase in nonclassical, depressor RAS components in females and/or an upregulation of pressor components in males. For example, vasoconstriction in response to Ang II is greater in male spontaneously hypertensive rat (SHR) model and is associated with an increase in vascular AT1R gene expression [181]. In females, expression of AT2R is upregulated and promotes decreased blood pressure in response to low-dose Ang II [182]. Sometimes sex differences are not found under normal conditions but are stimulated in response to disease. While there is no sex difference in AT1R or AT2R mRNA in femoral arteries at baseline, wire injury increases AT2R expression to a greater extent in females and mediates vascular protection [183]. Hypertensive mRen2.Lewis male rats have greater circulating angiotensinogen, Ang I, and Ang II but less Ang-(1–7) compared to females, along with greater renin and ACE activity [184]. All of these sex differences are absent in the normotensive Lewis control rats. Many results from animal models are recapitulated in clinical studies; renin is lower in women versus men, [185] while circulating Ang-(1–7) is higher in women [186].

Animal Models

One of the most commonly used techniques for inducing hypertension in rodent models is infusion of Ang II, which induces an increase in blood pressure via its effects on vasoconstriction, sodium reuptake, and sympathetic activation. This infusion can be used in conjunction with mouse KO models and dietary alterations such as increased salt to determine how these additional factors influence the hypertensive response and the pressure-induced damage. Interestingly, many animal models display sex differences in the hypertensive response to Ang II infusion. In C57BL/6J mice, infusion of Ang II increases mean arterial pressure in males by 40 mm Hg but increases pressure in females by only 10 mm Hg [187]. A similar although smaller sex difference is found in infused MF1 mice [188]. Male Sprague-Dawley rats infused with Ang II also show a significantly greater pressor response in comparison to their female counterparts [182].

The SHR model is the most widely used genetic model of hypertension and was created by selectively breeding Wistar Kyoto (WKY) rats with higher blood pressures for many generations. Males develop hypertension more quickly than females and reach a higher systolic blood pressure [189]. The protection in females is associated with greater endothelium-dependent vasodilation [190] and an attenuated calcium response to Ang II in isolated vascular SMCs [191].

Sexual dimorphism is present in more than one animal model of salt-sensitive hypertension. The Dahl salt-sensitive (SS) rat was produced through selective breeding of Sprague-Dawley rats that displayed salt sensitivity. Increases or decreases in dietary sodium in both male and female Dahl SS rats elevate blood pressure and induce tissue damage. However, high salt-induced hypertension in males develops faster and is associated with a reduction in longevity [192]. When challenged with low salt, the blood pressure in females never reaches that of the males [193]. DOCA-salt hypertension is another model of salt sensitivity induced by coadministration of the steroid deoxycorticosterone acetate and a 1% salt diet [194]. Both the Dahl SS and DOCA-salt models are considered low-renin models because they suppress renal production of renin [195], but the DOCA-salt model significantly activates the central RAS [196]. DOCA-salt has a greater pressor effect in males versus females, but central infusion of an AT2R antagonist increases pressure in females and eliminates this sex difference [197].

The mRen2.Lewis congenic strain was obtained by a backcross of the transgenic Sprague-Dawley rat possessing a copy of the mouse renin 2 gene with the Lewis strain to obtain an inbred population after nine generations [198]. Insertion of the mouse renin 2 gene results in extrarenal expression of renin and is confirmed by measurement of mouse plasma renin concentration after addition of exogenous angiotensinogen at pH 8.5 [199]. Systolic blood pressure in heterozygous female mRen2.Lewis rats is more than 50 mm Hg lower than that of male littermates [184]. Removal of circulating estrogen via OVX abolishes the gender difference in blood pressure and also increases the expression of many RAS components [198]. Conversely, replacement of circulating estrogens reverses these effects. This model represents an Ang II-dependent model of hypertension that displays sex differences in blood pressure and the RAS.

Mechanisms

While the benefits of hormone therapy in postmenopausal women remain controversial, it is clear that endogenous estrogens exert protective effects in the cardiovascular system. Premenopausal women have a lower cardiovascular risk compared to men [143], and early menopause [200] or bilateral oophorectomy [201] increases lifetime cardiovascular risk. Estrogen acutely decreases peripheral resistance in menopausal women and dilates isolated resistance arteries [202,203]. Studies utilizing ERαKO and ERβKO mice show that these receptors do not solely mediate estradiol-induced vasorelaxation [172,204]. Within minutes, sublingual estradiol decreases peripheral resistance in menopausal women [202,203]. Many sex differences in animal models are attributed to estrogen because removal of the majority of circulating estrogens via surgical OVX increases blood pressure, while treatment with 17-β-estradiol restores normal blood pressure. OVX of female mRen2.Lewis rats at 5 weeks of age increases systolic blood pressure by more than 40 mm Hg, which can be reversed by estradiol replacement at the time of surgery [198]. Similarly, OVX exacerbates hypertension in Dahl SS rats fed on a normal-salt [205], high-salt [206], and low-salt diet [207]. Estradiol replacement blocks the increase in blood pressure [208] and reduces the associated vascular oxidative damage [208,209].

The ability of estrogens to counteract vascular dysfunction and hypertension most likely is mediated by activation of one or more of its three receptors, ERα, ERβ, and GPER. The role of each of these receptors in conferring protection from hypertension differs depending on the animal strain and model being tested. In OVX mRen2.Lewis female rats, selective GPER activation lowers blood pressure and induces vasodilation to the same extent as treatment with E2 [210,211]. ERαKO mice show an exacerbated pressor response to Ang II infusion, similar to that seen in OVX mice [212]. In aldosterone-induced hypertension, ERβ receptors in the brain mediate the protective effects of estrogen [213], and both male and female ERβKO have increased blood pressure and vascular dysfunction [214]. The potential cardioprotective role of a recently discovered membrane-bound variant of ERα is yet to be determined [215].

Androgens also contribute to the development and maintenance of high blood pressure. The most interesting research has emerged from the SHR model, where OVX does not increase pressure in females but castration reduces pressure in males [189]. In OVX female SHR, testosterone administration elevates pressure, and the pressure increases seen in aging animals are accompanied by a naturally occurring increase in circulating testosterone [216]. In women, the postmenopausal decrease in estrogen is followed by an increase in testosterone, which follows the timeline for the onset of postmenopausal hypertension [77,217]. Furthermore, the characteristic hyperandrogenemia in women diagnosed with polycystic ovary syndrome is associated with increased systolic and diastolic blood pressure [218]. These animal and clinical studies suggest that an increase in androgens and/or the ratio of estrogen to androgen may influence hypertensive status.

Determining the relative contribution of sex hormones versus sex chromosomes to hypertension has historically been a difficult task, but two rodent models were created to answer this question. The first model was created by switching the Y chromosomes between a substrain of SHR, the SHR stroke-prone (SHRSP) model, and the control WKY strain [219]. In salt-loaded animals, transfer of the Y chromosome from WKY rats to SHRSP animals decreases blood pressure, while the Y chromosome from an SHRSP animal has a tendency to increase pressure in male WKY rats. These studies implicate the Y chromosome as an important promoter of hypertension in males.

A mouse model consisting of four genotypes—XX and XY females and XX and XY males—was created more recently to separate chromosomal versus gonadal contributions to disease [220]. When mice are gonadectomized and infused with Ang II, blood pressure is greater in mice carrying two X chromosomes compared to mice with XY chromosomes, regardless of whether they are male or female [188]. The increased pressure induced by female XX chromosomes is in contrast to the observed protection from hypertension in intact female control mice. Furthermore, castrated XX and XY males have increased levels of circulating Ang II compared to OVX XX and XY females, indicating that the presence of gonadal hormones before gonadectomy at 42–45 days of age impacts the development of hypertension. The divergent results obtained from these two animal models emphasize that the role of sex chromosomes in sexual dimorphism in hypertension is still unclear.

Management

Treatment for hypertension includes lifestyle modifications, such as a low-sodium diet, increased physical activity, and decreased alcohol intake and cigarette smoking [221]. Drugs prescribed for this disease include diuretics, sympatholytics, vasodilators, and drugs that affect the RAS [221]. For patients with severe or uncontrolled hypertension, treatment usually involves a combination of drugs from two or more of these classes. However, the influence of sex and sex hormones on the efficacy of these therapies has only recently received attention. Interestingly, only 44.8% of women achieve blood pressure control compared to 51.1% of men [222]. Moreover, the biggest gender gap emerges in the older population, where men over 65 years of age far exceed women in blood pressure control [223]. There are sex differences in the type of antihypertensive prescribed, with women more likely to take a diuretic or beta-blocker and men more commonly receiving an ACEi or calcium channel blocker [224]. There are underlying sex differences in pharmacokinetics and adverse effects of antihypertensive drugs [225]. Some of these sex differences in treatment effect are mirrored in animal models; for example, candesartan produces a greater blood pressure lowering effect in male versus female SHR [226]. However, additional work is needed on the mechanisms that contribute to sexual dimorphism in the response to antihypertensives.

Emerging environmental factors that increase blood pressure must also be considered in terms of sex. Exogenous chemicals that alter physiological estrogen signaling are called "environmental estrogens" and include naturally occurring and synthetic compounds such as phytoestrogens (genistein), pesticides (DDT, atrazine), pharmaceuticals (diethylstilbestrol), heavy metals (lead, nickel), and industrial compounds (BPA, phthalates, fire retardants) [227]. Because endogenous estrogens are protective in the cardiovascular system, it is not surprising that these compounds induce adverse cardiovascular effects in females. Lead toxicity is associated with hypertension during pregnancy [228] and phthalate exposure correlates with higher blood pressure in women [229]. Bisphenol A (BPA) is an environmental estrogen found in plastic that is being investigated for its pressure effects in both correlative and causative studies [230–232]. Although women are exposed to exogenous estrogens throughout their lifetime, their effects may become even more relevant in the later stages of life when endogenous estrogens are low.

PULMONARY ARTERIAL HYPERTENSION AND RIGHT VENTRICULAR DYSFUNCTION

Pulmonary arterial hypertension (PAH) is a condition characterized by progressive pulmonary vessel remodeling and elevations in pulmonary vascular resistance that lead to right ventricular (RV) dysfunction and eventually RV failure. PAH has several origins but is always associated with a poor prognosis. The disease is defined by a mean systolic artery pressure of at least 25 mm Hg at rest with a pulmonary capillary wedge pressure of <15 mm Hg and can be further categorized based on a pulmonary vascular resistance >3 Wood units [233]. Therapeutic options for PAH remain limited and are focused on delaying progression of the disease rather than offering a cure. How female sex hormones influence the diseased small circulation and the RV is critical to our understanding of the pathophysiology of PAH and to development of new therapeutic strategies. Here, we provide a brief review of the sex differences in the epidemiology and pathophysiology of PAH, and discuss the effects of estrogen within the cardiopulmonary system and on treatment of PAH.

Epidemiology

Our knowledge of the epidemiology of PAH and indicators of patient prognosis has vastly improved over the past three decades, with information coming from both observational patient registries and clinical trial data. PAH is a rare disease with an estimated prevalence of 7–52 cases per million [234–237]. The annual incidence of idiopathic PAH in Europe and the United States is 1–2 cases per million [238]. Despite the introduction of many PAH-specific therapies, treatment algorithms [239], and better patient screening [240], mortality associated with PAH remains high, with 1-, 3-, and 5-year survival rates of 85%, 68%, and 57%, respectively, as reported in the US REVEAL (Registry to Evaluate Early and Long-term Pulmonary Arterial Hypertension Disease Management) registry [241,242]. Similar survival rates have been reported by a French registry, including 87% and 67% survival rates at 1 and 3 years, respectively [243]. PAH is primarily a disease of women [244] and, depending on the classification, the female-to-male ratio can be as large as 4:1 [245]. Yet, when age is accounted for, the estimated 5-year survival rate from the time of diagnosis is about 10% higher in women than men older than 60 years of age [244].

Etiology

According to the clinical classification of PAH from the Fifth World Symposium (Nice, France 2013), PAH can be idiopathic, hereditable/familial, or associated with other medical conditions such as connective tissue disease, HIV infection, portal hypertension, sickle cell disease, and congenital heart disease. PAH has also been associated with drug and toxin exposures, particularly the ingestion of appetite suppressing drugs, fenfluramine and dexfenfluramine, and denatured grapeseed oil [246]. Thus, the correct diagnosis of patients is important, as the prognosis and management for each subcategory is different. Initial diagnosis of PAH can be difficult as patients often present with nonspecific symptoms such as breathlessness, fatigue, peripheral edema, chest pain during exertion, lightheadedness, syncope, and abdominal distension [247].

Pathophysiology, Mechanisms, and Treatment Strategies

The pathophysiological characteristics of PAH in general include pulmonary arterial endothelial cell (EC) dysfunction, pulmonary artery EC and SMC proliferation, vasoconstriction, and in situ thrombosis [248]. Many factors have been identified that contribute to vasoconstriction and vascular remodeling. These fall into several interrelated and overlapping categories, as follows: vasoactive factors, calcium signaling molecules, inflammatory mediators, growth factors, bone morphogenetic protein receptor 2 (BMPR2) gene mutations, and metabolic dysfunction [249]. Over the past two decades, three major pathways—prostacyclin, endothelin, and NO pathways—are recognized as playing key roles in the development and progression of PAH [250–252].

These three pathways are targeted by PAH-specific therapies that fall into three drug classes: prostacyclin (PGI_2) analogues, endothelin receptor antagonists, and phosphodiesterase-5 inhibitors (iPDE5) [253]. Some of these therapies improve symptoms, survival, functional class, and time to clinical worsening. The use of these agents in clinical practice has been associated with outcome improvements among patients with PAH compared with historical data; however, room for improvement exists [234,241,252,254]. Pulmonary arterial vessel relaxation is one of the main goals of PAH treatment. Thus, new therapeutic approaches that induce NO release and relaxation through different pathways, such as iPDE5 and PGI_2 analogues, are emerging that may improve the quality of life of patients with PAH [253].

Despite PAH being more frequently observed in females, a personalized medicine approach is not yet possible. Recently, it has been reported that women with PAH exhibit a greater clinical benefit from ET-1 receptor antagonists than men [255,256]. As men often present with significantly higher mean pulmonary arterial pressures than women [244], this could provide an explanation as to why women respond better to treatment (as their phenotype is less severe). There may also be differences in circulating ET-1 levels between women and men, with men having higher ET-1 compared to women [255]. This heterogeneity in treatment response may reflect pathophysiological differences between sexes or distinct disease phenotypes.

It is presumed that the female sex hormones, primarily estrogens, are involved in both the development of the disease process as well as the sex-specific response to treatment. The pathophysiologic mechanisms of how sex-based differences might occur in PAH and RV failure are complex, and involve acute and chronic effects of estrogens and androgens, as well as variability in estrogen metabolism, which have been reviewed extensively elsewhere [255,257,258]. Evidence for a role of estrogens in the pathogenesis of PAH includes the presence of a high level of CYP119A1 in the medial layer of pulmonary arteries from PAH patients [255]. CYP119A1, a member of the cytochrome P-450 family, synthesizes estrogens through the aromatization of androgens. Preclinical studies show a unique proliferative phenotype in female pulmonary artery smooth muscle cells (PASMC) (the hypoxic mouse and Sugen 5416/hypoxic rat models) caused by elevations in aromatase and reductions in BMPR2 expression. A mutation in the BMPR2 gene promotes cell proliferation or prevents cell death, resulting in an overgrowth of cells in the smallest arteries throughout the lungs.

Moreover, CYP119A1 inhibition with the aromatase inhibitor anastrozole led to lowering of circulating estrogen levels and reductions in pulmonary vascular resistance in female rats and mice with PAH, suggesting that the therapeutic effects of anastrozole may be related to a decrease in estrogen synthesis by the PASMCs. In contrast, no therapeutic effects of anastrozole were observed in male rats and mice; plasma estrogen was below the level of detection and unaffected by the CYP119A1 inhibitor [245].

Given that endogenous estrogens are cardioprotective, as discussed in earlier sections of this chapter, there is concern that treatment with aromatase inhibitors may facilitate RV dysfunction. Interestingly, anastrozole had no detrimental effects on right or left heart function in female and male rats [245]. Also, clinical data do not support a link between aromatase antagonism and CVD or PAH risk, or a deleterious effect on lipid metabolism in humans [259]. In fact, it has been proposed that the increased capacity of female PASMCs to produce estrogen locally via aromatase contributes to a reduction in the BMPR2 signaling axis, and the subsequent promotion of cell proliferation and pulmonary vascular disease in women. Thus, aromatase inhibitors may have a therapeutic effect on PAH in females [245].

Despite the fact that RV function is a major determinant of survival in PAH patients [243,254,260,261], most research in the field has focused on the impact of sex hormones on the pulmonary vasculature [258]. Although it is generally accepted that vasoconstriction is important in the early stages of the disease, the major factor responsible for the high pulmonary vascular resistance in severe, established PAH is the formation of occlusive neointimal and plexiform lesions in small, peripheral pulmonary arteries [258]. Ultimately, the increase in pulmonary vascular resistance results in maladaptive RV remodeling and failure, characterized by systolic and diastolic dysfunction, alterations in mitochondrial bioenergetics, ischemia, inflammation, oxidative stress, proapoptotic signaling, and cardiomyocyte death [258,262–264]. Thus, in addition to the increased pulmonary arterial pressure during the establishment of PAH, the majority of deaths are caused by failure of the RV [258,265].

Considering that, compared to men, RV systolic function is superior in both healthy women as well as in women with PAH, and given the fact that RV EF correlates with estrogen levels in healthy women receiving postmenopausal hormone replacement [266], it is conceivable that estrogen also has direct, ER-mediated RV cardioprotective effects [258]. There are studies that clearly demonstrate RV protective properties of estrogen; even in studies suggesting that estrogen worsens proliferative processes in the pulmonary vasculature, RV protective properties were also observed [267].

Regarding the role of the ERs on RV function, Frump et al. [268] showed significant correlation between RV ERα expression in sugen/hypoxia-induced PAH in male and female Sprague-Dawley rats and hemodynamic (RV systolic pressure, cardiac output), structural (RV hypertrophy), and biochemical alterations, which strongly suggests a role of this classical ER subtype in the modulation of RV function during the establishment of this pulmonary artery disease process. These investigators also showed a small contribution of ERβ in the attenuation of experimentally induced RV dysfunction (caused by high pulmonary arterial pressure and RV overload). What role the new estrogen receptor GPER may have in limiting the adverse effects of PAH on RV function and structure remains unclear.

The contradictory evidence that estrogen plays a role in the pathogenesis of PAH, while also having a cardioprotective effect, has led to the concept of the "estrogen paradox" [255]. Fig. 5.1 summarizes one mechanistic hypothesis for the estrogen paradox, in which mitogenic pathways induced by high levels of estrogen in the pulmonary circulation lead to RV dysfunction, while the cardioprotective effects of circulating ovarian estrogens (in premenopausal women) work to counteract the increased RV afterload. Upregulation of CYP119A1 produces high levels of estrogen in the pulmonary system via conversion of androsterone/testosterone, and this local estrogen activates mitogenic pathways in healthy pulmonary arteries to cause PASMC proliferation; increase in arteriole wall thickness, pulmonary artery pressures, and RV afterload; remodel pulmonary vessels; and predispose the RV to remodeling and dysfunction. In contrast, circulating ovarian estrogens activate ERs on cardiac cells, including cardiomyocytes and cardiofibroblasts, to counteract these remodeling processes and preserve RV function. Future studies will need to use experimental models that recapitulate the clinical presentation of PAH in order to understand the local tissue compartment- and context-specific effects of estrogen and its receptors. This may help to elucidate the molecular mechanisms of sex differences in PAH and facilitate the development of novel targeted therapies for PAH patients of both sexes.

Hormone Therapy and CVD Risk

In CHD, HF, and hypertension, extensive observational and experimental data suggest that estrogen is cardioprotective. However, due to the findings of the Women's Health Initiative trial, neither estrogen

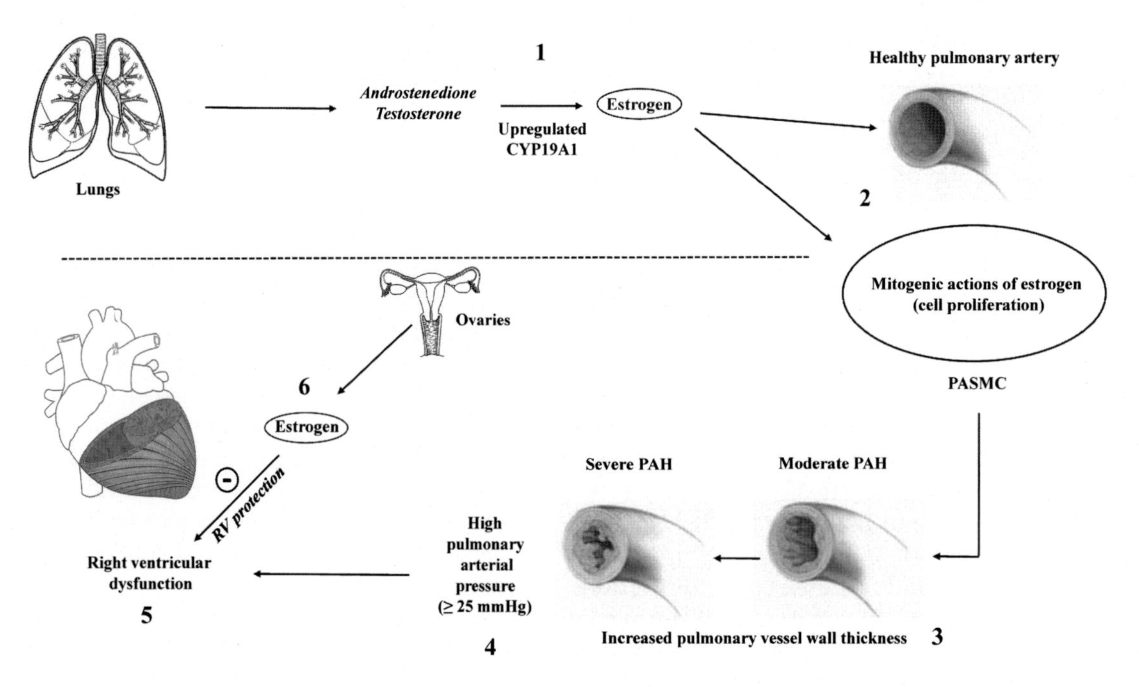

FIGURE 5.1 **Mechanistic basis for the estrogen paradox in PAH.** Overproduction of estrogen induces cell proliferation in pulmonary arteries and long-term RV dysfunction, but estrogen released from ovaries may be cardioprotective. (1) Upregulated CYP119A1 produces high levels of estrogen in the pulmonary system upon conversion of androsterone/testosterone. (2) Estrogen activates mitogenic pathways in healthy pulmonary arteries, causing PASMC proliferation. (3) Remodeling of pulmonary vessels occurs upon estrogen-mediated increases in the wall thickness, which induces PAH. (4) High pulmonary arterial pressure caused by hypertrophied pulmonary arterioles. (5) Long-term RV dysfunction related to persistent increased afterload. (6) Estrogen is synthesized in the ovaries and it can activate its receptors present on cardiomyocytes, thus regulating the cardiac function. CYP19A1: cytochrome P-450 aromatase enzyme; PASMC: pulmonary arterial smooth muscle cell; PAH: pulmonary arterial hypertension.

therapy alone nor in combination with progestin is recommended for primary or secondary prevention of CVD in postmenopausal women [269,270]. In brief, 16,608 postmenopausal women with an intact uterus, aged from 59 to 79 years, were randomized to receive placebo or combination of conjugated equine estrogens (CEE) and progestin. The study was stopped after a mean follow-up of 6.5 years and the landmark report demonstrated that the risks of combined hormone therapy (increased risk of heart disease, stroke, blood clots, and breast cancer) exceeded the benefits (reduced risk of hip fracture, and colorectal cancer) [269]. In a parallel study conducted in 10,739 women with prior hysterectomy, patients were randomized to either CEE alone or placebo. The results of the estrogen-only arm showed an increased risk of stroke but no effect on CHD or breast cancer [270]. Stratifying the rates of CHD events by age makes apparent that younger postmenopausal women on hormone therapy show a dramatic reduction in coronary risk [271]. Similarly, the risk of stroke is reduced in the 50–59-year-old postmenopausal cohort. Evidence continues to emerge that estrogen alone has a more benign risk/benefit profile than combined therapy [272]. Furthermore, concerns about hormone therapy in women, as a whole, may not apply to young women. The possibility that the coronary vasculature and central nervous systems derive benefit if estrogen is initiated during a critical window of opportunity, near the time of menopause versus later in life, is being addressed in the ELITE trial [273]. While hormone therapy should not be used with the intention of lowering the possibility of heart disease in healthy women with early or surgical menopause or with significant vasomotor symptoms, short-term estrogen replacement improves quality of life without increasing CVD risk.

CONCLUSION

It is increasingly recognized that biological differences in the mechanisms regulating cardiovascular function in women and men critically impact the expression, clinical presentation, and outcome of chronic diseases such a CHD, hypertension, and HF. That sex differences exist in CVD has been borne out in numerous epidemiological and clinical studies that have explored the prevalence and presentation of heart disease in women, as well as robust preclinical studies demonstrating estrogen's diverse effects on the vasculature (via both genomic and nongenomic mechanisms), including endothelial cell function and growth, SMC migration and proliferation, production of eNOS and NO, modulation of the local RAS, and inflammatory events. Some of these sex differences may also be linked to sexual dimorphism of myocardial growth processes and myocardial calcium handling, as well as to myocardial remodeling that occurs with advanced age and the loss of ovarian estrogens following the menopause. While our knowledge of the mechanisms that underlie sex differences in CVD is still developing, increasing information regarding the signaling pathways and cellular targets of the sex hormones and their receptors (particularly estrogens) in both the vasculature and in the heart will help in the development of viable targets to improve pharmacological, gene-based, and cell-based therapies. Increasing information about the contribution of specific gene variants to the susceptibility of sex-specific heart disease, their interaction with other genes, the environment, and specific drug treatments will be important in order to realize the promise of personalized medicine. There is no doubt that continued study in gender- and sex-specific areas of cardiovascular research will enhance our understanding of the spectrum of CVD, lead to better preventive strategies, management, and overall outcomes.

References

[1] Hayward CS, Kelly RP, Collins P. The roles of gender, the menopause and hormone replacement on cardiovascular function. Cardiovasc Res 2000;46:28–49.

[2] Argiento P, Vanderpool RR, Mulè M, Russo MG, D'Alto M, Bossone E, et al. Exercise stress echocardiography of the pulmonary circulation: limits of normal and sex differences. Chest 2012;142:1158–65.

[3] Lerner DJ, Kannel WB. Patterns of coronary heart disease morbidity and mortality in the sexes: a 26-year follow-up of the Framingham population. Am Heart J 1986;111:383–90.

[4] Lloyd-Jones DM, Larson MG, Beiser A, Levy D. Lifetime risk of developing coronary heart disease. Lancet 1999;353:89–92.

[5] Wenger NK. You've come a long way, baby: cardiovascular health and disease in women: problems and prospects. Circulation 2004;109:558–60.

[6] Alter DA, Naylor CD, Austin PC, Tu JV. Biology or bias: practice patterns and long-term outcomes for men and women with acute myocardial infarction. J Am Coll Cardiol 2002;39:1909–16.

[7] Shaw LJ, Bairey Merz CN, Pepine CJ, Reis SE, Bittner V, Kelsey SF, WISE Investigators, et al. Insights from the NHLBI-Sponsored Women's Ischemia Syndrome Evaluation (WISE) Study: Part I: gender differences in traditional and novel risk factors, symptom evaluation, and gender-optimized diagnostic strategies. J Am Coll Cardiol 2006;47(Suppl. 3):S4–20.

[8] Huxley RR, Woodward M. Cigarette smoking as a risk factor for coronary heart disease in women compared with men: a systematic review and meta-analysis of prospective cohort studies. Lancet 2011;378:1297–305.

[9] Bolego C, Poli A, Paoletti R. Smoking and gender. Cardiovasc Res 2002;53:568–76.

[10] Lockyer M. Women with diabetes at greater risk of CHD than men. Practitioner 2014;258:7.

[11] Peters SA, Huxley RR, Woodward M. Diabetes as risk factor for incident coronary heart disease in women compared with men: a systematic review and meta-analysis of 64 cohorts including 858,507 individuals and 28,203 coronary events. Diabetologia 2014;57:1542–51.

[12] Blöndal M, Ainla T, Marandi T, Baburin A, Eha J. Sex-specific outcomes of diabetic patients with acute myocardial infarction who have undergone percutaneous coronary intervention: a register linkage study. Cardiovasc Diabetol 2012;11:96.

[13] Champney KP, Veledar E, Klein M, Samady H, Anderson D, Parashar S, et al. Sex-specific effects of diabetes on adverse outcomes after percutaneous coronary intervention: trends over time. Am Heart J 2007;153:970–8.

[14] Sowers JR. Diabetes mellitus and cardiovascular disease in women. Arch Intern Med 1998;158:617–21.

[15] Steinberg HO, Paradisi G, Cronin J, Crowde K, Hempfling A, Hook G, et al. Type II diabetes abrogates sex differences in endothelial function in premenopausal women. Circulation 2000;101:2040–6.

[16] White RE, Gerrity R, Barman SA, Han G. Estrogen and oxidative stress: a novel mechanism that may increase the risk for cardiovascular disease in women. Steroids 2010;75 788–93.

[17] Appelman Y, van Rijn BB, Ten Haaf ME, Boersma E, Peters SA. Sex differences in cardiovascular risk factors and disease prevention. Atherosclerosis 2015; [in press]. http://dx.doi.org/10.1016/j.atherosclerosis.2015.01.027. [Epub ahead of print].

[18] Solomon CG, Hu FB, Dunaif A, Rich-Edwards JE, Stampfer MJ, Willett WC, et al. Menstrual cycle irregularity and risk for future cardiovascular disease. J Clin Endocrinol Metab 2002;87:2013–17.

[19] Orio Jr F, Palomba S, Spinelli L, Cascella T, Tauchmanovà L, Zullo F, et al. The cardiovascular risk of young women with polycystic ovary syndrome: an observational, analytical, prospective case-control study. J Clin Endocrinol Metab 2004;89:3696–701.

[20] Bassuk SS, Manson JE. Oral contraceptives and menopausal hormone therapy: relative and attributable risks of cardiovascular disease, cancer, and other health outcomes. Ann Epidemiol 2015;25:193–200.

[21] Mosca L, Benjamin EJ, Berra K, Bezanson JL, Dolor RJ, Lloyd-Jones DM, Newby LK, American Heart Association, et al. Effectiveness-based guidelines for the prevention of cardiovascular disease in women—2011 update: a guideline from the American Heart Association. J Am Coll Cardiol 2011;57:1404–23.

[22] Udell JA, Lu H, Redelmeier DA. Long-term cardiovascular risk in women prescribed fertility therapy. J Am Coll Cardiol 2013;62:1704–12.

[23] Pepine CJ, Abrams J, Marks RG, Morris JJ, Scheidt SS, Handberg E. Characteristics of a contemporary population with angina pectoris. TIDES Investigators. Am J Cardiol 1994;74:226–31.

[24] DeVon HA, Zerwic JJ. Symptoms of acute coronary syndromes: are there gender differences? A review of the literature. Heart Lung 2002;31:235–45.

[25] de Torbal A, Boersma E, Kors JA, van Herpen G, Deckers JW, van der Kuip DA, et al. Incidence of recognized and unrecognized myocardial infarction in men and women aged 55 and older: the Rotterdam Study. Eur Heart J 2006;27:729–36.

[26] Shlipak MG, Elmouchi DA, Herrington DM, Lin F, Grady D, Hlatky MA, Heart and Estrogen/progestin Replacement Study Research Group, et al. The incidence of unrecognized myocardial infarction in women with coronary heart disease. Ann Intern Med 2001;134:1043–7.

[27] Jónsdóttir LS, Sigfusson N, Sigvaldason H, Thorgeirsson G. Incidence and prevalence of recognised and unrecognised myocardial infarction in women. The Reykjavik Study. Eur Heart J 1998;19:1011–18.

[28] Wiviott SD, Cannon CP, Morrow DA, Murphy SA, Gibson CM, McCabe CH, et al. Differential expression of cardiac biomarkers by gender in patients with unstable angina/non-ST-elevation myocardial infarction: a TACTICS-TIMI 18 (Treat Angina with Aggrastat and determine Cost of Therapy with an Invasive or Conservative Strategy-Thrombolysis In Myocardial Infarction 18) substudy. Circulation 2004;109:580–6.

[29] Sbarouni E, Georgiadou P, Voudris V. Gender-specific differences in biomarkers responses to acute coronary syndromes and revascularization procedures. Biomarkers 2011;16:457–65.

[30] Buchthal SD, den Hollander JA, Merz CN, Rogers WJ, Pepine CJ, Reichek N, et al. Abnormal myocardial phosphorus-31nuclear magnetic resonance spectroscopy in women with chest pain but normal coronary angiograms. N Engl J Med 2000;342:829–35.

[31] Gurevitz O, Jonas M, Boyko V, Rabinowitz B, Reicher-Reiss H. Clinical profile and long-term prognosis of women < or = 50 years of age referred for coronary angiography for evaluation of chest pain. Am J Cardiol 2000;85:806–9.

[32] Sullivan AK, Holdright DR, Wright CA, Sparrow JL, Cunningham D, Fox KM. Chest pain in women: clinical, investigative, and prognostic features. BMJ 1994;308:883–6.

[33] Bairey Merz CN, Shaw LJ, Reis SE, Bittner V, Kelsey SF, Olson M, WISE Investigators, Insights from the NHLBI-Sponsored Women's Ischemia Syndrome Evaluation (WISE) Study, et al. Part II: gender differences in presentation, diagnosis, and outcome with regard to gender-based pathophysiology of atherosclerosis and macrovascular and microvascular coronary disease. J Am Coll Cardiol 2006;47(Suppl. 3):S21–9.

[34] Wessel TR, Arant CB, McGorray SP, Sharaf BL, Reis SE, Kerensky RA, NHLBI Women's Ischemia Syndrome Evaluation (WISE), et al. Coronary microvascular reactivity is only partially predicted by atherosclerosis risk factors or coronary artery disease in women evaluated for suspected ischemia: results from the NHLBI Women's Ischemia Syndrome Evaluation (WISE). Clin Cardiol 2007;30:69–74.

[35] Yahagi K, Davis HR, Arbustini E, Virmani R. Sex differences in coronary artery disease: pathological observations. Atherosclerosis 2015;239:260–7.

[36] Lansky AJ, Ng VG, Maehara A, Weisz G, Lerman A, Mintz GS, et al. Gender and the extent of coronary atherosclerosis, plaque composition, and clinical outcomes in acute coronary syndromes. JACC Cardiovasc Imaging 2012;5(Suppl. 3):S62–72.

[37] Knowlton AA, Lee AR. Estrogen and the cardiovascular system. Pharmacol Ther 2012;135:54–70.

[38] Cossette É, Cloutier I, Tardif K, DonPierre G, Tanguay JF. Estradiol inhibits vascular endothelial cells pro-inflammatory activation induced by C-reactive protein. Mol Cell Biochem 2013;373:137–47.

[39] Ruiz-Sanz JI, Navarro R, Martínez R, Martín C, Lacort M, Matorras R, et al. 17beta-estradiol affects in vivo the low density lipoprotein composition, particle size, and oxidizability. Free Radic Biol Med 2001;31:391–7.

[40] Knopp RH, Paramsothy P, Retzlaff BM, Fish B, Walden C, Dowdy A, et al. Sex differences in lipoprotein metabolism and dietary response: basis in hormonal differences and implications for cardiovascular disease. Curr Cardiol Rep 2006;8:452–9.

[41] Resanovic I, Rizzo M, Zafirovic S, Bjelogrlic P, Perovic M, Savic K, et al. Anti-atherogenic effects of 17β-estradiol. Horm Metab Res 2013;45:701–8.

[42] Fadini GP, de Kreutzenberg S, Albiero M, Coracina A, Pagnin E, Baesso I, et al. Gender differences in endothelial progenitor cells and cardiovascular risk profile: the role of female estrogens. Arterioscler Thromb Vasc Biol 2008;28:997–1004.

[43] Murphy E, Steenbergen C. Estrogen regulation of protein expression and signaling pathways in the heart. Biol Sex Differ 2014;5:6.

[44] Deschamps AM, Murphy E, Sun J. Estrogen receptor activation and cardioprotection in ischemia reperfusion injury. Trends Cardiovasc Med 2010;20:73–8.

[45] Bae S, Zhang L. Gender differences in cardioprotection against ischemia/reperfusion injury in adult rat hearts: focus on Akt and protein kinase C signaling. J Pharmacol Exp Ther 2005;315:1125–35.

[46] Wang M, Baker L, Tsai BM, Meldrum KK, Meldrum DR. Sex differences in the myocardial inflammatory response to ischemia-reperfusion injury. Am J Physiol Endocrinol Metab 2005;288:E321–6.

[47] Brown DA, Lynch JM, Armstrong CJ, Caruso NM, Ehlers LB, Johnson MS, et al. Susceptibility of the heart to ischaemia-reperfusion injury and exercise-induced cardioprotection are sex-dependent in the rat. J Physiol 2005;564:619–30.

[48] Johnson MS, Moore RL, Brown DA. Sex differences in myocardial infarct size are abolished by sarcolemmal KATP channel blockade in rat. Am J Physiol Heart Circ Physiol 2006;290: H2644–7.

[49] Cross HR, Murphy E, Steenbergen C. Ca(2 +) loading and adrenergic stimulation reveal male/female differences in susceptibility to ischemia-reperfusion injury. Am J Physiol Heart Circ Physiol 2002;283:H481–9.

[50] Sun J, Picht E, Ginsburg KS, Bers DM, Steenbergen C, Murphy E. Hypercontractile female hearts exhibit increased S-nitrosylation of the L-type Ca2 + channel alpha1 subunit and reduced ischemia/reperfusion injury. Circ Res 2006;98:403–11.

[51] Wu JC, Nasseri BA, Bloch KD, Picard MH, Scherrer-Crosbie M. Influence of sex on ventricular remodeling after myocardial infarction in mice. J Am Soc Echocardiogr 2003;16:1158–62.

[52] Fang L, Gao XM, Moore XL, Kiriazis H, Su Y, Ming Z, et al. Differences in inflammation, MMP activation and collagen damage account for gender difference in murine cardiac rupture following myocardial infarction. J Mol Cell Cardiol 2007;43:535–44.

[53] Chen Q, Williams R, Healy CL, Wright CD, Wu SC, O'Connell TD. An association between gene expression and better survival in female mice following myocardial infarction. J Mol Cell Cardiol 2010;49:801–11.

[54] Cavasin MA, Tao Z, Menon S, Yang XP. Gender differences in cardiac function during early remodeling after acute myocardial infarction in mice. Life Sci 2004;75:2181–92.

[55] Shioura KM, Geenen DL, Goldspink PH. Sex-related changes in cardiac function following myocardial infarction in mice. Am J Physiol Regul Integr Comp Physiol 2008;295:R528–34.

[56] Cavasin MA, Sankey SS, Yu AL, Menon S, Yang XP. Estrogen and testosterone have opposing effects on chronic cardiac remodeling and function in mice with myocardial infarction. Am J Physiol Heart Circ Physiol 2003;284:H1560–9.

[57] Cavasin MA, Tao ZY, Yu AL, Yang XP. Testosterone enhances early cardiac remodeling after myocardial infarction, causing rupture and degrading cardiac function. Am J Physiol Heart Circ Physiol 2006;290:H2043–50.

[58] Desrois M, Sidell RJ, Gauguier D, Davey CL, Radda GK, Clarke K. Gender differences in hypertrophy, insulin resistance and ischemic injury in the aging type 2 diabetic rat heart. J Mol Cell Cardiol 2004;37:547–55.
[59] Gabel SA, Walker VR, London RE, Steenbergen C, Korach KS, Murphy E. Estrogen receptor beta mediates gender differences in ischemia/reperfusion injury. J Mol Cell Cardiol 2005;38:289–97.
[60] Wang M, Wang Y, Weil B, Abarbanell A, Herrmann J, Tan J, et al. Estrogen receptor beta mediates increased activation of PI3K/Akt signaling and improved myocardial function in female hearts following acute ischemia. Am J Physiol Regul Integr Comp Physiol 2009;296:R972–8.
[61] Wang M, Crisostomo PR, Markel T, Wang Y, Lillemoe KD, Meldrum DR. Estrogen receptor beta mediates acute myocardial protection following ischemia. Surgery 2008;144:233–8.
[62] Nikolic I, Liu D, Bell JA, Collins J, Steenbergen C, Murphy E. Treatment with an estrogen receptor-beta-selective agonist is cardioprotective. J Mol Cell Cardiol 2007;42:769–80.
[63] Lin J, Steenbergen C, Murphy E, Sun J. Estrogen receptor-beta activation results in S-nitrosylation of proteins involved in cardioprotection. Circulation 2009;120:245–54.
[64] Wang M, Crisostomo P, Wairiuko GM, Meldrum DR. Estrogen receptor-alpha mediates acute myocardial protection in females. Am J Physiol Heart Circ Physiol 2006;290:H2204–9.
[65] Jeanes HL, Tabor C, Black D, Ederveen A, Gray GA. Oestrogen-mediated cardioprotection following ischaemia and reperfusion is mimicked by an oestrogen receptor (ER)alpha agonist and unaffected by an ER beta antagonist. J Endocrinol 2008;197:493–501.
[66] Babiker FA, Lips DJ, Delvaux E, Zandberg P, Janssen BJ, Prinzen F, et al. Oestrogen modulates cardiac ischaemic remodelling through oestrogen receptor-specific mechanisms. Acta Physiol (Oxf) 2007;189:23–31.
[67] Bopassa JC, Eghbali M, Toro L, Stefani E. A novel estrogen receptor GPER inhibits mitochondria permeability transition pore opening and protects the heart against ischemia-reperfusion injury. Am J Physiol Heart Circ Physiol 2010;298: H16–23.
[68] Deschamps AM, Murphy E. Activation of a novel estrogen receptor, GPER, is cardioprotective in male and female rats. Am J Physiol Heart Circ Physiol 2009;297:H1806–13.
[69] Han G, White RE. G-protein-coupled estrogen receptor as a new therapeutic target for treating coronary artery disease. World J Cardiol 2014;6:367–75.
[70] O'Donoghue M, Boden WE, Braunwald E, Cannon CP, Clayton TC, de Winter RJ, et al. Early invasive vs conservative treatment strategies in women and men with unstable angina and non-ST-segment elevation myocardial infarction: a meta-analysis. JAMA 2008;300:71–80.
[71] Boersma E, Harrington RA, Moliterno DJ, White H, Théroux P, Van de Werf F, et al. Platelet glycoprotein IIb/IIIa inhibitors in acute coronary syndromes: a meta-analysis of all major randomised clinical trials. Lancet 2002;359:189–98.
[72] Gulati M, Shaw LJ, Bairey Merz CN. Myocardial ischemia in women: lessons from the NHLBI WISE study. Clin Cardiol 2012;35:141–8.
[73] Solinas E, Nikolsky E, Lansky AJ, Kirtane AJ, Morice MC, Popma JJ, et al. Gender-specific outcomes after sirolimus-eluting stent implantation. J Am Coll Cardiol 2007;50:2111–16.
[74] Lansky AJ, Costa RA, Mooney M, Midei MG, Lui HK, Strickland W, TAXUS-IV Investigators, et al. Gender-based outcomes after paclitaxel-eluting stent implantation in patients with coronary artery disease. J Am Coll Cardiol 2005;45:1180–5.
[75] Puskas JD, Edwards FH, Pappas PA, O'Brien S, Peterson ED, Kilgo P, et al. Off-pump techniques benefit men and women and narrow the disparity in mortality after coronary bypass grafting. Ann Thorac Surg 2007;84:1447–54.
[76] Lundberg G, King S. Coronary revascularization in women. Clin Cardiol 2012;35:156–9.
[77] Mozaffarian D, Benjamin EJ, Go AS, Arnett DK, Blaha MJ, Cushman M, American Heart Association Statistics Committee and Stroke Statistics Subcommittee, et al. Heart disease and stroke statistics—2015 update: a report from the American Heart Association. Circulation 2015;131:e29–e322.
[78] Bui AL, Horwich TB, Fonarow GC. Epidemiology and risk profile of heart failure. Nat Rev Cardiol 2011;8:30–41.
[79] Ho KK, Anderson KM, Kannel WB, Grossman W, Levy D. Survival after the onset of congestive heart failure in Framingham Heart Study subjects. Circulation 1993;88 107–15.
[80] Upadhya B, Taffet GE, Cheng CP, Kitzman DW. Heart failure with preserved ejection fraction in the elderly: scope of the problem. J Mol Cell Cardiol 2015; [in press]. http://dx.doi.org/10.1016/j.yjmcc.2015.02.025. [Epub ahead of print]
[81] Roger VL, Weston SA, Redfield MM, Hellermann-Homan JP, Killian J, Yawn BP, et al. Trends in heart failure incidence and survival in a community-based population. JAMA 2004;292:344–50.
[82] Bibbins-Domingo K, Lin F, Vittinghoff E, Barrett-Connor E, Hulley SB, Grady D, et al. Predictors of heart failure among women with coronary disease. Circulation 2004;110:1424–30.
[83] Heiat A, Gross CP, Krumholz HM. Representation of the elderly, women, and minorities in heart failure clinical trials. Arch Intern Med 2002;162:1682–8.
[84] Ghali JK, Krause-Steinrauf HJ, Adams KF, Khan SS, Rosenberg YD, Yancy CW, et al. Gender differences in advanced heart failure: insights from the BEST study. J Am Coll Cardiol 2003;42:2128–34.
[85] Shin JJ, Hamad E, Murthy S, Piña IL. Heart failure in women. Clin Cardiol 2012;35:172–7.
[86] Grigioni F, Barbieri A, Russo A, Reggianini L, Bonatti S, Potena L, et al. Prognostic stratification of women with chronic heart failure referred for heart transplantation: relevance of gender as compared with gender-related characteristics. J Heart Lung Transplant 2006;25:648–52.
[87] Sharma K, Kass DA. Heart failure with preserved ejection fraction: mechanisms, clinical features, and therapies. Circ Res 2014;115:79–96.
[88] Schneider B, Athanasiadis A, Stöllberger C, Pistner W, Schwab J, Gottwald U, et al. Gender differences in the manifestation of tako-tsubo cardiomyopathy. Int J Cardiol 2013;166:584–8.
[89] Tranter MH, Wright PT, Sikkel MB, Lyon AR. Takotsubo cardiomyopathy: the pathophysiology. Heart Fail Clin 2013;9:187–96.
[90] Garg R, Yusuf S. Overview of randomized trials of angiotensin-converting enzyme inhibitors on mortality and morbidity in patients with heart failure. Collaborative Group on ACE Inhibitor Trials. JAMA 1995;273:1450–6.
[91] Flather MD, Yusuf S, Køber L, Pfeffer M, Hall A, Murray G, et al. Long-term ACE-inhibitor therapy in patients with heart failure or left-ventricular dysfunction: a systematic overview of data from individual patients. ACE-Inhibitor Myocardial Infarction Collaborative Group. Lancet 2000;355:1575–81.
[92] Pfeffer MA, Swedberg K, Granger CB, Held P, McMurray JJ, Michelson EL, CHARM Investigators and Committees, et al. Effects of candesartan on mortality and morbidity in patients with chronic heart failure: the CHARM-Overall programme. Lancet 2003;362:759–66.

[93] Consensus recommendations for the management of chronic heart failure. On behalf of the membership of the advisory council to improve outcomes nationwide in heart failure. Am J Cardiol 1999;83(2A):1A–38A.

[94] The Cardiac Insufficiency Bisoprolol Study II (CIBIS-II): a randomised trial. Lancet 1999;353:9–13.

[95] Katz AM. Angiotensin II: hemodynamic regulator or growth factor? J Mol Cell Cardiol 1990;22:739–47.

[96] Cohn JN, Ferrari R, Sharpe N. Cardiac remodeling—concepts and clinical implications: a consensus paper from an international forum on cardiac remodeling. Behalf of an International Forum on Cardiac Remodeling. J Am Coll Cardiol 2000;35:569–82.

[97] Hummel SL, Kitzman DW. Update on heart failure with preserved ejection fraction. Curr Cardiovasc Risk Rep 2013;7:495–502.

[98] Piro M, Della Bona R, Abbate A, Biasucci LM, Crea F. Sex-related differences in myocardial remodeling. J Am Coll Cardiol 2010;55:1057–65.

[99] Weinberg EO, Thienelt CD, Katz SE, Bartunek J, Tajima M, Rohrbach S, et al. Gender differences in molecular remodeling in pressure overload hypertrophy. J Am Coll Cardiol 1999;34:264–73.

[100] Douglas PS, Katz SE, Weinberg EO, Chen MH, Bishop SP, Lorell BH. Hypertrophic remodeling: gender differences in the early response to left ventricular pressure overload. J Am Coll Cardiol 1998;32:1118–25.

[101] Bhuiyan MS, Shioda N, Fukunaga K. Ovariectomy augments pressure overload-induced hypertrophy associated with changes in Akt and nitric oxide synthase signaling pathways in female rats. Am J Physiol Endocrinol Metab 2007;293: E1606–14.

[102] Cui YH, Tan Z, Fu XD, Xiang QL, Xu JW, Wang TH. 17 beta-estradiol attenuates pressure overload-induced myocardial hypertrophy through regulating caveolin-3 protein in ovariectomized female rats. Mol Biol Rep 2011;38:4885–92.

[103] Tagashira H, Bhuiyan S, Shioda N, Fukunaga K. Distinct cardioprotective effects of 17β-estradiol and dehydroepiandrosterone on pressure overload-induced hypertrophy in ovariectomized female rats. Menopause 2011;18:1317–26.

[104] Li J, Jubair S, Janicki JS. Estrogen inhibits mast cell chymase release to prevent pressure overload-induced adverse cardiac remodeling. Hypertension 2015;65:328–34.

[105] Levick SP, Meléndez GC, Plante E, McLarty JL, Brower GL, Janicki JS. Cardiac mast cells: the centrepiece in adverse myocardial remodelling. Cardiovasc Res 2011;89:12–19.

[106] Urata H, Boehm KD, Philip A, Kinoshita A, Gabrovsek J, Bumpus FM, et al. Cellular localization and regional distribution of an angiotensin II-forming chymase in the heart. J Clin Invest 1993;91:1269–81.

[107] Ahmad S, Varagic J, Westwood BM, Chappell MC, Ferrario CM. Uptake and metabolism of the novel peptide angiotensin-(1-12) by neonatal cardiac myocytes. PLoS One 2011;6:e15759.

[108] Trask AJ, Jessup JA, Chappell MC, Ferrario CM. Angiotensin-(1-12) is an alternate substrate for angiotensin peptide production in the heart. Am J Physiol Heart Circ Physiol 2008;294: H2242–7.

[109] Skavdahl M, Steenbergen C, Clark J, Myers P, Demianenko T, Mao L, et al. Estrogen receptor-beta mediates male-female differences in the development of pressure overload hypertrophy. Am J Physiol Heart Circ Physiol 2005;288:H469–76.

[110] Fliegner D, Schubert C, Penkalla A, Witt H, Kararigas G, Dworatzek E, et al. Female sex and estrogen receptor-beta attenuate cardiac remodeling and apoptosis in pressure overload. Am J Physiol Regul Integr Comp Physiol 2010;298: R1597–606.

[111] van Eickels M, Grohé C, Cleutjens JP, Janssen BJ, Wellens HJ, Doevendans PA. 17beta-estradiol attenuates the development of pressure-overload hypertrophy. Circulation 2001;104 1419–23.

[112] Patten RD, Pourati I, Aronovitz MJ, Alsheikh-Ali A, Eder S, Force T, et al. 17 Beta-estradiol differentially affects left ventricular and cardiomyocyte hypertrophy following myocardial infarction and pressure overload. J Card Fail 2008;14:245–53.

[113] Witt H, Schubert C, Jaekel J, Fliegner D, Penkalla A, Tiemann K, et al. Sex-specific pathways in early cardiac response to pressure overload in mice. J Mol Med (Berl) 2008;86:1013–24.

[114] Karatas A, Hegner B, de Windt LJ, Luft FC, Schubert C, Gross V, et al. Deoxycorticosterone acetate-salt mice exhibit blood pressure-independent sexual dimorphism. Hypertension 2008;51:1177–83.

[115] Nematbakhsh M, Khazaei M. The effect of estrogen on serum nitric oxide concentrations in normotensive and DOCA salt hypertensive ovariectomized rats. Clin Chim Acta 2004;344:53–7.

[116] Arias-Loza PA, Hu K, Dienesch C, Mehlich AM, König S, Jazbutyte V, et al. Both estrogen receptor subtypes, alpha and beta, attenuate cardiovascular remodeling in aldosterone salt-treated rats. Hypertension 2007;50:432–8.

[117] Dent MR, Tappia PS, Dhalla NS. Gender differences in cardiac dysfunction and remodeling due to volume overload. J Card Fail 2010;16:439–49.

[118] Gardner JD, Brower GL, Janicki JS. Gender differences in cardiac remodeling secondary to chronic volume overload. J Card Fail 2002;8:101–7.

[119] Gardner JD, Murray DB, Voloshenyuk TG, Brower GL, Bradley JM, Janicki JS. Estrogen attenuates chronic volume overload induced structural and functional remodeling in male rat hearts. Am J Physiol Heart Circ Physiol 2010;298: H497–504.

[120] Dent MR, Tappia PS, Dhalla NS. Gender differences in apoptotic signaling in heart failure due to volume overload. Apoptosis 2010;15:499–510.

[121] Dent MR, Tappia PS, Dhalla NS. Gender related alterations of β-adrenoceptor mechanisms in heart failure due to arteriovenous fistula. J Cell Physiol 2012;227:3080–7.

[122] Chancey AL, Gardner JD, Murray DB, Brower GL, Janicki JS. Modulation of cardiac mast cell-mediated extracellular matrix degradation by estrogen. Am J Physiol Heart Circ Physiol 2005;289:H316–21.

[123] Queirós AM, Eschen C, Fliegner D, Kararigas G, Dworatzek E, Westphal C, et al. Sex- and estrogen-dependent regulation of a miRNA network in the healthy and hypertrophied heart. Int J Cardiol 2013;169:331–8.

[124] Pedram A, Razandi M, Narayanan R, Dalton JT, McKinsey TA, Levin ER. Estrogen regulates histone deacetylases to prevent cardiac hypertrophy. Mol Biol Cell 2013;24:3805–18.

[125] Gürgen D, Hegner B, Kusch A, Catar R, Chaykovska L, Hoff U, et al. Estrogen receptor-beta signals left ventricular hypertrophy sex differences in normotensive deoxycorticosterone acetate-salt mice. Hypertension 2011;57:648–54.

[126] Gürgen D, Kusch A, Klewitz R, Hoff U, Catar R, Hegner B, et al. Sex-specific mTOR signaling determines sexual dimorphism in myocardial adaptation in normotensive DOCA-salt model. Hypertension 2013;61:730–6.

[127] Zhao Z, Wang H, Jessup JA, Lindsey SH, Chappell MC, Groban L. Am J Physiol Heart Circ Physiol. 2014;306(5):H628–40. Available from: http://dx.doi.org/10.1152/ajpheart.00859.2013.

[128] Wang H, Jessup JA, Lin MS, Chagas C, Lindsey SH, Groban L. Activation of GPR30 attenuates diastolic dysfunction and left ventricle remodelling in oophorectomized mRen2.Lewis rats. Cardiovasc Res 2012;94:96–104.
[129] Jessup JA, Zhang L, Chen AF, Presley TD, Kim-Shapiro DB, Chappell MC, et al. Neuronal nitric oxide synthase inhibition improves diastolic function and reduces oxidative stress in ovariectomized mRen2.Lewis rats. Menopause 2011;18:698–708.
[130] Jessup JA, Zhang L, Presley TD, Kim-Shapiro DB, Wang H, Chen AF, et al. Tetrahydrobiopterin restores diastolic function and attenuates superoxide production in ovariectomized mRen2.Lewis rats. Endocrinology 2011;152:2428–36.
[131] Jessup JA, Wang H, MacNamara LM, Presley TD, Kim-Shapiro DB, Zhang L, et al. Estrogen therapy, independent of timing, improves cardiac structure and function in oophorectomized mRen2.Lewis rats. Menopause 2013;20:860–8.
[132] Zhao Z, Wang H, Lin M, Groban L. GPR30 decreases cardiac chymase/angiotensin II by inhibiting local mast cell number. Biochem Biophys Res Commun 2015;459:131–6.
[133] Delbeck M, Golz S, Vonk R, Janssen W, Hucho T, Isensee J, et al. Impaired left-ventricular cardiac function in male GPR30-deficient mice. Mol Med Rep 2011;4:37–40.
[134] Shekelle PG, Rich MW, Morton SC, Atkinson CS, Tu W, Maglione M, et al. Efficacy of angiotensin-converting enzyme inhibitors and beta-blockers in the management of left ventricular systolic dysfunction according to race, gender, and diabetic status: a meta-analysis of major clinical trials. J Am Coll Cardiol 2003;41:1529–38.
[135] Rathore SS, Wang Y, Krumholz HM. Sex-based differences in the effect of digoxin for the treatment of heart failure. N Engl J Med 2002;347:1403–11.
[136] Pitt B, Williams G, Remme W, Martinez F, Lopez-Sendon J, Zannad F, et al. The EPHESUS trial: eplerenone in patients with heart failure due to systolic dysfunction complicating acute myocardial infarction. Eplerenone Post-AMI Heart Failure Efficacy and Survival Study. Cardiovasc Drugs Ther 2001;15:79–87.
[137] Pitt B, Zannad F, Remme WJ, Cody R, Castaigne A, Perez A, et al. The effect of spironolactone on morbidity and mortality in patients with severe heart failure. Randomized Aldactone Evaluation Study Investigators. N Engl J Med 1999;341:709–17.
[138] Goldenberg I, Moss AJ, Hall WJ, Foster E, Goldberger JJ, Santucci P, MADIT-CRT Executive Committee, et al. Predictors of response to cardiac resynchronization therapy in the Multicenter Automatic Defibrillator Implantation Trial with Cardiac Resynchronization Therapy (MADIT-CRT). Circulation 2011;124:1527–36.
[139] Yoon K, Kwack SJ, Kim HS, Lee BM. Estrogenic endocrine-disrupting chemicals: molecular mechanisms of actions on putative human diseases. J Toxicol Environ Health B Crit Rev 2014;17:127–74.
[140] Farley TA, Dalal MA, Mostashari F, Frieden TR. Deaths preventable in the U.S. by improvements in use of clinical preventive services. Am J Prev Med 2010;38:600–9.
[141] Wang G, Fang J, Ayala C. Hypertension-associated hospitalizations and costs in the United States, 1979–2006. Blood Press 2014;23:126–33.
[142] James PA, Oparil S, Carter BL, Cushman WC, Dennison-Himmelfarb C, Handler J, et al. 2014 evidence-based guideline for the management of high blood pressure in adults: report from the panel members appointed to the Eighth Joint National Committee (JNC 8). JAMA 2014;311:507–20.
[143] Go AS, Mozaffarian D, Roger VL, Benjamin EJ, Berry JD, Blaha MJ, American Heart Association Statistics Committee and Stroke Statistics Subcommittee, et al. Executive summary: heart disease and stroke statistics—2014 update: a report from the American Heart Association. Circulation 2014;129:399–410.
[144] Boden-Albala B, Sacco RL. Lifestyle factors and stroke risk: exercise, alcohol, diet, obesity, smoking, drug use, and stress. Curr Atheroscler Rep 2000;2:160–6.
[145] Agarwal A, Williams GH, Fisher ND. Genetics of human hypertension. Trends Endocrinol Metab 2005;16:127–33.
[146] Doris PA, Fornage M. The transcribed genome and the heritable basis of essential hypertension. Cardiovasc Toxicol 2005;5:95–108.
[147] Schubert R, Mulvany MJ. The myogenic response: established facts and attractive hypotheses. Clin Sci (Lond) 1999;96:313–26.
[148] Miller Jr FJ, Dellsperger KC, Gutterman DD. Myogenic constriction of human coronary arterioles. Am J Physiol 1997;273:H257–64.
[149] Osol G, Halpern W. Myogenic properties of cerebral blood vessels from normotensive and hypertensive rats. Am J Physiol 1985;249:H914–21.
[150] Izzard AS, Bund SJ, Heagerty AM. Myogenic tone in mesenteric arteries from spontaneously hypertensive rats. Am J Physiol 1996;270:H1–6.
[151] Gros R, Van Wert R, You X, Thorin E, Husain M. Effects of age, gender, and blood pressure on myogenic responses of mesenteric arteries from C57BL/6 mice. Am J Physiol Heart Circ Physiol 2002;282:H380–8.
[152] Wellman GC, Bonev AD, Nelson MT, Brayden JE. Gender differences in coronary artery diameter involve estrogen, nitric oxide, and Ca(2+)-dependent K+ channels. Circ Res 1996;79:1024–30.
[153] Huang A, Sun D, Koller A, Kaley G. Gender difference in myogenic tone of rat arterioles is due to estrogen-induced, enhanced release of NO. Am J Physiol 1997;272:H1804–9.
[154] Geary GG, Krause DN, Duckles SP. Estrogen reduces myogenic tone through a nitric oxide-dependent mechanism in rat cerebral arteries. Am J Physiol 1998;275:H292–300.
[155] Furchgott RF, Zawadzki JV. The obligatory role of endothelial cells in the relaxation of arterial smooth muscle by acetylcholine. Nature 1980;288:373–6.
[156] Marsh N, Marsh A. A short history of nitroglycerine and nitric oxide in pharmacology and physiology. Clin Exp Pharmacol Physiol 2000;27:313–19.
[157] Hermann M, Flammer A, Lüscher TF. Nitric oxide in hypertension. J Clin Hypertens (Greenwich) 2006;8(12 Suppl. 4):17–29.
[158] Hashimoto M, Akishita M, Eto M, Ishikawa M, Kozaki K, Toba K, et al. Modulation of endothelium-dependent flow-mediated dilatation of the brachial artery by sex and menstrual cycle. Circulation 1995;92:3431–5.
[159] Benjamin EJ, Larson MG, Keyes MJ, Mitchell GF, Vasan RS, Keaney JF, et al. Clinical correlates and heritability of flow-mediated dilation in the community: the Framingham Heart Study. Circulation 2004;109:613–19.
[160] Forte P, Kneale BJ, Milne E, Chowienczyk PJ, Johnston A, Benjamin N, et al. Evidence for a difference in nitric oxide biosynthesis between healthy women and men. Hypertension 1998;32:730–4.
[161] Barber DA, Miller VM. Gender differences in endothelium-dependent relaxations do not involve NO in porcine coronary arteries. Am J Physiol 1997;273:H2325–32.

[162] Graham DA, Rush JW. Cyclooxygenase and thromboxane/prostaglandin receptor contribute to aortic endothelium-dependent dysfunction in aging female spontaneously hypertensive rats. J Appl Physiol (1985) 2009;107:1059–67.
[163] Leblanc AJ, Chen B, Dougherty PJ, Reyes RA, Shipley RD, Korzick DH, et al. Divergent effects of aging and sex on vasoconstriction to endothelin in coronary arterioles. Microcirculation 2013;20:365–76.
[164] Villar IC, Hobbs AJ, Ahluwalia A. Sex differences in vascular function: implication of endothelium-derived hyperpolarizing factor. J Endocrinol 2008;197:447–62.
[165] Li Z, Duckles SP. Influence of gender on vascular reactivity in the rat. J Pharmacol Exp Ther 1994;268:1426–31.
[166] Lindsey SH, da Silva AS, Silva MS, Chappell MC. Reduced vasorelaxation to estradiol and G-1 in aged female and adult male rats is associated with GPR30 downregulation. Am J Physiol Endocrinol Metab 2013;305:E113–18.
[167] El-Mas MM, El-Gowilly SM, Gohar EY, Ghazal AR. Sex and hormonal influences on the nicotine-induced attenuation of isoprenaline vasodilations in the perfused rat kidney. Can J Physiol Pharmacol 2009;87:539–48.
[168] Kneale BJ, Chowienczyk PJ, Brett SE, Coltart DJ, Ritter JM. Gender differences in sensitivity to adrenergic agonists of forearm resistance vasculature. J Am Coll Cardiol 2000;36:1233–8.
[169] Hart EC, Charkoudian N, Wallin BG, Curry TB, Eisenach J, Joyner MJ. Sex and ageing differences in resting arterial pressure regulation: the role of the β-adrenergic receptors. J Physiol 2011;589:5285–97.
[170] Hart EC, Joyner MJ, Wallin BG, Charkoudian N. Sex, ageing and resting blood pressure: gaining insights from the integrated balance of neural and haemodynamic factors. J Physiol 2012;590:2069–79.
[171] Jiang CW, Sarrel PM, Lindsay DC, Poole-Wilson PA, Collins P. Endothelium-independent relaxation of rabbit coronary artery by 17 beta-oestradiol in vitro. Br J Pharmacol 1991;104:1033–7.
[172] Cruz MN, Douglas G, Gustafsson JA, Poston L, Kublickiene K. Dilatory responses to estrogenic compounds in small femoral arteries of male and female estrogen receptor-beta knockout mice. Am J Physiol Heart Circ Physiol 2006;290:H823–9.
[173] Naderali EK, Smith SL, Doyle PJ, Williams G. Vasorelaxant effects of oestradiols on guinea pigs: a role for gender differences. Eur J Clin Invest 2001;31:215–20.
[174] Costarella CE, Stallone JN, Rutecki GW, Whittier FC. Testosterone causes direct relaxation of rat thoracic aorta. J Pharmacol Exp Ther 1996;277:34–9.
[175] Perusquía M, Navarrete E, González L, Villalón CM. The modulatory role of androgens and progestins in the induction of vasorelaxation in human umbilical artery. Life Sci 2007;81:993–1002.
[176] Yıldırım E, Erol K. The effects of testosterone on isolated sheep coronary artery. Anadolu Kardiyol Derg 2011;11:343–50.
[177] Sullivan JC. Sex and the renin-angiotensin system: inequality between the sexes in response to RAS stimulation and inhibition. Am J Physiol Regul Integr Comp Physiol 2008;294: R1220–6.
[178] Komukai K, Mochizuki S, Yoshimura M. Gender and the renin-angiotensin-aldosterone system. Fundam Clin Pharmacol 2010;24:687–98.
[179] Hilliard LM, Sampson AK, Brown RD, Denton KM. The "his and hers" of the renin-angiotensin system. Curr Hypertens Rep 2013;15:71–9.
[180] Chappell MC, Marshall AC, Alzayadneh EM, Shaltout HA, Diz DI. Update on the Angiotensin converting enzyme 2-Angiotensin (1–7)-MAS receptor axis: fetal programing, sex differences, and intracellular pathways. Front Endocrinol (Lausanne) 2014;4:201.
[181] Silva-Antonialli MM, Tostes RC, Fernandes L, Fior-Chadi DR, Akamine EH, Carvalho MH, et al. A lower ratio of AT1/AT2 receptors of angiotensin II is found in female than in male spontaneously hypertensive rats. Cardiovasc Res 2004;62:587–93.
[182] Sampson AK, Moritz KM, Jones ES, Flower RL, Widdop RE, Denton KM. Enhanced angiotensin II type 2 receptor mechanisms mediate decreases in arterial pressure attributable to chronic low-dose angiotensin II in female rats. Hypertension 2008;52:666–71.
[183] Okumura M, Iwai M, Ide A, Mogi M, Ito M, Horiuchi M. Sex difference in vascular injury and the vasoprotective effect of valsartan are related to differential AT2 receptor expression. Hypertension 2005;46:577–83.
[184] Pendergrass KD, Pirro NT, Westwood BM, Ferrario CM, Brosnihan KB, Chappell MC. Sex differences in circulating and renal angiotensins of hypertensive mRen(2). Lewis but not normotensive Lewis rats. Am J Physiol Heart Circ Physiol 2008;295:H10–20.
[185] Schunkert H, Danser AH, Hense HW, Derkx FH, Kürzinger S, Riegger GA. Effects of estrogen replacement therapy on the renin-angiotensin system in postmenopausal women. Circulation 1997;95:39–45.
[186] Sullivan JC, Rodriguez-Miguelez P, Zimmerman MA, Harris RA. Differences in angiotensin (1–7) between men and women. Am J Physiol Heart Circ Physiol 2015; [in press]. http://dx.doi.org/10.1152/ajpheart.00897.2014. [Epub ahead of print].
[187] Xue B, Pamidimukkala J, Hay M. Sex differences in the development of angiotensin II-induced hypertension in conscious mice. Am J Physiol Heart Circ Physiol 2005;288: H2177–84.
[188] Ji H, Zheng W, Wu X, Liu J, Ecelbarger CM, Watkins R, et al. Sex chromosome effects unmasked in angiotensin II-induced hypertension. Hypertension 2010;55:1275–82.
[189] Reckelhoff JF, Zhang H, Granger JP. Testosterone exacerbates hypertension and reduces pressure-natriuresis in male spontaneously hypertensive rats. Hypertension 1998;31:435–9.
[190] Kauser K, Rubanyi GM. Gender difference in endothelial dysfunction in the aorta of spontaneously hypertensive rats. Hypertension 1995;25:517–23.
[191] Loukotová J, Kunes J, Zicha J. Gender-dependent difference in cell calcium handling in VSMC isolated from SHR: the effect of angiotensin II. J Hypertens 2002;20:2213–19.
[192] Knudsen KD, Dahl LK, Thompson K, Iwai J, Heine M, Leitl G. Effects of chronic excess salt ingestion. Inheritance of hypertension in the rat. J Exp Med 1970;132:976–1000.
[193] Dahl LK, Knudsen KD, Ohanian EV, Muirhead M, Tuthill R. Role of the gonads in hypertension-prone rats. J Exp Med 1975;142:748–59.
[194] Selye H, Hall CE, Rowley EM. Malignant hypertension produced by treatment with desoxycorticosterone acetate and sodium chloride. Can Med Assoc J 1943;49:88–92.
[195] Makrides SC, Mulinari R, Zannis VI, Gavras H. Regulation of renin gene expression in hypertensive rats. Hypertension 1988;12:405–10.
[196] Itaya Y, Suzuki H, Matsukawa S, Kondo K, Saruta T. Central renin-angiotensin system and the pathogenesis of DOCA-salt hypertension in rats. Am J Physiol 1986;251:H261–8.
[197] Dai SY, Peng W, Zhang YP, Li JD, Shen Y, Sun XF. Brain endogenous angiotensin II receptor type 2 (AT2-R) protects against DOCA/salt-induced hypertension in female rats. J Neuroinflammation 2015;12:1–11.

[198] Chappell MC, Gallagher PE, Averill DB, Ferrario CM, Brosnihan KB. Estrogen or the AT1 antagonist olmesartan reverses the development of profound hypertension in the congenic mRen2. Lewis rat. Hypertension 2003;42:781–6.
[199] Bohlender J, Ménard J, Edling O, Ganten D, Luft FC. Mouse and rat plasma renin concentration and gene expression in (mRen2)27 transgenic rats. Am J Physiol 1998;274 H1450–6.
[200] Lisabeth LD, Beiser AS, Brown DL, Murabito JM, Kelly-Hayes M, Wolf PA. Age at natural menopause and risk of ischemic stroke: the Framingham heart study. Stroke 2009;40:1044–9.
[201] Rivera CM, Grossardt BR, Rhodes DJ, Brown Jr RD, Roger VL, Melton LJ, et al. Increased cardiovascular mortality after early bilateral oophorectomy. Menopause 2009;16:15–23.
[202] Leonardo F, Medeirus C, Rosano GM, Pereira WI, Sheiban I, Gebara O, et al. Effect of acute administration of estradiol 17 beta on aortic blood flow in menopausal women. Am J Cardiol 1997;80:791–3.
[203] Pines A, Fisman EZ, Drory Y, Shapira I, Averbuch M, Eckstein N, et al. The effects of sublingual estradiol on left ventricular function at rest and exercise in postmenopausal women: an echocardiographic assessment. Menopause 1998;5:79–85.
[204] Freay AD, Curtis SW, Korach KS, Rubanyi GM. Mechanism of vascular smooth muscle relaxation by estrogen in depolarized rat and mouse aorta. Role of nuclear estrogen receptor and Ca2 + uptake. Circ Res 1997;81:242–8.
[205] Harrison-Bernard LM, Schulman IH, Raij L. Postovariectomy hypertension is linked to increased renal AT1 receptor and salt sensitivity. Hypertension 2003;42:1157–63.
[206] Brinson KN, Rafikova O, Sullivan JC. Female sex hormones protect against salt-sensitive hypertension but not essential hypertension. Am J Physiol Regul Integr Comp Physiol 2014;307:R149–57.
[207] Hinojosa-Laborde C, Craig T, Zheng W, Ji H, Haywood JR, Sandberg K. Ovariectomy augments hypertension in aging female Dahl salt-sensitive rats. Hypertension 2004;44:405–9.
[208] Sasaki T, Ohno Y, Otsuka K, Suzawa T, Suzuki H, Saruta T. Oestrogen attenuates the increases in blood pressure and platelet aggregation in ovariectomized and salt-loaded Dahl salt-sensitive rats. J Hypertens 2000;18:911–17.
[209] Zhang L, Fujii S, Kosaka H. Effect of oestrogen on reactive oxygen species production in the aortas of ovariectomized Dahl salt-sensitive rats. J Hypertens 2007;25:407–14.
[210] Lindsey SH, Cohen JA, Brosnihan KB, Gallagher PE, Chappell MC. Chronic treatment with the G protein-coupled receptor 30 agonist G-1 decreases blood pressure in ovariectomized mRen2.Lewis rats. Endocrinology 2009;150:3753–8.
[211] Lindsey SH, Carver KA, Prossnitz ER, Chappell MC. Vasodilation in response to the GPR30 agonist G-1 is not different from estradiol in the mRen2.Lewis female rat. J Cardiovasc Pharmacol 2011;57:598–603.
[212] Xue B, Pamidimukkala J, Lubahn DB, Hay M. Estrogen receptor-alpha mediates estrogen protection from angiotensin II-induced hypertension in conscious female mice. Am J Physiol Heart Circ Physiol 2007;292:H1770–6.
[213] Xue B, Zhang Z, Beltz TG, Johnson RF, Guo F, Hay M, et al. Estrogen receptor-β in the paraventricular nucleus and rostroventrolateral medulla plays an essential protective role in aldosterone/salt-induced hypertension in female rats. Hypertension 2013;61:1255–62.
[214] Zhu Y, Bian Z, Lu P, Karas RH, Bao L, Cox D, et al. Abnormal vascular function and hypertension in mice deficient in estrogen receptor beta. Science 2002;295:505–8.
[215] Lee LM, Cao J, Deng H, Chen P, Gatalica Z, Wang ZY. ER-alpha36, a novel variant of ER-alpha, is expressed in ER-positive and -negative human breast carcinomas. Anticancer Res 2008;28:479–83.
[216] Fortepiani LA, Zhang H, Racusen L, Roberts II LJ, Reckelhoff JF. Characterization of an animal model of postmenopausal hypertension in spontaneously hypertensive rats. Hypertension 2003;41:640–5.
[217] Laughlin GA, Barrett-Connor E, Kritz-Silverstein D, von Mühlen D. Hysterectomy, oophorectomy, and endogenous sex hormone levels in older women: the Rancho Bernardo Study. J Clin Endocrinol Metab 2000;85:645–51.
[218] Chen MJ, Yang WS, Yang JH, Chen CL, Ho HN, Yang YS. Relationship between androgen levels and blood pressure in young women with polycystic ovary syndrome. Hypertension 2007;49:1442–7.
[219] Negrín CD, McBride MW, Carswell HV, Graham D, Carr FJ, Clark JS, et al. Reciprocal consomic strains to evaluate y chromosome effects. Hypertension 2001;37:391–7.
[220] De Vries GJ, Rissman EF, Simerly RB, Yang LY, Scordalakes EM, Auger CJ, et al. A model system for study of sex chromosome effects on sexually dimorphic neural and behavioral traits. J Neurosci 2002;22:9005–14.
[221] Brunton LL, editor. Goodman and Gilman's the pharmacological basis of therapeutics. New York, NY: McGraw-Hill; 2006.
[222] Gu Q, Burt VL, Paulose-Ram R, Dillon CF. Gender differences in hypertension treatment, drug utilization patterns, and blood pressure control among US adults with hypertension: data from the National Health and Nutrition Examination Survey 1999–2004. Am J Hypertens 2008;21:789–98.
[223] Daugherty SL, Masoudi FA, Ellis JL, Ho PM, Schmittdiel JA, Tavel HM, et al. Age-dependent gender differences in hypertension management. J Hypertens 2011;29:1005–11.
[224] Ljungman C, Kahan T, Schiöler L, Hjerpe P, Hasselström J, Wettermark B, et al. Gender differences in antihypertensive drug treatment: results from the Swedish Primary Care Cardiovascular Database (SPCCD). J Am Soc Hypertens 2014;8:882–90.
[225] Seeland U, Regitz-Zagrosek V. Sex and gender differences in cardiovascular drug therapy. Handb Exp Pharmacol 2012;(214):211–36.
[226] Zimmerman MA, Harris RA, Sullivan JC. Female spontaneously hypertensive rats are more dependent on ANG (1-7) to mediate effects of low-dose AT1 receptor blockade than males. Am J Physiol Renal Physiol 2014;306:F1136–42.
[227] Yoon SS, Gu Q, Nwankwo T, Wright JD, Hong Y, Burt V. Trends in blood pressure among adults with hypertension: United States, 2003 to 2012. Hypertension 2015;65:54–61.
[228] Yazbeck C, Thiebaugeorges O, Moreau T, Goua V, Debotte G, Sahuquillo J, et al. Maternal blood lead levels and the risk of pregnancy-induced hypertension: the EDEN Cohort Study. Environ Health Perspect 2009;117:1526–30.
[229] Shiue I, Hristova K. Higher urinary heavy metal, phthalate and arsenic concentrations accounted for 3-19% of the population attributable risk for high blood pressure: US NHANES, 2009–2012. Hypertens Res 2014;37:1075–81.
[230] Shankar A, Teppala S. Urinary bisphenol A and hypertension in a multiethnic sample of US adults. J Environ Public Health 2012;2012:481641.
[231] Bae S, Hong YC. Exposure to bisphenol A from drinking canned beverages increases blood pressure: randomized crossover trial. Hypertension 2015;65:313–19.
[232] Bae S, Kim JH, Lim YH, Park HY, Hong YC. Associations of bisphenol A exposure with heart rate variability and blood pressure. Hypertension 2012;60:786–93.
[233] Rosenthal JL, Jacob MS. Biomarkers in pulmonary arterial hypertension. Curr Heart Fail Rep 2014;11:477–84.

[234] Ling Y, Johnson MK, Kiely DG, Condliffe R, Elliot CA, Gibbs JS, et al. Changing demographics, epidemiology, and survival of incident pulmonary arterial hypertension: results from the pulmonary hypertension registry of the United Kingdom and Ireland. Am J Respir Crit Care Med 2012;186:790–6.

[235] Humbert M, Sitbon O, Chaouat A, Bertocchi M, Habib G, Gressin V, et al. Pulmonary arterial hypertension in France: results from a national registry. Am J Respir Crit Care Med 2006;173:1023–30.

[236] Peacock AJ, Murphy NF, McMurray JJ, Caballero L, Stewart S. An epidemiological study of pulmonary arterial hypertension. Eur Respir J 2007;30:104–9.

[237] Frost AE, Badesch DB, Barst RJ, Benza RL, Elliott CG, Farber HW, et al. The changing picture of patients with pulmonary arterial hypertension in the United States: how REVEAL differs from historic and non-US Contemporary Registries. Chest 2011;139:128–37.

[238] Gaine SP, Rubin LJ. Primary pulmonary hypertension. Lancet 1998;352:719–25.

[239] Galiè N, Corris PA, Frost A, Girgis RE, Granton J, Jing ZC, et al. Updated treatment algorithm of pulmonary arterial hypertension. J Am Coll Cardiol 2013;62(Suppl. 25):D60–72.

[240] Humbert M, Yaici A, de Groote P, Montani D, Sitbon O, Launay D, et al. Screening for pulmonary arterial hypertension in patients with systemic sclerosis: clinical characteristics at diagnosis and long-term survival. Arthritis Rheum 2011;63:3522–30.

[241] Benza RL, Miller DP, Barst RJ, Badesch DB, Frost AE, McGoon MD. An evaluation of long-term survival from time of diagnosis in pulmonary arterial hypertension from the REVEAL Registry. Chest 2012;142:448–56.

[242] McGoon MD, Miller DP. REVEAL: a contemporary US pulmonary arterial hypertension registry. Eur Respir Rev 2012;21:8–18.

[243] Humbert M, Sitbon O, Yaïci A, Montani D, O'Callaghan DS, Jaïs X, French Pulmonary Arterial Hypertension Network, et al. Survival in incident and prevalent cohorts of patients with pulmonary arterial hypertension. Eur Respir J 2010;36 549–55.

[244] Shapiro S, Traiger GL, Turner M, McGoon MD, Wason P, Barst RJ. Sex differences in the diagnosis, treatment, and outcome of patients with pulmonary arterial hypertension enrolled in the registry to evaluate early and long-term pulmonary arterial hypertension disease management. Chest 2012;141:363–73.

[245] Mair KM, Wright AF, Duggan N, Rowlands DJ, Hussey MJ, Roberts S, et al. Sex-dependent influence of endogenous estrogen in pulmonary hypertension. Am J Respir Crit Care Med 2014;190:456–67.

[246] Lai YC, Potoka KC, Champion HC, Mora AL, Gladwin MT. Pulmonary arterial hypertension: the clinical syndrome. Circ Res 2014;115:115–30.

[247] Purcell H. Moving forward in pulmonary hypertension. Br J Cardiol 2009;16(Suppl. 1):S2–3.

[248] Alencar AKN, Barreiro EJ, Sudo RT, Zapata-Sudo G, Adenosine A. Receptor as a target of treatment for pulmonary arterial hypertension. Clin Res Pulmonol 2014;2(2):1021.

[249] Schermuly RT, Ghofrani HA, Wilkins MR, Grimminger F. Mechanisms of disease: pulmonary arterial hypertension. Nat Rev Cardiol 2011;8:443–55.

[250] Humbert M, Sitbon O, Simonneau G. Treatment of pulmonary arterial hypertension. N Engl J Med 2004;351:1425–36.

[251] Sitbon O, Morrell N. Pathways in pulmonary arterial hypertension: the future is here. Eur Respir Rev 2012;21:321–7.

[252] Galiè N, Ghofrani AH. New horizons in pulmonary arterial hypertension therapies. Eur Respir Rev 2013;22:503–14.

[253] Baliga RS, MacAllister RJ, Hobbs AJ. New perspectives for the treatment of pulmonary hypertension. Br J Pharmacol 2011;163:125–40.

[254] Humbert M, Sitbon O, Chaouat A, Bertocchi M, Habib G, Gressin V, et al. Survival in patients with idiopathic, familial, and anorexigen-associated pulmonary arterial hypertension in the modern management era. Circulation 2010;122:156–63.

[255] Mair KM, Johansen AK, Wright AF, Wallace E, MacLean MR. Pulmonary arterial hypertension: basis of sex differences in incidence and treatment response. Br J Pharmacol 2014;171:567–79.

[256] Gabler NB, French B, Strom BL, Liu Z, Palevsky HI, Taichman DB, et al. Race and sex differences in response to endothelin receptor antagonists for pulmonary arterial hypertension. Chest 2012;141:20–6.

[257] Martin YN, Pabelick CM. Sex differences in the pulmonary circulation: implications for pulmonary hypertension. Am J Physiol Heart Circ Physiol 2014;306:H1253–64.

[258] Lahm T, Tuder RM, Petrache I. Progress in solving the sex hormone paradox in pulmonary hypertension. Am J Physiol Lung Cell Mol Physiol 2014;307:L7–26.

[259] Perez EA. Safety profiles of tamoxifen and the aromatase inhibitors in adjuvant therapy of hormone-responsive early breast cancer. Ann Oncol 2007;18(Suppl. 8):viii26–35.

[260] Benza RL, Miller DP, Gomberg-Maitland M, Frantz RP, Foreman AJ, Coffey CS, et al. Predicting survival in pulmonary arterial hypertension: insights from the Registry to Evaluate Early and Long-Term Pulmonary Arterial Hypertension Disease Management (REVEAL). Circulation 2010;122:164–72.

[261] van de Veerdonk MC, Kind T, Marcus JT, Mauritz GJ, Heymans MW, Bogaard HJ, et al. Progressive right ventricular dysfunction in patients with pulmonary arterial hypertension responding to therapy. J Am Coll Cardiol 2011;58 2511–19.

[262] Bogaard HJ, Abe K, Vonk Noordegraaf A, Voelkel NF. The right ventricle under pressure: cellular and molecular mechanisms of right-heart failure in pulmonary hypertension. Chest 2009;135:794–804.

[263] Haddad F, Doyle R, Murphy DJ, Hunt SA. Right ventricular function in cardiovascular disease, part II: pathophysiology, clinical importance, and management of right ventricular failure. Circulation 2008;117:1717–31.

[264] Voelkel NF, Quaife RA, Leinwand LA, Barst RJ, McGoon MD, Meldrum DR, National Heart, Lung, and Blood Institute Working Group on Cellular and Molecular Mechanisms of Right Heart Failure, et al. Right ventricular function and failure: report of a National Heart, Lung, and Blood Institute working group on cellular and molecular mechanisms of right heart failure. Circulation 2006;114:1883–91.

[265] Tonelli AR, Arelli V, Minai OA, Newman J, Bair N, Heresi GA, et al. Causes and circumstances of death in pulmonary arterial hypertension. Am J Respir Crit Care Med 2013;188:365–9.

[266] Ventetuolo CE, Ouyang P, Bluemke DA, Tandri H, Barr RG, Bagiella E, et al. Sex hormones are associated with right ventricular structure and function: The MESA-right ventricle study. Am J Respir Crit Care Med 2011;183:659–67.

[267] White K, Dempsie Y, Nilsen M, Wright AF, Loughlin L, MacLean MR. The serotonin transporter, gender, and 17β oestradiol in the development of pulmonary arterial hypertension. Cardiovasc Res 2011;90:373–82.

[268] Frump AL, Goss KN, Vayl A, Albrecht M, Fisher AJ, Tursunova R, et al. Estradiol improves right ventricular function in rats with severe angioproliferative pulmonary hypertension: effects of endogenous and exogenous sex hormones. Am J Physiol Lung Cell Mol Physiol 2015; [in press]. Available from: http://dx.doi.org/10.1152/ajplung.00006.2015. [Epub ahead of print].

[269] Rossouw JE, Anderson GL, Prentice RL, LaCroix AZ, Kooperberg C, Stefanick ML, Writing Group for the Women's Health Initiative Investigators, et al. Risks and benefits of estrogen plus progestin in healthy postmenopausal women: principal results from the Women's Health Initiative randomized controlled trial. JAMA 2002;288:321–33.

[270] Anderson GL, Limacher M, Assaf AR, Bassford T, Beresford SA, Black H, Women's Health Initiative Steering Committee, et al. Effects of conjugated equine estrogen in postmenopausal women with hysterectomy: the Women's Health Initiative randomized controlled trial. JAMA 2004;291:1701–12.

[271] Rossouw JE, Prentice RL, Manson JE, Wu L, Barad D, Barnabei VM, et al. Postmenopausal hormone therapy and risk of cardiovascular disease by age and years since menopause. JAMA 2007;297:1465–77.

[272] North American Menopause Society. Estrogen and progestogen use in postmenopausal women: 2010 position statement of The North American Menopause Society. Menopause 2010;17:242–55.

[273] Hodis HN, Mack WJ, Shoupe D, Azen SP, Stanczyk FZ, Huang-Levine J, et al. Testing the menopausal hormone therapy timing hypothesis: the Early vs. Late Intervention Trial with Estradiol. Circulation 2014;130:A13283.

C H A P T E R

6

Sex Differences in Pulmonary Anatomy and Physiology: Implications for Health and Disease

Venkatachalem Sathish[1,2] *and Y.S. Prakash*[1,2]

[1]Department of Physiology and Biomedical Engineering, Mayo Clinic, Rochester, MN, United States
[2]Department of Anesthesiology, Mayo Clinic, Rochester, MN, United States

INTRODUCTION

Intrinsic sex differences in lung anatomy and physiology are apparent at all life stages from the time of embryonic lung development through adulthood and aging [1–6]. In both males and females, prepubertal, postpubertal, and the intervening childbearing years for women are windows into the modulatory roles of sex steroids (estrogens, progesterone, and testosterone) in normal lung structure/function, and the pathophysiology of a range of pulmonary diseases. For example, the American Lung Association highlights the increasingly higher burden of chronic obstructive pulmonary disease (COPD) among women (37% more prevalent) that cannot be explained by differences in smoking alone [7]. Existing data also suggest that adolescent girls are more susceptible to cigarette smoke [8], which may contribute to this sex disparity in COPD, and may also help explain sex differences in lung cancer [9]. Similarly, several clinical studies have shown an age-related variation in asthma, such that prepubertal boys are more likely to have asthma, while following puberty, women show greater asthma which decreases during the postmenopausal years [2,10–16]. Furthermore, asthmatic women show variations in symptom severity with their menstrual cycle, and during pregnancy, suggesting that female sex hormones may modulate asthma pathophysiology [2,4,17–19]. Overall, accumulating epidemiological and clinical data highlight the importance of sex differences in the lung, and suggest a regulatory role for sex steroids. Accordingly, understanding these concepts becomes important both in terms of appreciating sex differences in normal pulmonary physiology, and in disease development and its eventual therapy.

SEX DIFFERENCES IN THE PULMONARY SYSTEM

Sex Differences in the Developing Lung

Human fetal lung development originates at a very early stage of gestation, and although humans are born with a fixed number of conducting airways, alveolar development continues after birth up to 2–4 years of age [20–26]. Even during fetal development, sex differences in the lung are noted. Female fetuses display smaller airways and less number of bronchi compared to males; however, lung maturation rates are faster in female fetuses [27,28]. Furthermore, maternal and fetal sex steroids have regulatory effects on lung development (eg, see recent reviews) [1–5]. Estradiol is produced by the placenta and testosterone is produced by fetal testes after sex differentiation [29]. Dynamic changes in circulating sex steroid levels and also intrinsic production of sex steroids by fetuses can contribute to critical phases of lung development and to sex differences [30,31]. The saccular phase extending into the late gestation period involves production of testosterone by fetal testes, which delays surfactant production in the male lung [31]. On the other hand, female fetuses produce lung surfactant earlier and are believed to preserve patency of small airways [32,33]. Studies in human fetal lung show that branching morphogenesis of the lung involves androgens and

Sex Differences in Physiology
DOI: http://dx.doi.org/10.1016/B978-0-12-802388-4.00006-9

androgen receptors (ARs) [31,34]. Similarly, estrogen receptors (ERs) are expressed in the early stages of developing lung [35,36]. Distinct from androgens, estrogens have stimulatory effects on lung maturation [37]. In addition, both ERα and ERβ seem to be crucial in alveolar formation, with genetic alteration of ERs resulting in a diminished quantity of alveoli in female fetuses [38,39]. Thus, at birth, males had bigger lungs than females with more respiratory bronchioles [27]. However, female airways and lung parenchyma grow more proportionately (concept of dysanapsis) compared to male airways throughout childhood and adolescence [40–42]. Briefly, dysanapsis means that disproportionate growth of airways and lung parenchyma leads to the variability in the maximal flows below 70% of vital capacity (VC). Hence, larger lungs in the male would have longer conducting airways and would thus be at a disadvantage during expiration [43]. Because of this comparative growth in females, the overall specific airway resistance is lesser in females, and enables higher forced expiratory flow rates [44], providing an airflow advantage in prepubertal girls. The relevance of these intrinsic and steroid-induced differences between male and female fetuses and infants, and in lung development, lies in the interesting clinical findings that premature male infants are more susceptible to respiratory distress syndrome [45] and bronchopulmonary dysplasia (alveolar simplification and vascular dysmorphogenesis). Such early establishment of sex differences in the structure and function of the lung may also influence respiratory diseases later in life, particularly in the context of postpubertal changes in sex steroid levels.

Sex Differences in the Adult Lung

Irrespective of age, the lungs of women are smaller compared to men when normalized for height. Despite these anatomical differences, expiratory flow rates are in fact higher in women compared to men [46], which may reflect the pattern of airway versus alveolar growth. During childhood and adolescence, the airways versus lung parenchyma of females grow proportionally, but in males, conducting airways grow slower than the remainder of the lung resulting in disproportionately fewer alveoli for the number of airways (although males may actually have more airways to begin with). Accordingly, such dysanaptic growth results in overall narrower airways in males and contributes to higher airway resistance, and lesser forced expiratory flow rates [42]. With puberty, the greater expiratory flow rate is not as obvious in women, although the forced expiratory volume 1/forced vital capacity (FEV1/FVC) ratio tends to be greater in women [47]. In women, the processes of pregnancy induce changes in lung anatomy that should result in altered function. The gravid uterus elevates the diaphragm and compresses the lungs within the thoracic cavity, reducing total lung capacity and functional residual capacity (FRC) [48,49]. Yet, there is no detected decrease in FVC, FEV1, or VC, which may be partly due to concurrent increases in inspiratory capacity and ~50% decline in pulmonary resistance [49]. With aging, independent of any accompanying disease, there are age-related changes such as decreased elastic recoil of the larger airways and increased fibrosis that result in decreased maximal expiratory flow rate. But even here, such changes are thought to be greater in men compared to women, and also occur more slowly in women [2,50–52].

Thus, overall, with relatively larger airway sizes, and slower detrimental changes throughout life, the airways of females are naturally suited to function better. Yet, these anatomical and functional advantages may be offset (at least clinically) by the paradoxical greater incidence of asthma in boys versus girls and more in women versus men [11,12,15,16,53], suggesting a modulating role for sex steroids: an area that is not well understood.

SEX STEROID SIGNALING IN THE LUNG

Sex steroid signaling and effects are complex, and cell-, tissue- and context-dependent. Although primarily gonadally derived, local production of sex steroid that occurs in many tissues including the lung [54,55] further complicates interpretation. The signaling of sex steroids has been reviewed elsewhere [56–60]. Relevant to the lung, the classical ERs (ERα and ERβ), progesterone receptors (PR-A and PR-B), and AR have all been found in different lung elements [61–70], but their signaling is not well-established. In this regard, the complexity and context-specific nature of sex steroid receptor signaling may be relevant in eventually understanding whether and how estrogens (in particular) and other hormones are beneficial or detrimental in lung disease. For example, ERα is a superior transcriptional activator than ERβ, and ERβ can inhibit ERα activation [71,72]. ERα and ERβ can form homo- and, in some cases, heterodimers in the nucleus. ERα also can heterodimerize with AR, which modifies transcriptional activity [73]. Both PR-A and PR-B bind progesterone with comparable affinities, but their transcriptional activities diverge considerably [60,74]. Testosterone and its most active metabolite, 5α-dihydrotestosterone (DHT) bind to AR leading to nuclear localization of the receptor, predominantly with homodimerization, or with ERα, for example [75,76]. In addition, steroid receptors have multiple Ser/Thr phosphorylation sites on their N-terminal domain which may be responsible for ligand-independent activation or may augment activation in the presence of low levels of ligand. Although these many nuances of sex steroid signaling are largely unexplored in the lung, these processes suggest that the effects of estrogen and progesterone in women, testosterone in men, or even estrogen in aging males (due to the effect of aromatase) may be quite complex.

In addition to classical steroid hormone receptor activation that is largely genomic and requires longer durations [77,78], rapid nongenomic sex steroid signaling occurring in seconds to minutes have been observed, which may or may not involve classical steroid receptors, but is quite relevant to the lung. Rapid signaling can involve membrane-localized or cytoplasmic receptors, as well as the newly recognized G-protein coupled receptors GPR30 [58–60,77,79–81]. A major aspect of rapid nongenomic sex steroid signaling is intracellular Ca^{2+} regulation [58,82]. Here, estrogen can activate the PLC-DAG-IP_3 signaling cascade, leading to increases in $[Ca^{2+}]_i$, for example in osteoblasts [83], or in the case of progesterone by opening L-type Ca^{2+} channels [84]. Both PLC-dependent mechanisms and L-type Ca^{2+} channel modulation can occur in the case of testosterone effects in osteoblasts and macrophages [85,86]. However, sex steroid effects in the cardiovascular system predominantly involve inhibition of L-type Ca^{2+} channels and plasma membrane ion fluxes [58,82,87], with decreased vascular contractility [88–91]. A similar effect may be involved in the lung (see below).

Sex steroid receptor signaling can activate a myriad of intracellular signaling pathways including mitogen activated protein kinases (MAPKs), tyrosine kinases, and lipid kinases, with further modulation of steroid receptors themselves [92–94]. Estradiol has been shown to activate ERK 1/2, p38, and JNK pathways [59,60,95,96] which are also relevant to nongenomic effects mediated via phosphorylation. PR-B displays crosstalk with ERs and priming ER to activate Src-Ras-ERK pathways [97]. PRs also activate p42 MAPK and PI3K in *Xenopus* oocytes [93]. ERα activation leads to PI3K-Akt-eNOS signaling cascade and production of the vasodilator nitric oxide (NO) in endothelial cells [59,98]. AR activation can involve the c-Src, Raf-1, and ERK-2 pathways resulting in downstream involvement of MAPK [99,100].

IMPACT OF SEX STEROIDS ON PULMONARY STRUCTURE AND FUNCTION

Sex Steroid Effects in the Epithelium

Sex steroid signaling plays a significant role in the maintenance and function of airway epithelium throughout the lifespan. Airway epithelium is the first line of defense for environmental and inflammatory stimuli in the lung and is involved in the pathophysiology of lung diseases such as asthma, COPD, and fibrosis. ERs and PRs are expressed in the nasal mucosa and suggest a role for female sex steroids in modulating nasal epithelial function [101]. For example, studies show increased nasal edema and mucus secretion during different stages of the menstrual cycle [102–104]. Some other studies suggest the significance of estrogen signaling in the nasal mucosa in the context of rhinitis, further supported by clinical data showing augmented rhinitis symptoms in women using oral contraceptives [105]. Experimental animal studies show thickened nasal mucosa after exogenous estrogen treatment [106,107], and female sex steroids may augment expression of histamine receptors in human nasal epithelial cells and mucosal microvascular endothelial cells [108].

A major role for estrogen signaling in bronchial epithelial cells (BECs) may be in regulating NO in the airway (Fig. 6.1). Previous studies in vascular endothelium have established that estrogens can facilitate dissociation of eNOS from caveolae lipid rafts, leading to NO production and vasodilation [109,110]. Exhaled NO is used as an indicator of airway inflammation [111,112], and BECs are the major source of NO in the airway. Acute exposure to estradiol increases conversion of $^{[3H]}$L-arginine to $^{[3H]}$L-citrulline through eNOS activation in immortalized epithelial cell lines [113]. We and others have shown constitutive expression of ERα and ERβ in human airway epithelial cells [63,113,114], and their involvement in increasing NO production via eNOS [63]. Importantly, we showed in epithelium-intact human bronchial rings that E_2 can induce rapid relaxation of agonist-induced bronchoconstriction: an effect that substantially involves the epithelium [63]. A recent study also showed that nongenomic estrogen effects in the epithelium can involve inhibition of basal stromal interaction molecule 1 (STIM1) phosphorylation and reduction in store-operated Ca^{2+} entry (SOCE) [115]. Estrogen also increases posttranslational modification of mucins [116]. Limited studies also suggest that estrogens can promote proliferation of immortalized BECs [62]. PR expression is limited to the proximal region of the cilia of airway epithelia, with PR-B being more abundant. Progesterone decreases ciliary beat frequency, while coadministration of 17β-estradiol interestingly inhibits this effect [117].

The differences in the sex steroid effects have also been observed in the alveolar epithelial cells. For example, androgens are involved in the maturation of alveolar type II epithelial cells and the consequent production of surfactant in male fetuses [118–120]. The expressions of ERs, along with PR and AR have

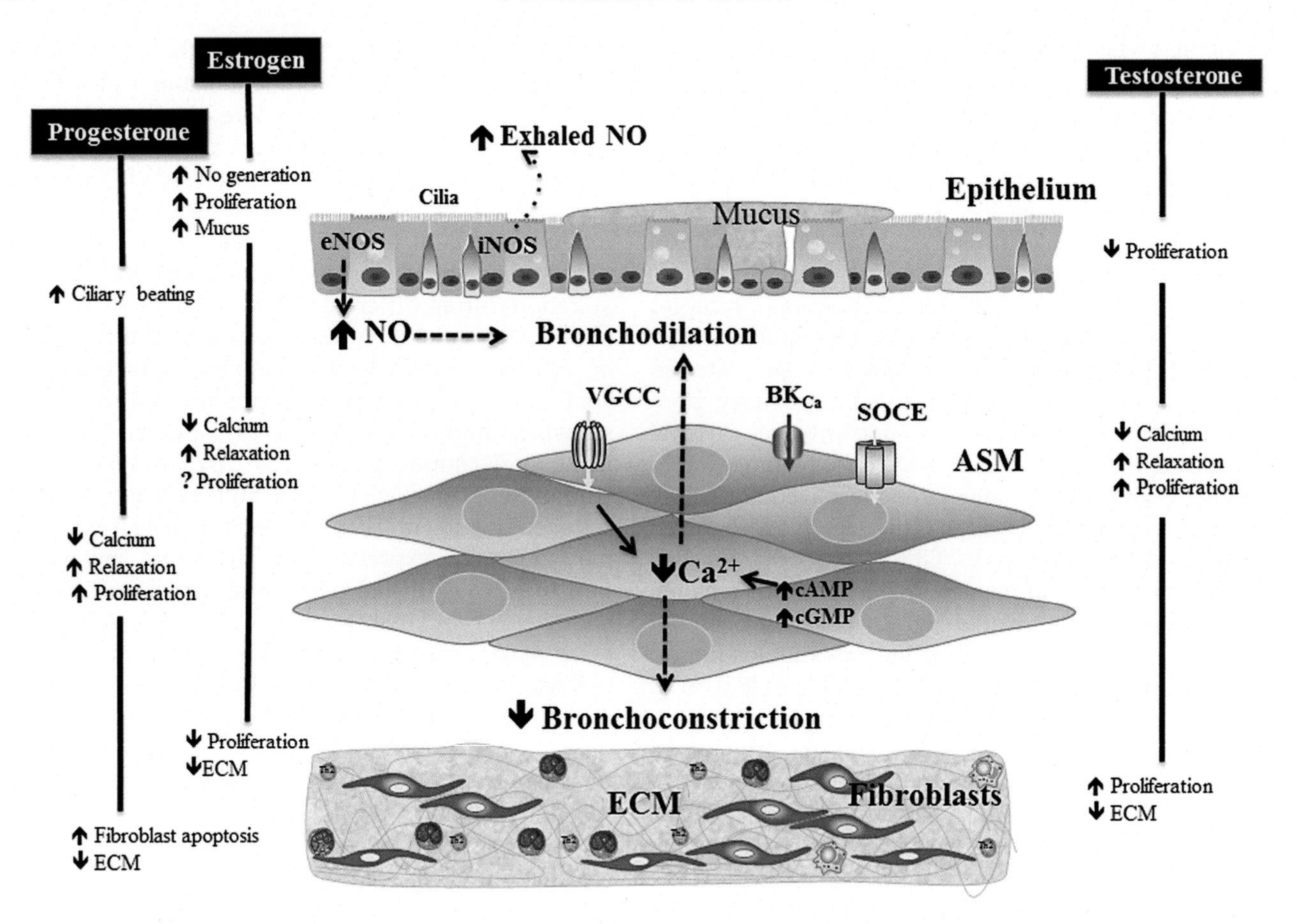

FIGURE 6.1 Schematic summarizing sex differences in pulmonary physiology at the cellular level, and the effects of sex steroids (estrogen, progesterone, and testosterone) on different cell types.

been observed in the alveolar epithelial layer of lung biopsies [70,121,122]. Additionally, the enzymes aromatase and 17b-hydroxysteroid dehydrogenase are also observed in the alveolar epithelium [123,124], demonstrating the local effects of sex steroids and their metabolites in the alveolar epithelium.

Alveolar fluid clearance maintains gas (O_2 and CO_2) exchange across the alveolar epithelium. Many conditions and diseases affecting alveolar fluid clearance (eg, cystic fibrosis, pulmonary edema, and acute lung injury) also appear to display differences between males and females [2,125–130]. For example, males exhibit greater incidence and severity of preterm infant respiratory distress syndrome [131–133] and studies have shown that treatment with E_2 and P4 reduces the sex disparity in infants [134,135].

Maintaining appropriate fluid levels at the alveolar epithelium requires robust regulation of Na^+ reabsorption and Cl^- secretion, without which the hydrostatic pressure across the pulmonary capillaries would flood the alveolar space within minutes. In rats, alveolar fluid clearance in females exceeds that of males and is sensitive to amiloride, an inhibitor of the epithelial sodium channel (ENaC) [136]. Several studies show that E_2 increases Na^+ reabsorption via the ENaC and its associated nonselective cation channels (NSCs) in alveolar epithelial cells and that E_2 decreases Cl^- secretion via the cystic fibrosis transmembrane receptor (CFTR) [115,137–143], suggesting that sex hormones are responsible for the observed sex difference in alveolar fluid clearance.

E_2, at physiological concentrations <1 nM, appears to predominantly act via nongenomic mechanisms to alter alveolar ion transport. The open probability of highly Na^+-selective ENaC and the amiloride-sensitive NSC increased in response to 0.7 nM E_2 within minutes and this effect was inhibited by G-1, a G-protein coupled ER antagonist [139]. Other studies have shown that, E_2 (1–3 nM), through inhibition of CFTR and other Cl^- secreting channels, reduced alveolar surface volume in human adult airway epithelial monolayers and had an even greater effect in cells from CF patients [137,140]. Consistent with these results, females with CF display changes in lung function across the menstrual cycle [144].

Studies on the sex steroid effects in other disease states such as emphysema are limited. Some animal studies have suggested that administration of progesterone or medroxyprogesterone is able to reverse emphysematic changes in lung airspaces [145,146]. Other studies also reported a similar positive effect of progesterone on papain-induced emphysema [147,148]. Progesterone administration to emphysema patients also improved symptoms and reduced hypercapnia [149]. Interestingly, the mechanisms underlying these effects were not established and thus, it is not known whether these effects can be attributed to sex steroid signaling in alveolar epithelium per se.

Sex Steroid Effects in Airway Smooth Muscle

Diseases such as asthma are characterized by airway hyperresponsiveness, inflammation, and remodeling [150–153]. A modulatory role for sex steroids in airway smooth muscle (ASM) structure function has been shown in a limited number of studies over several years, but focus on this area is now increasing with renewed interest regarding asthma in women. Foster et al. studied the effects of different sex steroids on potentiating isoprenaline-induced relaxation of isolated pig bronchus [154] and found estradiol to be the strongest potentiator of the bronchodilatory response to β-adrenoreceptor stimulation [154], although the role of ASM per se is not clear from this study. Previous in vivo studies using ER knockout mice or ovariectomized mice with estrogen replacement therapy suggest that sex steroids, particularly estrogen, promote bronchodilation [40,155–158], but a potential caveat in these reports is the use of nonphysiological concentrations of sex steroids. Nonetheless, some studies in ovariectomized rats suggest an epithelium-dependent mechanism for estrogen-induced bronchodilation [155], while others suggest epithelium-independent mechanisms [159–161]. Here, estradiol-induced prostaglandin synthesis and cGMP-mediated relaxation may be involved [161–163]. We recently reported that physiologically relevant concentrations of estradiol (1–10 nM) increase cAMP in human ASM cells, and potentiate β-adrenergic effect on intracellular Ca^{2+}, thus promoting bronchodilation [164,165].

While in vitro and limited in vivo work suggest estrogens can have bronchodilatory functions in the airway, further examination regarding sex differences in the context of asthma, or estrogen effects in the context of inflammation are less clear. Here, the complex effects of sex steroids on the immune system and other resident cell types make interpretation quite difficult. For example, numerous studies using murine models of allergic asthma show sex differences in airway reactivity and airway inflammation, however the data are contradictory [13,40,158,166–169]. Briefly, C57BL/6 male mice show more airway hyperresponsiveness compared to females, suggesting a "protective" role for estrogen [13], whereas female mice display more airway inflammation [170,171]. Additional animal studies suggest that estradiol markedly reduces carbachol-induced airway constriction through NO-cGMP-PKG signaling, which opens calcium-activated potassium channels [167]. Matsubara and colleagues found that male mice alone exhibit methacholine-induced airway hyperresponsiveness, and dose-dependent administration of estrogen in males diminishes airway hyperresponsiveness [158]. Carey et al. found that ERα knockout mice exhibit greater airway responsiveness to methacholine, connected to M2 muscarinic receptor dysfunction [172]. While these studies could suggest a bronchodilatory, protective effect of estrogens at the level of ASM, a fundamental limitation is that clinical data still show that women have greater asthma than men: a discrepancy that is not yet well-explained mechanistically.

The physiological and biochemical mechanisms by which estrogens directly influence ASM are still under exploration. We have shown that physiologically relevant concentrations of estrogens (~100 pM) markedly decrease Ca^{2+} responses to bronchoconstrictor agonist (ACh and histamine) in human ASM cells [64]. We further showed full-length expression ERα and ERβ in ASM, and that reductions in $[Ca^{2+}]_i$ are largely mediated by ERα, involving inhibition of L-type Ca^{2+} channels and reduction in SOCE [64,165] (Fig. 6.1). In addition to these nongenomic effects, limited studies show genomic signaling, for example, 2-methoxyestradiol (2-ME, estrogen metabolite) and its analogs have an inhibitory effect on human ASM cell proliferation [67].

In addition to estrogen, progesterone also potentiates the relaxing properties of isoprenaline in preconstricted pig bronchus [154], and is more potent than testosterone, although less effective than estradiol. Progesterone and 5β-pregnenalone prevents agonist-induced contraction of guinea pig trachea via inhibition of Ca^{2+} influx [173]. However, in vivo, in ovalbumin-sensitized male mice for example, progesterone exacerbates airway hyperresponsiveness [174].

Physiological concentrations of testosterone relax preconstricted rabbit tracheal smooth muscle in a dose-dependent manner, but this effect appears to be epithelium-dependent [175], and surprisingly does not appear to involve AR. On the other hand, the androgen 5α-DHT produces substantial relaxation in male guinea pig and bovine tracheal rings [176] via epithelium-independent mechanisms and involves voltage-dependent calcium channel inhibition [176,177].

Airway diseases such as asthma involve increases in smooth muscle proliferation and hyperplasia, which can further contribute to increased airway reactivity. Studies regarding sex steroid role of ASM proliferation and extracellular matrix (ECM) production are very limited. Stewart et al. showed that physiologic concentrations of testosterone, estradiol, and progesterone had no significant effect on thrombin-induced cell proliferation in human ASM [178]. However, testosterone and dehyroepiandrosterone have been shown to significantly inhibit rat tracheal smooth muscle proliferation [179]. In contrast to this observation, increased ASM proliferation by estrogen and testosterone has also been reported [180]. In ongoing studies, we have found that ER activation leads to inhibition of human ASM proliferation in the context of inflammation, and may involve differential ERα versus ERβ activation [181]. In fetal ASM cells, the estrogen metabolizing enzyme CYP1a1 may contribute to estradiol-induced inhibition of cell proliferation [182], and thus local estrogen metabolism may be important.

Sex Steroid Effects in Fibroblasts

Studies on sex steroid receptor expression and function in lung fibroblasts are also limited and conflicting. However, interest in this area of research is rapidly growing due to the potential role of sex steroids in pulmonary fibrosis. Male mice show decreased lung function and increased fibrosis in bleomycin models of pulmonary fibrosis [183]. Castration restores lung function, whereas DHT replacement therapy worsens it, indicating the possible detrimental effect of androgens [183]. Ovariectomy significantly reduces lung collagen deposition in bleomycin-induced lung fibrosis in female rats [184], but other studies suggest that fibrosis induced by bleomycin is resistant to treatment with 2-ME [185], while some show a protective role for sex steroids in pulmonary fibrosis [186,187]. Antiproliferative effects of estrogens may involve the Raf1-ERK-MAPK pathway [188], and a role for ERβ [189].

Sex Steroid Effects in Airway Nerves

The important role of nerves in the airway include defensive reflexes such as sneezing and coughing in response to noxious stimuli in the upper airways, cholinergic, parasympathetic nerve-mediated bronchoconstriction, and noncholinergic, parasympathetic nerve-mediated bronchodilation [190]. Although sex differences in ventilatory responses with age, variations with menstrual cycle and in pregnancy have all been reported [191–194], sex differences and sex steroid modulation of innervation and neuronal pathways are less clear. Neurosteriods secreted by the peripheral nervous system [195,196] can modulate neurons and glia cells, for example, progesterone is a known respiratory stimulant in pregnancy [197], in patients with breathing disorders [198–200], and in sleep-disordered breathing in postmenopausal women [201–203]. Some studies show that testosterone supplementation in hypogondal men increases obstructive sleep apnea [193] and modulates ventilation, while others show the opposite [193,204]. An additional source of nerve-related pulmonary regulation comes from nonadrenergic, noncholinergic (NANC) innervation involving both excitatory neurotransmitters (neurokinin A, substance P, calcitonin gene-related peptides, etc.) and inhibitory neurotransmitters (vasoactive intestinal peptide and NO) [205–211]. While neurally derived NO can modulate urogenitary [212,213] and gastrointestinal [214] smooth muscles, there are no data on how estrogen influences NANC in regards to ASM function. However, increased nasal cholinergic nerve activity has been noted in pregnant women [215]. Isolated tracheas from ERα knockout mice showed increased ACh release, suggesting the contribution of estrogen signaling in the neurally derived signals [172].

Sex Steroid Effects in Immune Cells

Sex steroids are thought to be involved in regulating multiple immune cells such as dendritic cells, lymphocytes, regulatory T-cells, and B lymphocytes, all of which may be relevant to lung function, particularly in the context of recruitment of immune cells in asthma and modulating of the inflammatory environment. Sex differences in immune responses and their regulation have been reported in several studies [216–218]. Increased female prevalence for autoimmune diseases (systemic lupus erythematosus, rheumatoid arthritis, and multiple sclerosis), and a higher ratio for allergic asthma in females collectively suggest that female sex hormones could play a major role in immunologic airway inflammation. Women usually show greater humoral responses, but in men both humoral and cell-mediated immune responses are relatively suppressed by sex steroids [81]. However, it is important to note that sex steroid effects on immune cells are likely dependent on steroid concentration, timing, duration, and context of exposure [219].

SEX DIFFERENCES IN DISEASES OF THE LUNG

As mentioned earlier, clinical and epidemiological studies suggest significant sex differences in the natural history of asthma. The primary observation is that prepubertal boys have a higher incidence of asthma compared to girls, whereas following puberty and throughout the reproductive years, women have more asthma [1,2,16], with severity modulated by life events such as the menstrual cycle and pregnancy [220]. Although sex differences in asthma are reduced in the later stages of life, some reports suggest increased asthma prevalence in aging men [221,222]. In addition to asthma, COPD incidence also appears to be increasing in women [223,224], with increased susceptibility of women to cigarette smoke, who typically show greater conducting airway disease compared to parenchymal changes in males. What is not known is how biological susceptibility, age-related variations with hormonal milieu, environmental exposures, and socioeconomic factors could play a role in these differences.

The link between sex steroids and airway reactivity is certainly not straightforward. For example, during early menarche, girls are at higher risk for asthma [221]. However, girls with Turner's syndrome and low serum levels of estrogen also exhibit greater airway responsiveness, while exogenous estrogen administration reduces airway hyperresponsiveness in these patients. Although cyclical variations in estrogens and progesterone levels during the menstrual cycle can impact asthma symptoms, some studies note that women with moderate asthma have reduced premenstrual asthma exacerbations with oral contraceptive therapy, which reduce large variations in circulating hormones [225,226]. A recent study investigated changes in bronchial reactivity of asthmatic women during the luteal and follicular phases of the menstrual cycle and suggested a link between Peoria-menstrual asthma and testosterone levels in sputum samples [227]. Asthma also complicates 8–13% of pregnancies [228], but although estrogen and progesterone concentrations increase progressively and peak through the third trimester [16], acute asthmatic exacerbations are observed in only about 10% of these women, while ~30% display improvement, another 30% exhibit worsening, and no changes are observed in the remainder [229]. It is also important to note here that, fetal sex may influence asthma during pregnancy [230,231].

In contrast to human studies, animal studies more tend to show that estrogen protects against airway hyperresponsiveness [158,167]. In addition, spontaneous airway hyperresponsiveness has been reported in ERα knockout mice [172]. Other animal studies suggest that androgens also play a protective role, while estrogens are pro-inflammatory [168,170,171,232]. Exogenous progesterone aggravates allergic airway disease in mice [174].

In terms of COPD, clinical studies suggest that women may be biologically more vulnerable to lung damage triggered by tobacco smoke and environmental pollutants [7]. Women develop COPD after smoking less number of cigarettes in their lifetime [8,233], and demonstrate racial and sex differences in manifestation of COPD [234], as well as proteomic differences in their bronchoalveolar lavage suggesting differences in putative molecular pathways [235].

In contrast to asthma and COPD, increased prevalence of pulmonary fibrosis is observed in males compared to females [236] with faster progression, and lower survival [237,238]. However, animal data suggest increased mortality and fibrosis in female rats compared to males, and ovariectomy is protective whereas estrogen exacerbates disease [184]. Some studies do show that androgens play an exacerbating role in the diminished lung function induced by bleomycin [183,239].

IMPLICATIONS OF SEX DIFFERENCES IN THE LUNG

In this current chapter, we have highlighted the potential importance of sex differences in lung structure and function along with the possible influence of sex steroids and their signaling in normal and diseased lung (Table 6.1). Clinical implications of sex differences in lung structure and function have been integrated

TABLE 6.1 Sex Differences in the Pulmonary Diseases

Disease	Prevalence and/or Severity	References
Childhood asthma	M>F	[2,10–12,14–16, 240,241]
Adult asthma	F>M	
COPD: bronchitis	F>M	[7,145,184,242,243]
COPD: emphysema	M>F	
Preterm infant respiratory distress syndrome and bronchopulmonary dysplasia	M>F	[244–246]
Pulmonary hypertension	F>M	[182,247–251]
Cystic fibrosis	M=F	[144,184,252–254]
Pulmonary fibrosis	M>F	[183,236,255]
Childhood allergic rhinitis	M>F	[256–259]
Adult allergic rhinitis	M=F	
Lung cancer: adenocarcinoma	F>M	[121,122,260,261]
Lung cancer: squamous cell	M>F	

into some medical settings such as the use of sex-specific values of lung function adjusted for size and age, predicted values for common variables in lung function tests, etc. Although sex differences are clearly present, and steroids likely influence airway structure and function, the substantial mechanistic discrepancy (where studied) and lack of information and clarity regarding protective versus detrimental roles of sex steroids are also clear.

Sex steroids can influence lung structure and function throughout life. Variations in sex steroid levels in fetal life can impact structural development of the lung with substantial postnatal consequences, represented by early childhood diseases such as respiratory distress syndrome and bronchopulmonary dysplasia, and childhood asthma. It is also important to note maternal smoking effect in utero, and childhood tobacco smoke exposure, especially considering increased numbers of smoking women and higher susceptibility to cigarette smoke effect. The role of sex steroids in women during life stages such as menstrual cycle and pregnancy remain to be clarified, but have implications for premenstrual asthma, and modulation of asthma in the course of pregnancy. In this regard, the possible dual effects of estrogens on inflammation versus bronchoconstriction and bronchodilation remain to be clarified. While estrogens and progesterone sometimes yield similar effects in the lung, they may also produce opposing effects that would color interpretation regarding the beneficial versus detrimental roles of sex steroids. Importantly, the relative concentrations, specific receptor expression, and the context in which exposure occurs, as well as interactions between sex steroids may all determine the overall effect of sex steroids in women. Added to that complexity is tissue metabolism and conversion of sex steroids, for example, testosterone to estradiol via aromatase in aged men.

CONCLUSION

In conclusion, intrinsic sex differences are important to lung structure and function throughout life. Sex steroids play modulating roles that are incompletely understood. The implications of intrinsic and steroid-induced differences in the lung lie in normal pulmonary physiology across the age, at specific life stages such as puberty, pregnancy, menopause, and aging, and in diseases such as asthma, COPD, and fibrosis. Understanding these concepts is important not only to appreciating intrinsic sex differences, but also for the development of sex-specific markers and therapeutic strategies in lung disease.

Acknowledgment

Supported by Flight Attendants Medical Research Institute (VS), NIH grants HL123494 (VS), HL056470, and HL088029 (YSP).

References

[1] Sathish V, Martin YN, Prakash YS. Sex steroid signaling: implications for lung diseases. Pharmacol Ther 2015;150:94–108.
[2] Townsend EA, Miller VM, Prakash YS. Sex differences and sex steroids in lung health and disease. Endocr Rev 2012;33:1–47.
[3] Martin YN, Pabelick CM. Sex differences in the pulmonary circulation: implications for pulmonary hypertension. Am J Physiol Heart Circ Physiol 2014;306:H1253–64.
[4] Tam A, Morrish D, Wadsworth S, Dorscheid D, Man S, Sin D. The role of female hormones on lung function in chronic lung diseases. BMC Womens Health 2011;11:24.
[5] Verma MK, Miki Y, Sasano H. Sex steroid receptors in human lung diseases. J Steroid Biochem Mol Biol 2011;127:216–22.
[6] Dimitropoulou C, Drakopanagiotakis F, Catravas JD. Estrogen as a new therapeutic target for asthma and chronic obstructive pulmonary disease. Drug News Perspect 2007;20:241–52.
[7] Ala D. Taking her breath away: the rise of COPD in women. Am Lung Assoc 2013;1–26. Retrieved from: <http://www.lung.org/our-initiatives/research/lung-health-disparities/the-rise-of-copd-in-women.html>.
[8] Gold DR, Wang X, Wypij D, Speizer FE, Ware JH, Dockery DW. Effects of cigarette smoking on lung function in adolescent boys and girls. N Engl J Med 1996;335:931–7.
[9] McDuffie HH, Klaassen DJ, Dosman JA. Female-male differences in patients with primary lung cancer. Cancer 1987;59:1825–30.
[10] Becklake MR, Kauffmann F. Gender differences in airway behaviour over the human life span. Thorax 1999;54:1119–38.
[11] Bjornson CL, Mitchell I. Gender differences in asthma in childhood and adolescence. J Gend Specif Med 2000;3:57–61.
[12] Caracta CF. Gender differences in pulmonary disease. Mt Sinai J Med 2003;70:215–24.
[13] Card JW, Carey MA, Bradbury JA, DeGraff LM, Morgan DL, Moorman MP, et al. Gender differences in murine airway responsiveness and lipopolysaccharide-induced inflammation. J Immunol 2006;177:621–30.
[14] Carey MA, Card JW, Voltz JW, Germolec DR, Korach KS, Zeldin DC. The impact of sex and sex hormones on lung physiology and disease: lessons from animal studies. Am J Physiol Lung Cell Mol Physiol 2007;293:L272–8.
[15] Jensen-Jarolim E, Untersmayr E. Gender-medicine aspects in allergology. Allergy 2008;63:610–15.
[16] Melgert BN, Ray A, Hylkema MN, Timens W, Postma DS. Are there reasons why adult asthma is more common in females? Curr Allergy Asthma Rep 2007;7:143–50.
[17] Kim S, Kim J, Park SY, Um HY, Kim K, Kim Y, et al. Effect of pregnancy in asthma on health care use and perinatal outcomes. J Allergy Clin Immunol 2015;136(5):1215–23. e1.
[18] Murphy VE, Gibson PG. Premenstrual asthma: prevalence, cycle-to-cycle variability and relationship to oral contraceptive use and menstrual symptoms. J Asthma 2008;45:696–704.
[19] Murphy VE, Clifton VL, Gibson PG. Asthma exacerbations during pregnancy: incidence and association with adverse pregnancy outcomes. Thorax 2006;61:169–76.
[20] Dezateux C, Stocks J. Lung development and early origins of childhood respiratory illness. Br Med Bull 1997;53:40–57.
[21] Merkus PJ, ten Have-Opbroek AA, Quanjer PH. Human lung growth: a review. Pediatr Pulmonol 1996;21:383–97.

[22] Stocks J, Dezateux C, Hoo AF, Rabbette PS, Costeloe K, Wade A. Delayed maturation of Hering-Breuer inflation reflex activity in preterm infants. Am J Respir Crit Care Med 1996;154:1411–17.
[23] Koos BJ, Rajaee A. Fetal breathing movements and changes at birth. Adv Exp Med Biol 2014;814:89–101.
[24] Stocks J, Hislop A, Sonnappa S. Early lung development: lifelong effect on respiratory health and disease. Lancet Respir Med 2013;1:728–42.
[25] Reyburn B, Martin RJ, Prakash YS, MacFarlane PM. Mechanisms of injury to the preterm lung and airway: implications for long-term pulmonary outcome. Neonatology 2012;101:345–52.
[26] Britt Jr RD, Faksh A, Vogel E, Martin RJ, Pabelick CM, Prakash YS. Perinatal factors in neonatal and pediatric lung diseases. Expert Rev Respir Med 2013;7:515–31.
[27] Thurlbeck WM. Postnatal human lung growth. Thorax 1982;37:564–71.
[28] Thurlbeck WM, Angus GE. Growth and aging of the normal human lung. Chest 1975;67:3S–6S.
[29] Abramovich DR. Human sexual differentiation–in utero influences. J Obstet Gynaecol Br Commonw 1974;81:448–53.
[30] Pasqualini JR. Enzymes involved in the formation and transformation of steroid hormones in the fetal and placental compartments. J Steroid Biochem Mol Biol 2005;97:401–15.
[31] Seaborn T, Simard M, Provost PR, Piedboeuf B, Tremblay Y. Sex hormone metabolism in lung development and maturation. Trends Endocrinol Metab 2010;21:729–38.
[32] Fleisher B, Kulovich MV, Hallman M, Gluck L. Lung profile: sex differences in normal pregnancy. Obstet Gynecol 1985;66:327–30.
[33] Torday JS, Nielsen HC. The sex difference in fetal lung surfactant production. Exp Lung Res 1987;12:1–19.
[34] Kimura Y, Suzuki T, Kaneko C, Darnel AD, Akahira J, Ebina M, et al. Expression of androgen receptor and 5alpha-reductase types 1 and 2 in early gestation fetal lung: a possible correlation with branching morphogenesis. Clin Sci (Lond) 2003;105: 709–13.
[35] Takeyama J, Suzuki T, Inoue S, Kaneko C, Nagura H, Harada N, et al. Expression and cellular localization of estrogen receptors alpha and beta in the human fetus. J Clin Endocrinol Metab 2001;86:2258–62.
[36] Brandenberger AW, Tee MK, Lee JY, Chao V, Jaffe RB. Tissue distribution of estrogen receptors alpha (ER-alpha) and beta (ER-beta) mRNA in the midgestational human fetus. J Clin Endocrinol Metab 1997;82:3509–12.
[37] Beyer C, Kuppers E, Karolczak M, Trotter A. Ontogenetic expression of estrogen and progesterone receptors in the mouse lung. Biol Neonate 2003;84:59–63.
[38] Massaro D, Massaro GD. Estrogen receptor regulation of pulmonary alveolar dimensions: alveolar sexual dimorphism in mice. Am J Physiol Lung Cell Mol Physiol 2006;290:L866–70.
[39] Massaro GD, Mortola JP, Massaro D. Estrogen modulates the dimensions of the lung's gas-exchange surface area and alveoli in female rats. Am J Physiol 1996;270:L110–14.
[40] Carey MA, Card JW, Voltz JW, Germolec DR, Korach KS, Zeldin DC. The impact of sex and sex hormones on lung physiology and disease: lessons from animal studies. Am J Physiol Lung Cell Mol Physiol 2007;293:L272–8.
[41] Hoffstein V. Relationship between lung volume, maximal expiratory flow, forced expiratory volume in one second, and tracheal area in normal men and women. Am Rev Respir Dis 1986;134:956–61.
[42] Pagtakhan RD, Bjelland JC, Landau LI, Loughlin G, Kaltenborn W, Seeley G, et al. Sex differences in growth patterns of the airways and lung parenchyma in children. J Appl Physiol 1984;56:1204–10.
[43] Mead J. Dysanapsis in normal lungs assessed by the relationship between maximal flow, static recoil, and vital capacity. Am Rev Respir Dis 1980;121:339–42.
[44] Doershuk CF, Fisher BJ, Matthews LW. Specific airway resistance from the perinatal period into adulthood. Alterations in childhood pulmonary disease. Am Rev Respir Dis 1974;109:452–7.
[45] Martin JA, Hamilton BE, Sutton PD, Ventura SJ, Menacker F, Kirmeyer S. Births: final data for 2004. Natl Vital Stat Rep 2006;55:1–101.
[46] Hibbert M, Lannigan A, Raven J, Landau L, Phelan P. Gender differences in lung growth. Pediatr Pulmonol 1995;19:129–34.
[47] Schrader PC, Quanjer PH, Olievier IC. Respiratory muscle force and ventilatory function in adolescents. Eur Respir J 1988;1:368–75.
[48] Garcia-Rio F, Pino-Garcia JM, Serrano S, Racionero MA, Terreros-Caro JG, Alvarez-Sala R, et al. Comparison of helium dilution and plethysmographic lung volumes in pregnant women. Eur Respir J 1997;10:2371–5.
[49] Gee JB, Packer BS, Millen JE, Robin ED. Pulmonary mechanics during pregnancy. J Clin Invest 1967;46:945–52.
[50] Bode FR, Dosman J, Martin RR, Ghezzo H, Macklem PT. Age and sex differences in lung elasticity, and in closing capacity in nonsmokers. J Appl Physiol 1976;41:129–35.
[51] Pelzer AM, Thomson ML. Effect of age, sex, stature, and smoking habits on human airway conductance. J Appl Physiol 1966;21:469–76.
[52] Brody JS. Cell-to-cell interactions in lung development. Pediatr Pulmonol 1985;1:S42–8.
[53] Carey MA, Card JW, Voltz JW, Arbes Jr SJ, Germolec DR, Korach KS, et al. It's all about sex: gender, lung development and lung disease. Trends Endocrinol Metab 2007;18:308–13.
[54] Payne AH, Hales DB. Overview of steroidogenic enzymes in the pathway from cholesterol to active steroid hormones. Endocr Rev 2004;25:947–70.
[55] Lahm T, Crisostomo PR, Markel TA, Wang M, Weil BR, Novotny NM, et al. The effects of estrogen on pulmonary artery vasoreactivity and hypoxic pulmonary vasoconstriction: potential new clinical implications for an old hormone. Crit Care Med 2008;36:2174–83.
[56] Gillies GE, McArthur S. Estrogen actions in the brain and the basis for differential action in men and women: a case for sex-specific medicines. Pharmacol Rev 2010;62:155–98.
[57] Lamont KR, Tindall DJ. Androgen regulation of gene expression. Adv Cancer Res 2010;107:137–62.
[58] Miller VM, Duckles SP. Vascular actions of estrogens: functional implications. Pharmacol Rev 2008;60:210–41.
[59] Simoncini T, Genazzani AR. Non-genomic actions of sex steroid hormones. Eur J Endocrinol 2003;148:281–92.
[60] Edwards DP. Regulation of signal transduction pathways by estrogen and progesterone. Annu Rev Physiol 2005;67:335–76.
[61] Couse J, Lindzey J, Grandien K, Gustafsson J, Korach K. Tissue distribution and quantitative analysis of estrogen receptor-alpha (ERalpha) and estrogen receptor-beta (ERbeta) messenger ribonucleic acid in the wild-type and ERalpha-knockout mouse. Endocrinology 1997;138:4613–21.
[62] Ivanova MM, Mazhawidza W, Dougherty SM, Klinge CM. Sex differences in estrogen receptor subcellular location and activity in lung adenocarcinoma cells. Am J Respir Cell Mol Biol 2010;42:320–30.
[63] Townsend EA, Meuchel LW, Thompson MA, Pabelick CM, Prakash YS. Estrogen increases nitric-oxide production in human bronchial epithelium. J Pharmacol Exp Ther 2011;339:815–24.

[64] Townsend EA, Thompson MA, Pabelick CM, Prakash YS. Rapid effects of estrogen on intracellular $Ca2^+$ regulation in human airway smooth muscle. Am J Physiol Lung Cell Mol Physiol 2010;298:L521–30.

[65] Jia S, Zhang X, He DZZ, Segal M, Berro A, Gerson T, et al. Expression and Function of a Novel Variant of Estrogen Receptor–α36 in Murine Airways. Am J Respir Cell Mol Biol 2011;45:1084–9.

[66] Dahlman-Wright K, Cavailles V, Fuqua SA, Jordan VC, Katzenellenbogen JA, Korach KS, et al. International union of pharmacology. LXIV. Estrogen receptors. Pharmacol Rev 2006;58:773–81.

[67] Hughes RA, Harris T, Altmann E, McAllister D, Vlahos R, Robertson A, et al. 2-Methoxyestradiol and analogs as novel antiproliferative agents: analysis of three-dimensional quantitative structure-activity relationships for DNA synthesis inhibition and estrogen receptor binding. Mol Pharmacol 2002;61:1053–69.

[68] Abraham Z, Lourdes A, Christian G, Martha O, Maria C. Expression and regulation of estrogen, progesterone and androgen receptors in airway smooth muscle cells in an allergic asthma model. Am J Respir Crit Care Med 2011;183: A3651.

[69] Wilson CM, McPhaul MJ. A and B forms of the androgen receptor are expressed in a variety of human tissues. Mol Cell Endocrinol 1996;120:51–7.

[70] Mikkonen L, Pihlajamaa P, Sahu B, Zhang FP, Janne OA. Androgen receptor and androgen-dependent gene expression in lung. Mol Cell Endocrinol 2010;317:14–24.

[71] Zhang Z, Teng CT. Estrogen receptor-related receptor alpha 1 interacts with coactivator and constitutively activates the estrogen response elements of the human lactoferrin gene. J Biol Chem 2000;275:20837–46.

[72] Hall JM, McDonnell DP. The estrogen receptor beta-isoform (ERbeta) of the human estrogen receptor modulates ERalpha transcriptional activity and is a key regulator of the cellular response to estrogens and antiestrogens. Endocrinol 1999;140:5566–78.

[73] Bennett NC, Gardiner RA, Hooper JD, Johnson DW, Gobe GC. Molecular cell biology of androgen receptor signalling. Int J Biochem Cell Biol 2010;42:813–27.

[74] Giangrande PH, McDonnell DP. The A and B isoforms of the human progesterone receptor: two functionally different transcription factors encoded by a single gene. Recent Prog Horm Res 1999;54:291–313 discussion 313–294.

[75] Lee YF, Shyr CR, Thin TH, Lin WJ, Chang C. Convergence of two repressors through heterodimer formation of androgen receptor and testicular orphan receptor-4: a unique signaling pathway in the steroid receptor superfamily. Proc Natl Acad Sci USA 1999;96:14724–9.

[76] Zhou ZX, Wong CI, Sar M, Wilson EM. The androgen receptor: an overview. Recent Prog Horm Res 1994;49:249–74.

[77] Foradori CD, Weiser MJ, Handa RJ. Non-genomic actions of androgens. Front Neuroendocrinol 2008;29:169–81.

[78] Groner B, Ponta H, Beato M, Hynes NE. The proviral DNA of mouse mammary tumor virus: its use in the study of the molecular details of steroid hormone action. Mol Cell Endocrinol 1983;32:101–16.

[79] Papadopoulou N, Papakonstanti EA, Kallergi G, Alevizopoulos K, Stournaras C. Membrane androgen receptor activation in prostate and breast tumor cells: molecular signaling and clinical impact. IUBMB Life 2009;61:56–61.

[80] Heldring N, Pike A, Andersson S, Matthews J, Cheng G, Hartman J, et al. Estrogen receptors: how do they signal and what are their targets. Physiol Rev 2007;87:905–31.

[81] Gilliver SC. Sex steroids as inflammatory regulators. J Steroid Biochem Mol Biol 2010;120:105–15.

[82] Tofovic SP. Estrogens and development of pulmonary hypertension - interaction of estradiol metabolism and pulmonary vascular disease. J Cardiovasc Pharmacol 2010;56:696–708.

[83] Le Mellay V, Grosse B, Lieberherr M. Phospholipase C beta and membrane action of calcitriol and estradiol. J Biol Chem 1997;272:11902–7.

[84] Grosse B, Kachkache M, Le Mellay V, Lieberherr M. Membrane signalling and progesterone in female and male osteoblasts. I. Involvement of intracellular $Ca(2^+)$, inositol trisphosphate, and diacylglycerol, but not cAMP. J Cell Biochem 2000;79:334–45.

[85] Benten WP, Lieberherr M, Stamm O, Wrehlke C, Guo Z, Wunderlich F. Testosterone signaling through internalizable surface receptors in androgen receptor-free macrophages. Mol Biol Cell 1999;10:3113–23.

[86] Lieberherr M, Grosse B. Androgens increase intracellular calcium concentration and inositol 1,4,5-trisphosphate and diacylglycerol formation via a pertussis toxin-sensitive G-protein. J Biol Chem 1994;269:7217–23.

[87] Valverde MA, Rojas P, Amigo J, Cosmelli D, Orio P, Bahamonde MI, et al. Acute activation of Maxi-K channels (hSlo) by estradiol binding to the beta subunit. Science 1999;285:1929–31.

[88] Nakajima T, Kitazawa T, Hamada E, Hazama H, Omata M, Kurachi Y. 17beta-Estradiol inhibits the voltage-dependent L-type $Ca2^+$ currents in aortic smooth muscle cells. Eur J Pharmacol 1995;294:625–35.

[89] Minshall RD, Pavcnik D, Browne DL, Hermsmeyer K. Nongenomic vasodilator action of progesterone on primate coronary arteries. J Appl Physiol 2002;92:701–8.

[90] Crews JK, Khalil RA. Gender-specific inhibition of $Ca2^+$ entry mechanisms of arterial vasoconstriction by sex hormones. Clin Exp Pharmacol Physiol 1999;26:707–15.

[91] Deenadayalu VP, White RE, Stallone JN, Gao X, Garcia AJ. Testosterone relaxes coronary arteries by opening the large-conductance, calcium-activated potassium channel. Am J Physiol Heart Circ Physiol 2001;281:H1720–7.

[92] Nazareth LV, Weigel NL. Activation of the human androgen receptor through a protein kinase A signaling pathway. J Biol Chem 1996;271:19900–7.

[93] Bagowski CP, Myers JW, Ferrell Jr JE. The classical progesterone receptor associates with p42 MAPK and is involved in phosphatidylinositol 3-kinase signaling in Xenopus oocytes. J Biol Chem 2001;276:37708–14.

[94] Kato S, Endoh H, Masuhiro Y, Kitamoto T, Uchiyama S, Sasaki H, et al. Activation of the estrogen receptor through phosphorylation by mitogen-activated protein kinase. Science 1995;270:1491–4.

[95] Endoh H, Sasaki H, Maruyama K, Takeyama K, Waga I, Shimizu T, et al. Rapid activation of MAP kinase by estrogen in the bone cell line. Biochem Biophys Res Commun 1997;235:99–102.

[96] Pearson G, Barry P, Timmins C, Stickley J, Hocking M. Changes in the profile of paediatric intensive care associated with centralisation. Intensive Care Med 2001;27:1670–3.

[97] Migliaccio A, Piccolo D, Castoria G, Di Domenico M, Bilancio A, Lombardi M, et al. Activation of the Src/p21ras/Erk pathway by progesterone receptor via cross-talk with estrogen receptor. EMBO J 1998;17:2008–18.

[98] Simoncini T, Genazzani AR, Liao JK. Nongenomic mechanisms of endothelial nitric oxide synthase activation by the selective estrogen receptor modulator raloxifene. Circulation 2002;105:1368–73.

[99] Kousteni S, Bellido T, Plotkin LI, O'Brien CA, Bodenner DL, Han L, et al. Nongenotropic, sex-nonspecific signaling through the estrogen or androgen receptors: dissociation from transcriptional activity. Cell 2001;104:719–30.

[100] Migliaccio A, Castoria G, Di Domenico M, de Falco A, Bilancio A, Lombardi M, et al. Steroid-induced androgen receptor-oestradiol receptor beta-Src complex triggers prostate cancer cell proliferation. EMBO J 2000;19:5406–17.

[101] Shirasaki H, Watanabe K, Kanaizumi E, Konno N, Sato J, Narita S, et al. Expression and localization of steroid receptors in human nasal mucosa. Acta Otolaryngol 2004;124:958–63.

[102] Haeggstrom A, Ostberg B, Stjerna P, Graf P, Hallen H. Nasal mucosal swelling and reactivity during a menstrual cycle. ORL J Otorhinolaryngol Relat Spec 2000;62:39–42.

[103] Paulsson B, Gredmark T, Burian P, Bende M. Nasal mucosal congestion during the menstrual cycle. J Laryngol Otol 1997;111:337–9.

[104] Ellegard E, Karlsson G. Nasal congestion during the menstrual cycle. Clin Otolaryngol Allied Sci 1994;19:400–3.

[105] Pelikan Z. Possible immediate hypersensitivity reaction of the nasal mucosa to oral contraceptives. Ann Allergy 1978;40:211–19.

[106] Mortimer H, Wright RP, Collip JB. The effect of the administration of oestrogenic hormones on the nasal mucosa of the monkey (*Macaca mulatta*). Can Med Assoc J 1936;35:503–13.

[107] Taylor M. An experimental study of the influence of the endocrine system on the nasal respiratory mucosa. J Laryngol Otol 1961;75:972–7.

[108] Hamano N, Terada N, Maesako K, Ikeda T, Fukuda S, Wakita J, et al. Expression of histamine receptors in nasal epithelial cells and endothelial cells—the effects of sex hormones. Int Arch Allergy Immunol 1998;115:220–7.

[109] Kim KH, Moriarty K, Bender JR. Vascular cell signaling by membrane estrogen receptors. Steroids 2008;73:864–9.

[110] Sud N, Wiseman DA, Black SM. Caveolin 1 is required for the activation of endothelial nitric oxide synthase in response to 17beta-estradiol. Mol Endocrinol 2010;24:1637–49.

[111] Mandhane PJ, Hanna SE, Inman MD, Duncan JM, Greene JM, Wang HY, et al. Changes in exhaled nitric oxide related to estrogen and progesterone during the menstrual cycle. Chest 2009;136:1301–7.

[112] Jatakanon A, Lim S, Kharitonov SA, Chung KF, Barnes PJ. Correlation between exhaled nitric oxide, sputum eosinophils, and methacholine responsiveness in patients with mild asthma. Thorax 1998;53:91–5.

[113] Kirsch EA, Yuhanna IS, Chen Z, German Z, Sherman TS, Shaul PW. Estrogen acutely stimulates endothelial nitric oxide synthase in H441 human airway epithelial cells. Am J Respir Cell Mol Biol 1999;20:658–66.

[114] Ivanova M, Mazhawidza W, Dougherty S, Minna J, Klinge C. Activity and intracellular location of estrogen receptors alpha and beta in human bronchial epithelial cells. Mol Cell Endocrinol 2009;305:12–21.

[115] Sheridan JT, Gilmore RC, Watson MJ, Archer CB, Tarran R. 17beta-Estradiol inhibits phosphorylation of stromal interaction molecule 1 (STIM1) protein: implication for store-operated calcium entry and chronic lung diseases. J Biol Chem 2013;288:33509–18.

[116] Tam A, Wadsworth S, Dorscheid D, Man SF, Sin DD. Estradiol increases mucus synthesis in bronchial epithelial cells. PLoS One 2014;9:e100633.

[117] Jain R, Ray JM, Pan J-h, Brody SL. Sex hormone—dependent regulation of cilia beat frequency in airway epithelium. Am J Respir Cell Mol Biol 2012;46:446–53.

[118] Dammann CE, Ramadurai SM, McCants DD, Pham LD, Nielsen HC. Androgen regulation of signaling pathways in late fetal mouse lung development. Endocrinology 2000;141:2923–9.

[119] Provost PR, Simard M, Tremblay Y. A link between lung androgen metabolism and the emergence of mature epithelial type II cells. Am J Respir Crit Care Med 2004;170:296–305.

[120] Nielsen HC. Androgen receptors influence the production of pulmonary surfactant in the testicular feminization mouse fetus. J Clin Invest 1985;76:177–81.

[121] Marquez-Garban DC, Chen HW, Fishbein MC, Goodglick L, Pietras RJ. Estrogen receptor signaling pathways in human non-small cell lung cancer. Steroids 2007;72:135–43.

[122] Ishibashi H, Suzuki T, Suzuki S, Niikawa H, Lu L, Miki Y, et al. Progesterone receptor in non-small cell lung cancer—a potent prognostic factor and possible target for endocrine therapy. Cancer Res 2005;65:6450–8.

[123] Plante J, Simard M, Rantakari P, Cote M, Provost PR, Poutanen M, et al. Epithelial cells are the major site of hydroxysteroid (17beta) dehydrogenase 2 and androgen receptor expression in fetal mouse lungs during the period overlapping the surge of surfactant. J Steroid Biochem Mol Biol 2009;117:139–45.

[124] Marquez-Garban DC, Chen HW, Goodglick L, Fishbein MC, Pietras RJ. Targeting aromatase and estrogen signaling in human non-small cell lung cancer. Ann N Y Acad Sci 2009;1155:194–205.

[125] Bastarache JA, Ong T, Matthay MA, Ware LB. Alveolar fluid clearance is faster in women with acute lung injury compared to men. J Crit Care 2011;26:249–56.

[126] Klein SL, Passaretti C, Anker M, Olukoya P, Pekosz A. The impact of sex, gender and pregnancy on 2009 H1N1 disease. Biol Sex Differ 2010;1:5.

[127] Olesen HV, Pressler T, Hjelte L, Mared L, Lindblad A, Knudsen PK, et al. Gender differences in the Scandinavian cystic fibrosis population. Pediatr Pulmonol 2010;45:959–65.

[128] Serfling RE, Sherman IL, Houseworth WJ. Excess pneumonia-influenza mortality by age and sex in three major influenza A2 epidemics, United States, 1957–58, 1960 and 1963. Am J Epidemiol 1967;86:433–41.

[129] Stephenson A, Hux J, Tullis E, Austin PC, Corey M, Ray J. Higher risk of hospitalization among females with cystic fibrosis. J Cyst Fibros 2011;10:93–9.

[130] Wolk KE, Lazarowski ER, Traylor ZP, Yu EN, Jewell NA, Durbin RK, et al. Influenza A virus inhibits alveolar fluid clearance in BALB/c mice. Am J Respir Crit Care Med 2008;178:969–76.

[131] Elsmen E, Hansen Pupp I, Hellstrom-Westas L. Preterm male infants need more initial respiratory and circulatory support than female infants. Acta Paediatr 2004;93:529–33.

[132] Farrell EE, Silver RK, Kimberlin LV, Wolf ES, Dusik JM. Impact of antenatal dexamethasone administration on respiratory distress syndrome in surfactant-treated infants. Am J Obstet Gynecol 1989;161:628–33.

[133] Kovar J, Waddell BJ, Sly PD, Willet KE. Sex differences in response to steroids in preterm sheep lungs are not explained by glucocorticoid receptor number or binding affinity. Pediatr Pulmonol 2001;32:8–13.

[134] Trotter A, Ebsen M, Kiossis E, Meggle S, Kueppers E, Beyer C, et al. Prenatal estrogen and progesterone deprivation impairs alveolar formation and fluid clearance in newborn piglets. Pediatr Res 2006;60:60–4.

[135] Trotter A, Maier L, Grill HJ, Kohn T, Heckmann M, Pohlandt F. Effects of postnatal estradiol and progesterone replacement in extremely preterm infants. J Clin Endocrinol Metab 1999;84:4531–5.

[136] Kooijman EE, Kuzenko SR, Gong D, Best MD, Folkesson HG. Phosphatidylinositol 4,5-bisphosphate stimulates alveolar epithelial fluid clearance in male and female adult rats. Am J Physiol Lung Cell Mol Physiol 2011;301:L804–11.
[137] Coakley RD, Sun H, Clunes LA, Rasmussen JE, Stackhouse JR, Okada SF, et al. 17beta-Estradiol inhibits $Ca2^{+}$-dependent homeostasis of airway surface liquid volume in human cystic fibrosis airway epithelia. J Clin Invest 2008;118:4025–35.
[138] Fanelli T, Cardone RA, Favia M, Guerra L, Zaccolo M, Monterisi S, et al. Beta-oestradiol rescues DeltaF508CFTR functional expression in human cystic fibrosis airway CFBE41o-cells through the up-regulation of NHERF1. Biol Cell 2008;100:399–412.
[139] Greenlee MM, Mitzelfelt JD, Yu L, Yue Q, Duke BJ, Harrell CS, et al. Estradiol activates epithelial sodium channels in rat alveolar cells through the G protein-coupled estrogen receptor. Am J Physiol Lung Cell Mol Physiol 2013;305:L878–89.
[140] Saint-Criq V, Kim SH, Katzenellenbogen JA, Harvey BJ. Nongenomic estrogen regulation of ion transport and airway surface liquid dynamics in cystic fibrosis bronchial epithelium. PLoS One 2013;8:e78593.
[141] Ito Y, Sato S, Son M, Kondo M, Kume H, Takagi K, et al. Bisphenol A inhibits Cl(-) secretion by inhibition of basolateral K+ conductance in human airway epithelial cells. J Pharmacol Exp Ther 2002;302:80–7.
[142] Laube M, Kuppers E, Thome UH. Modulation of sodium transport in alveolar epithelial cells by estradiol and progesterone. Pediatr Res 2011;69:200–5.
[143] Sweezey N, Tchepichev S, Gagnon S, Fertuck K, O'Brodovich H. Female gender hormones regulate mRNA levels and function of the rat lung epithelial Na channel. Am J Physiol 1998;274:C379–86.
[144] Johannesson M, Ludviksdottir D, Janson C. Lung function changes in relation to menstrual cycle in females with cystic fibrosis. Respir Med 2000;94:1043–6.
[145] Ino T, Aviado DM. Cardiopulmonary effects of progestational agents in emphysematous rats. Chest 1971;59:659–66.
[146] Ito H, Aviado DM. Prevention of pulmonary emphysema in rats by progesterone. J Pharmacol Exp Ther 1968;161: 197–204.
[147] Giles RE, Finkel MP, Williams JC, Winbury MM. The therapeutic effect of progesterone in papain-induced emphysema. Proc Soc Exp Biol Med 1974;147:489–93.
[148] Giles RE, Williams JC, Finkel MP, Winbury MM. Progesterone antagonism of papain emphysema: role of sex, estrogens and serum antitrypsin. Proc Soc Exp Biol Med 1973;144:487–91.
[149] Tyler JM. The effect of progesterone on the respiration of patients with emphysema and hypercapnia. J Clin Invest 1960;39:34–41.
[150] Holgate ST. A brief history of asthma and its mechanisms to modern concepts of disease pathogenesis. Allergy Asthma Immunol Res 2010;2:165–71.
[151] Girodet PO, Ozier A, Bara I, Tunon de Lara JM, Marthan R, Berger P. Airway remodeling in asthma: new mechanisms and potential for pharmacological intervention. Pharmacol Ther 2011;130:325–37.
[152] Gil FR, Lauzon AM. Smooth muscle molecular mechanics in airway hyperresponsiveness and asthma. Can J Physiol Pharmacol 2007;85:133–40.
[153] Dekkers BG, Maarsingh H, Meurs H, Gosens R. Airway structural components drive airway smooth muscle remodeling in asthma. Proc Am Thorac Soc 2009;6:683–92.
[154] Foster PS, Goldie RG, Paterson JW. Effect of steroids on beta-adrenoceptor-mediated relaxation of pig bronchus. Br J Pharmacol 1983;78:441–5.
[155] Degano B, Prevost MC, Berger P, Molimard M, Pontier S, Rami J, et al. Estradiol decreases the acetylcholine-elicited airway reactivity in ovariectomized rats through an increase in epithelial acetylcholinesterase activity. Am J Respir Crit Care Med 2001;164:1849–54.
[156] Dimitropoulou C, Drakopanagiotakis F, Chatterjee A, Snead C, Catravas JD. Estrogen replacement therapy prevents airway dysfunction in a murine model of allergen-induced asthma. Lung 2009;187:116–27.
[157] Riffo-Vasquez Y, Ligeiro de Oliveira AP, Page CP, Spina D, Tavares-de-Lima W. Role of sex hormones in allergic inflammation in mice. Clin Exp Allergy 2007;37:459–70.
[158] Matsubara S, Swasey CH, Loader JE, Dakhama A, Joetham A, Ohnishi H, et al. Estrogen determines sex differences in airway responsiveness after allergen exposure. Am J Respir Cell Mol Biol 2008;38:501–8.
[159] Degano B, Mourlanette P, Valmary S, Pontier S, Prevost MC, Escamilla R. Differential effects of low and high-dose estradiol on airway reactivity in ovariectomized rats. Respir Physiol Neurobiol 2003;138:265–74.
[160] Kouloumenta V, Hatziefthimiou A, Paraskeva E, Gourgoulianis K, Molyvdas PA. Sexual dimorphism in airway responsiveness to sex hormones in rabbits. Am J Physiol Lung Cell Mol Physiol 2007;293:L516.
[161] Pang JJ, Xu XB, Li HF, Zhang XY, Zheng TZ, Qu SY. Inhibition of beta-estradiol on trachea smooth muscle contraction in vitro and in vivo. Acta Pharmacol Sin 2002;23: 273–7.
[162] Assem ES, Wan BY, Peh KH, Pearce FL. Effect of genistein on agonist-induced airway smooth muscle contraction. Inflamm Res 2006;55(Suppl. 1):S13–14.
[163] Lin AH, Leung GP, Leung SW, Vanhoutte PM, Man RY. Genistein enhances relaxation of the spontaneously hypertensive rat aorta by transactivation of epidermal growth factor receptor following binding to membrane estrogen receptors-alpha and activation of a G protein-coupled, endothelial nitric oxide synthase-dependent pathway. Pharmacol Res 2011;63:181–9.
[164] Townsend EA, Sathish V, Thompson MA, Pabelick CM, Prakash YS. Estrogen effects on human airway smooth muscle involve cAMP and protein kinase A. Am J Physiol Lung Cell Mol Physiol 2012;303:L923–8.
[165] Sathish V, Freeman MR, Long E, Thompson MA, Pabelick CM, Prakash YS. Cigarette smoke and estrogen signaling in human airway smooth muscle. Cell Physiol Biochem 2015;36:1101–15.
[166] Card JW, Voltz JW, Ferguson CD, Carey MA, DeGraff LM, Peddada SD, et al. Male sex hormones promote vagally mediated reflex airway responsiveness to cholinergic stimulation. Am J Physiol Lung Cell Mol Physiol 2007;292:L908–14.
[167] Dimitropoulou C, White RE, Ownby DR, Catravas JD. Estrogen reduces carbachol-induced constriction of asthmatic airways by stimulating large-conductance voltage and calcium-dependent potassium channels. Am J Respir Cell Mol Biol 2005;32:239–47.
[168] Hayashi T, Adachi Y, Hasegawa K, Morimoto M. Less sensitivity for late airway inflammation in males than females in BALB/c mice. Scand J Immunol 2003;57:562–7.
[169] Melgert BN, Postma DS, Kuipers I, Geerlings M, Luinge MA, van der Strate BW, et al. Female mice are more susceptible to the development of allergic airway inflammation than male mice. Clin Exp Allergy 2005;35:1496–503.
[170] Corteling R, Trifilieff A. Gender comparison in a murine model of allergen-driven airway inflammation and the response to budesonide treatment. BMC Pharmacol 2004;4:4.

[171] Seymour BW, Friebertshauser KE, Peake JL, Pinkerton KE, Coffman RL, Gershwin LJ. Gender differences in the allergic response of mice neonatally exposed to environmental tobacco smoke. Dev Immunol 2002;9:47–54.
[172] Carey MA, Card JW, Bradbury JA, Moorman MP, Haykal-Coates N, Gavett SH, et al. Spontaneous airway hyperresponsiveness in estrogen receptor-alpha-deficient mice. Am J Respir Crit Care Med 2007;175:126–35.
[173] Perusquia M, Hernandez R, Montano LM, Villalon CM, Campos MG. Inhibitory effect of sex steroids on guinea-pig airway smooth muscle contractions. Comp Biochem Physiol C Pharmacol Toxicol Endocrinol 1997;118:5–10.
[174] Hellings PW, Vandekerckhove P, Claeys R, Billen J, Kasran A, Ceuppens JL. Progesterone increases airway eosinophilia and hyper-responsiveness in a murine model of allergic asthma. Clin Exp Allergy 2003;33:1457–63.
[175] Kouloumenta V, Hatziefthimiou A, Paraskeva E, Gourgoulianis K, Molyvdas PA. Non-genomic effect of testosterone on airway smooth muscle. Br J Pharmacol 2006;149: 1083–91.
[176] Bordallo J, de Boto MJ, Meana C, Velasco L, Bordallo C, Suarez L, et al. Modulatory role of endogenous androgens on airway smooth muscle tone in isolated guinea-pig and bovine trachea; involvement of beta2-adrenoceptors, the polyamine system and external calcium. Eur J Pharmacol 2008;601:154–62.
[177] Montaño LM, Espinoza J, Flores-Soto E, Chávez J, Perusquía M. Androgens are bronchoactive drugs that act by relaxing airway smooth muscle and preventing bronchospasm. J Endocrinol 2014;222:1–13.
[178] Stewart AG, Fernandes D, Tomlinson PR. The effect of glucocorticoids on proliferation of human cultured airway smooth muscle. Br J Pharmacol 1995;116:3219–26.
[179] Dashtaki R, Whorton AR, Murphy TM, Chitano P, Reed W, Kennedy TP. Dehydroepiandrosterone and analogs inhibit DNA binding of AP-1 and airway smooth muscle proliferation. J Pharmacol Exp Ther 1998;285:876–83.
[180] Stamatiou R, Paraskeva E, Papagianni M, Molyvdas PA, Hatziefthimiou A. The mitogenic effect of testosterone and 17beta-estradiol on airway smooth muscle cells. Steroids 2011;76:400–8.
[181] Sathish V, Freeman MR, Manlove LJ, Thompson MA, Pabelick CM, Prakash YS. Estrogen Receptor Beta (ERb) Blunts Inflammation-Induced Human Airway Smooth Muscle Proliferation And Remodeling. Am J Respir Crit Care Med 2014;189:A5318.
[182] Martin YN, Manlove L, Dong J, Carey WA, Thompson MA, Pabelick CM, et al. Hyperoxia-induced changes in estradiol metabolism in postnatal airway smooth muscle. Am J Physiol Lung Cell Mol Physiol 2015;308:L141–6.
[183] Voltz JW, Card JW, Carey MA, Degraff LM, Ferguson CD, Flake GP, et al. Male sex hormones exacerbate lung function impairment after bleomycin-induced pulmonary fibrosis. Am J Respir Cell Mol Biol 2008;39:45–52.
[184] Gharaee-Kermani M, Hatano K, Nozaki Y, Phan SH. Gender-based differences in bleomycin-induced pulmonary fibrosis. Am J Pathol 2005;166:1593–606.
[185] Langenbach SY, Wheaton BJ, Fernandes DJ, Jones C, Sutherland TE, Wraith BC, et al. Resistance of fibrogenic responses to glucocorticoid and 2-methoxyestradiol in bleomycin-induced lung fibrosis in mice. Can J Physiol Pharmacol 2007;85:727–38.
[186] Kondo H, Kasuga H, Noumura T. Effects of various steroids on in vitro lifespan and cell growth of human fetal lung fibroblasts (WI-38). Mech Ageing Dev 1983;21:335–44.
[187] Tofovic SP, Zhang X, Jackson EK, Zhu H, Petrusevska G. 2-methoxyestradiol attenuates bleomycin-induced pulmonary hypertension and fibrosis in estrogen-deficient rats. Vascul Pharmacol 2009;51:190–7.
[188] Flores-Delgado G, Bringas P, Buckley S, Anderson KD, Warburton D. Nongenomic estrogen action in human lung myofibroblasts. Biochem Biophys Res Commun 2001;283:661–7.
[189] Morani A, Barros RP, Imamov O, Hultenby K, Arner A, Warner M, et al. Lung dysfunction causes systemic hypoxia in estrogen receptor beta knockout (ERbeta-/-) mice. Proc Natl Acad Sci USA 2006;103:7165–9.
[190] Canning BJ, Fischer A. Neural regulation of airway smooth muscle tone. Respir Physiol 2001;125:113–27.
[191] da Silva SB, de Sousa Ramalho Viana E, de Sousa MB. Changes in peak expiratory flow and respiratory strength during the menstrual cycle. Respir Physiol Neurobiol 2006;150:211–19.
[192] Driver HS, McLean H, Kumar DV, Farr N, Day AG, Fitzpatrick MF. The influence of the menstrual cycle on upper airway resistance and breathing during sleep. Sleep 2005;28:449–56.
[193] Saaresranta T, Polo O. Hormones and breathing. Chest 2002;122:2165–82.
[194] White DP, Douglas NJ, Pickett CK, Weil JV, Zwillich CW. Sexual influence on the control of breathing. J Appl Physiol Respir Environ Exerc Physiol 1983;54:874–9.
[195] Mellon SH, Griffin LD. Neurosteroids: biochemistry and clinical significance. Trends Endocrinol Metab 2002;13:35–43.
[196] Behan M, Wenninger JM. Sex steroidal hormones and respiratory control. Respir Physiol Neurobiol 2008;164:213–21.
[197] Slatkovska L, Jensen D, Davies GA, Wolfe LA. Phasic menstrual cycle effects on the control of breathing in healthy women. Respir Physiol Neurobiol 2006;154:379–88.
[198] Tatsumi K. Effects of female hormones on respiratory and cardiovascular regulation]. Nihon Kokyuki Gakkai Zasshi 1999;37:359–67.
[199] Kimura H, Tatsumi K, Kunitomo F, Okita S, Tojima H, Kouchiyama S, et al. Obese patients with sleep apnea syndrome treated by progesterone. Tohoku J Exp Med 1988;156 (Suppl.):151–7.
[200] Dempsey JA, Skatrud JB. A sleep-induced apneic threshold and its consequences. Am Rev Respir Dis 1986;133:1163–70.
[201] Young T, Finn L, Austin D, Peterson A. Menopausal status and sleep-disordered breathing in the Wisconsin Sleep Cohort Study. Am J Respir Crit Care Med 2003;167:1181–5.
[202] Shahar E, Redline S, Young T, Boland LL, Baldwin CM, Nieto FJ, et al. Hormone replacement therapy and sleep-disordered breathing. Am J Respir Crit Care Med 2003;167:1186–92.
[203] Bixler EO, Vgontzas AN, Lin HM, Ten Have T, Rein J, Vela-Bueno A, et al. Prevalence of sleep-disordered breathing in women: effects of gender. Am J Respir Crit Care Med 2001;163:608–13.
[204] Kapsimalis F, Kryger MH. Gender and obstructive sleep apnea syndrome, part 2: mechanisms. Sleep 2002;25:499–506.
[205] Joos GF, Germonpre PR, Pauwels RA. Neural mechanisms in asthma. Clin Exp Allergy 2000;30(Suppl. 1):60–5.
[206] Linden A. NANC neural control of airway smooth muscle tone. Gen Pharmacol 1996;27:1109–21.
[207] Mackay TW, Hulks G, Douglas NJ. Non-adrenergic, noncholinergic function in the human airway. Respir Med 1998;92:461–6.
[208] Belvisi MG, Ward JK, Mitchell JA, Barnes PJ. Nitric oxide as a neurotransmitter in human airways. Arch Int Pharmacodyn Ther 1995;329:97–110.

[209] Lammers JW, Barnes PJ, Chung KF. Nonadrenergic, noncholinergic airway inhibitory nerves. Eur Respir J 1992;5:239–46.

[210] van der Velden VH, Hulsmann AR. Autonomic innervation of human airways: structure, function, and pathophysiology in asthma. Neuroimmunomodulation 1999;6:145–59.

[211] Ward JK, Barnes PJ, Springall DR, Abelli L, Tadjkarimi S, Yacoub MH, et al. Distribution of human i-NANC bronchodilator and nitric oxide-immunoreactive nerves. Am J Respir Cell Mol Biol 1995;13:175–84.

[212] Andronowska A, Chrusciel M. Influence of estradiol-17beta and progesterone on nitric oxide (NO) production in the porcine endometrium during first half of pregnancy. Reprod Biol 2008;8:43–55.

[213] Scott PA, Tremblay A, Brochu M, St-Louis J. Vasorelaxant action of 17-estradiol in rat uterine arteries: role of nitric oxide synthases and estrogen receptors. Am J Physiol Heart Circ Physiol 2007;293:H3713–19.

[214] Shah S, Nathan L, Singh R, Fu YS, Chaudhuri G. E2 and not P4 increases NO release from NANC nerves of the gastrointestinal tract: implications in pregnancy. Am J Physiol Regul Integr Comp Physiol 2001;280:R1546–54.

[215] Toppozada H, Michaels L, Toppozada M, El-Ghazzawi I, Talaat M, Elwany S. The human respiratory nasal mucosa in pregnancy. An electron microscopic and histochemical study. J Laryngol Otol 1982;96:613–26.

[216] Da Silva JA. Sex hormones and glucocorticoids: interactions with the immune system. Ann N Y Acad Sci 1999;876:102–17 discussion 117–108.

[217] Whitacre CC, Reingold SC, O'Looney PA. A gender gap in autoimmunity. Science 1999;283:1277–8.

[218] Huber SA, Pfaeffle B. Differential Th1 and Th2 cell responses in male and female BALB/c mice infected with coxsackievirus group B type 3. J Virol 1994;68:5126–32.

[219] Straub RH. The complex role of estrogens in inflammation. Endocr Rev 2007;28:521–74.

[220] Haggerty C, Ness R, Kelsey S, Waterer G. The impact of estrogen and progesterone on asthma. Ann Allergy Asthma Immunol 2003;90:284–91.

[221] Postma DS. Gender differences in asthma development and progression. Gend Med 2007;4(Suppl. B):S133–46.

[222] McCallister JW, Mastronarde JG. Sex differences in asthma. J Asthma 2008;45:853–61.

[223] Mannino DM, Homa DM, Akinbami LJ, Ford ES, Redd SC. Chronic obstructive pulmonary disease surveillance–United States, 1971–2000. MMWR Surveill Summ 2002;51:1–16.

[224] Troisi RJ, Speizer FE, Rosner B, Trichopoulos D, Willett WC. Cigarette smoking and incidence of chronic bronchitis and asthma in women. Chest 1995;108:1557–61.

[225] Dratva J, Schindler C, Curjuric I, Stolz D, Macsali F, Gomez FR, et al. Perimenstrual increase in bronchial hyperreactivity in premenopausal women: results from the population-based SAPALDIA 2 cohort. J Allergy Clin Immunol 2010;125:823–9.

[226] Forbes L, Jarvis D, Burney P. Do hormonal contraceptives influence asthma severity? Eur Respir J 1999;14:1028–33.

[227] Matteis M, Polverino F, Spaziano G, Roviezzo F, Santoriello C, Sullo N, et al. Effects of sex hormones on bronchial reactivity during the menstrual cycle. BMC Pulm Med 2014;14:108.

[228] Kwon HL, Belanger K, Bracken MB. Asthma prevalence among pregnant and childbearing-aged women in the United States: estimates from national health surveys. Ann Epidemiol 2003;13:317–24.

[229] Schatz M, Dombrowski MP, Wise R, Thom EA, Landon M, Mabie W, et al. Asthma morbidity during pregnancy can be predicted by severity classification. J Allergy Clin Immunol 2003;112:283–8.

[230] Dodds L, Armson BA, Alexander S. Use of asthma drugs is less among women pregnant with boys rather than girls. BMJ 1999;318:1011.

[231] Beecroft N, Cochrane GM, Milburn HJ. Effect of sex of fetus on asthma during pregnancy: blind prospective study. BMJ 1998;317:856–7.

[232] Ligeiro de Oliveira AP, Oliveira-Filho RM, da Silva ZL, Borelli P, Tavares de Lima W. Regulation of allergic lung inflammation in rats: interaction between estradiol and corticosterone. Neuroimmunomodulation 2004;11:20–7.

[233] Gillum R. Frequency of attendance at religious services and cigarette smoking in American women and men: the Third National Health and Nutrition Examination Survey. Prev Med 2005;41:607–13.

[234] Kamil F, Pinzon I, Foreman MG. Sex and race factors in early-onset COPD. Curr Opin Pulm Med 2013;19:140–4 110.1097/MCP.1090b1013e32835d32903b.

[235] Kohler M, Sandberg A, Kjellqvist S, Thomas A, Karimi R, Nyrén S, et al. Gender differences in the bronchoalveolar lavage cell proteome of patients with chronic obstructive pulmonary disease. J Allergy Clin Immunol 2013;131:743–51 e749.

[236] Meltzer EB, Noble PW. Idiopathic pulmonary fibrosis. Orphanet J Rare Dis 2008;3:8.

[237] Han MK, Murray S, Fell CD, Flaherty KR, Toews GB, Myers J, et al. Sex differences in physiological progression of idiopathic pulmonary fibrosis. Eur Respir J 2008;31:1183–8.

[238] Olson AL, Swigris JJ, Lezotte DC, Norris JM, Wilson CG, Brown KK. Mortality from pulmonary fibrosis increased in the United States from 1992 to 2003. Am J Respir Crit Care Med 2007;176:277–84.

[239] Redente EF, Jacobsen KM, Solomon JJ, Lara AR, Faubel S, Keith RC, et al. Age and sex dimorphisms contribute to the severity of bleomycin-induced lung injury and fibrosis. Am J Physiol Lung Cell Mol Physiol 2011;301:L510–18.

[240] Myers TR. Pediatric asthma epidemiology: incidence, morbidity, and mortality. Respir Care Clin N Am 2000;6:1–14.

[241] Postma DS. Gender differences in asthma development and progression. Gend Med 2007;4(Suppl. B):S133–46.

[242] Camp PG, Coxson HO, Levy RD, Pillai SG, Anderson W, Vestbo J, et al. Sex differences in emphysema and airway disease in smokers. Chest 2009;136:1480–8.

[243] Martinez FJ, Curtis JL, Sciurba F, Mumford J, Giardino ND, Weinmann G, et al. Sex differences in severe pulmonary emphysema. Am J Respir Crit Care Med 2007;176:243–52.

[244] Farstad T, Bratlid D, Medbo S, Markestad T, Norwegian Extreme Prematurity Study Group. Bronchopulmonary dysplasia - prevalence, severity and predictive factors in a national cohort of extremely premature infants. Acta Paediatr 2011;100:53–8.

[245] Anadkat JS, Kuzniewicz MW, Chaudhari BP, Cole FS, Hamvas A. Increased risk for respiratory distress among white, male, late preterm and term infants. J Perinatol 2012; 32:780–5.

[246] Perelman RH, Palta M, Kirby R, Farrell PM. Discordance between male and female deaths due to the respiratory distress syndrome. Pediatrics 1986;78:238–44.

[247] Humbert M, Sitbon O, Chaouat A, Bertocchi M, Habib G, Gressin V, et al. Pulmonary arterial hypertension in France: results from a national registry. Am J Respir Crit Care Med 2006;173:1023–30.

[248] Runo JR, Loyd JE. Primary pulmonary hypertension. Lancet 2003;361:1533–44.

[249] Robles AM, Shure D. Gender issues in pulmonary vascular disease. Clin Chest Med 2004;25:373–7.

[250] Rich S, Dantzker DR, Ayres SM, Bergofsky EH, Brundage BH, Detre KM, et al. Primary pulmonary hypertension. A national prospective study. Ann Intern Med 1987;107:216–23.

[251] Loyd JE, Butler MG, Foroud TM, Conneally PM, Phillips III JA, Newman JH. Genetic anticipation and abnormal gender ratio at birth in familial primary pulmonary hypertension. Am J Respir Crit Care Med 1995;152:93–7.

[252] Gurwitz D, Corey M, Francis PW, Crozier D, Levison H. Perspectives in cystic fibrosis. Pediatr Clin North Am 1979;26:603–15.

[253] Davis PB. The gender gap in cystic fibrosis survival. J Gend Specif Med 1999;2:47–51.

[254] Rosenfeld M, Davis R, FitzSimmons S, Pepe M, Ramsey B. Gender gap in cystic fibrosis mortality. Am J Epidemiol 1997;145:794–803.

[255] Gribbin J, Hubbard RB, Le Jeune I, Smith CJ, West J, Tata LJ. Incidence and mortality of idiopathic pulmonary fibrosis and sarcoidosis in the UK. Thorax 2006;61:980–5.

[256] Bertelsen RJ, Instanes C, Granum B, Lodrup Carlsen KC, Hetland G, Carlsen KH, et al. Gender differences in indoor allergen exposure and association with current rhinitis. Clin Exp Allergy 2010;40:1388–97.

[257] Senthilselvan A, Rennie D, Chenard L, Burch LH, Babiuk L, Schwartz DA, et al. Association of polymorphisms of toll-like receptor 4 with a reduced prevalence of hay fever and atopy. Ann Allergy Asthma Immunol 2008;100:463–8.

[258] PausJenssen ES, Cockcroft DW. Sex differences in asthma, atopy, and airway hyperresponsiveness in a university population. Ann Allergy Asthma Immunol 2003;91:34–7.

[259] Fagan JK, Scheff PA, Hryhorczuk D, Ramakrishnan V, Ross M, Persky V. Prevalence of asthma and other allergic diseases in an adolescent population: association with gender and race. Ann Allergy Asthma Immunol 2001;86:177–84.

[260] Albain KS, Belani CP, Bonomi P, O'Byrne KJ, Schiller JH, Socinski M. PIONEER: a phase III randomized trial of paclitaxel poliglumex versus paclitaxel in chemotherapy-naive women with advanced-stage non-small-cell lung cancer and performance status of 2. Clin Lung Cancer 2006;7:417–19.

[261] Recchia AG, Musti AM, Lanzino M, Panno ML, Turano E, Zumpano R, et al. A cross-talk between the androgen receptor and the epidermal growth factor receptor leads to p38MAPK-dependent activation of mTOR and cyclinD1 expression in prostate and lung cancer cells. Int J Biochem Cell Biol 2009;41:603–14.

CHAPTER

7

Sex Differences in Renal Physiology and Pathophysiology

Carolyn M. Ecelbarger

Department of Medicine, Center for the Study of Sex Differences in Health, Aging, and Disease, Georgetown University, Washington, DC, United States

KIDNEY ANATOMY, DEVELOPMENT, AND OVERALL FUNCTION

Kidney Weight, Volume, and Structure

The kidneys, in humans, are a pair of organs lying in the retroperitoneal space in the lower back region. An adult kidney is about 4–5 inches or approximately 11 cm long and shaped like a bean. In most mammalian species, kidney size is either slightly larger in males or not different between the sexes (when normalized by body surface area) [1]. However, androgens have been demonstrated to play a permissive role in the growth of the kidney, especially compensatory growth following unilateral nephrectomy (removal of one kidney) [2]. The kidney consists of four gross regions: (1) the cortex, or outer zone, where bulk reabsorption occurs for numerous substances in the proximal tubule (PT) (water, glucose, salts, peptides), thick ascending limb (TAL) (salts), and distal tubule (water, salts); (2) the outer stripe of the outer medulla, containing the straight PT, a portion of the TAL, and cortical collecting duct (CCD); (3) the inner stripe of the outer medulla, containing thin descending limb, TAL, and CD; and (4) the inner medulla, containing CD, and thin limbs. The medulla, as a whole, is essential in fine-tuning of the final concentration of the urine via regulated transport and reabsorption. Each human kidney contains about 1 million nephrons, or independent filtering units. Blood enters the kidneys via branches of the ascending aorta of the vasculature, that is, the right and left renal arteries. Approximately 20% of cardiac output enters the renal arteries, or in other words, is filtered at each pass. Therefore, reabsorption is a primary role of the kidney, otherwise, an excessive amount and volume of vital nutrients and water would be lost in the urine. The renal arteries form additional branches into progressively smaller-dimensioned arteries, and eventually into the afferent arterioles in the glomeruli, where filtering occurs. The glomerulus is attached to the epithelial-cell lined renal tubule where urine is generated by active and passive reabsorption and secretion of a number of substances. Blood is filtered of larger mass proteins by passing across several cell layers which constitute the barrier. The resulting ultrafiltrate is then modified to become urine as it passes from the PT, through the loop of Henle, into the distal tubule, CD, and eventually bladder. Urine can become concentrated or diluted (relative to plasma) depending on the fluid intake of the individual. Regulation of the concentration of urine is discussed in greater detail in the section regarding vasopressin control.

Kidney lengths and volumes measured by sonography were found to correlate with height in both sexes in a study conducted in Copenhagen evaluating adult human subjects ranging in age from 30 to 70 years [1]. In this study, the researchers demonstrated that men had longer kidneys (on average), as compared to women; however, the data were not normalized by height. In a study in rats, Oudar et al. [3] reported increased volume of the cortex in male animals, as compared to females, while females had a greater relative volume of medulla. The increased cortical volume in the male rats was due to increased PT length and development. Dihydrotestosterone treatment of female

Sex Differences in Physiology
DOI: http://dx.doi.org/10.1016/B978-0-12-802388-4.00007-0

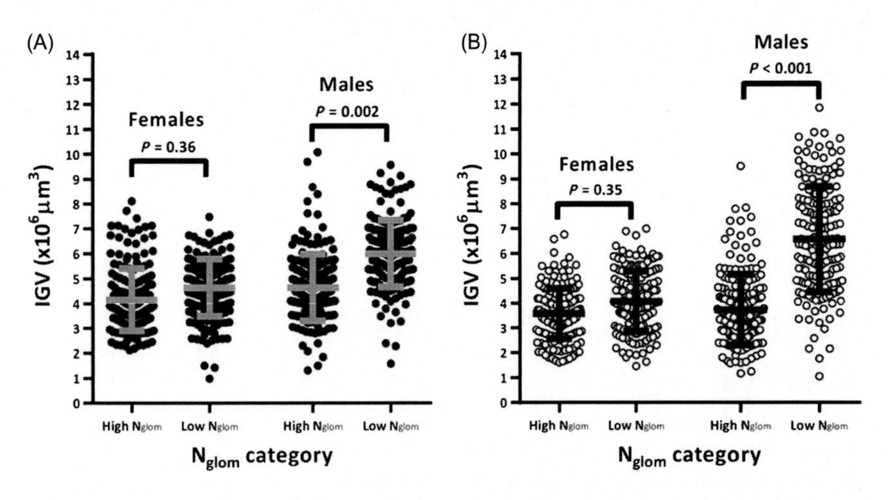

FIGURE 7.1 Correlation between nephron number and glomerular size in men and women. Sex comparisons of individual glomerular volume (IGV) by total nephron number in (A) African-American and (B) Caucasian subjects. The size of the glomerulus increased in males (both races) when glomerular number was reduced. *Source: From Puelles VG, Douglas-Denton RN, Zimanyi MA, Armitage JA, Hughson MD, Kerr PG, et al. Glomerular hypertrophy in subjects with low nephron number: contributions of sex, body size and race. Nephrol Dial Transplant 2014;29:1686–95. with copyright permission.*

rats led to an increase in PT development, supporting the view that renal PT growth and enlargement was an androgen-mediated effect. Glomerular volume was similar between the sexes. Females had about a 20% larger volume of TALs, especially in short-looped nephrons [3]. The impact of these sex differences on renal function including urinary concentrating capacity has not been evaluated. Furthermore, whether similar differences exist in human kidneys is not clear.

Nephron Number and Fetal Reprogramming

The total number of nephrons in a kidney is predetermined at birth or shortly thereafter. The number of nephrons, however can clearly influence overall kidney function, and in particular, influence susceptibility to hypertension (high blood pressure (BP)). One study conducted on autopsy showed that women had approximately 14% ($P < 0.001$) lower numbers of glomeruli than men in a sampling of both Caucasian and African-American subjects [4]. In comparison, a study in rats [5] showed approximate equal numbers of glomeruli between the sexes despite smaller kidney weights in females. Fetal malnutrition including low-protein, very low-calorie, or high-sodium diets consumed by the mother can also result in kidneys of lower volume, that is, reduced nephron number. This process is known as reprogramming, and in addition to affecting the structure/function of a number of major organs, it results in small for gestational age (SGA) babies. Moreover, SGA babies, later in life, have increased susceptibility to a variety of metabolic and cardiovascular diseases, including hypertension and type 2 diabetes. In the majority of studies reported, males are found to be more susceptible to pathologies associated with low nephron numbers, for example, to the development of hypertension [6]. Puelles et al. [7] showed that low nephron numbers in male, but not female, humans correlated with an increase in glomerular size (Fig. 7.1) and the propensity for eventual renal injury and failure. The authors speculated to some extent on why females did not show compensatory glomerular hypertrophy with low nephron numbers. They suggested that metabolic demand may influence hypertrophy of the glomerulus, and larger body surface area (such as in males) would likely be associated with greater demand [7]. Moreover, recent studies suggest that genetic programming alterations due to subpar prenatal environments have the capacity to transcend generations, and the penetrance of this transcendence may depend upon sex. Gallo et al. [8] demonstrated reduced nephron numbers and elevated BP in the F2 generation of male (but not female) rats in which the "grandmother" underwent bilateral uterine vessel ligation prior to pregnancy. Bilateral uterine vessel ligation is a model for uterine-placental insufficiency, or a state in which the fetus does not receive an adequate blood supply, and therefore nutrients. In their particular study, this sex difference could not be explained by differences in the renin-angiotensin system (RAS) [8], which is discussed in greater detail in a later section. In another study, the prenatal influence of corticosterone, a stress hormone released by the adrenal cortex, was evaluated with respect to nephron number and BP in mice [9]. Corticosterone (a steroid) was administered to pregnant female mice for 60 h beginning mid-pregnancy, that is, embryonic day 12.5 (gestation period for mice is about 21 days). A number of sex differences were found in the expression of genes in the kidneys at embryonic day 14.5. After birth, corticosterone treatment led to a reduction in body weight, kidney weight, and nephron numbers (about 33% lower) in both males and females; however, altered cardiovascular function was only observed in the male mice. Surprisingly, the "alteration" in cardiovascular function in this study was found to be hypotension (low BP) rather than the expected hypertension.

Renal Hemodynamics

Renal blood flow (RBF) and glomerular filtration rate (GFR) are two commonly utilized hemodynamic measures to estimate "renal function." RBF and GFR define the flow rate of blood and urine, respectively, through the kidney. Both are reflections of cardiac output and renovascular resistance. GFR can be estimated by measuring the concentrations of substances in the plasma and in the urine that are both freely filtered by the glomerulus and not reabsorbed to any extent along the renal tubule. Absolute values of both GFR and RBF are generally higher in men due to a larger body size (body surface area). However, using radiolabeled I^{125} iothalamate, Levey et al. [10] found no sex difference in GFR in human subjects when values were normalized for body surface area. In the clinic, algorithms are often used to estimate GFR from plasma creatinine, an endogenous breakdown product of muscle that is produced and should be excreted at a fairly constant rate day-to-day. Common algorithms used to estimate GFR from plasma creatinine are: (1) the Cockcroft and Gault equation [11], (2) the Chronic Kidney Disease Epidemiology Collaboration (CKD-EPI) model [12], and (3) the Modification of Diet in Renal Disease Study (MDRD) model [13]. These algorithms include corrections for body weight, race, age, and sex. In these models, GFR has been reported to be about 15% lower in women, than men, even when normalized for body surface area [11]. Thus, a correction factor (multiplicand) of about 0.85 for females is normally included when converting plasma creatinine into estimated GFR in most models. Note in the below Cockcroft-Gault equation, if plasma creatinine concentration was inserted for "Cr," the numerator would be adjusted down by a factor of 0.85 if female, and also adjusted down with increased age.

$$\text{Cockcroft} - \text{Gault} - \text{CrCl} = (140 - \text{age}) \times (\text{Wt in kg}) \times (0.85 \text{ if female})/(72 \times \text{Cr})$$

The flow of blood in the kidney has also been shown to be modified by sex. In both sexes, the kidney regulates glomerular pressure via tubuloglomerular feedback (TGF). TGF is a process whereby sodium concentrations are monitored by macula densa cells, that is, specialized cells found in the post-TAL portion of the renal tubule. Sodium enters these cells through the bumetanide-sensitive Na-K-2Cl cotransporter (NKCC2) in the apical membrane. If sodium concentrations are high, the macula densa cells release chemical signals to the afferent arteriole of that tubule (which is in close physical proximity to the macula densa) leading to vasoconstriction, which then reduces GFR and therefore the sodium load. This allows for the BP of the kidney to be modulated independently of the systemic BP. Loss or attenuation of autoregulation of TGF can lead to high GFR and RBF. This is known as hyperfiltration and is damaging and can cause detachment of podocytes, that is, specialized cells that line the glomerulus and form the barrier to larger and charged molecules. This is the early process of CKD, which is described in greater detail later.

A handful of studies have examined sex differences in RBF in the kidney. One study by Munger et al. [5] conducted in rats, led to the conclusion that male kidneys are relatively "vasodilated," as compared to those of females. The investigators determined that renal vascular resistance (RVR) was higher in female rats. RVR was calculated using acquired BP, hematocrit, and paraminohippuric acid (PAH) clearance data. This relatively high RVR, especially preglomerular, could very well be protective of the kidney over the long term. In their study, the authors made two other key observations: (1) ovariectomy in the females reduced RVR, suggesting a role for ovarian steroids in this effect and (2) the acute response to cyclooxygenase inhibitors paradoxically produced vasodilation in females and vasoconstriction in males, suggesting sex differences in the regulation of the arachidonic acid cascade, which produces a number of vasoactive substances.

Another study [14] conducted in humans assessed the modulatory effect of endogenous gonadal steroids and oral contraceptives on GFR and effective renal plasma flow (ERPF) under a low-salt (LS) or high-salt (HS) diet. While under most conditions, dietary salt did not affect these two hemodynamic measures, GFR was found to be significantly increased by HS in subjects on oral contraceptives. Furthermore, ERPF was increased during the luteal phase (high estrogen) relative to the follicular phase of the menstrual cycle. These studies support modulation of hemodynamics and perhaps the predisposition to "salt-sensitivity" by female sex steroids.

WATER HOMEOSTASIS AND AQUAPORINS

Kidney Regulation of Water Homeostasis

As mentioned earlier, a major role of the kidney is to regulate final urine concentration (water content) in order to maintain plasma homeostasis. That is, the plasma osmolality in nearly all mammalian species is tightly controlled to $\sim$290 mOsm/kg $\cdot$ H_2O despite large day-to-day and species-to-species differences in water intake. In humans, assuming an average GFR of around 90–120 mL/min, a high-volume of fluid (both sodium and water) needs to be reabsorbed at various sites along the renal tubule to avoid becoming rapidly dehydrated. About 65% of the filtered load of water is reabsorbed in the PT following the concentration gradient as sodium chloride (NaCl) is reabsorbed by the active action of the basolateral sodium, potassium, and adenosine triphosphatase (Na-K-ATPase) pump. Another ~15% is reabsorbed in the thin descending limb. The thin and TALs are impermeable to water, allowing for the development of the cortical-medullary osmotic gradient, that is, the interstitium of the medulla contains greater amounts of solutes than that of the cortex. The remaining approximately 20% of the water, thus enters the distal tubule, that is, the distal convoluted tubule (DCT), connecting tubule (CNT), and CD, where its fate is determined by the absence or presence of the circulatory peptide hormone, vasopressin (or antidiuretic hormone, ADH). Higher relative levels of vasopressin will increase water reabsorption, and lower levels will reduce it. Sex differences in a number of the renal processes that regulate water balance have been reported and are expanded on below.

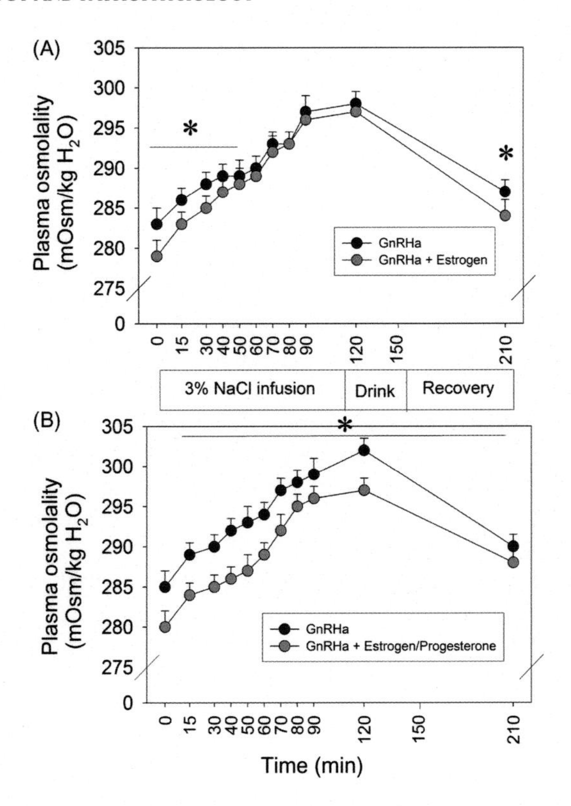

FIGURE 7.2 Plasma osmolality in response to hypertonic saline in women treated with (A) estrogen or (B) estrogen plus progesterone. Gonadotropin-releasing hormone analogue (GnRHa) was administered to block endogenous sex steroid release. *Source: Adapted from Stachenfeld NS, Silva C, Keefe DL, Kokoszka CA, Nadel ER. Effects of oral contraceptives on body fluid regulation. J Appl Physiol 1999;87:1016–25.*

Vasopressin and V2 Receptors

Vasopressin is a 9-amino acid peptide synthesized in the hypothalamus and released by the posterior pituitary in response to an increase in plasma osmolality or a decrease in blood volume, both indicators of relative dehydration. Sex differences exist in regard to the regulation of secretion and sensitivity to arginine vasopressin (AVP, type found in mammals). AVP release, like a number of hormones, is regulated in circadian fashion [15]. Adult women have been shown to have an attenuated circadian rhythm in AVP secretion which likely influences incidence of nocturia (waking up at night more than 2× to void the bladder), which has been shown to be higher in women, as compared to men [16,17]. Stachenfeld and colleagues have conducted a number of studies in the broad area of the influence of ovarian steroids on water handling by the kidney. In one study [18], they showed that plasma osmolality was lower in women treated for 4 days with estrogen or estrogen plus progesterone during a hypertonic saline challenge (Fig. 7.2). These results suggested that the set point for plasma osmolality was affected by sex steroids, and therefore either renal AVP sensitivity or the threshold for AVP production/release could be different. Indeed, other studies by this group showed, in general, that during states of high-estrogen activity, for example, during the mid-luteal phase of the menstrual cycle and with oral contraceptives, there was a relative reduction in the osmotic threshold for AVP release [19]. Additionally, they reported increased NaCl retention due to estradiol. However, free water retention was not altered in females treated with estradiol despite greater circulating AVP. This effect, they interpreted, was perhaps due to relatively reduced sensitivity of the kidney to AVP. When comparing men to women, however, they showed resting plasma AVP levels were higher in men, and men had a greater sensitivity to infused hypertonic saline with regard to osmotic AVP release [20,21].

AVP acts on two major types of receptors: (1) V1 (V1a and V1b), which are primarily associated with vasoconstriction in the vasculature (to increase BP) and (2) V2, found in the kidney distal tubule (CD principal cells) and TAL and associated with water reabsorptive mechanisms. Both receptors act in concert to maintain normal BP when blood volume drops due to low water intake, hemorrhage, or some other real or perceived loss of intravascular volume. V1a receptors may also be expressed in intercalated cells of the kidney. Deep-sequencing of microdissected tubules from rat revealed fairly robust expression of V1a in CNT and CCD [22]. The V2 receptor gene (V2R) is encoded on the X-chromosome, and therefore disorders associated with its mutation, for example, nephrogenic diabetes insipidus, occur with greater frequency in males. This is due to the fact that females would usually have one normal X-chromosome to compensate for the one harboring the mutation. This is true of nearly all

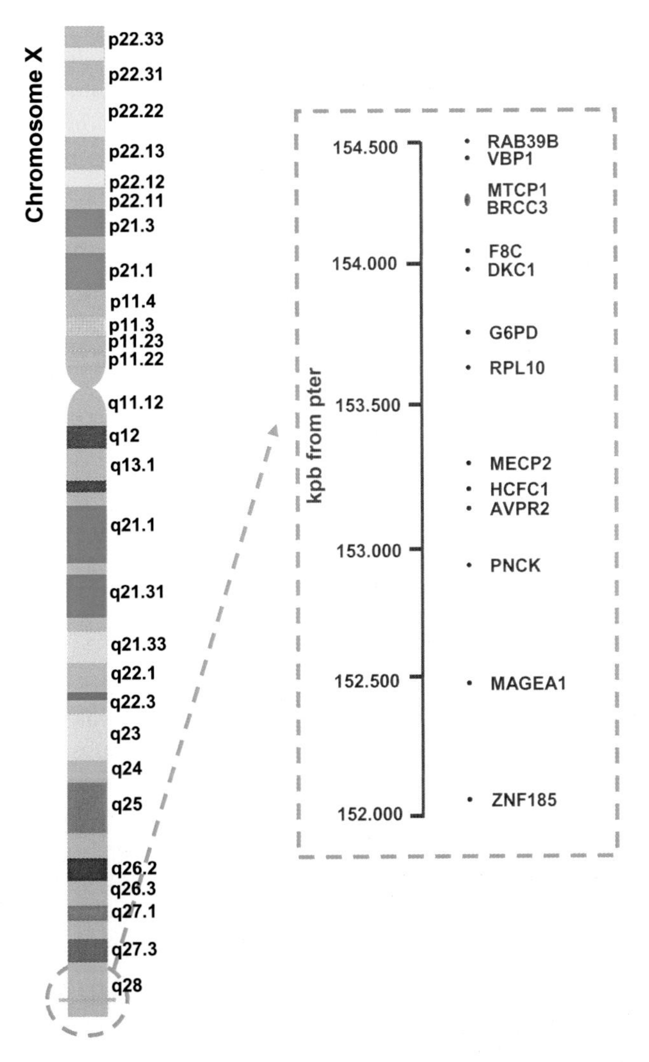

FIGURE 7.3 The V2R gene, AVPR2, and adjacent genes' locations on the X-chromosome in the Xq28 region. *Source: Adapted from Juul KV, Bichet DG, Nielsen S, Norgaard JP. The physiological and pathophysiological functions of renal and extrarenal vasopressin V2 receptors. Am J Physiol Renal Physiol 2014;306:F931–40.*

X-linked disorders. Moreover, the V2R gene is located on a region of the X-chromosome putatively more susceptible to escape from X-inactivation [23,24] (Fig. 7.3, see more regarding this topic in a later portion of the chapter). This would predict higher levels of V2R in female kidney. This has recently been shown to be true in rats in a study by Liu et al. [25] where female rats had greater mRNA and protein expression of V2R, and greater sensitivity to administered desmopressin (V2R-receptor-select agonist).

Regulation of Renal Aquaporins

Water reabsorption along the renal tubule is generally passive (does not directly require ATP/energy) to occur, but rather follows a concentration gradient created by the active (ATP utilizing) reabsorption of NaCl. Therefore, a primary determinant of water reabsorption by the renal tubule is the tubule's permeability. One might expect a lipid bilayer, which constitutes the outer membrane of epithelial cells lining the renal tubule, to be fairly hydrophobic, and not allow efficient water entry/exit. This is useful, for example, in the TAL, for generating the cortico-medullary osmotic gradient to allow for eventual fine-tuning of water reabsorption. However, in the PT and distal tubule (including CD), numerous intramembrane hydrophilic proteins known as water channels or "aquaporins" increase the porosity of the lipid bilayer allowing for H_2O molecules to be reabsorbed from the luminal fluid across the apical membrane, into the cell, and then across the basolateral membrane and back into the circulation. Aquaporins (AQPs), as a class, have six-membrane-spanning regions, with charged amino acids (hydrophilic) forming the actual water channel and intracellular amino and carboxy tails.

AQP1 is the isoform found expressed in PT and thin limbs. Most reports suggest that this water channel is constitutively expressed at high levels allowing for reabsorption of a large number of water molecules in an unregulated fashion. A recent study by Herak-Kramberger et al. [26] demonstrated that male rats had greater expression of AQP1 in kidney (80% protein, 40% higher mRNA) as compared to females. Gonadectomy appeared to reduce expression in both sexes. It is unclear from this study if the number of channels is higher in males due to the greater length or size of the PT in male animals (as discussed in the early portion of this chapter), or if the number of channels per cell is different.

AQP2 is the apical water channel expressed in the late DCT, CNT, and CD (cortical through inner medullary). AQP2 appears to be the most highly expressed mRNA transcript in the CCD, as assessed by deep-sequencing in rat [22]. Recent studies also show AQP2 expression in bladder, reproductive tissues, and brain. AQP2 transcription is upregulated by AVP [27]; therefore this provides one mode for increased permeability of the CD during periods of high AVP. A second mechanism involves translocation (trafficking) of existing AQP2 molecules from cytosolic sites into the apical membrane [27]. This mode of regulation, which may involve phosphorylation, can rapidly (in a matter of minutes) alter the water permeability of the membrane and increase water reabsorption [28]. A handful of studies have reported sex or sex hormone differences in the regulation of AQP2. Prepubertal levels of urinary AQP2 (a marker for renal levels) have been shown to be higher than postpubertal levels in a study conducted on healthy children of different ages [29]. Sharma et al. [30] reported a greater baseline protein expression of AQP2 in the kidney of female mice. However, in this study, females also experienced a greater downregulation of AQP2 on a high-fructose diet, as compared to male mice [30]. Kim et al. [31] demonstrated increased AQP2 expression in rat bladder in response to estradiol repletion in ovariectomized rats. Moreover, in support of greater expression in females, and with estradiol, Zou et al. [32] recently identified an estrogen-response element in the AQP2 gene promoter region.

AQP3 and AQP4 are found on the basolateral membranes of the renal CD, and mediate conductance of water molecules across this membrane. Similar to AQP2, both of these channels are expressed in a number of other tissues, in addition to kidney. Similar to AQP2, AQP3, but not AQP4, has been demonstrated to be upregulated transcriptionally by vasopressin [33]. Less has been reported regarding sex differences in the expression of these channels. Similar to what was observed for AQP2, estradiol repletion of ovariectomized rats led to increased expression of AQP3 in bladder [31]. AQP4 is the most highly expressed water channel in brain tissue and has an important role in protecting against edema. Estradiol has been shown to increase the expression of AQP4 in brain [34].

Hyponatremia

Inappropriate regulation of water reabsorption by the kidney can lead to either hyper or hyponatremia (high or low serum Na+). Hyponatremia is one of the most common morbidities observed in the clinic and may be associated with a greater number of adverse outcomes in women [35,36]. The syndrome of inappropriate ADH secretion (SIADH) is one cause of hyponatremia, that is, circulating AVP (ADH) levels are too high (relative to serum sodium). Causes of SIADH include certain medications, tumors, postoperative stress, CNS disturbances, and pulmonary disorders [37]. In this condition, renal AQP2 levels and apical localization increase inappropriately and too much water is reabsorbed, resulting in dilutional hyponatremia. Hyponatremia coupled with high AVP normally will elicit a "desensitization" of the renal tubule to AVP, over time, a process known as "vasopressin escape." Verbalis and colleagues [25] demonstrated impaired vasopressin escape in female, as compared to male, rats. That is, the females had higher urine osmolality and reduced urine volume in the model of desmopressin (V2R agonist) infusion plus high water intake, a common animal model of vasopressin escape [38]. Higher expression of V2R (as discussed above) might explain the susceptibility of females to hyponatremia given any level of circulating vasopressin. In addition to the elderly, another population susceptible to dilutional hyponatremia is endurance athletes. A study by Ayus et al. [39] demonstrated increased incidence of cerebral edema in female marathon runners as compared to their male counterparts. However, a recent study conducted with 50 adolescent marathon runners to test susceptibility in young runners (20 females and 30 males, 13–17 years of age) showed only a slight reduction in plasma sodium after the marathon in all subjects (about 2 mEq/L) with no sex differences [40]. Another study evaluating 96 subjects who had recently completed a marathon showed lower starting body weights, higher water consumption, and reduced sweat rates statistically explained serum sodium levels in their study, but sex was not an independent predictor [41].

BP CONTROL AND ELECTROLYTE HOMEOSTASIS

Sex Differences in BP

The kidney plays a vital role in maintaining plasma homeostasis and BP not only via regulated water reabsorption, but also through fine-tuning of the reabsorption of sodium, potassium, and chloride, the major electrolytes of the body that determine blood volume. Premenopausal women have consistently been demonstrated to have lower BP than young men, on average. Nonetheless, sex differences in BP have been shown to be attenuated at menopause [42]. In fact, some studies show higher BP in postmenopausal women when compared to age-matched men [43]. These studies implicate an important role for sex steroids, in particular estradiol, as a major determinant of sex differences in BP. In fact, several studies have shown a relative rightward shift in the pressure-natriuresis relationship in men as compared to women (Fig. 7.4). That is, males may require higher BPs to excrete the same amount of sodium, than do women. The pressure-natriuresis relationship is thought to be regulated by differences in the RAS and nitric oxide (NO) and/or oxidative stress. In addition to the RAS, a fairly large number of hormones have been shown to affect renal sodium reabsorption, and therefore, are implicated in BP control, for example, vasopressin, dopamine, insulin, and endothelin (ET). Sex differences have been reported in a number of these systems.

Major Sodium Transporters, Exchangers, and Channel Subunits

Despite lower BP in females, many sodium transporters and channels of the renal tubule have been shown to be upregulated either directly by estradiol or in the female sex. Sodium is reabsorbed at a number of sites

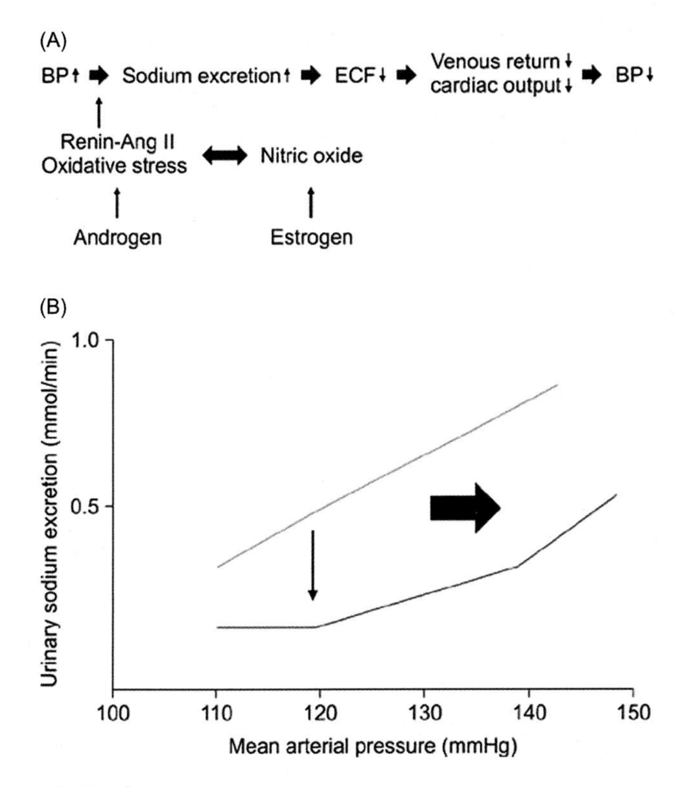

FIGURE 7.4 Potential mechanisms underlying basic sex differences in BP. (A) Androgens have been shown to activate the RAS and increase oxidative stress, while estrogens increase NO. (B) These two conditions may have opposite effects on the pressure-natriuretic curve, that is, the BP which elicits a given degree of natriuresis. A rightward shift (red line) indicates reduced sensitivity of natriuresis to a given rise in BP. *Source: From Kim JM, Kim TH, Lee HH, Lee SH, Wang T. Postmenopausal hypertension and sodium sensitivity. J Menopausal Med 2014;20:1–6.*

along the renal tubule, including the PT, TAL, and in the distal tubule (CNT through CD). The regulation of specific transporters and channel subunits between the sexes and by sex hormones has been evaluated in a number of studies [44–48]. For example, in the TAL, the bumetanide-sensitive Na-K-2Cl cotransporter (NKCC2, Slc12a1) has been reported to be decreased in abundance by estradiol repletion to ovariectomized rats [44]. In contrast, another major sodium-reabsorptive protein of the same family, but expressed solely in the DCT, the thiazide-sensitive sodium chloride cotransporter (NCC, Slc12a3), has been shown to be increased on the apical membrane in female, relative to male, rats. Furthermore, estradiol repletion to ovariectomized rats similarly led to increased apical localization of this transporter [45]. It is interesting to note that this protein is also strongly upregulated by aldosterone, and some studies suggest aldosterone and estrogen may have shared activity at the G-protein-coupled estrogen receptor (GPER-1) [49] (discussed in a later section). Nonetheless, in another study conducted in rats, estradiol repletion to ovariectomized rats decreased NCC whole-cell abundance [44]. The sodium hydrogen exchanger (NHE3, Slc9a3) a major apical exchanger for sodium reabsorption in the PT and TAL was demonstrated to be downregulated in the epididymis of mice lacking the estrogen receptor-alpha, although renal levels of NHE3 were not reported [50]. The epithelial sodium channel subunits (ENaC, Scnn1a, Scnn1b, Scnn1g) expressed in the late DCT2, CNT, and the CD, and responsible for fine-tuning of sodium reabsorption in these segments, have been shown to be differentially regulated between the sexes [48] and by sex hormones [44,46]. Female sex and, specifically, estrogen have been shown to upregulate mRNA expression for one or more of the ENaC subunits [51]. Therefore, the overall effect of the sex steroids on the expression, not to mention the activity, of sodium transporters and channels along the renal tubule is still not clear. Similar to other studies involving sex steroid replacement, dose and timing of replacement is likely an important factor determining overall expression and activity.

Renal NO

A great number of the sex differences in renal function relating to BP control have been postulated to be mediated either directly or indirectly by sex differences in renal oxidative stress or NO production [52–56]. NO is a potent vasodilator and antinatriuretic agent produced by serial monooxygenation of L-arginine. The major source of NO in the kidney is via enzymatic production from one of three NO synthase (NOS) isoforms, that is, NOS1–3. All three isoforms have been localized to renal vasculature and tubule segments [57,58]. The influence of sex or sex steroids on the expression and activity of the NOS isoforms in kidney has been controversial. Many, but not all, studies showed increased expression and/or activity of NOS isoforms in the kidneys of females versus males, in rodent models. Likewise, while estradiol has been fairly clearly shown to activate NOS isoforms in vitro, the complex in vivo milieu precludes a consistent association between sex or sex steroid status and NOS activity and expression. For example, female spontaneously hypertensive rats (SHRs) have been shown to have higher NOS activity in the renal medulla as compared to males [59]. Ovariectomy reduced renal NOS activity and the protein expression of the NOS1 isoform in this strain [60]. Treatment of the high BP in these rats with hydrochlorothiazide and reserpine (antihypertensive medications), not only abolished BP differences, but also the sex difference in NOS1 expression. This suggested that the relatively higher NOS1 expression and activity in intact females was in response to the hypertension, and may have been responsible for sex and OVX differences observed in this parameter. In contrast, another study showed that ovariectomy increased NOS1 medullary gene expression and protein in female mRen2.Lewis hypertensive rats [61]. Strain differences and degree of hypertension may play a role in NOS responsivity. In the obese Zucker rat (OZR), a model for metabolic syndrome with modest hypertension, female rats were shown to have greater activity of cortical but not medullary NOS as compared to male OZR [55].

Endothelins

ETs are 21-amino acid peptides that influence both vasculature contractility and sodium reabsorption in the kidney by binding primarily to endothelin A (ETA) or endothelin B (ETB) receptors [62]. ETB receptors expressed in the TAL and CD tubules have been reported to be responsible for many natriuretic actions of ET-1 [62–64]. Using the Spotting-Lethal (sl) rat which lacks endogenous ETB receptors, Kawanishi et al. [63] demonstrated a role for the ETB receptor in the sex difference in BP in the deoxycorticosterone acetate (DOCA) salt model. In this model, a mineralocorticoid receptor agonist (similar to aldosterone) is infused into the rats in combination with the feeding of a high-NaCl diet. The model results in sodium retention and occasionally hypertension. In this study, the investigators found that wild-type (WT) female rats had a markedly lower BP response to DOCA-HS than did WT male rats; however, in the ETB knockout (KO) rats (sl) this sex difference was abolished [63]. Nonetheless, a recent study by Nakano and Pollock [62] suggested that ETA receptors may mediate natriuresis in response to ET-1 in female rats. They showed that female, but not male rats, display a natriuretic response to acutely infused ET-1. The natriuresis was enhanced in female rats with genetic deletion or KO of ETB receptors, and was also inhibited by an ETA receptor antagonist, ovariectomy, or an antagonist of NOS1. The authors went on to conclude that this sex-differential response to ET-1 might play a role in premenopausal BP sex differences observed in humans. Male rats in the above study were found to have a reduction in medullary blood flow with ET infusion, while females were not. This likely contributed to the sex difference in natriuresis [62]. A more recent study by members of the same group [65] estimated no differences between the sexes in the ETB receptor in the inner medulla in rats, but that male rats had greater activity of ETA receptor at least at this renal site. They utilized a select agonist of the ETB receptor (ET-3) in binding studies and computed the difference between total binding (ET-1) and binding with ET-3 to deduce ETA receptor binding. However, using a deep-sequencing approach on microdissected tubules to evaluate mRNA expression in male rats, Lee et al. [22] found evidence for expression of ETB, but not ETA, in microdissected inner medullary collecting duct (IMCD) [22]. ET provides an example of the complexity underlying control of sodium homeostasis and BP by a number of hormonal systems.

Dopamine

Dopamine (3, 4-dihydroxyphenethylamine), also known as prolactin-inhibiting hormone, is a member of the catecholamine family. Dopamine can be synthesized in the kidney independent of renal nerves, and therein its actions are primarily natriuretic. Therefore, a reduction in dopamine levels or impaired receptor action can result in salt-sensitive hypertension [66]. Similar to several other hormones, the actions of dopamine are dependent upon the expression of dopamine receptors. In the kidney, there are a number of dopamine receptor isoforms (DR) expressed along the length of the renal tubule, denoted D1R–D5R, and studies using KO mice have shown that lack any of the five receptor subtypes can lead to hypertension [67]. While estrogen effects and sex differences in dopamine receptor signaling and expression have been fairly extensively studied in brain and more specifically in regard to behavioral modifications [68,69], less has been reported for kidney. A study by Wang and associates [70] examined the renal expression of dopamine receptors in obese and lean Zucker rats, a model for metabolic syndrome, and salt-sensitive hypertension. They found that obese rats (on a HS diet) had approximately 50% lower D1R protein levels (by western blotting), as compared to lean rats. Female lean rats had the highest D1R levels, about 30% higher than the lean males, and 280% higher than obese females. Another receptor subtype significantly affected by sex in this study was D3R which was, in contrast, significantly increased in the obese rats. Female OZRs had 240% higher D3R than lean males. Furthermore, female lean Zucker rats were resistant to the HS diet with regard to BP changes, while all other groups of rats, that is, lean males, obese males, and obese females showed a significant rise in BP when placed on the high-NaCl diet [55].

The RAS Overview

Some of the more robust sex differences in hormonal systems that regulate BP have been described for the RAS. As discussed earlier, renin secretion is increased in response to low sodium detected in the macula densa cells residing in the juxtaglomerular apparatus of the kidney. Renin acts as an enzyme to convert angiotensinogen to angiotensin I [71]. Angiotensin I is converted to the potent vasoconstrictor, angiotensin II (Ang II), an 8-amino acid peptide, by angiotensin converting enzyme (ACE). Ang II binds to either AT1 or AT2 receptors. However, Ang 7 (also known as Ang 1-7) can be produced via cleavage by neprilysin of Ang I and endopeptidase (ACE2). Ang 7 has been shown to have vasodilatory actions in the vasculature through the MAS receptor. Finally, Ang 7 can be converted to Ala1-Ang 7, which has been shown to bind to the Mrg (Mas-related G-protein coupled) receptor.

Angiotensinogen, ACE, and Renin

There exists evidence for both a systemic and organ-based (including kidney) RAS. Both renal and systemic RAS play central roles in the regulation of BP. Since many of the studies examining sex differences have been conducted in hypertensive subjects, there is some disagreement as to whether any or all of the components of the RAS are differentially regulated between the sexes in nonhypertensive control subjects. For example, Miller et al. [72] did not find baseline differences between sexes in plasma renin, ACE, or Ang II levels in healthy, young, human subjects. Likewise, Pendergrass et al. [73] did not find differences in plasma angiotensinogen or Ang II in normotensive Lewis rats. However, Ellison et al. [74] found an increase with puberty in the expression of the angiotensinogen in male WKY rats and strikingly higher levels in males, than females in adulthood. These relatively higher levels were abolished with castration in the males. Another study conducted in normotensive Wistar rats demonstrated that ovariectomy led to increased plasma ACE activity and that 17β estradiol (E_2) replacement abrogated this effect [75]. Strain differences may play a role in some of these discrepancies. In hypertensive strains, there is greater agreement that components of the RAS may be upregulated in males. For example, Pendergrass found elevation of plasma ACE and Ang II in male versus female mRen(2) hypertensive rats [73]. These transgenic rats have high activity of the RAS under normal conditions as they overexpress renin. Surprisingly, some studies have shown that estradiol itself may increase some intrarenal components of the RAS. In a study in healthy young women, those with higher circulating estradiol levels had greater renin activity [72]. Furthermore, the promoter region of the angiotensinogen gene has been shown to be responsive to estradiol [76].

Angiotensin II Sensitivity and Receptors

Male mice treated with Ang II via subcutaneous infusion have been demonstrated to develop rapid and robust hypertension, while this effect, although still present, is markedly attenuated in females [47,77]. For example, Tiwari et al. demonstrated about a 20 mm Hg difference in BP between male and female C57Bl6 mice over a 7-day course of Ang II infusion [47]. Age did not significantly affect this response. These results suggest increased sensitivity of males, as compared to females, to similar doses of Ang II infusion. In contrast, a study by Miller et al. [72] showed increased sensitivity of women to infused Ang II, at least with regard to changes in GFR. In their study, women had a significant reduction in GFR, while men showed no change in response to much lower doses of Ang II (0.5–2.5 ng/kg/min). One possible explanation given for this response was relatively increased preglomerular vasoconstriction in the females to a given dose of Ang II. Another factor which may explain sex differences in Ang II sensitivity is the receptor subtype expression in any given tissue. There are two main subtypes of Ang II receptors, that is, AT1 and AT2. AT1 receptors are Gq/11-coupled receptors that are coupled to the activation of phospholipase C. In kidney, binding studies show that AT1R are expressed in the glomerulus, as well as in the proximal and distal tubule. mRNA for AT1R (Agtr1a) was highest in TAL in rats [22]. Receptor binding studies, as well as immunoblotting and immunohistochemistry have demonstrated increased AT1R in the kidney of male rodents, as compared to females, and that ovariectomy increases these numbers in female rodents [75,78]. In contrast, a number of investigators have reported that AT2 receptors are increased in females [79–81]. These receptors are coupled to NO generation and vasodilation. Armando et al. [82] demonstrated that estrogen replacement upregulated renal AT2R in ovariectomized mice. Therefore, it may be that that the ratio of AT1/AT2 receptors in any given tissue may be an important factor in determining whether vasoconstriction or for that matter, sodium reabsorption occurs. Nevertheless, other studies have not observed higher levels of AT2R in female versus male rodents [83]. Some evidence suggests that AT2R renal expression may be restricted to vascular, rather than tubular, cells in the kidney [22,82].

ACE2, Ang 7, MAS Receptor

A number of sex or sex-steroid-mediated differences in the relative activity and/or expression level of the components of the "protective" RAS axis, consisting of ACE2, Ang 7, and the MAS receptor, have been described. Ang 7 (also referred to as Ang 1-7), the product of ACE2 enzymatic cleavage, binds to the MAS receptor, which is coupled to NO production [84]. Sullivan and colleagues [85] described abrogation of sex differences in Ang II-induced BP rises in the SHR by antagonism of the Mas receptor. In another study, relatively higher ACE2 activity in female C57Bl6 mice was shown to be protective against obesity-induced hypertension [86]. However, Pendergrass and colleagues [73] demonstrated increased BP, renal Ang II, as well as, ACE2 activity in male mren2.Lewis rats. In contrast, nephrolysin activity and Ang 1-7 expression were reduced in the males.

Aldosterone

Another hormone which is directly affected by the RAS is the mineralocorticoid hormone, aldosterone. Aldosterone is normally produced in the adrenal medulla in response to Ang II stimulation. Roesch et al. [87] reported that estradiol attenuated Ang II stimulated aldosterone secretion in ovariectomized rats. Aldosterone is perhaps the most influential hormone mediating antinatriuresis in the kidney by acting on the distal nephron and CD. It does so primarily by activation and increased expression of the thiazide-sensitive NCC [88,89] of the DCT and ENaC [90] in the CD. Direct infusion of aldosterone into a rat or mouse can increase BP in sensitive models, especially when coupled with a high-NaCl diet [91]. Little has been reported with regard to sex differences in aldosterone sensitivity of the kidney. However, Michaelis et al. [92] reported increased spironolactone (MR antagonist) sensitivity of BP in gonadectomized male, versus female, Wistar rats treated with aldosterone and high-NaCl diet. Intact males and females were not examined in their study. Therefore, these findings suggest sex differences that are independent of circulating sex steroids, and are putatively due to either sex chromosomal effects or developmental effects of the steroids (see later section on "Biological Mechanisms Underlying Sex Differences in the Kidney"). In addition, although the classic MR receptor is a steroid receptor, which acts by binding directly to DNA and influencing gene transcription, aldosterone has also been reported to bind to a G-protein coupled (fast-acting) receptor. In fact, GPR30 (G-protein coupled receptor 30), also known as GPER-1, a "fast-acting estrogen receptor" is reportedly involved in MR signaling, at least in the vasculature [93]. Therefore, the interaction between estrogen and mineralocorticoid receptor signaling may potentially influence sex differences in sodium balance and BP.

METABOLIC INFLUENCES ON THE KIDNEY

Sugar Reabsorption in PT

Depending upon the cell type, the renal tubule has the capacity to reabsorb and metabolize sugar molecules for energy (ATP production), as well as synthesize glucose (gluconeogenesis) [94]. These activities vary dramatically in the different cell types of the renal tubule. The PT, for example, is highly enriched in oxidative enzymes and does not utilize glucose to any large extent for energy. Free fatty acids are the preferred energy source for PT [94,95]. In contrast, the medulla is an obligate user of glucose for ATP synthesis via glycolysis [95]. Most filtered glucose is reabsorbed into the general circulation by the PT by apically localized sodium-glucose coupled cotransporters, that is, SGLT1 (Slc5a1) and SGLT2 (Slc5a2). There have been reports of elevated SGLT2 in the kidneys of female versus male rats, however the opposite finding (males > females) was reported for mice [96]. A novel class of SGLT2 antagonists has recently emerged as an additional therapeutic strategy to treat high circulating glucose levels in type 2 diabetes [97]. The drugs work by antagonizing PT (especially the S1 segment) uptake of glucose, and therefore allowing it to be excreted in the urine, since no other cell type in the renal tubule is capable of reabsorbing glucose to any large extent. These medications, for example, phlorizin, canagliflozin, and dapagliflozin have been reported to be equally efficacious in men and women; however, women may have greater incidence of genitourinary tract infections while on these drugs [97].

Gluconeogenesis and Fructose

The PT is also capable of gluconeogenesis, or the synthesis of "new glucose." Little work has been reported on sex differences in glucose synthesis in the kidney; however, in liver, the other major site of gluconeogenesis in the body, enzyme transcripts involved in gluconeogenesis have been shown to be relatively upregulated in male, as compared to female, rats [98]. Fructose, another monosaccharide, is also reabsorbed in PT. Sharma et al. [30] recently demonstrated increased expression of the fructose transporter (GLUT5, Slc2a5) in kidneys from male, as compared to female, mice fed high-levels of dietary fructose for 12 weeks. In this study, male mice also had greater upregulation of the first enzyme in fructose metabolism, that is, ketohexokinase, suggesting male mice may be better able to adapt and metabolize fructose in the renal PT, as compared to females. In addition, fructose metabolism has been demonstrated to lead to elevated plasma levels of uric acid due to rapid utilization of ATP at the cellular level by the action of ketohexokinase, an enzyme which does not seem to be rate-limiting, as compared to hexokinase, the enzyme involved in the first step in metabolism for glucose. Uric acid is a by-product of ATP utilization and serum levels have been shown to be higher in men. High plasma uric acid levels are associated with cardiovascular disease. The kidney reabsorbs uric acid via transporters. A recent study by Samimi et al. [99] demonstrated higher baseline levels of plasma uric acid in men, but a positive correlation between serum uric acid levels and BP only in women.

PHOSPHORUS, CALCIUM, VITAMIN D, AND RENAL STONES

Phosphorus

The kidney has an important role in whole-body phosphorus homeostasis. The bulk of inorganic phosphorus (Pi) reabsorption from the filtrate occurs in the PT (~65%), through the sodium-coupled cotransporter, NaPi-2a (gene symbol Slc34a1) or NaPi-2c (Slc34a3). Another 10–20% Pi is reabsorbed distally, and the remainder excreted [100]. Parathyroid hormone (PTH) is the major hormonal regulator of phosphorus reabsorption in PT and is inhibitory. Also inhibitory are fibroblast growth factor 23 (FGF23), a bone-derived hormone, in complex with its receptor (FGFR) and Klotho [101]. High levels of serum Pi have been associated with cardiovascular and renal disease, primarily due to calcification of soft tissue; therefore proper regulation of Pi reabsorption at the level of the PT is essential.

A number of sex differences have been reported for the renal regulation of Pi reabsorption. Moreover, hypophosphatemia has been associated with hormone replacement therapy in postmenopausal women, suggesting a direct effect of sex steroids [102]. Faroqui et al. [103] demonstrated that estrogen treatment of ovariectomized rats decreased NaPi-2a protein levels in the kidney and led to hyperphosphaturia and hypophosphatemia. This could potentially play a role in the hypophosphatemia observed occasionally in women on HRT [104]. Furthermore, estrogen has also been associated with a reduction in the circulating level of PTH, and PTH levels have been shown to be elevated postmenopausally [105]. These two effects might be expected to have opposite effects on PT Pi reabsorption. This suggests that the rise in PTH postmenopausally may be a response to higher circulating levels of phosphorus, thus an attempt to reduce phosphorus retention at the level of the kidney. The downregulation of PTH release from the parathyroid by estrogen may be an indirect effect with estrogen increasing the release of FGF23, and FGF23 inhibiting parathyroid release through FGFR localized on the parathyroid gland [106]. On the balance, estrogen seems to be protective against excessive phosphorus reabsorption in PT, which may partially explain reduced cardiovascular-renal disease in women premenopausally. Nonetheless, postmenopausal women on HRT might be increasingly susceptible to hypophosphatemia and bone loss due to lower than acceptable Pi-reabsorption in the PT.

Calcium

Like phosphorus, the kidney is vital in maintaining plasma calcium concentrations in the narrow range of 8.5–10.5 mg/dL [107]. The kidney reabsorbs about 60% of Ca^{2+} in the PT, 15% in the TAL, and 10–15% in the DCT through CNT. Fine-tuning of Ca^{2+} reabsorption occurs in the DCT and CNT, under control of PTH and activated vitamin D. Ca^{2+} is taken across the apical membrane through Ca^{2+} channels, that is, transient receptor potential vanilloid 5 and/or 6 (TRPV5 and 6). Calbindin-D_{28K} binds to Ca^{2+} in the cytoplasm so that intracellular levels of ionized Ca^{2+} are not affected. Basolateral extrusion of Ca^{2+} occurs either through $Na+/Ca^{2+}$ exchanger (NCX1) or the plasma membrane Ca^{2+}ATPase (PMCA). Dong et al. [108] examined the expression of mRNA transcripts for proteins involved in Ca^{2+} reabsorption in the distal tubule in aging female rats. They found that ovariectomy markedly reduced the expression of TRPV5, calbindin-28, and the β-subunit of PMCA under high or low-Ca^{2+} diets, with little effect of the diets themselves. Therefore a reduction in the expression of these proteins in kidney after menopause could conceivably contribute to bone loss associated with menopause.

Vitamin D

Another major player in the regulation of body phosphorus and calcium that acts primarily through the kidney is vitamin D. Vitamin D_3 (cholecalciferol) and vitamin D_2 (ergocalciferol) are steroid hormones, which are obtained through the diet or from synthesis in the skin. Vitamin D is converted to 25-hydroxycholecalciferol (calcidiol) by the liver. Calcidiol is taken up by the PT through a process known as endocytosis. This process involves megalin. In the PT, calcidiol is converted into its active form, 1,25-dihydroxycholecalciferol (calcitriol) by the enzyme, 25-hydroxyvitamin D_3 1-alpha-hydroxylase (CYP27B1). Calcitriol circulates in the blood and, as expected, is important in bone formation and remodeling, but is also active in a variety of other systems including the kidney, neuromuscular, and the immune systems. Most studies have not shown clear sex differences in the circulating levels of calcidiol or calcitriol, although calcidiol levels are more routinely measured. Hagenfeldt and Berlin [109] reported CYP27B1 activity, as measured by isotope-dilution mass spectrometry, to be on average 75% higher ($P < 0.05$) in preparations made from human male as compared to female kidney cortex. Samples were obtained from kidneys removed due to tumors. A significant inverse correlation between CYP27B1 activity and serum phosphorus was found, but no correlation was found between enzyme activity and serum calcium or calcitriol levels. CYP27B1 is also downregulated by the FGF23/FGFR/Klotho complex.

CKD and Dysregulation of Calcium and Phosphorus Homeostasis

Patients with CKD (discussed in greater detail below) are highly susceptible to disorders of calcium and phosphorus homeostasis. Predictably, impaired reabsorptive activity in the renal tubule can lead to increased serum levels of P in particular. CKD patients have been shown to be at risk for lower calcitriol levels, probably due to impaired PT function. Moreover, vitamin D has been shown to be protective of the kidney, with genetic deletion of the receptor leading to severe renal disease [110]. Treatment by calcitriol or activated vitamin D analogs has been shown to be highly protective against a number of renal diseases. Ovariectomy in rats with CKD has been shown to increase intact PTH, an effect that was attenuated with estrogen replacement [106]. Martin et al. [111] demonstrated a correlation between high serum P and risk of carotid atheromatous plaque in human CKD patients undergoing carotid ultrasound. In men, serum P within the normal range was associated with increased risk, whereas in women, the risk was increased only in women showing high serum P, suggesting higher sensitivity in men to the same levels of serum P. In this study, mean serum P was also significantly higher in women as compared to men (3.88 vs 3.62 mg/dL). This effect was independent of PTH.

Renal Stones

The incidence of renal stone formation (nephrolithiasis) is 30–70% higher in men than in women [112]. The mechanisms underlying this sex difference are not entirely clear; however, the incidence in women is apparently on the rise [112]. Jayachandran and colleagues [113] examined urinary extracellular vesicles (ECV) collected from stone-forming and control men and women in an effort to elucidate sex differences in the etiology of this disorder. Control women excreted ECV with a significantly greater number of exosomal markers, adhesion molecules, inflammatory factors, and markers associated with specific renal tubular sites, than did ECV from control men. However, the ECV from stone-forming women were similar to that of control and stone-forming men. Therefore, they concluded that the ECV profile in women, but not men, could potentially be a prognostic indicator of propensity to form stones [113]. Another study showed that stone-forming women may be at increased risk of developing hypertension as compared to stone-forming men [114].

ACID-BASE HOMEOSTASIS AND RENAL ELIMINATION OF TOXINS AND DRUGS

Organic Anion and Cation Transporters

Organic anion transporters (OATs) and organic cation transporters (OCTs) are members of the superfamily of transporters, Slc (solute carriers). Most are in the families Slc21 (also known as Slc0) and Slc22, both important in the secretion of common drugs, toxins, and nutrients [115]. One major site of expression for both OATs and OCTs is the renal PT, although they are found in most barrier epithelium. Many pharmaceuticals, including antibiotics, nonsteroidal anti-inflammatory agents, diuretics, and antivirals are not freely filtered by the glomerulus (as they are bound to albumin), but enter the pericapillary circulatory system. They are then transported across the basolateral membrane of the PT cells by one or more members of this diverse family of transporters for eventual secretion and elimination in the urine. Sex differences in activity and expression of OATs and OCTs have been described [116]; however, species differences have also been noted [117]. Cerrutti and colleagues [116] demonstrated delayed clearance of PAH, a reference substance secreted by the broader class of OATs, and furosemide, a loop diuretic also secreted by OATs, in female as compared to male rats after infusion of the substances. In contrast, using rabbits, Groves et al. [117] did not find any sex differences in organic anion or cation transport in PT. In some cases, secretion of drugs or medications into the tubule lumen, for example, furosemide, which binds and inhibits the sodium-potassium-2 chloride cotransporter (NKCC2) of the TAL, is necessary for efficacy. Therefore, differences in the secretion profile can be expected to affect and potentially explain sex differences that have been noted in the pharmacokinetics of several common medications [118]. Microarray studies conducted by Sabolic et al. [119] showed a 63-fold higher level of expression of OATp1 in male rats and a 32-fold higher level of OAT2 in female rats, as compared to the opposite sex, respectively.

Renal Metabolism and Elimination of Toxins

A number of sex differences have been documented with regard to the susceptibility of the kidney to a variety of toxic substances. Many of these differences can be ascribed to differences in drug metabolizing enzymes primarily in the liver including those of the cytochrome P450 (CYP) class, but other differences in the renal elimination or metabolism of toxins between the sexes have been described. For example, male rats have been shown to be more sensitive to the nephrocarcinogenic effects of hexachlorobenzene (HCB) and dimethyl methylphosphonate, two widely used chemicals in industry. In a study in rats, this particular male-specific tendency correlated with increased renal expression of a small molecular weight drug-binding protein, alpha 2u-globulin. Male rats treated with HCB had over 11-fold higher levels of this protein in renal PT than did untreated male rats [120]. Impaired lysosomal degradation of alpha 2u-globulin results in protein droplets in the PT, cell death, and sloughing [120,121]. Accelerated cell death and cycling in these male rats led to renal tumor formation. Alpha 2u-globulin is under the control of androgens and this likely contributes to greater expression. Gautier et al. [122] measured urine biomarkers in response to injury by gentamicin in Sprague-Dawley rats and reported numerous sex differences highlighting the need to test both sexes. Male rats have also been shown to be more susceptible to renal carcinogen, ochratoxin [123]. Ochratoxin is a mycotoxin produced from fungi and found in moldy cereals and grains. Greater toxicity in male animals has been proposed to be due to differences in OAT activity between the sexes [123], although the mechanism underlying nephrocarcinosis due to ochratoxin A is not completely understood.

SPECIAL CONDITIONS AND RENAL PATHOLOGY

Renal Aging

Age is a major determinant of renal anatomy and functionality [124]. Renal function, that is, GFR, has been demonstrated to decline at a rate of around 0.5–1.0 mL/min/year in human subjects [124]. Various estimates have been put forth using different age and health-status humans. The rate of decline appears to accelerate as age advances. One manifestation of renal aging is an increase in proteinuria due to a loss in the integrity of the size/charge barrier activity of the glomerulus. Other manifestations include a loss of nephron numbers, increased interstitial fibrosis and inflammation, and rarefaction of capillary density. Although only a handful of studies have looked at "healthy aging," that is, in subjects not affected with chronic diseases such as hypertension, diabetes, or the metabolic syndrome, most studies show a more rapid decline in renal function with age in males [125–127]. The majority of studies show androgens exacerbate the rate of decline [125], and estrogens may be protective [128,129]. Maric and colleagues [129] demonstrated some of the protective effects of estrogen appeared to be due to an increase in NOS expression and activity in the kidney. As discussed previously, NO is a potent vasodilator, produced locally by a number of cell types including endothelial and epithelial, of the kidney, and found to lower BP and improve blood flow [130]. Fortepiani and colleagues [125] studied the role of testosterone specifically in the changes in renal function and BP in aging rats. Castration led to a reduction in BP, proteinuria, and increased GFR and RBF in SHR. Loria et al. [127] showed an increased degree of proteinuria in the aging male rat kidney, relative to female, and this sex difference was more pronounced with an early developmental insult, in their case, angiotensin receptor blockade (which reduces total nephron number).

One recently considered factor associated with renal aging are epigenetic modifications, that is, alterations in gene expression due to inhibitory factors such as competitive binding by micro ribonucleic acids (miRNA). A study conducted in F344 rats over-the-life-span revealed that over 2 dozen miRNAs in the kidney were "sex-biased" in expression at one age or another (Fig. 7.5) [131]. For example, miR-208 and miR-501 were over fivefold greater in females, while miR-130b was greater in males. "Male-biased" miRNA expression patterns in the aging kidney correlated with the male predominance of fibrosis and mononuclear cell infiltration (a marker for inflammation).

Chronic Kidney Disease

CKD is a progressive disorder defined as the gradual loss of kidney function over time. CKD, in general, occurs secondarily to other chronic diseases including, but not limited to, type 2 diabetes mellitus, hypertension, and polycystic kidney disease (PKD). It is diagnosed via imaging, histology, proteinuria, and/or a reduction in GFR below 60 mL/min/1.73 m^2 for more than 3 months [132]. As many as 26 million American

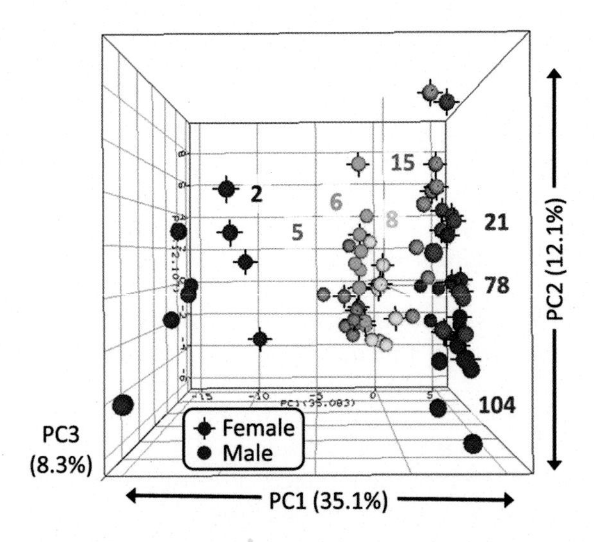

FIGURE 7.5 Age and sex differences in miRNA expression profile in kidneys from male and female rats—174 miRNA were differentially expressed (1.5-fold difference in expression and $P < 0.05$). Each sphere represents the expression profile of one animal plotted in 3-dimensional space according to the top three principal components. Spheres of the same color are animals of the same group (weeks of age). Spheres with black vertices indicated females while those without represent males ($n = 5$ or 4). *Source: From Kwekel JC, Vijay V, Desai VG, Moland CL, Fuscoe JC. Age and sex differences in kidney microRNA expression during the life span of F344 rats. Biol Sex Differ 2015;6:1.*

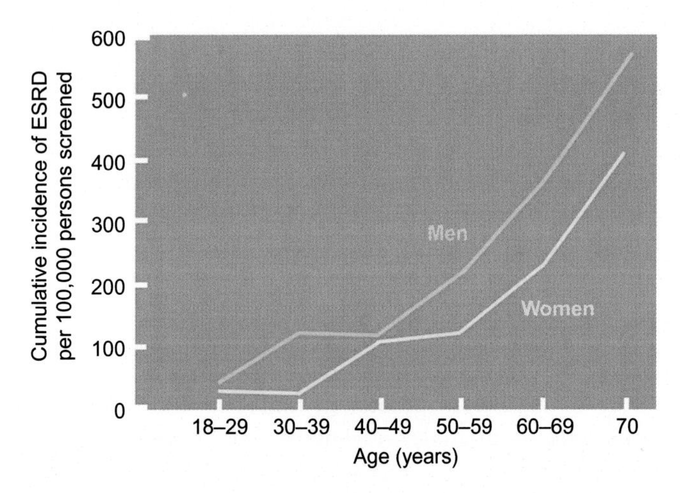

FIGURE 7.6 The cumulative incidence of ESRD is higher in men at nearly all ages, although the rate of increase in women parallels that of men in the postmenopausal years. *Source: From Iseki K. Gender differences in chronic kidney disease. Kidney Int 2008;74:415–17.*

adults have CKD, and millions of others are at risk for developing it due to the presence of comorbidities. When all classes of CKD are combined, men in general, exhibit greater renal injury and progress to end-stage renal disease (ESRD) at a more rapid rate than women, at least prior to menopause (Fig. 7.6). After menopause, however, the rate of development has been demonstrated to be equal or even higher in women, highlighting the protective effects of ovarian steroids [133]. Estrogen has been shown to have several protective effects on the kidney that may reduce or attenuate renal damage in models of CKD. Ovariectomy in diabetic mice was shown to increase the risk of glomerular disease (glomerulosclerosis) [134]. Estrogen therapy to postmenopausal women was shown to reduce albuminuria [135]. Estrogen has also been demonstrated to reduce markers of inflammation, fibrosis (scarring of the kidney), and the production of the vasodilator, NO in animal models. Taken together, a variety of actions of estradiol that are protective against hypertension are also protective of kidney function.

Nonetheless, the underlying etiology of CKD is ultimately an important determinant of the rate of progression, for example, in diabetic kidney disease, the most common cause of CKD, several studies have clearly shown a loss of any protective advantage to females. Yu and Lyles [136] conducted a cross-sectional analysis in a managed-care setting in which they examined the prevalence and severity of diabetic complications in both young and old diabetic men and women. They found that women, especially those greater than 60 years of age, had greater incidence of advanced renal disease, that is, estimated GFR < 30 mL/min/1.73 m^2.

Testosterone deficiency and hypogonadism are also associated with CKD in men [137]. In fact, low serum testosterone in this population may explain the anemia and osteoporosis associated with CKD. Nonetheless, because testosterone has also been associated with several negative factors relating to progression of CKD, including oxidative stress, high BP, inflammation, and enhanced cell proliferation, there has been some controversy as to whether it would be beneficial or perhaps counter-productive to restore or replace low testosterone levels in this population.

Polycystic Kidney Disease

PKD is a fairly common genetic disorder which can lead to CKD and renal failure. The most common form of PKD (autosomal dominant PKD, ADPKD) is caused by mutations in polycystin 1 or 2, which are found on chromosomes 16 and 4, respectively. PKD1 is a more commonly found mutation. Mutations in these proteins cause proliferative changes in kidney structure whereby, the kidney develops internal cysts, which enlarge, secrete fluid, and ultimately detrimentally affect function. ADPKD has an incidence of approximately 1:1000 live births. Approximately 10% of ESRD patients treated with hemodialysis may have originally been diagnosed with ADPKD. While the incidence of PKD is generally thought to be the same between the sexes, male sex is a risk factor for the progression of ADPKD. Male patients generally begin hemodialysis at a younger age. Using animal models, androgens have been shown to potentiate renal cell proliferation and cyst enlargement through ERK1/2- (extracellular signal-related kinase) dependent and independent signaling [138]. In another study, mTOR (mammalian target of rapamycin) inhibition was shown to attenuate cyst formation and renal enlargement in male, but not in female Han:SPRD-cy rats (a rodent model of PKD) [139]. MTORC1 (mTOR complex 1) signaling was reduced in both males and females, but only male mice showed an inhibition of mTORC2 signaling with rapamycin. MTORC2 signaling is associated with serine phosphorylation of Akt. While rapamycin has classically been considered primarily an mTORC1 antagonist, it may in actuality have sex differential effects on mTORC2.

Ureteral Obstruction

Ureteral obstruction (UO) or a blockage of urine flow through the kidney has been associated with renal damage primarily as a result of enhanced fibrosis (renal scar development). Nephrolithiasis is one cause of UO, and others include trauma, prostate enlargement, and certain cancers. When flow is restored, the kidney usually is able to recover to some extent, although the chronicity of blockade is largely an important determinant of reversibility. In rodents, unilateral UO is often modeled by clamping or tying off one of the ureters for a set period of time, for example, 24 h, removing the opposite kidney, and then relieving the clamp. In one study, following this model, male animals were found to be more susceptible to certain indices of renal injury including reduced creatinine clearance (GFR), tubular dilatation, and cellular apoptosis, than were female rats; however, urinary concentrating ability was not affected by sex in this study [140].

BIOLOGICAL MECHANISMS UNDERLYING SEX DIFFERENCES IN THE KIDNEY

Sex Steroids Versus Gene Dosage

Similar to other systems in the body, sex differences in kidney function are the result of either sex steroid influences, which are clearly remarkably different between the sexes and at different points in the lifespan of both men and women, and the sex chromosomal complement (SCC) (XX vs XY). The majority of sex differences in the kidney that have been described in the literature can be attributed to the presence or absence of sex steroids. Estrogen, progesterone, and testosterone have all been shown to directly regulate renal function and affect sensitivity to renal disease, as fully discussed in previous sections. These sex steroids can either have permanent, that is, developmental or organizational effects that are generally considered irreversible, such as the number of nephrons in the kidney at birth, or acute, that is, reversible effects, such as you might observe during phases of the menstrual cycle, for example, regulation of sodium reabsorption. Other potentially malleable effects of sex steroids would include the activity and expression key receptors in the kidneys, such as those involved in mediating RAS effects or NO enzymes. Many changes in these systems or receptors have been described over-the-lifespan and are modified with alterations in sex steroid milieu.

X-Inactivation and the Kidney

The SCC, as discussed elsewhere, has the capacity to affect the phenotype of the kidney in at least two major ways. First, there are some genes on the Y-chromosome that do not have a similar substitute on the X, for example, the Sry (sex-determining region on the Y-chromosome). Second, the X-associated gene dosage may vary between the sexes due to incomplete silencing of one X in female, that is, genes that escape X-cell inactivation (XCI). Females are normally a cellular "mosaic" with regard to which X-chromosome is expressed. One of the pair of X-chromosomes is silenced in each cell, seemingly at random, leading the average female to express approximately 50% paternal- and 50% maternal-X-inherited genes at the organism level. This provides a means to normalize X-gene dosage, and generally provides protection against X-linked diseases. However, there is "skewing" of this pattern in some individuals due to random chance, with some persons expressing as much as 80% one lineage and only 20% the other. The influence of XCI on kidney function has not been yet fully elucidated; however, a recent study [141] conducted in a line of mice in which the Sry (Sex-determining gene on the Y-chromosome) was transferred to an autosome demonstrated increased BP in XX versus XY mice during Ang II infusion. This study demonstrated at least one cardiovascular effect of the SCC that was independent of sex steroids. Furthermore, as discussed earlier (see Fig. 7.3), female rats have been shown to have higher renal mRNA expression of the vasopressin V2R [25], whose gene is found near the end of the long arm of the X-chromosome.

Another instance pertinent to the kidney in which the XCI has been found to be relevant is during renal transplantation. Aging, in general, is associated with increased XCI skewing possibly due to stem-cell senescence [142]. Interestingly, XCI skewing has been proposed to be beneficial if it occurs in the recipient of a kidney transplant, but detrimental if it occurs in the donor [143]. Simmonds et al. [143] found that XCI skewing was associated with less allograft rejection at least in the first year in the recipient, but a greater chance for rejection in the donor. All-in-all this might suggest, at least with regard to women, the best match would be a young donor and an older recipient. Moreover, donor-recipient sex mismatch has also been associated with poor renal graft survival [144]. Potentially, some of these same mechanisms underlie this effect.

Sex Steroid Receptors and the Kidney

Ovarian hormones and testosterone are steroids, and their classic actions have been defined as occurring through nuclear receptors. Nuclear receptors are generally associated with more chronic effects, "genomic" effects, such as changes in DNA transcription. Both androgen and estrogen steroid receptors have been localized to various sites along the renal tubule [119]. Androgen receptors have been shown to be expressed in PT as well as CCD as assessed by real-time RT-PCR on microdissected tubules [145]. Early studies characterizing renal binding sites for estrogen with a radiolabeled ligand (^{3}H-estradiol) found the greatest labeling in the PT S1 and S2 segments, but with smaller degrees of labeling in the distal tubule [146]. Classic steroid estrogen receptors exist as two isoforms, that is, ERα and ERβ, which are derived from two different genes. Both isoforms exist in the kidney, but precise tubular locations have not been clearly elucidated. Lane and colleagues [147] have studied the effects of ERα KO in mice on renal growth. They've shown that, unlike WT mice, ERα KO mice do not show sexual dimorphism of kidney weights, that is, in WT animals, male kidneys are usually larger for body size. The ERα KO mice also have reduced compensatory growth of the kidney in response to streptozotocin-induced diabetes [147].

In addition to the nuclear receptors, evidence for rapidly-acting, membrane-associated estrogen receptors in kidney have been reported [148]. Both ERα and the GPER-1 have been localized to membranes and may mediate some of the rapid, nongenomic actions of estradiol in kidney. GPER-1 is coupled to Gs and adenyl cyclase activation. Some [49], but not all, studies [149] suggest aldosterone is a ligand for GPER-1 in addition to estradiol. Female mice with whole-body GPER-1 KO have been shown to have hyperglycemia, impaired glucose tolerance, high BP, and reduced body growth. Cheng et al. [149] localized GPER-1 using immunofluorescence to the DCTs and loop of Henle in kidney. GPER-1 therefore, could potentially play a role in the redistribution of the thiazide-sensitive NCC observed with estradiol and in female rodents [45]. It is unclear whether GPER-1 KO mice have any renal phenotype associated with GPER-1 absence.

SUMMARY

At a gross level, determinants of renal functionality are similar in males and females, for example, renal function can be estimated by creatinine clearance or even plasma creatinine in both sexes; however, adjustments to the standard formulas do need to be made. Moreover, finer anatomical differences do exist, and regulation of a variety of renal activities including reabsorption of a number of vital nutrients has been demonstrated to be under the control of sex steroids or even the sex chromosomes. Therefore, the continued study of both sexes in health and disease is essential to our development of a comprehensive understanding of those factors that control the kidney.

Acknowledgments

The author would like to thank Marcus Byrd, B.S. and Amber Ecelbarger in artistic assistance with original artwork, as well as, redrawing/adapting several existing figures from the literature.

References

[1] Emamian SA, Nielsen MB, Pedersen JF, Ytte L. Kidney dimensions at sonography: correlation with age, sex, and habitus in 665 adult volunteers. AJR Am J Roentgenol 1993;160:83–6.

[2] Zeier M, Schonherr R, Amann K, Ritz E. Effects of testosterone on glomerular growth after uninephrectomy. Nephrol Dial Transplant 1998;13:2234–40.

[3] Oudar O, Elger M, Bankir L, Ganten D, Ganten U, Kriz W. Differences in rat kidney morphology between males, females and testosterone-treated females. Ren Physiol Biochem 1991;14:92–102.

[4] Hoy WE, Bertram JF, Denton RD, Zimanyi M, Samuel T, Hughson MD. Nephron number, glomerular volume, renal disease and hypertension. Curr Opin Nephrol Hypertens 2008;17:258–65.

[5] Munger K, Baylis C. Sex differences in renal hemodynamics in rats. Am J Physiol 1988;254:F223–31.

[6] Lankadeva YR, Singh RR, Tare M, Moritz KM, Denton KM. Loss of a kidney during fetal life: long-term consequences and lessons learned. Am J Physiol Renal Physiol 2014;306:F791–800.

[7] Puelles VG, Douglas-Denton RN, Zimanyi MA, Armitage JA, Hughson MD, Kerr PG, et al. Glomerular hypertrophy in subjects with low nephron number: contributions of sex, body size and race. Nephrol Dial Transplant 2014;29:1686–95.

[8] Gallo LA, Tran M, Cullen-McEwen LA, Denton KM, Jefferies AJ, Moritz KM, et al. Transgenerational programming of fetal nephron deficits and sex-specific adult hypertension in rats. Reprod Fertil Dev 2014;26:1032–43.

[9] O'Sullivan L, Cuffe JS, Koning A, Singh RR, Paravicini TM, Moritz KM. Excess prenatal corticosterone exposure results in albuminuria, sex-specific hypotension and altered heart rate responses to restraint stress in aged adult mice. Am J Physiol Renal Physiol 2015;308:F1065–73.

[10] Levey AS, Bosch JP, Lewis JB, Greene T, Rogers N, Roth D. A more accurate method to estimate glomerular filtration rate from serum creatinine: a new prediction equation. Modification of Diet in Renal Disease Study Group. Ann Intern Med 1999;130:461–70.

[11] Cockcroft DW, Gault MH. Prediction of creatinine clearance from serum creatinine. Nephron 1976;16:31–41.

[12] Levey AS, Stevens LA, Schmid CH, Zhang YL, Castro III AF, Feldman HI, et al. A new equation to estimate glomerular filtration rate. Ann Intern Med 2009;150:604–12.

[13] Stevens LA, Levey AS. Use of the MDRD study equation to estimate kidney function for drug dosing. Clin Pharmacol Ther 2009;86:465–7.

[14] Pechere-Bertschi A, Maillard M, Stalder H, Brunner HR, Burnier M. Blood pressure and renal haemodynamic response to salt during the normal menstrual cycle. Clin Sci (Lond) 2000;98:697–702.

[15] Bailey M, Silver R. Sex differences in circadian timing systems: implications for disease. Front Neuroendocrinol 2014;35:111–39.

[16] Kamperis K, Hansen MN, Hagstroem S, Hvistendahl G, Djurhuus JC, Rittig S. The circadian rhythm of urine production, and urinary vasopressin and prostaglandin E2 excretion in healthy children. J Urol 2004;171:2571–5.

[17] Hvistendahl GM, Frokiaer J, Nielsen S, Djurhuus JC. Gender differences in nighttime plasma arginine vasopressin and delayed compensatory urine output in the elderly population after desmopressin. J Urol 2007;178:2671–6.

[18] Stachenfeld NS, Keefe DL. Estrogen effects on osmotic regulation of AVP and fluid balance. Am J Physiol Endocrinol Metab 2002;283:E711–21.

[19] Stachenfeld NS, Silva C, Keefe DL, Kokoszka CA, Nadel ER. Effects of oral contraceptives on body fluid regulation. J Appl Physiol 1999;87:1016–25.

[20] Stachenfeld NS. Sex hormone effects on body fluid regulation. Exerc Sport Sci Rev 2008;36:152–9.

[21] Stachenfeld NS, Splenser AE, Calzone WL, Taylor MP, Keefe DL. Sex differences in osmotic regulation of AVP and renal sodium handling. J Appl Physiol (1985) 2001;91:1893–901.

[22] Lee JW, Chou CL, Knepper MA. Deep sequencing in microdissected renal tubules identifies nephron segment-specific transcriptomes. J Am Soc Nephrol 2015;26:2669–77.

[23] Juul KV, Bichet DG, Nielsen S, Norgaard JP. The physiological and pathophysiological functions of renal and extrarenal vasopressin V2 receptors. Am J Physiol Renal Physiol 2014;306:F931–40.

[24] Carrel L, Willard HF. X-inactivation profile reveals extensive variability in X-linked gene expression in females. Nature 2005;434:400–4.

[25] Liu J, Sharma N, Zheng W, Ji H, Tam H, Wu X, et al. Sex differences in vasopressin V(2) receptor expression and vasopressin-induced antidiuresis. Am J Physiol Renal Physiol 2011;300: F433–40.

[26] Herak-Kramberger CM, Breljak D, Ljubojevic M, Matokanovic M, Lovric M, Rogic D, et al. Sex-dependent expression of water channel AQP1 along the rat nephron. Am J Physiol Renal Physiol 2015;308:F809–21.

[27] Wilson JL, Miranda CA, Knepper MA. Vasopressin and the regulation of aquaporin-2. Clin Exp Nephrol 2013;17:751–64.

[28] Kortenoeven ML, Fenton RA. Renal aquaporins and water balance disorders. Biochim Biophys Acta 2014;1840:1533–49.

[29] Mahler B, Kamperis K, Ankarberg-Lindgren C, Frokiaer J, Djurhuus JC, Rittig S. Puberty alters renal water handling. Am J Physiol Renal Physiol 2013;305:F1728–35.

[30] Sharma N, Li J, Ecelbarger CM. Increased in renal proximal tubule GLUT5 and ketohexokinase in male mice, but not female mice, in response to high-fructose feeding may contribute to sex differences in renal responses. FASEB J 2014;28:1135.3.

[31] Kim SO, Song SH, Hwang EC, Oh KJ, Ahn K, Jung SI, et al. Changes in aquaporin (AQP)2 and AQP3 expression in ovariectomized rat urinary bladder: potential implication of water permeability in urinary bladder. World J Urol 2012;30:207–12.

[32] Zou LB, Zhang RJ, Tan YJ, Ding GL, Shi S, Zhang D, et al. Identification of estrogen response element in the aquaporin-2 gene that mediates estrogen-induced cell migration and invasion in human endometrial carcinoma. J Clin Endocrinol Metab 2011;96:E1399–408.

[33] Ecelbarger CA, Nielsen S, Olson BR, Murase T, Baker EA, Knepper MA, et al. Role of renal aquaporins in escape from vasopressin-induced antidiuresis in rat. J Clin Invest 1997;99:1852–63.

[34] Shin JA, Choi JH, Choi YH, Park EM. Conserved aquaporin 4 levels associated with reduction of brain edema are mediated by estrogen in the ischemic brain after experimental stroke. Biochim Biophys Acta 2011;1812:1154–63.

[35] Fraser CL, Arieff AI. Epidemiology, pathophysiology, and management of hyponatremic encephalopathy. Am J Med 1997;102:67–77.

[36] Ayus JC, Wheeler JM, Arieff AI. Postoperative hyponatremic encephalopathy in menstruant women. Ann Intern Med 1992;117:891–7.

[37] Ecelbarger CA, Murase T, Tian Y, Nielsen S, Knepper MA, Verbalis JG. Regulation of renal salt and water transporters during vasopressin escape. Prog Brain Res 2002;139:75–84.

[38] Verbalis JG, Murase T, Ecelbarger CA, Nielsen S, Knepper MA. Studies of renal aquaporin-2 expression during renal escape from vasopressin-induced antidiuresis. Adv Exp Med Biol 1998;449:395–406.

[39] Ayus JC, Varon J, Arieff AI. Hyponatremia, cerebral edema, and noncardiogenic pulmonary edema in marathon runners. Ann Intern Med 2000;132:711–14.

[40] Traiperm N, Gatterer H, Burtscher M. Plasma electrolyte and hematological changes after marathon running in adolescents. Med Sci Sports Exerc 2013;45:1182–7.

[41] Roberts WO. Risk factors for developing hyponatremia in marathon running. Clin J Sport Med 2008;18:550–1.

[42] Kim JM, Kim TH, Lee HH, Lee SH, Wang T. Postmenopausal hypertension and sodium sensitivity. J Menopausal Med 2014;20:1–6.

[43] Wenner MM, Stachenfeld NS. Blood pressure and water regulation: understanding sex hormone effects within and between men and women. J Physiol 2012;590:5949–61.

[44] Heo NJ, Son MJ, Lee JW, Jung JY, Kim S, Oh YK, et al. Effect of estradiol on the expression of renal sodium transporters in rats. Climacteric 2013;16:265–73.

[45] Verlander JW, Tran TM, Zhang L, Kaplan MR, Hebert SC. Estradiol enhances thiazide-sensitive NaCl cotransporter density in the apical plasma membrane of the distal convoluted tubule in ovariectomized rats. J Clin Invest 1998;101:1661–9.

[46] Kienitz T, Allolio B, Strasburger CJ, Quinkler M. Sex-specific regulation of ENaC and androgen receptor in female rat kidney. Horm Metab Res 2009;41:356–62.

[47] Tiwari S, Li L, Riazi S, Halagappa VK, Ecelbarger CM. Sex differences in adaptive down-regulation of pre-macula densa sodium transporters with Ang II infusion in mice. Am J Physiol Renal Physiol 2010;298(1):F187–95. Available from: http://dx.doi.org/10.1152/ajprenal.00088.2009.

[48] Tiwari S, Li L, Riazi S, Halagappa VK, Ecelbarger CM. Sex and age result in differential regulation of the renal thiazide-sensitive NaCl cotransporter and the epithelial sodium channel in angiotensin II-infused mice. Am J Nephrol 2009;30:554–62.
[49] Feldman RD, Gros R. Unraveling the mechanisms underlying the rapid vascular effects of steroids: sorting out the receptors and the pathways. Br J Pharmacol 2011;163:1163–9.
[50] Joseph A, Hess RA, Schaeffer DJ, Ko C, Hudgin-Spivey S, Chambon P, et al. Absence of estrogen receptor alpha leads to physiological alterations in the mouse epididymis and consequent defects in sperm function. Biol Reprod 2010;82:948–57.
[51] Gambling L, Dunford S, Wilson CA, McArdle HJ, Baines DL. Estrogen and progesterone regulate alpha, beta, and gammaENaC subunit mRNA levels in female rat kidney. Kidney Int 2004;65:1774–81.
[52] Dantas AP, Franco Mdo C, Silva-Antonialli MM, Tostes RC, Fortes ZB, Nigro D, et al. Gender differences in superoxide generation in microvessels of hypertensive rats: role of NAD(P)H-oxidase. Cardiovasc Res 2004;61:22–9.
[53] Ji H, Pesce C, Zheng W, Kim J, Zhang Y, Menini S, et al. Sex differences in renal injury and nitric oxide production in renal wrap hypertension. Am J Physiol Heart Circ Physiol 2005;288: H43–7.
[54] Reckelhoff JF. Gender differences in the regulation of blood pressure. Hypertension 2001;37:1199–208.
[55] Riazi S, Madala Halagappa VK, Dantas AP, Hu X, Ecelbarger CA. Sex differences in renal nitric oxide synthase (NOS), NAD (P)H oxidase, and blood pressure in obese Zucker rats. Gend Med 2007;4:1–16.
[56] Roesch DM, Shi M, Verbalis JG, Ecelbarger CA, Sandberg K. Sex differences in the renal and cardiovascular responses to aldosterone: role of nitric oxide. FASEB J 2005;20:A1194.
[57] Mattson DL, Wu F. Control of arterial blood pressure and renal sodium excretion by nitric oxide synthase in the renal medulla. Acta Physiol Scand 2000;168:149–54.
[58] Mattson DL, Wu F. Nitric oxide synthase activity and isoforms in rat renal vasculature. Hypertension 2000;35:337–41.
[59] Sullivan JC, Pardieck JL, Hyndman KA, Pollock JS. Renal NOS activity, expression, and localization in male and female spontaneously hypertensive rats. Am J Physiol Regul Integr Comp Physiol 2010;298:R61–9.
[60] Sasser JM, Brinson KN, Tipton AJ, Crislip GR, Sullivan JC. Blood pressure, sex, and female sex hormones influence renal inner medullary nitric oxide synthase activity and expression in spontaneously hypertensive rats. Am J Physiol 2010;298(1): F187–95.
[61] Yamaleyeva LM, Gallagher PE, Vinsant S, Chappell MC. Discoordinate regulation of renal nitric oxide synthase isoforms in ovariectomized mRen2.Lewis rats. Am J Physiol Regul Integr Comp Physiol 2007;292:R819–26.
[62] Nakano D, Pollock DM. Contribution of endothelin A receptors in endothelin 1-dependent natriuresis in female rats. Hypertension 2009;53:324–30.
[63] Kawanishi H, Hasegawa Y, Nakano D, Ohkita M, Takaoka M, Ohno Y, et al. Involvement of the endothelin ET(B) receptor in gender differences in deoxycorticosterone acetate-salt-induced hypertension. Clin Exp Pharmacol Physiol 2007;34:280–5.
[64] Plato CF, Pollock DM, Garvin JL. Endothelin inhibits thick ascending limb chloride flux via ET(B) receptor-mediated NO release. Am J Physiol Renal Physiol 2000;279:F326–33.
[65] Jin C, Speed JS, Hyndman KA, O'Connor PM, Pollock DM. Sex differences in ET-1 receptor expression and Ca2 + signaling in the IMCD. Am J Physiol Renal Physiol 2013;305:F1099–104.
[66] Banday AA, Lokhandwala MF. Dopamine receptors and hypertension. Curr Hypertens Rep 2008;10:268–75.
[67] Cuevas S, Villar VA, Jose PA, Armando I. Renal dopamine receptors, oxidative stress, and hypertension. Int J Mol Sci 2013;14:17553–72.
[68] Trainor BC. Stress responses and the mesolimbic dopamine system: social contexts and sex differences. Horm Behav 2011;60:457–69.
[69] Sanchez MG, Bourque M, Morissette M, Di Paolo T. Steroids-dopamine interactions in the pathophysiology and treatment of CNS disorders. CNS Neurosci Ther 2010;16:e43–71.
[70] Wang X, Li F, Jose PA, Ecelbarger CM. Reduction of renal dopamine receptor expression in obese Zucker rats: role of sex and angiotensin II. Am J Physiol Renal Physiol 2010;299: F1164–70.
[71] Chappell MC, Marshall AC, Alzayadneh EM, Shaltout HA, Diz DI. Update on the Angiotensin converting enzyme 2-angiotensin (1-7)-MAS receptor axis: fetal programing, sex differences, and intracellular pathways. Front Endocrinol (Lausanne) 2014;4:201.
[72] Miller JA, Anacta LA, Cattran DC. Impact of gender on the renal response to angiotensin II. Kidney Int 1999;55:278–85.
[73] Pendergrass KD, Pirro NT, Westwood BM, Ferrario CM, Brosnihan KB, Chappell MC. Sex differences in circulating and renal angiotensins of hypertensive mRen(2). Lewis but not normotensive Lewis rats. Am J Physiol Heart Circ Physiol 2008;295:H10–20.
[74] Ellison KE, Ingelfinger JR, Pivor M, Dzau VJ. Androgen regulation of rat renal angiotensinogen messenger RNA expression. J Clin Invest 1989;83:1941–5.
[75] Dean SA, Tan J, O'Brien ER, Leenen FH. 17beta-estradiol downregulates tissue angiotensin-converting enzyme and ANG II type 1 receptor in female rats. Am J Physiol Regul Integr Comp Physiol 2005;288:R759–66.
[76] Gordon MS, Chin WW, Shupnik MA. Regulation of angiotensinogen gene expression by estrogen. J Hypertens 1992;10 361–6.
[77] Ji H, Zheng W, Wu X, Liu J, Ecelbarger CM, Watkins R, et al. Sex chromosome effects unmasked in angiotensin II-induced hypertension. Hypertension 2010;55:1275–82.
[78] Rogers JL, Mitchell AR, Maric C, Sandberg K, Myers A, Mulroney SE. Effect of sex hormones on renal estrogen and angiotensin type 1 receptors in female and male rats. Am J Physiol Regul Integr Comp Physiol 2007;292:R794–9.
[79] Mirabito KM, Hilliard LM, Kett MM, Brown RD, Booth SC, Widdop RE, et al. Sex- and age-related differences in the chronic pressure-natriuresis relationship: role of the angiotensin type 2 receptor. Am J Physiol Renal Physiol 2014;307: F901–7.
[80] Hilliard LM, Chow CL, Mirabito KM, Steckelings UM, Unger T, Widdop RE, et al. Angiotensin type 2 receptor stimulation increases renal function in female, but not male, spontaneously hypertensive rats. Hypertension 2014;64:378–83.
[81] Mirabito KM, Hilliard LM, Head GA, Widdop RE, Denton KM. Pressor responsiveness to angiotensin II in female mice is enhanced with age: role of the angiotensin type 2 receptor. Biol Sex Differ 2014;5:13.
[82] Armando I, Jezova M, Juorio AV, Terron JA, Falcon-Neri A, Semino-Mora C, et al. Estrogen upregulates renal angiotensin II AT(2) receptors. Am J Physiol Renal Physiol 2002;283:F934–43.
[83] Schneider MP, Wach PF, Durley MK, Pollock JS, Pollock DM. Sex differences in acute ANG II-mediated hemodynamic responses in mice. Am J Physiol Regul Integr Comp Physiol 2010;299:R899–906.
[84] Chappel MC, Ferrario CM. ACE and ACE2: their role to balance the expression of angiotensin II and angiotensin-(1-7). Kidney Int 2006;70:8–10.

[85] Sullivan JC, Bhatia K, Yamamoto T, Elmarakby AA. Angiotensin (1-7) receptor antagonism equalizes angiotensin II-induced hypertension in male and female spontaneously hypertensive rats. Hypertension 2010;56:658–66.
[86] Gupte M, Thatcher SE, Boustany-Kari CM, Shoemaker R, Yiannikouris F, Zhang X, et al. Angiotensin converting enzyme 2 contributes to sex differences in the development of obesity hypertension in C57BL/6 mice. Arterioscler Thromb Vasc Biol 2012;32:1392–9.
[87] Roesch DM, Tian Y, Zheng W, Shi M, Verbalis JG, Sandberg K. Estradiol attenuates angiotensin-induced aldosterone secretion in ovariectomized rats. Endocrinology 2000;141:4629–36.
[88] Ellison DH, Velazquez H, Wright FS. Thiazide-sensitive sodium chloride cotransport in early distal tubule. Am J Physiol 1987;253:F546–54.
[89] Kim GH, Masilamani S, Turner R, Mitchell C, Wade JB, Knepper MA. The thiazide-sensitive Na-Cl cotransporter is an aldosterone-induced protein. Proc Natl Acad Sci USA 1998;95:14552–7.
[90] Blazer-Yost BL, Liu X, Helman SI. Hormonal regulation of ENaCs: insulin and aldosterone. Am J Physiol 1998;274: C1373–9.
[91] Riazi S, Khan O, Hu X, Ecelbarger CA. Aldosterone infusion with high-NaCl diet increases blood pressure in obese but not lean Zucker rats. Am J Physiol Renal Physiol 2006;291: F597–605.
[92] Michaelis M, Hofmann PJ, Gotz F, Bartel C, Kienitz T, Quinkler M. Sex-specific effects of spironolactone on blood pressure in gonadectomized male and female Wistar rats. Horm Metab Res 2012;44:291–5.
[93] Gros R, Ding Q, Davis M, Shaikh R, Liu B, Chorazyczewski J, et al. Delineating the receptor mechanisms underlying the rapid vascular contractile effects of aldosterone and estradiol. Can J Physiol Pharmacol 2011;89:655–63.
[94] Mitrakou A. Kidney: its impact on glucose homeostasis and hormonal regulation. Diabetes Res Clin Pract 2011;93(Suppl. 1):S66–72.
[95] Gerich JE. Role of the kidney in normal glucose homeostasis and in the hyperglycaemia of diabetes mellitus: therapeutic implications. Diabet Med 2010;27:136–42.
[96] Vrhovac I, Balen Eror D, Klessen D, Burger C, Breljak D, Kraus O, et al. Localizations of Na-D-glucose cotransporters SGLT1 and SGLT2 in human kidney and of SGLT1 in human small intestine, liver, lung, and heart. Pflugers Arch 2014;467:1881–98.
[97] Kim GW, Chung SH. Clinical implication of SGLT2 inhibitors in type 2 diabetes. Arch Pharm Res 2014;37:957–66.
[98] Gustavsson C, Yassin K, Wahlstrom E, Cheung L, Lindberg J, Brismar K, et al. Sex-different hepaticglycogen content and glucose output in rats. BMC Biochem 2010;11:38.
[99] Samimi A, Ramesh S, Turin TC, MacRae JM, Sarna MA, Reimer RA, et al. Serum uric acid level, blood pressure, and vascular angiotensin II responsiveness in healthy men and women. Physiol Rep 2014;2(12).
[100] Prasad N, Bhadauria D. Renal phosphate handling: physiology. Indian J Endocrinol Metab 2013;17:620–7.
[101] Quarles LD. Role of FGF23 in vitamin D and phosphate metabolism: implications in chronic kidney disease. Exp Cell Res 2012;318:1040–8.
[102] Castelo-Branco C, Martinez de Osaba MJ, Pons F, Gonzalez-Merlo J. The effect of hormone replacement therapy on post-menopausal bone loss. Eur J Obstet Gynecol Reprod Biol 1992;44:131–6.
[103] Faroqui S, Levi M, Soleimani M, Amlal H. Estrogen downregulates the proximal tubule type IIa sodium phosphate cotransporter causing phosphate wasting and hypophosphatemia. Kidney Int 2008;73:1141–50.
[104] Uemura H, Irahara M, Yoneda N, Yasui T, Genjida K, Miyamoto KI, et al. Close correlation between estrogen treatment and renal phosphate reabsorption capacity. J Clin Endocrinol Metab 2000;85:1215–19.
[105] Khosla S, Atkinson EJ, Melton III LJ, Riggs BL. Effects of age and estrogen status on serum parathyroid hormone levels and biochemical markers of bone turnover in women: a population-based study. J Clin Endocrinol Metab 1997;82:1522–7.
[106] Carrillo-Lopez N, Roman-Garcia P, Rodriguez-Rebollar A, Fernandez-Martin JL, Naves-Diaz M, Cannata-Andia JB. Indirect regulation of PTH by estrogens may require FGF23. J Am Soc Nephrol 2009;20:2009–17.
[107] Jeon US. Kidney and calcium homeostasis. Electrolyte Blood Press 2008;6:68–76.
[108] Dong XL, Zhang Y, Wong MS. Estrogen deficiency-induced Ca balance impairment is associated with decrease in expression of epithelial Ca transport proteins in aged female rats. Life Sci 2014;96:26–32.
[109] Hagenfeldt Y, Berlin T. The human renal 25-hydroxyvitamin D3-1 alpha-hydroxylase: properties studied by isotope-dilution mass spectrometry. Eur J Clin Invest 1992;22:223–8.
[110] Bover J, Egido J, Fernandez-Giraldez E, Praga M, Solozabal-Campos C, Torregrosa JV, et al. Vitamin D, vitamin D receptor and the importance of its activation in patients with chronic kidney disease. Nefrologia 2015;35:28–41.
[111] Martin M, Valls J, Betriu A, Fernandez E, Valdivielso JM. Association of serum phosphorus with subclinical atherosclerosis in chronic kidney disease. Sex makes a difference. Atherosclerosis 2015;241:264–70.
[112] Trinchieri A. Epidemiology of urolithiasis: an update. Clin Cases Miner Bone Metab 2008;5:101–6.
[113] Jayachandran M, Lugo G, Heiling H, Miller VM, Rule AD, Lieske JC. Extracellular vesicles in urine of women with but not without kidney stones manifest patterns similar to men: a case control study. Biol Sex Differ 2015;6:2.
[114] Gillen DL, Coe FL, Worcester EM. Nephrolithiasis and increased blood pressure among females with high body mass index. Am J Kidney Dis 2005;46:263–9.
[115] Pelis RM, Wright SH. SLC22, SLC44, and SLC47 transporters—organic anion and cation transporters: molecular and cellular properties. Curr Top Membr 2014;73:233–61.
[116] Cerrutti JA, Quaglia NB, Brandoni A, Torres AM. Effects of gender on the pharmacokinetics of drugs secreted by the renal organic anions transport systems in the rat. Pharmacol Res 2002;45:107–12.
[117] Groves CE, Suhre WB, Cherrington NJ, Wright SH. Sex differences in the mRNA, protein, and functional expression of organic anion transporter (Oat) 1, Oat3, and organic cation transporter (Oct) 2 in rabbit renal proximal tubules. J Pharmacol Exp Ther 2006;316:743–52.
[118] Seeland U, Regitz-Zagrosek V. Sex and gender differences in cardiovascular drug therapy. Handb Exp Pharmacol 2012;211–36.
[119] Sabolic I, Asif AR, Budach WE, Wanke C, Bahn A, Burckhardt G. Gender differences in kidney function. Pflugers Arch 2007;455:397–429.
[120] Bouthillier L, Greselin E, Brodeur J, Viau C, Charbonneau M. Male rat specific nephrotoxicity resulting from subchronic administration of hexachlorobenzene. Toxicol Appl Pharmacol 1991;110:315–26.
[121] Blumbach K, Pahler A, Deger HM, Dekant W. Biotransformation and male rat-specific renal toxicity of diethyl ethyl- and dimethyl methylphosphonate. Toxicol Sci 2000;53:24–32.

[122] Gautier JC, Gury T, Guffroy M, Masson R, Khan-Malek R, Hoffman D, et al. Comparison between male and female Sprague-Dawley rats in the response of urinary biomarkers to injury induced by gentamicin. Toxicol Pathol 2014;42 1105–16.

[123] Zepnik H, Volkel W, Dekant W. Toxicokinetics of the mycotoxin ochratoxin A in F 344 rats after oral administration. Toxicol Appl Pharmacol 2003;192:36–44.

[124] Bolignano D, Mattace-Raso F, Sijbrands EJ, Zoccali C. The aging kidney revisited: a systematic review. Ageing Res Rev 2014;14:65–80.

[125] Fortepiani LA, Yanes L, Zhang H, Racusen LC, Reckelhoff JF. Role of androgens in mediating renal injury in aging SHR. Hypertension 2003;42:952–5.

[126] Erdely A, Greenfeld Z, Wagner L, Baylis C. Sexual dimorphism in the aging kidney: effects on injury and nitric oxide system. Kidney Int 2003;63:1021–6.

[127] Loria A, Reverte V, Salazar F, Saez F, Llinas MT, Salazar FJ. Sex and age differences of renal function in rats with reduced ANG II activity during the nephrogenic period. Am J Physiol Renal Physiol 2007;293:F506–10.

[128] Lee WL, Cheng MH, Tarng DC, Yang WC, Lee FK, Wang PH. The benefits of estrogen or selective estrogen receptor modulator on kidney and its related disease-chronic kidney disease-mineral and bone disorder: osteoporosis. J Chin Med Assoc 2013;76:365–71.

[129] Maric C, Xu Q, Sandberg K, Hinojosa-Laborde C. Age-related renal disease in female Dahl salt-sensitive rats is attenuated with 17 beta-estradiol supplementation by modulating nitric oxide synthase expression. Gend Med 2008;5:147–59.

[130] Cowley Jr AW, Abe M, Mori T, O'Connor PM, Ohsaki Y, Zheleznova NN. Reactive oxygen species as important determinants of medullary flow, sodium excretion, and hypertension. Am J Physiol Renal Physiol 2015;308:F179–97.

[131] Kwekel JC, Vijay V, Desai VG, Moland CL, Fuscoe JC. Age and sex differences in kidney microRNA expression during the life span of F344 rats. Biol Sex Differ 2015;6:1.

[132] Iseki K. Gender differences in chronic kidney disease. Kidney Int 2008;74:415–17.

[133] Jafar TH, Schmid CH, Stark PC, Toto R, Remuzzi G, Ruggenenti P, et al. The rate of progression of renal disease may not be slower in women compared with men: a patient-level meta-analysis. Nephrol Dial Transplant 2003;18:2047–53.

[134] Doublier S, Lupia E, Catanuto P, Elliot SJ. Estrogens and progression of diabetic kidney damage. Curr Diabetes Rev 2011;7:28–34.

[135] Agarwal M, Selvan V, Freedman BI, Liu Y, Wagenknecht LE. The relationship between albuminuria and hormone therapy in postmenopausal women. Am J Kidney Dis 2005;45:1019–25.

[136] Yu MK, Lyles CR, Bent-Shaw LA, Young BA, Pathways A. Risk factor, age and sex differences in chronic kidney disease prevalence in a diabetic cohort: the pathways study. Am J Nephrol 2012;36:245–51.

[137] Dousdampanis P, Trigka K, Fourtounas C, Bargman JM. Role of testosterone in the pathogenesis, progression, prognosis and comorbidity of men with chronic kidney disease. Ther Apher Dial 2014;18:220–30.

[138] Nagao S, Kusaka M, Nishii K, Marunouchi T, Kurahashi H, Takahashi H, et al. Androgen receptor pathway in rats with autosomal dominant polycystic kidney disease. J Am Soc Nephrol 2005;16:2052–62.

[139] Belibi F, Ravichandran K, Zafar I, He Z, Edelstein CL. mTORC1/2 and rapamycin in female Han:SPRD rats with polycystic kidney disease. Am J Physiol Renal Physiol 2011;300:F236–44.

[140] Quinlan MR, Cronin P, Daly PJ, Watson RW, Manucha W, Docherty NG, et al. A gender comparison of postobstructive injury in the rat kidney. Kidney Blood Press Res 2010;33:266–73.

[141] Ji H, Zheng W, Wu X, Liu J, Ecelbarger CM, Watkins R, et al. Sex chromosome effects unmasked in angiotensin II-induced hypertension. Hypertension 2010;55:1275–82.

[142] Hatakeyama C, Anderson CL, Beever CL, Penaherrera MS, Brown CJ, Robinson WP. The dynamics of X-inactivation skewing as women age. Clin Genet 2004;66:327–32.

[143] Simmonds MJ, Benavente D, Brand OJ, Moore J, Ball S, Ferro CJ, et al. Skewing of female X-chromosome inactivation: an epigenetic risk factor for kidney transplantation outcome. Transplantation 2013;95:e25–8.

[144] Zhou JY, Cheng J, Huang HF, Shen Y, Jiang Y, Chen JH. The effect of donor-recipient gender mismatch on short- and long-term graft survival in kidney transplantation: a systematic review and meta-analysis. Clin Transplant 2013;27:764–71.

[145] Boulkroun S, Le Moellic C, Blot-Chabaud M, Farman N, Courtois-Coutry N. Expression of androgen receptor and androgen regulation of NDRG2 in the rat renal collecting duct. Pflugers Arch 2005;451:388–94.

[146] Davidoff M, Caffier H, Schiebler TH. Steroid hormone binding receptors in the rat kidney. Histochemistry 1980;69:39–48.

[147] Lane PH. Estrogen receptors in the kidney: lessons from genetically altered mice. Gend Med 2008;5(Suppl. A):S11–18.

[148] Bjornstrom L, Sjoberg M. Mechanisms of estrogen receptor signaling: convergence of genomic and nongenomic actions on target genes. Mol Endocrinol 2005;19:833–42.

[149] Cheng SB, Dong J, Pang Y, LaRocca J, Hixon M, Thomas P, et al. Anatomical location and redistribution of G protein-coupled estrogen receptor-1 during the estrus cycle in mouse kidney and specific binding to estrogens but not aldosterone. Mol Cell Endocrinol 2014;382:950–9.

CHAPTER

8

Sex Differences in Gastrointestinal Physiology and Diseases: From Endogenous Sex Hormones to Environmental Endocrine Disruptor Agents

Eric Houdeau

Intestinal Development, Xenobiotics & ImmunoToxicology, Research Centre in Food Toxicology (INRA Toxalim UMR 1331), Toulouse, France

Besides their fundamental role in the development and maintenance of reproductive function, sex steroids show a variety of biological effects in the stomach and small and large bowels that participate to maintain homeostasis of the gastrointestinal (GI) tract, and affect susceptibility and/or magnitude of GI disorders. The significance of sex steroid receptors in the digestive tract has received particular attention since the late 70s, for example, when the presence of estrogen receptors (ER) in primary colon malignancies suggested that some large bowel cancers could be endocrine dependent, possibly sharing common etiological factors with breast cancer [1]. With the discovery in 1996 of ERβ [2], a new nuclear receptor for estradiol with dominant functions opposite to those of other ER (renamed ERα) [3,4], many studies have reevaluated the importance of estrogen signaling in GI physiology and pathophysiology. Differences have been shown in the ERα/ERβ expression ratio along the GI tract that determine the regional sensitivity to circulating estrogens and the resulting global actions according to the gender and age of individuals. Hence, novel estrogen-dependent effects have emerged in the control of epithelial cell proliferation [5], intestinal permeability to luminal agents [6], and fluid movements [7], as well as inflammatory responses [8,9] or central nervous targets involved in pain processes [10,11]. A sex bias in GI functions is supported by sexually dimorphic gene expression in the small intestine, suggesting sex dimorphic phenotypes and functions [12]. Estrogen-mediated influences on GI functions occur through genomic (ie, transcriptional effects mediated by nuclear receptors) and nongenomic responses (ie, independent of gene transcription), the latter requiring the activation of membrane-associated G protein-coupled estrogen receptor termed GPR30 or GPER. Membrane GPR30 is found in the intestine [7], as well as in the central nervous system, and is involved in the process of peripheral pain perception together with nuclear ERs [13,14]. Receptors for progesterone (PR-A and B isoforms) and androgen (AR-A and B) are also expressed along the GI tract [15]. Both PR and AR have been found in fetal tissues as observed for ER [16,17], suggesting an important role for sex hormones in the architectural and functional development of the GI tract. In females, a prominent PR expression into the colonic musculature has been associated with a sex-related difference in gastric emptying and intestinal motility, leading to the hypothesis of a progesterone-evoked "block" of muscle contractility in slow transit constipation [18]. In males, androgens at puberty affect the GI microbiota composition, which becomes less diverse than the female commensal flora, with consequences on immune system activity [19,20]. Today, most experimental findings are relevant to the clinical debate for sex-related symptomatology and prevalence in epidemiology, trying to decipher many confounding factors between "true" sex-linked biological effects and gender-related social determinants in our modern societies. This chapter presents the current state of knowledge regarding the role of estrogen, progesterone, and androgen signaling

Sex Differences in Physiology
DOI: http://dx.doi.org/10.1016/B978-0-12-802388-4.00008-2

pathways along the digestive tract, along with the influence of hormonal status on disease susceptibility and activity. We will discuss whether sex differences in GI functions are linked to sexually dimorphic expression of hormone receptors in responsive tissues or to the sole difference in circulating sex steroids between males and females. A further understanding of the complex role of these hormones in GI homeostasis and its dysfunctions is necessary to dissect out modulatory mechanisms in the brain–gut axis. This includes sex differences in stress responses and regulatory mechanisms of visceral pain perception, the latter encompassing neuronal system activities from afferent nerve sources and local nerve interactions with the mucosal immune system [21,22]. Finally, this chapter discusses how these receptors may also act as sensors of endocrine disruptors present in the diet. Indeed, part of these molecules shows estrogen-like activity (eg, bisphenols, pesticides, plant-derived phytoestrogens), evokes chronic disturbances in GI physiology in rodents, with immunotoxic effects after developmental exposure, a long-lasting perturbation appearing predominantly in females [23,24].

ESTROGEN, ANDROGEN, AND PROGESTERONE RECEPTOR EXPRESSION ALONG THE GI TRACT

Tissue Distribution of Estrogen Receptors

Before the discovery of ERβ by Kuiper et al. [2], an earlier study had already reported various ER-like DNA sequences and different molecular weights for ER-like proteins along the rat intestine, suggesting variant genes encoding for different ERs in the GI tract [25]. With the distinction made by Gustafsson's team [2], it is now well established that biological effects of estrogens are mediated by two receptors, namely ERα and ERβ, which are products of distinct genes located on different chromosomes [3]. They belong to the superfamily of nuclear receptors (including PR and AR as well) that translate hormone stimuli into genomic modifications, enhancing or repressing the transcriptional expression of various genes [26]. A dominant function for ERβ appears to activate estrogen signaling pathways often with opposite actions on ERα-mediated transcriptional responses [3]. This suggests an intricate interplay among ER genomic activities when ERα and ERβ are coexpressed in tissues and cells. In this context, because the endogenous ligand 17β-estradiol binds equally to ERα and ERβ, the resulting actions for estradiol in organ physiopathology are closely linked to the ERα/ERβ expression ratio in target tissues. Furthermore, local determination of ERα/ERβ levels appears crucial for toxicity testing of estrogenic endocrine disruptors that circulate in the environment, some exhibiting better binding activity for either ERα or ERβ depending on their chemical structure, with effects closely linked to the dominant ER type expressed in exposed tissues.

In the esophagus, mapping the distribution of ERα and ERβ proteins has received poor attention in humans. By using immunohistochemistry, Taylor and Al-Azzawi [27] showed ERα and ERβ costaining in the stratified squamous epithelium, while ERβ proteins predominated in the mucous glands (Table 8.1). In the human stomach, the distribution pattern of ER types has gained attention based on clinical data indicating a sexual dimorphism in gastric acid secretion. Both ERα and ERβ have been detected in the human normal gastric mucosa as observed in rodent tissues (Table 8.1). In the female rat using RT-PCR and Northern blot analysis, ERβ mRNAs were found expressed in greater abundance than ERα transcripts, with the highest level found in the fundic mucosa [29]. Immunolocalization showed ERβ and ERα proteins in fundic parietal cells involved in acid secretion, as well as in epithelial cells in the progenitor zone [30]. In humans, higher levels of ERβ mRNA have also been reported in fetal gastric tissues [16,17]. Because these authors reported a prominent ERβ mRNA expression along the whole fetal GI tract, regardless of the gestational age, it is suggested that the effects of estrogens during GI development are predominantly mediated through ERβ rather than ERα. In adult rat gastric tissues, the demonstration that both ER proteins are often found colocalized in parietal cells [30] also suggests that circulating estrogens may directly control the secretory activity of these cells through a balanced effect between ERβ and ERα signaling pathways.

In the duodenum, studies report conflicting observations depending on the species and hormonal status. ERα mRNAs have been found abundantly expressed in duodenal tissues from ovariectomized (OVX) rats [29], whereas ERβ transcripts were reported as the only represented ER transcript in this region in female mice without gonadectomy [34]. This finding has been confirmed using Western blot analysis in mouse neonate as well as in adults, irrespective of sex. One may speculate that interspecies differences could explain these discrepancies between rat and mice studies, and consistent with this idea, a greater abundance of ERα transcripts has been also reported in the bovine duodenum [37] as observed in female rats [29] in contrast to mice [34] (Table 8.1). On the other hand, it is also conceivable that the ERα/ERβ expression ratio in the small intestine may depend

TABLE 8.1 Tissue Distribution of Sex Hormone Receptors in the Upper Gastrointestinal Tract and Main Physiological Effects

Region	Species	Sex	Model	Tissue/cells	Methods	Main findings/effects	Refs.
Esophagus	Human	ns	Adult normal tissues	Stratified epithelium, mucous glands	IHC	• ERα + ERβ • ERβ only	[27]
		Female	Fetus (20 weeks)	ns	RT-PCR	• ERβ > ERα	[16,17]
	Rat	Female/male			IHC	• ERα in female only	[28]
Stomach	Rat	Female	OVX±17β-estradiol	Fundus and antrum	RT-PCR	• ERβ > ERα in fundic mucosa/inverse expression ratio in antrum	[29]
				Fundic parietal cells (acid secretion)	IHC	• ERα and ERβ coexpressed • E2 dose-dependently decreases gastric acid secretion	[30]
	Human	Female/male	Fetus (13–20 weeks)	ns	RT-PCR	• ERβ > ERα	[16,17]
	Mouse	ns	WT vs GPR30-lacZ mice	Chief cells (secret hydrolytic enzymes) in fundic glands	Gene reporter assay	• Local GPER activity	[31]
Duodenum	Rat	Female	OVX	ns	RT-PCR	• ERα > ERβ	[29]
			Adults/isolated duodenal segments±17β-estradiol	Epithelial cells	Ligand activity	• Rapid increase of Ca^{2+} absorption (ie, nongenomic activity)	[32,33]
	Mouse	Female/male	Neonates vs adults	Epithelial cells (gastroduodenal junction)	RT-PCR, WB, IHC	• ERβ >> ERα/no sex difference • Decrease in ERβ expression during development • ERβ-positive cells expand throughout the duodenal villi after weaning	[34]
		Male	Adults±17β-estradiol	Smooth muscle cells	Ligand activity	• E2 mediates rapid/nongenomic (GPER-like) decrease of spontaneous contractility	[35]
		Female	OVX±17β-estradiol	Smooth muscle cells	IHC	• PR in duodenal musculature • No effect of E2 treatment on PR expression	[36]
	Cow	Female	Heifer	ns	RT-PCR	• ERα + ERβ/no sex difference • Higher AR and PR expression than in distal gut	[37]
	Human	Female/male	Healthy volunteers (20–29 vs 60–69 years); duodenal luminal administration of 17β-estradiol	Proximal duodenal mucosa, epithelial cells (villous and crypt cells)	WB, IHC	• No sex difference in ER expression • E2 stimulates bicarbonate (HCO_3^-) secretion in all ages and sex	[38]
Jejunum/ileum	Mouse	Female/male	OVX±17β-estradiol	Lamina propria, enteric neurons	WB, IHC	• ERβ > ERα • ERs expressed in nerve cells of Auerbach and Meissner plexuses	[39]
		Female	OVX±17β-estradiol	Smooth muscle cells	IHC	• PR in jejunal musculature • No effect of E2 treatment on PR expression	[36]
		ns	Duodenal vs ileal muscle strips ± testosterone or 5α-DHT in organ bath	Smooth muscle cells	Ligand activity	• Androgens potentiate extracellular Ca^{2+} and ACh-induced ileal contractility (nongenomic mechanism)/no effect on duodenal musculature	[40]

(*Continued*)

TABLE 8.1 (Continued)

Region	Species	Sex	Model	Tissue/cells	Methods	Main findings/effects	Refs.
	Rat	Male	Ileal strips in organ bath±17β-estradiol	Smooth muscle cells	Ligand activity	• E2 decreases ACh-induced contractions	[41]
	Human	Female/male	Fetus (13–20 weeks)	ns	RT-PCR	• ERβ > ERα	[17]
	Cow	Female	Heifer	ns	RT-PCR	• High ERα/ERβ ratio in the jejunum/inverse relation in the ileum • AR and low PR expression	[37]

ER, estrogen receptors (alpha and beta); PR, progesterone receptors (A and B); AR, androgen receptors (A and B); OVX, ovariectomy; E2, estradiol; WT, wild type; GPER (or Gpr30), G protein-coupled estrogen receptor; ACh, acetylcholine; RT-PCR, reverse transcription polymerase chain reaction; WB, Western blotting; IHC, immunohistochemistry; ns, not specified.

upon marked differences in the circulating levels of sex hormones between cyclic and gonadectomized females in the above rodent studies. In human biopsies, Tuo et al. [38] showed that both ERα and ERβ proteins were expressed at similar levels in the duodenal mucosa of healthy male and female volunteers, and primarily found in epithelial cells.

ERα and ERβ protein distribution has been studied in the small intestine and colon using Western blot analysis and immunohistochemistry in adult female and male mice [39]. These authors showed ERβ as the more prominent nuclear ER along the gut, with higher expression in colonic tissues compared to the small bowel (Tables 8.1 and 8.2). Such a regionalization is consistent with the very recent study of Choijookhuu et al. [34] in the same species. However in their report, authors showed that the jejunum and the ileum are devoid of ERβ transcripts, which are expressed only in the colon, and that the whole intestine appears negative for ERα mRNAs. Apart from this inconsistency, a dominant ERβ expression in colonic mucosa has been consistently observed in humans, both women and men [48], and during fetal life [17] (Table 8.2). In the human normal colonic mucosa, a gradient from moderate to strong ERβ immunostaining was noticed from the villi apex to crypt epithelial cells, with a distribution pattern similar in the proximal and distal segments [49]. This indicates that ERβ is preferentially expressed in stem cells located at the base of the colonic crypts, ie, in progenitor and proliferative cells that differentiate into all epithelial cell lineages for epithelial renewal [5,42]. In women as well as men, ERβ staining was also reported in deeper human colonic mucosa including endothelial cells, vascular smooth muscle, lymphocytes, and enteric neurons of the Auerbach and Meissner plexuses, as well as in the smooth musculature of the colonic wall [49]. Depending on the study, either there is no evidence for sex differences in the colonic amount and cell distribution of ERβ in normal mucosa [49], or there is higher expression in ERβ mRNA levels (2×) in women [48] (Table 8.2). Of note, the presence of ER in enteric neurons has been also reported in rats and in mice, where both ERα and ERβ proteins were present within the nucleus and the cytoplasm, with some ERα staining reaching the nerve fibers [30,39]. Interestingly, Kawano et al. [39] showed immune cell-specific distribution of ERα in submucosal macrophages along the whole intestine, the number of which fluctuated according to the reproductive cycle period, but in the colon only (Table 8.2). More precisely, the highest number of ERα-positive macrophages was found between two successive sexual cycles, ie, from diestrus to proestrus stages when plasma progesterone falls and estradiol increases in rodents [61]. Importantly, such an estrous cycle-mediated change has been mimicked after 17β-estradiol treatment in OVX mice, suggesting a main role for estradiol in the modulation of macrophage density in the colon during the reproductive cycle. In addition to macrophages, ERα and ERβ have also been reported expressed in a variety of immune cells, including lymphocytes and dendritic cells that compose the gut-associated lymphoid tissue (GALT) [62].

Finally, ERα proteins have been found to be abundantly expressed in the anorectal region in both male and female rats, but no information is available on ERβ. Interestingly, ERα immunostaining was mainly detected in the nuclei of basal cells of squamous epithelium of the anus connected to the anorectal junction, and the number of ERα-positive cells sharply decreased with age in males only [28] (Table 8.2).

Beside these classical nuclear ERs, nongenomic regulatory pathways should be also taken into account when observing an estrogenic modulation of GI functions. However, to our knowledge, there is no report mapping the expression of a plasma membrane pool of GPER along the GI tract, except one study based on a mutant mouse model (ie, Lac-Z reporter in

TABLE 8.2 Tissue Distribution of Sex Hormone Receptors in the Lower Gastrointestinal Tract and Main Physiological Effects

Region	Species	Sex	Model	Tissue/cells	Methods	Main findings/effects	Refs.
Colon	Mouse	Female/male	Young adults OVX±17β-estradiol	Epithelial cells, macrophages	WB, IHC	• ERβ > ERα • ERα-positive macrophages vary with estrous cycle stages and estrogen levels	[39]
		Female	Neonates vs cyclic adults	Mucosa	RT-PCR	• ERβ >> ERα: stable expression during development	[34]
		Female/male	WT vs ERβ$^{-/-}$ mice	Epithelium, mucosa	Phenotype	• ERβ involved in epithelial cell renewal/cell-to-cell junctions (barrier integrity)/mucin secretion by goblet cells • ERβ drives eosinophil and lymphocyte mucosal infiltration	[5,42]
		Female	OVX±17β-estradiol vs pregnant mice, ERβ siRNA transfection	Epithelium, mucosa	WB, IHC, ISH	• ERβ expression in surface epithelial cells of the proximal colon increases in late pregnancy • E2 regulates the expression of NHE3 through ERβ ligand activity, hence drives Na^+ (and water) absorption	[43]
		Male	WT mice pretreated with P4 : in vivo model vs colonic strips in organ bath	Smooth muscle cells	Ligand activity	• P4 inhibits spontaneous and ACh-induced contractility • P4 reduces fecal output	[44]
			Muscle strips pretreated with testosterone or 5α-DHT in organ bath	Smooth muscle cells	Ligand activity	• Androgens induce Ca^{2+} sensitization in longitudinal muscle, hence potentiate spontaneous and ACh-induced contractility (nongenomic mechanism)	[45]
	Guinea pig	Male	Adults ± P4/P4 antagonist, colon segments in organ bath	Smooth muscle cells	Ligand activity	• P4 inhibits spontaneous and induced contractility through regulation of prostaglandins (PGE2/PGF2α) and G protein levels	[46,47]
	Human	Female/male	Midgestation fetus/adults	Epithelial cells, endothelial cells, GALT, enteric neurons	RT-PCR, WB, IHC	• ERβ >> ERα, higher ERβ levels in women than in men • Gradient of ERβ expression along colonic crypt • Regulation of epithelial cell proliferation • AR-B > AR-A	[15,17,30,39,48,49]
		Female	Adults with normal bowel transit vs chronic constipation (STC)	Smooth muscle cells	Ligand activity	• Overexpression of PR-A and PR-B in patients with STC compared to women with normal transit • PR-A mediates P4 inhibition of colonic muscle contractility through upregulation of Gs proteins that mediate relaxation/PR-B downregulates Gq proteins that mediate contraction	[18,46,50,51]
	Rat	Female	Estrous cycle, OVX±17β-estradiol or P4, selective ERβ/ERα agonists	Epithelium	Ligand activity	• Estrous cycle variation of colonic permeability, depending on estrogen levels • ERβ agonist decreases colonic permeability and reinforces gut barrier function through upregulation of TJ proteins/no effect of P4	[6]

(Continued)

TABLE 8.2 (Continued)

Region	Species	Sex	Model	Tissue/cells	Methods	Main findings/effects	Refs.
			Estrous cycle, OVX±17β-estradiol	Smooth muscle cells	Ligand activity	• Estrous cycle-related changes in OT-induced contraction, depending on estrogen levels • E2 increases OTR expression in enteric neurons	[52]
			Isolated distal colonic crypts±17β-estradiol/antagonist of nuclear ERs	Epithelium	Ligand activity	• Estrogen levels regulate water movements through nongenomic (GPER-like) control of Cl^- secretion/no effect of E2 in males	[53–56]
				Epithelium	Ligand activity	• E2 increases Ca^{2+} absorption in distal colon through GPER-like pathway	[57,58]
Anorectal junction	Rat	Female/male	Adults	Epithelium	IHC	• Number of ERα-positive cells decreases with age in males	[28]
		Female	OVX±17β-estradiol and/or P4	Mucosa	IHC, in vivo ligand activity	• P4 regulates collagen levels in synergy with E2, and E2 increases blood vessel density	[59]
	Human	Female/male	Normal continence	Smooth muscle cells, squamous epithelium	IHC	• AR coexpressed with ER and PR	[60]

ER, estrogen receptors (alpha and beta); PR, progesterone receptors (A and B); AR, androgen receptors (A and B); OVX, ovariectomy; E2, estradiol; P4, progesterone; WT, wild type; WB, Western blotting; IHC, immunohistochemistry; RT-PCR, reverse transcription polymerase chain reaction; ISH, in situ hybridization; NHE3, sodium/hydrogen exchanger-3; STC, slow transit constipation; TJ, tight junctions; ACh, acetylcholine; GALT, gut-associated lymphoid tissue; GPER, G protein-coupled estrogen receptor; OT/OTR, oxytocin/oxytocin receptors.

the GPER locus with subsequent X-galactosidase reporter assay for cellular localization), giving a specific signature for GPER-positive cells in gastric fundic glands [31] (Table 8.1). Nevertheless, the existence of GPER in other parts of the GI tract has been suggested from studies describing estrogen-dependent rapid signaling responses in rodent colonic tissues, and more recently in the sensory nerve circuitry controlling intestinal pain perception.

Progesterone and Androgen Receptors

Although the GI tract is mainly considered an estrogen-sensitive target, a number of studies also report that sex-related differences in circulating progesterone and androgen concentrations affect several GI functions. However, mapping of PR and AR expression in the GI tissues has been poorly documented. Progesterone receptors are represented by two protein isoforms, namely PR-A and PR-B, both originating from a single gene in rodents and humans. Using immunohistochemical analysis along the GI tract in OVX mice, PR proteins were only detected in the smooth muscle cells of the duodenum and the jejunum [36] (Table 8.1). Studies in women also reported PR mRNA expression in the musculature of normal colon [46,50,51], with increased expression in patients suffering from slow colonic transit and constipation (Table 8.2). In OVX mice, 17β-estradiol treatment did not affect PR levels in the GI tract, indicating that this pool of receptors in the intestinal muscle wall is not affected by plasma estrogen concentrations [36]. Altogether, the apparent low abundance of PR proteins in the GI organs compared to other sex steroid receptors is in line with the tissue distribution of PR mRNA found in bovine tissues, where PR expression also appears restricted to upper gut segments as noticed in mice studies and always with low quantities of transcripts compared to AR and ER mRNAs [37] (Table 8.1). It is however noteworthy that low density of receptors does not necessarily translate into small effects in target tissues. For instance, in the perianal region where PR expression is low compared to ER [37,60], a study in OVX rats with hormone replacement emphasized a greater effect of progesterone compared to 17β-estradiol on anal function, which involved progesterone-mediated changes in collagen levels, and a synergic activity with estrogen to control submucosal blood vessels density [59].

The distribution pattern of AR-A and AR-B has been examined in the human fetus and adult GI tissues [15], and in the bovine GI tract [37] (Tables 8.1 and 8.2).

In these studies, the AR-B isoform appeared to be the major receptor type compared to AR-A along the GI tract until the rectum with putative overexpression of AR-B proteins in the small gut compared to distal regions. Immunolocalization studies of the human anal canal clearly showed intense AR expression in the musculature of the internal sphincter in nearly all women and men patients with normal anal continence [60]. Interestingly, AR proteins in the human anal sphincter were reported to colocalize with abundant ER staining, and to a less extent with PR, at all ages and in both sexes (Table 8.2).

SEX HORMONE MODULATION OF GASTROINTESTINAL MOTILITY

The influence of hormonal status on GI transit has received longstanding attention to explain gender variability in motility patterns. Although controversies exist in human transit studies [63–67], the hypothesis of hormone "driven" motility in women has been supported by studies reporting changes in motility index (gastric emptying, intestinal transit) during the menstrual cycle or between pre- and postmenopausal women [68,69], and because GI motility is faster in men than in women [63,68–71], and that constipation is a common complication during pregnancy [72,73]. In addition, idiopathic chronic constipation associated with slow transit time in the colon is a common clinical observation in motility disorders, with epidemiology indicating the prevalence almost two to three times higher in women (starting at puberty) than in men [74]. These clinical findings suggest ovarian sex hormones as probable etiological factors. In the small bowel, several studies highlight a dominant estrogen-dependent modulatory effect on muscle excitability (Table 8.1). For instance, the synthetic estrogen, diethylstilbestrol, as well as 17β-estradiol, caused a rapid and dose-dependent nongenomic inhibition of duodenal spontaneous muscle activity in mice [35]. This effect was triggered by activation of K^+ channels and inhibition of L-type Ca^{2+} channels at the plasma membrane that drive Ca^{2+} entry in muscle cells necessary for contraction. An estrogen-induced smooth muscle relaxation has been also reported in the rat ileum [41]. Surprisingly, oxytocin (OT) receptors (OTR) have been shown to be expressed in the GI tract of humans [75] and rodents [76,77], with OT accelerating or inhibiting GI motility depending on species and region [77–79]. In the rat colon, enteric neurons of the myenteric plexus express OTR, and estradiol upregulates the amount of colonic OTR mRNA and protein, resulting in stimulated colonic motility [52] (Table 8.2). During the estrous cycle, this effect disappears in diestrus stage when estrogen plasma levels fall, while it is abolished in OVX rats without estradiol replacement, demonstrating a pure estrogen-dependent pathway, and specific to the follicular period [52].

The administration of progesterone in mice has been shown to decrease the contractile activity of circular muscle strips in the colon, and, concomitantly, fecal output [44] (Table 8.2). Progesterone treatment also decreased the basal motility index as well as neurotransmitter-induced contractions in the distal colon of guinea pig [47]. In women suffering from chronic constipation, the PR isoform B (PR-B) appears overexpressed in colonic muscle cells compared to healthy subjects, while no difference has been noticed in serum progesterone levels [50,51]. This phenotype correlates with changes in muscle responses to various agonists of G protein-coupled receptors, caused by decreased expression of $G\alpha_{q/11}$, which mediates contraction, and upregulation of $G\alpha_s$, which mediates relaxation [50] (Table 8.2). Importantly, most of these abnormalities in G protein expression have been reproduced in vitro after progesterone pretreatment of normal human colonic muscle cells [50], as well as in cells transfected with PR-B [51]. Of note, PR-B transfection made colonic muscle cells more sensitive to physiological progesterone concentrations than normal cells. Such an effect is consistent with the idea that constipation in women is primarily linked to overexpression of colonic PR-B, hence mediating hypersensitivity to circulating progesterone, and not to interindividual differences in plasma progesterone concentrations. This could also explain why the incidence of constipation in pregnancy is quite low (20–40%) despite the high progesterone levels across gravidity [80,81]. More recently, the same group of authors presented a more complete picture of the progesterone-mediated effects on colonic smooth musculature, showing that PR-B mediates downregulation of $G\alpha_{q/11}$, ie, the G protein linked to muscle contraction, while PR-A upregulates $G\alpha_s$ proteins that mediate relaxation [46,51] (Table 8.2).

Although ovarian hormones are thought to be the main players in sex differences in GI motility, recent studies report that androgens are also powerful nongenomic inducers of contractile activity in the gut [40,45]. In the mouse, in vitro pretreatment of intestinal muscle strips with testosterone (T) or its active metabolite 5α-dihydrotestosterone (DHT) acutely affected the frequency pattern of peristaltic activity in the longitudinal musculature of the small (except duodenal region) and large bowels and considerably increased the amplitude of muscle contractions in response to extracellular calcium ($CaCl_2$) or carbachol (CCh), an agonist of acetylcholine (Tables 8.1 and 8.2). These effects occur through a rapid, thus nongenomic, cascade of events in ileal and colonic muscle cells (activation of Rho kinase and concomitant increased phosphorylation of myosin light chain) leading to Ca^{2+} sensitization of the contractile machinery. Interestingly, the $CaCl_2$ or CCh-evoked mechanical tension increased by 400–900% in muscle cells pretreated with physiological concentrations of androgens (100 pM to 10 nM), indicating that T and DHT probably modulate GI motility [40,45]. This likely accounts for the commonly reported faster GI transit in males compared to females.

EPITHELIAL ION TRANSPORT AND GUT BARRIER FUNCTION

In the gut, one critical function of epithelial surfaces is to delineate the interface between the body compartment and intestinal lumen as a protective barrier against the diffusion of pathogens, toxins, and antigens [82], while also permitting nutrient and water absorption [83]. The epithelium is composed of absorptive enterocytes/colonocytes, mucus-producing goblet cells, Paneth cells that secrete antimicrobial peptides and growth factors, and enteroendocrine cells that produce various secretory peptides including cholecystokin in, a glucagon-like peptide involved in the regulation of appetite and digestion. The epithelium renews in 4–7 days in humans to ensure efficient absorptive and barrier functions throughout life [26,83,84]. An influence of sex hormones on intestinal epithelial functions has been commonly reported in animal studies with circulating estrogens as the main player.

Sex Hormone Modulation of Hydroelectrolytic Exchanges and Calcium Uptake

Sodium and Chloride Channels in Fluid Homeostasis

Ion channels in epithelial tissues are key factors in the maintenance of body fluid homeostasis, and estrogens play an important role in regulating hydroelectrolytic exchanges [85]. One of the main functional roles of the colon is water absorption in the body, able to absorb up to 80–90% of the lumen fluids, a mechanism governed by basolateral Na^+ absorption and apical Cl^- secretion in the epithelial cell layer [7,85]. Under normal conditions, this requires a fine balance between Na^+ entry and Cl^- efflux in enterocytes, since for example severe diarrheas are characterized by an exacerbated conductance of Cl^- channels (cystic fibrosis transmembrane conductance regulator, CFTR) at the luminal membrane, inducing rapid and excessive water secretion in the intestine, and the loss of body fluids with feces [86].

In a series of studies beginning in 2001, Harvey et al. [53–55] first described circulating estrogens attenuating the Cl^- secretion across the distal colonic epithelium during the rat estrous cycle. Exogenous 17β-estradiol mimicks this antisecretory effect and it occurs within 10 min of treatment, suggesting rapid nongenomic effects mediated by membrane ER (putatively GPER) rather than a classical transcriptional response mediated by nuclear ERα or ERβ (Table 8.2). Neither testosterone nor progesterone affected colonic Cl^- secretion, and 17β-estradiol administered to males did not mimic effects reported in females, indicating sex-specific regulation [53]. In females, the estrogen-evoked rapid inhibition of intestinal Cl^- secretion involved an intracellular cascade of events from activation of protein kinase C (PKCδ) to rapid phosphorylation of basolateral voltage-gated K^+ channels (ie, suppression of KCNQ1 activity with subsequent decrease in K^+ efflux that determines the rate of Cl^- secretion at apical membrane of colonocytes), with subsequent induction of tissue fluid retention [7,54,55]. Moreover, the sex-specific mode of action for estrogens correlated with sex differences in PKCδ expression in the distal colon found threefold higher in the female compared to male intestine [54]. Additionally, basal colonic PKCδ expression was reported to increase from the follicular period of the estrous cycle to the luteal phase, hence facilitating fluid retention after peak of plasma estrogen [56].

In addition to the nongenomic effects of estradiol, consistent with the idea of a membrane-associated (GPER-like) mode of action, there is a growing body of evidence for a modulatory role of ERβ in the control of colonic fluid movement in pregnant and nonpregnant rodents (Table 8.2). First, $ER\beta^{-/-}$ mice exhibit reduced expression of the colonic Na^+/H^+ exchanger (NHE3) that drives Na^+-dependent water absorption from the gut lumen, while 17β-estradiol treatment of wild-type OVX mice upregulates NHE3 [43]. Second, ERβ has also been demonstrated to play a key role in the control of paracellular permeability between epithelial cells in the colon of female rats, another major route for water and solutes, with estrous cycle-dependent permeability changes [6]. Whatever the study, authors emphasize that the body fluid-retaining effect of estrogens is consistent with fluid retention often reported in women of reproductive age or postmenopausal women under natural or artificial high estrogen states (ie, at mid-menstrual cycle, during pregnancy, with oral contraceptives or estrogen replacement).

Calcium Absorption

Calcium balance and homeostasis in the body is a function of dietary intake and intestinal absorption, and the modulation of intracellular Ca^{2+} pools is crucial for a wide range of physiological processes such as skeletal mineralization, cell proliferation, and muscle cell contraction. In animal studies, there is longstanding evidence for physiological modulation of Ca^{2+} transport by estrogens in the gut [87]. GI Ca^{2+} uptake first occurs in the upper intestine. Rapid 17β-estradiol effects (within 1–10 min, with maximal effect at 5 min) at physiological concentrations (~1 nM) have been reported in isolated rat duodenal cells (Table 8.1). This mechanism is initiated at the cell membrane and stimulates adenylate cyclase activity driving cAMP production and hence fast Ca^{2+} influx in enterocytes [32]. Authors showed that voltage-gated Ca^{2+} channels participated in at least 80% of the rapid Ca^{2+} uptake, an effect that involved the phospholipase C (PLC) signaling cascade, and was not reproduced by either progesterone or DHT, thus implying a specific and nongenomic effect of estradiol [33]. Moreover, a similar pathway for Ca^{2+} uptake has been reported in female rat distal colonic crypts (Table 8.2). This effect was shown to be sex specific and occurred through a Gαs protein-coupled receptor membrane (ie, GPER-like), with PKCδ and fatty acids involved in the estrogen effects [57,58]. Further studies by the same group using human T84 colonic cells have identified the transient receptor potential vanilloid 6 (TRPV6) channel as responsible for Ca^{2+} entry in response to estradiol, because siRNA targeting TRPV6 protein expression totally abolished the estrogen effect [88]. Of interest, TRPV6 is a major Ca^{2+} channel located at the apical and basolateral membranes of enterocytes in the small and large bowels. Its mRNA expression in the duodenum has been reported to be under the control of estradiol in mice and rats [89], consistent with TRPV6 channels participating in vivo in the rapid proabsorptive effect of circulating estrogens on intestinal Ca^{2+}.

Sex Hormone Modulation of Intestinal Barrier

Epithelial Cell Proliferation and Mucus Secretion

There is no information on a putative effect of androgen and progesterone on epithelial cell proliferation and differentiation in the gut. Nonetheless, it is unlikely that these hormones play a prominent role given the relatively poor expression of their receptors compared to ER, and that they are mainly found in nonepithelial tissues. Because ERβ is considered as the primary receptor mediating estrogen effects in the intestine, knockout mice have been produced, but observations have been limited to the colon [5,42]. The comparison of morphology, cell proliferation, and differentiation of colonic epithelium revealed that the number of proliferative cells is greater in mice lacking ERβ, with faster migration of epithelial cells toward the luminal surface than in wild-type littermates [42]. The latter observation suggests that estrogens acting on ERβ are able to slow down the cellular proliferation from crypt stem cells in the colon, in opposition to ERα, which appears to stimulate cell proliferation [3]. In addition, crypt shape and the number of cells per crypt-villus unit were not altered in $ER\beta^{-/-}$ mice, while these mice displayed disorganization of colon mucin secretion by goblet cells, a decrease in apoptosis, a loss of the differentiation marker cytokeratin 20 and of the cell-to-cell adhesion proteins α-catenin (adherens junctions) and plectin (desmosomes) [5,42] (Table 8.2). Altogether, these studies indicate that ERβ activity is a key factor in the organization and architectural maintenance of colonic epithelium and emphasizes an important role for local ERβ/ERα ratio in the control of barrier integrity [26].

It is of note that female rats in proestrus were reported to be more resistant to gut injury than to those in diestrus and that females were protected relative to males [90,91]. According to Sheth et al. [91], the female-specific protection may partly be explained by an estrous cycle-related fluctuation in the mucin content overlaying epithelial cells. For instance, estradiol has been shown to modulate mucus secretion in human bronchial epithelial cells through ERβ signaling pathways [92], and as mentioned above, disorganization of mucin localization has been observed in the colon of $ER\beta^{-/-}$ mice compared to wild-type mice [42]. The colonic lumen is a reservoir for a very large number of bacteria, including potential pathogens, where a thick mucus gel overlaying the epithelium prevents them from directly contacting the epithelial cell layer, hence limiting their entry downstream into the mucosa [93]. Consistent with the idea of ovarian hormones enhancing mucus barrier in females, recent in vitro studies using the mucus-producing human cell lineage HT29-MTX clearly showed that estradiol treatment increased the mucin content (+50%) as well as the mucus viscosity (twofold) and elasticity (eightfold) when compared to nonmucus-producing HT29 [94]. Both these findings suggest a better preservation of the physicochemical properties of the mucus barrier under estrogen dominance, that is, during the follicular period of the reproductive cycle, hence enhancing the protection of gut epithelial surfaces in female compared to male intestines.

Tight Junction Permeability

The epithelial barrier is primarily formed by plasma membranes of individual epithelial cells, but intercellular spaces between adjacent cells are also crucial [82,83]. Tight junction (TJ) transmembrane proteins located at the apical side of epithelial cells seal these intercellular spaces. They are intimately related to the perijunctional cytoskeleton through intracytoplasmic proteins, such as zonula occludens 1 (ZO-1), and TJ protein expression constitutes the rate-limiting step for paracellular transit [82,83]. Excessive paracellular permeability has been reported in gut barrier defects induced by acute or chronic stress conditions, leading to an excessive uptake of luminal antigens and bacterial products that initiate mucosal inflammatory responses and disease [95,96]. Accordingly, a disorganization of apical TJ protein complexes concomitant to a leaky gut epithelium is commonly observed in intestinal diseases [82]. In female rats, colonic paracellular permeability fluctuates across the estrous cycle. This effect is due to fluctuating levels of plasma estrogens, since it was reproduced in OVX rats with 17β-estradiol replacement, and abolished in the presence of an ER antagonist, while progesterone had no effect [6] (Table 8.2). Furthermore, the use of specific agonists for ERα and ERβ has demonstrated these effects to be mediated by estradiol binding to ERβ only, evoking upregulation of TJ transmembrane proteins that regulate paracellular spaces, namely occludin and junctional adhesion molecule (JAM)-A [6]. This role for ERβ in the maintenance of an efficient epithelial barrier has been confirmed in vitro using various human cell epithelial monolayers [6,97], or in rodent models of intestinal inflammation [97,98], while a decreased level of ERβ has been reported in colonic biopsies of patients suffering from inflammatory bowel disease (IBD) [97]. Finally, in the colon, these findings emphasize a multifaceted role of ERβ in strengthening the gut barrier, by reinforcing the TJ apical complex to control paracellular permeability and enhancing mucus secretion overlaying the epithelium, as noted earlier. Therefore, these two beneficial estrogen-dependent mechanisms act in concert to produce female-specific protection of the intestinal barrier function, and the reproductive cycle constitutes an important variable in these effects.

INFLUENCE OF SEX IN GASTROINTESTINAL DISEASES AND PAIN SENSITIVITY

Sex Differences in Gastroduodenal Ulcers

The gastroduodenal mucosa has multiple defense mechanisms to protect itself from gastric acid-induced tissue injury (eg, peptic ulcers). Epidemiological studies emphasize that the women to men ratio in peptic ulcer disease ranges from 1:2 to 1:3 [99,100]. In the stomach, the predominant ER expression in gastric cells first attracted attention to a role for estrogens in the control of acid secretion rates (Table 8.3). In OVX rats, estradiol treatment evokes a dose-dependent inhibition of gastric acid secretion and modulates the mRNA ratio of fundic trefoil factors in steady state conditions, a family of small peptides playing important roles in mucosal protection and restoration [29,30]. Interestingly, ERα and ERβ colocalize in parietal cells, and both ER types have been demonstrated to have equal protective roles in acetic acid-induced gastric ulcers in rats [101].

In the small bowel, epithelial cells lining the duodenum also need efficient defense against gastric acidification of the intestinal lumen. Duodenal bicarbonate (HCO_3^-) secretion is a primary defense mechanism for mucosal protection against gastric acids, and women are better protected from gastric acids than men by an estrogen-mediated stimulation of duodenal HCO_3^- secretion [38]. Furthermore, basal and acid-stimulated HCO_3^- responses in mice have been reported to be 1.5-fold higher in females than in males [133], as observed in women versus men (2.4-fold higher) [38]. In mice studies, authors showed that 17β-estradiol, but not progesterone, dose-dependently stimulated HCO_3^- secretion in both sexes, with greater responses in female mice than in males. In mouse duodenal cells, the proposed mechanisms involve a rapid mobilization of intracellular Ca^{2+} and of phosphatidyl-inositol 3 kinase (PI3K) activity, leading to a potentiation of prostaglandin (PGE2)-stimulated HCO_3^- secretion [102,103], with all effects mediated by ERα (Table 8.3). It is notable that basal and acid-stimulated HCO_3^- secretion in women paralleled the level of endogenous estradiol during the menstrual cycle and that young women (20–29 years old) had significant higher HCO_3^- secretion than post-menopausal subjects (60–69 years old) [38]. Consistent with a primary influence of circulating estrogens, healthy volunteers receiving duodenal luminal administration of 17β-estradiol had an increased HCO_3^- secretion. However, because a similar response was observed in men (Table 8.1), it is concluded that the sex difference in duodenal HCO_3^- secretion and associated protection primarily results from sex differences in circulating estrogens, not of a sex dimorphism in duodenal ER expression.

Clinical Manifestations of Hormonal Status in Irritable Bowel Syndrome and Inflammatory Bowel Diseases: Lessons from Animal Models

IBDs, that is, Crohn's disease and ulcerative colitis, are immunologically mediated pathologies characterized by chronic and relapsing inflammation of the GI

TABLE 8.3 Summary of Sex Hormone Effects in Gastrointestinal Diseases

Disease	Sex prevalence (Refs.)	Epidemiology and GI symptom expression (Refs.)	Main proposed mechanism(s) of action (Refs.)
Gastroduodenal/ peptic ulcer	Men > women [99,100]	• Sex difference in gastric acid production and duodenal HCO_3^- secretion promoting mucosal protection in women [38] • Basal and acid-stimulated HCO_3^- secretion higher in pre- vs postmenopausal women [38]	• Dose-dependent inhibition by estrogens on gastric acid secretion [30]: protection mediated by ERβ and ERα signaling pathways in acid-secreting parietal cells [101]/possible involvement of GPER ligand activity [31] • ERα ligand activity potentiates PGE2-induced duodenal HCO_3^- secretion [102,103]
IBD (CD and UC)	Women > men [104]	• Changes in GI symptoms across the menstrual cycle [105,106] • Postmenopausal HRT use increases the risk of UC [107], but alleviates IBD activity [108] • Women suffering from CD experience more severe symptoms than men [109,110] • Higher levels of immune reaction in women than men, likely predisposing women to IBD [111]	• E2 protects while P4 increases severity of experimental colitis in female rats: correlation with sex hormone regulation of colonic MIF expression controlling proinflammatory cytokine release [9] • Loss of estrogen-mediated immunoprotection (ERβ and ERα) on Treg function in a spontaneous murine model of IBD [112] • ERβ enhances intestinal barrier integrity [6], but decreased ERβ expression in colonic biopsies of IBD patients in relapse [97]
IBS	Women > men [113–116]	• Fluctuations in GI symptoms (bloating, abdominal pain, bowel habits) during the menstrual cycle [115,116] • Increased gut permeability correlates with GI symptoms in IBS patients [117,118] • Resident MC closely apposed to afferent nerve terminals in the gut of IBS patients [119] • Sex difference in stress response and central visceral pain processing as crucial factors for IBS prevalence in women [22,120]	• E2 decreases stress-induced IBS-like symptoms in rats through enhancement of gut barrier integrity [96]: possible ERβ-mediated mechanisms [6] • Pronociceptive effects of ERα vs antinociceptive ERβ-mediated responses in the spinal cord in a IBS-like visceral pain model [13,121], while membrane-associated GPER-like activity in afferent neurons (dorsal root ganglia) promotes hyperalgesia [14] • Sex difference in mucosal MC number and histamine release to neurotransmitter stimulation in the gut/MC reactivity under progesterone influence in rats [122] • Estrogens (ERβ and ERα) with multiple pathways along the brain–gut axis implicated in stress response, with consequences on gut barrier integrity and hypersensitivity (reviews in Refs. [22,120])
Gastric and colorectal cancer	Men > women [123–125]	• Women with ovariectomy have an increased risk of GC/cumulative years of menstrual cycling was inversely associated with GC risk [123] • Postmenopausal HRT use reduced the risk of colorecal malignancy [126]	• Antiproliferative role for ERβ in GI epithelium [5,42,127], but loss of ERβ expression during GC and colorectal cancer progression [49,128–130] • MIF expression promoting gastric and colon tumorigenesis [131,132] is decreased under estrogen dominance (rat colon) [9]

GI, gastrointestinal; ER, estrogen receptors; HCO_3^-, bicarbonate; GPER, G protein-coupled estrogen receptor; PGE2, prostaglandins; IBD, inflammatory bowel diseases; CD, Crohn's disease; UC, ulcerative colitis; HRT, hormonal replacement therapy; MIF, macrophage migration inhibitory factor; Treg, regulatory T cells; IBS, irritable bowel syndrome; MC, mast cells; GC, gastric cancer.

tract [134,135]. On the other hand, irritable bowel syndrome (IBS) is one of the most common functional GI disorders, characterized by abdominal pain and disturbance in bowel habits [136]. Sex differences in prevalence of IBS and IBD are commonly reported; two-thirds of IBS patients [113] and 54–62% of IBD patients [104] are women. Although these diseases are complex and involve multiple mechanisms of injury [134–136], a sex difference in these GI diseases has led to the question of whether endogenous sex hormones play a role among other social determinant factors, such as lifestyle, including diet, smoking and alcohol, as well as anxiety and depression, drug therapies, or the use of oral contraceptives or hormone replacement therapy [104,107,137,138]. To date, and supported by animals studies, clinical data highlight the influence of women's hormonal status (age at diagnosis, menstrual cycle period, pregnancy) on IBD or IBS activity, and the related symptoms [22,105,108,112,120,139].

Sex Hormones and Inflammation in IBD

Women of reproductive age suffering from IBD commonly report fluctuations in their GI symptoms across the menstrual cycle [105,106] and experience more severe disease manifestations compared with men patients, particularly those with Crohn's disease [109,110] (Table 8.3). In women without pathology, higher levels of immune activation and inflammation-associated gene expression have been found in gut mucosal samples compared to men, a sex difference that may predispose women to IBD [111]. It is also notable that a sex bias exists in gut microbiota after puberty [19,20] and that cells of the immune system influence the composition of the commensal flora [140]. IBD is considered to arise from an abnormal immune response to commensal bacteria, often observed with alterations in the composition and/or activity of the microbiota (intestinal dysbiosis) [140]. It is thought that sex hormones targeting one or several partners in this integrative loop linking the microbiota to gut barrier and immune cells might support a sex difference in the pathogenesis of IBD and/or activity.

However, rodent models of IBD mainly indicate an anti-inflammatory role for estrogens in the gut, linked to potent modulatory properties of innate immune responses and of T-helper 1 (Th-1)-mediated inflammation. Indeed, in experimental colitis, 17β-estradiol protects the intestine by reducing tissue damage (eg, ulceration, necrosis), mucosal neutrophil infiltration, and oxidative stress, along with proinflammatory cytokines, such as tumor necrosis factor (TNF)α and interleukin (IL)-1β, which participate in the onset of colitis and disease progression [8,9,141]. ERβ-selective compounds mimic the efficacy of estradiol treatment in rodent models of Crohn's disease, suggesting that ERβ is the key receptor mediating the anti-inflammatory effects of estrogens in the gut [98] (Table 8.3). It is also noteworthy that plant-derived phytoestrogens (ie, isoflavones), that structurally and functionally act as estrogen agonists with higher binding affinity for ERβ, reproduced these anti-inflammatory effects in rat models of IBD [142,143]. Altogether these studies emphasize that circulating estrogens have a protective role in active IBD rather than favor the disease prevalence in females.

Studies using female rats have indicated that the natural shift from estrogen to progesterone dominance during the estrous cycle evokes alternatively resistance then susceptibility of the colon to inflammatory stimuli [9]. One explanation for such an opposite response to inflammatory insult is based on the estrous-related variations in the intestinal content of macrophage migration inhibitory factor (MIF) in epithelial cells. The protein MIF is a proinflammatory cytokine regulated by sex steroids in reproductive and nonreproductive tissues [144–146] and is of central importance in the pathogenesis of colitis as MIF drives the TNFα and IL-1β cytokine network. By enhancing MIF and TNFα production, progesterone worsens the acute phase of colitis, while estrogen prevents acute inflammation by decreasing tissue MIF and IL-1β [9]. This suggests that natural changes in the plasma estrogen/progesterone balance regulate disease activity and enhance the risk of IBD relapse under progesterone dominance [9] (Table 8.3).

Another possible mechanism to explain sex differences in Crohn's disease has been recently proposed using a mouse strain (SAMP mice) that spontaneously develops IBD-like chronic inflammation [112]. Interestingly, female SAMP mice exhibit greater disease severity than their male counterparts, as observed in human patients. Female SAMP mice were found resistant to estrogen-mediated immunoprotection compared to SAMP males, due to a sex-specific loss in SAMP females of ERβ-mediated expansion and function of the regulatory T cells (Tregs) in the GALT [112] (Table 8.3). In the GALT, T cells routinely interact with commensal bacteria, and the Treg subset is important for gut homeostasis by dampening abnormal immune responses against the resident commensals and dietary antigens; one characteristic for IBD etiology is that the disease results from a breakdown of immune tolerance of the microbiota [147]. Goodman et al. [112] reported that Tregs in male SAMP mice are functionally more effective than in SAMP females at ameliorating chronic inflammation after estradiol treatment, suggesting that the higher severity of IBD in women patients may result from a progressive loss of the protective influence of endogenous estrogens upon tolerogenic mechanisms. Finally, in addition to imbalanced immune homeostasis, a common characteristic in IBD patients is an impaired epithelial barrier that correlates with the onset of colonic inflammation and increased intestinal permeability predictive of relapse [134,135]. One study shows that ERβ expression in normal colon mucosa from women is higher than in men [48], and ERβ activity is pivotal in the control of normal epithelial barrier function [6,42]. In Crohn's disease patients, decreased ERβ levels in colonic biopsies have been reported [97], so that the female susceptibility to IBD could also result from the loss of ERβ-mediated protection of epithelial barrier integrity (Table 8.3).

Sex Hormones, Epithelial Permeability and Visceral Pain in IBS

More women than men have IBS, but the importance of hormonal status on the disease occurrence has not been clearly established. One explanation is that women have higher rates of consultation in gastroenterology units than men [114], since they report more frequent symptoms such as abdominal pain, bloating, and changes in bowel habits (constipation and/or diarrhea), with amplification of GI symptoms from mid-menstrual cycle to the perimenses period [115,116]. A sex difference in stress response is also crucial in understanding IBS pathogenesis and symptomatology. Indeed, changes in hormonal status between pre- and postmenopausal women or during the menstrual cycle and pregnancy modulate the hypothalamo–pituitary–adrenal axis response to environmental stressors, with downstream effects on the intestinal barrier, GI motility, and visceral sensitivity [22] (Table 8.3). In IBS, increased intestinal permeability correlates with severity of symptoms [117,118], and estrogens alleviate most IBS-like symptoms in stressed female rats through reinforcement of epithelial permeability, and inhibition of the stress-induced increase in fecal proteolytic activity [96]. In parallel to these local effects, estrogens modulate neuronal pain processes in controlling afferent nerve responses to peripheral mechanical and chemical stimuli through the spinal cord and dorsal root ganglia (Table 8.3). Both ERα and ERβ, as well as GPER, are expressed in nerve cell bodies in the spinal cord and afferent neurons projecting to abdominal and pelvic regions [148,149]. In a rat model of IBS-like visceral pain, ERβ mediates antinociceptive effects, attenuating the response of visceroceptive dorsal horn neurons in the spinal cord to colorectal distension [121], while spinal ERα mediates estradiol-induced pronociceptive responses [13]. Estrogen also rapidly induces visceral hypersensitivity on membrane-associated GPER in afferent neurons located in the dorsal root ganglia [14]. In contrast to estrogens, the role for progesterone in IBS pathogenesis remains poorly investigated, and no information is available for androgen. In IBS, mast cell density increases in the intestinal mucosa, in close contact with afferent nerve terminals, and mast cell degranulation participates in intestinal hypersensitivity and increased gut permeability [119]. Interestingly, the intestinal mucosa of IBS patients displays a higher number of mast cells compared to healthy subjects, and progesterone in female rats modulates jejunal and colonic mast cell degranulation to afferent nerve stimulation [122] (Table 8.3).

Sex Hormones in Gastric and Colorectal Cancers

The male prevalence in gastric and colorectal cancers has been extensively debated, and female sex hormones, primarily estrogen, have been proposed to be protective in the two malignancies [123–125]. Epidemiology, animal and cell culture studies have been conducted in an attempt to decipher the protective role of endogenous estrogens, and the respective contribution of nuclear ERα and ERβ, as well as membrane GPERs in the development of tumors [128,150]. In gastric cancer, the presence of AR and PR has been also reported in addition to ER, but their expression levels were very low, except for ERβ, and all these receptors were reported downregulated in gastric tumors compared to adjacent normal tissues [128]. The expression of ERβ protein is also markedly decreased during colon cancer growth in humans [129,130,151]. Moreover, ERβ expression was inversely correlated with the hyperproliferative state of epithelium in adenomatous tissues [127,130]. This means that the protective role of estrogens in women is likely due to the antiproliferative effect of ERβ in normal GI epithelia, and a consequence of its predominant expression in gastric and colon tissues of healthy subjects (Table 8.3). Regarding progesterone signaling, no effect of progesterone treatment or PR gene knock out have been reported in rodent models of colorectal cancer [152].

Another interesting aspect is the link between anti-inflammatory activity of estrogens and cancer protection. Indeed, chronic inflammation is a well-established risk factor for GI cancers, and protumorigenic activity of the cytokine MIF has been demonstrated in several cancer models, including gastric and colon tumors [131,132]. MIF is constitutively expressed in gastric and colon epithelial cells, and chronic MIF treatment accelerates tumor progression [132]. These data, along with the demonstration that estradiol (or phytoestrogen-based diet) significantly reduced basal MIF content in GI tissues [9,143], further support the conclusion that the high estrogenic status in females protects the GI tract against tumorigenesis (Table 8.3). This is consistent with observational studies demonstrating that the use of hormone replacement therapy in postmenopausal women reduced the risk of colorectal cancer, although the significance depends on the timing of start and duration of treatment, and formulations of hormones [126].

AGE, GENDER, AND HORMONAL STATUS IN FOOD SAFETY: OPEN QUESTIONS FOR ENDOCRINE DISRUPTOR COMPOUNDS

Endocrine Disruptors: What Are They?

Endocrine disruptor compounds (EDCs) are environmental factors able to interfere with hormone-driven processes, mainly through agonist and/or antagonist activity on sex steroid receptors. A deviation from normal endocrine processes may lead to the development of chronic diseases. Age at exposure and gender are intrinsic variables, especially during the developmental period, with potential transgenerational effects [153]. Among these EDCs, estrogeno-mimetic chemicals—such as bisphenols (eg, BPA, BPS, BPF), phthalates (a family of plasticizers), certain pesticides or drugs—present in water, food, and many consumer products have been largely incriminated, due to their ability, even at low doses, to disrupt various signaling pathways evoked by estrogens in target cells [154,155]. Despite their natural origin, phytoestrogens such as soy isoflavones are also considered as potential EDCs when exposure starts during fetal life through the mother's diet [156], in contrast to their beneficial properties on gut barrier when they are consumed as a diet complement in adults [96,142,143]. Because oral exposure is a major route of contamination with EDCs, and the gut epithelium is the main GI tissue exposed through food, the influence of estrogeno-mimetic compounds on GI homeostasis and health, including dysregulation of immune system in the GALT, has attracted much attention in recent years [23,24,157].

The Perinatal Period Is a Vulnerable Window for EDC Exposure

The intestinal barrier function develops early in life. It is organized around a trophic epithelium in constant renewal and mucosal immune cells of GALT, incompetent at birth that must to develop in discriminating between pathogenic antigens and commensal bacteria that colonize the gut rapidly after birth [158]. Studies described throughout this chapter have emphasized the role of circulating estrogens and ER expression, mainly ERβ, as key factors in the development and maintenance of GI functions and GALT integrity. An EDC-mediated dysregulation of the maturing process of host defenses and immune tolerance may lead to the development of- or increase susceptibility to- chronic GI disorders, food intolerance, infectious diseases, or tumor development [24,157,159].

Analysis of the metabolic pathways of estrogenic EDCs is essential to predict their effects in the body. The estrogeno-mimetic properties may be altered through metabolism of the parental compound. For example, a first pass through the liver detoxifies bisphenol A (BPA, a chemical often used in food packaging) by adding a glucuronide conjugate, transforming the parent BPA into hormonally inactive BPA-glucuronide (BPA-Gluc) [159], and similar glucuronide conjugation occurs for phytoestrogens [160]. This process reduces bioavailability of EDCs for estrogen-sensitive targets into the body. However, before their absorption and conjugation in the liver, the free (ie, active) form of these compounds is present in the intestinal lumen and exhibits binding activity with ER expressed on both epithelial and immune cells. With respect to BPA, although differences exist in metabolic pathways between animals and humans, recent studies also report efficient deconjugation pathways in the body, particularly in the fetus, which are crucial for health authorities to consider in human risk assessment studies [159]. Studies performed in rodents have demonstrated transfer of BPA ingested by the pregnant mother to the fetus, then to the newborn during lactation [159]. For instance, in pregnant rats receiving a single oral dose of ^{14}C-labeled BPA at the 18th day of gravidity, the radioactivity is recovered in fetuses, primarily in the gut lumen, within 24 h after treatment. If the same BPA dosage is given orally to lactating rats, most of the radioactivity in neonates is also recovered in the intestine, within 24 h of feeding. A similar study in pregnant mice showed that placental transfer of BPA occurred within 30 min after administration to the mother, and 4% of the dose is found in fetal tissues, with nearly 50% in an unconjugated form, thus hormonally active during developmental stages of the GI tract and GALT [159].

Animal studies have reported effects on the intestinal epithelial barrier (eg, antimitotic effects, defect in calcium transport, epithelial permeability) and immune system responses (innate and adaptive immunity) to phytoestrogen [156,161,162] and/or BPA [23,24,157,163]. For instance, in vitro exposure of human Caco-2 epithelial cells to phytoestrogens decreased Ca^{2+} absorption in the presence of 17β-estradiol, and this effect disappeared in estrogen-depleted culture media, which suggests that phytoestrogen activities on intestinal epithelium are tightly linked to estrogenic status [162]. In piglets, the isoflavone genistein—a phytoestrogen displaying 10-fold higher affinity for ERβ than for ERα [164]—at a dose close to infant formula displayed potent antiproliferative effects in the jejunum and the ileum [161]. In one-generation studies after perinatal exposure to low doses of BPA (ie, transmaternal passage of BPA during pregnancy and lactational period), studies in rodent offspring at adulthood reported an upregulation of Th1-mediated responses and increased susceptibility to severe intestinal inflammation [23,163], weakening of host defenses with parasitic infection [157], and failure of oral tolerance together with impaired sensitization to dietary antigens [24]. A sex difference in BPA effects has been noticed, as dysregulated immune responses in inflammation have been primarily observed in females after perinatal exposure [23,24]. Additionally, BPA exposure at environmentally relevant concentrations promotes migration of human colon cancer cells via ERβ signaling pathways, suggesting a potential metastatic cancer risk [165]. In tissues where ERβ is the predominantly expressed ER, such as in the colon, a potent tissue response to low doses of BPA, despite its low ER binding activity (estimated to be 10,000-fold lower than estradiol [164]), may be related to better coupling of ERβ with its coactivators in the presence of xenoestrogens. This results in transcription of target genes that are estimated to be 500 times greater for the BPA–ERβ complex, in comparison to ERα [166–168].

SUMMARY AND CONCLUSIONS

Based on the findings highlighted in this chapter, it is clear that sex differences in GI physiology primarily result from sex-related hormonal status, rather than a dimorphism in the expression level of sex steroid receptors. Studies aimed at deciphering sex hormone effects on GI functions and disorders have demonstrated important roles for estrogens in epithelial cell proliferation and functions, immune cell activity in the GALT, and visceral pain processes. The involvement of estrogen in visceral pain processes is complex as estrogens drive both pro- and antinociceptive pathways that affect abdominal pain perception in women and quality of life in disease states. Estrogens are not the sole players since progesterone and androgens also affect GI motility, accelerating or slowing intestinal transit according to sex and age, menstrual cycle stage, or pregnancy. At puberty, androgens in males seem to cause profound changes in bacterial composition in the gut lumen, leading to sexually dimorphic microbiota. In conclusion, in addition to sex, it is essential to take into account differences in hormone status between pre- and postpubertal periods and between pre- and postmenopausal women to gain a better understanding of GI physiology, sex differences in epidemiology and symptomatology, as well as the vulnerability of GI development to estrogenic endocrine disruptors present in the environment.

Acknowledgments

The author wish to thank Drs Laurence Guzylack-Piriou, Sandrine Ménard, and Rafael Garcia-Villar for helpful comments on the manuscript, and Eric Gaultier for assistance in editing this chapter. In memoriam : Jean Fioramonti, PhD (1948–2015).

References

[1] Alford TC, Do HM, Geelhoed GW, Tsangaris NT, Lippman ME. Steroid hormone receptors in human colon cancers. Cancer 1979;43:980–4.
[2] Kuiper GG, Enmark E, Pelto-Huikko M, Nilsson S, Gustafsson JA. Cloning of a novel receptor expressed in rat prostate and ovary. Proc Natl Acad Sci USA 1996;93:5925–30.
[3] Matthews J, Gustafsson JA. Estrogen signaling: a subtle balance between ER alpha and ER beta. Mol Interv 2003;3:281–92.
[4] Böttner M, Thelen P, Jarry H. Estrogen receptor beta: tissue distribution and the still largely enigmatic physiological function. J Steroid Biochem Mol Biol 2014;139:245–51.
[5] Wada-Hiraike O, Imamov O, Hiraike H, Hultenby K, Schwend T, Omoto Y, et al. Role of estrogen receptor beta in colonic epithelium. Proc Natl Acad Sci USA 2006;103:2959–64.
[6] Braniste V, Leveque M, Buisson-Brenac C, Bueno L, Fioramonti J, Houdeau E. Oestradiol decreases colonic permeability through oestrogen receptor beta-mediated up-regulation of occludin and junctional adhesion molecule-A in epithelial cells. J Physiol 2009;587:3317–28.
[7] O'Mahony F, Thomas W, Harvey BJ. Novel female sex-dependent actions of oestrogen in the intestine. J Physiol 2009;587:5039–44.
[8] Harnish DC, Albert LM, Leathurby Y, Eckert AM, Ciarletta A, Kasaian M, et al. Beneficial effects of estrogen treatment in the HLA-B27 transgenic rat model of inflammatory bowel disease. Am J Physiol Gastrointest Liver Physiol 2004;286:G118–25.
[9] Houdeau E, Moriez R, Leveque M, Salvador-Cartier C, Waget A, Leng L, et al. Sex steroid regulation of macrophage migration inhibitory factor in normal and inflamed colon in the female rat. Gastroenterology 2007;132:982–93.
[10] Ji Y, Tang B, Traub RJ. Modulatory effects of estrogen and progesterone on colorectal hyperalgesia in the rat. Pain 2005;117:433–42.
[11] Ji Y, Tang B, Traub RJ. The visceromotor response to colorectal distention fluctuates with the estrous cycle in rats. Neuroscience 2008;154:1562–7.
[12] Huby RD, Glaves P, Jackson R. The incidence of sexually dimorphic gene expression varies greatly between tissues in the rat. PLoS One 2014;9:115792.
[13] Ji Y, Tang B, Traub RJ. Spinal estrogen receptor alpha mediates estradiol-induced pronociception in a visceral pain model in the rat. Pain 2011;152:1182–91.
[14] Lu CL, Hsieh JC, Dun NJ, Oprea TI, Wang PS, Luo JC, et al. Estrogen rapidly modulates 5-hydroxytrytophan-induced visceral hypersensitivity via GPR30 in rats. Gastroenterology 2009;137:1040–50.
[15] Wilson CM, McPhaul MJ. A and B forms of the androgen receptor are expressed in a variety of human tissues. Mol Cell Endocrinol 1996;120:51–7.
[16] Brandenberger AW, Tee MK, Lee JY, Chao V, Jaffe RB. Tissue distribution of estrogen receptors alpha (ER-α) and beta (ER-β) mRNA in the midgestational human fetus. J Clin Endocrinol Metab 1997;82:3509–12.
[17] Takeyama J, Suzuki T, Inoue S, Kaneko C, Nagura H, Harada N, et al. Expression and cellular localization of estrogen receptors alpha and beta in the human fetus. J Clin Endocrinol Metab 2001;86:2258–62.
[18] Guarino M, Cheng L, Cicala M, Ripetti V, Biancani P, Behar J. Progesterone receptors and serotonin levels in colon epithelial cells from females with slow transit constipation. Neurogastroenterol Motil 2011;23:575.
[19] Markle JG, Frank DN, Mortin-Toth S, Robertson CE, Feazel LM, Rolle-Kampczyk U, et al. Sex differences in the gut microbiome drive hormone-dependent regulation of autoimmunity. Science 2013;339:1084–8.

[20] Yurkovetskiy L, Burrows M, Khan AA, Graham L, Volchkov P, Becker L, et al. Gender bias in autoimmunity is influenced by microbiota. Immunity 2013;39:400–12.

[21] Mayer EA, Berman S, Chang L, Naliboff BD. Sex-based differences in gastrointestinal pain. Eur J Pain 2004;8:451–63.

[22] Mulak A, Taché Y, Larauche M. Sex hormones in the modulation of irritable bowel syndrome. World J Gastroenterol 2014;20:2433–48.

[23] Braniste V, Jouault A, Gaultier E, Polizzi A, Buisson-Brenac C, Leveque M, et al. Impact of oral bisphenol A at reference doses on intestinal barrier function and sex differences after perinatal exposure in rats. Proc Natl Acad Sci USA 2010;107:448–53.

[24] Ménard S, Guzylack-Piriou L, Leveque M, Braniste V, Lencina C, Naturel M, et al. Food intolerance at adulthood after perinatal exposure to the endocrine disruptor bisphenol A. FASEB J 2014;28:4893–900.

[25] Salih MA, Sims SH, Kalu DN. Putative intestinal estrogen receptor: evidence for regional differences. Mol Cell Endocrinol 1996;121:47–55.

[26] D'Errico I, Moschetta A. Nuclear receptors, intestinal architecture and colon cancer: an intriguing link. Cell Mol Life Sci 2008;65:1523–43.

[27] Taylor AH, Al-Azzawi F. Immunolocalisation of oestrogen receptor beta in human tissues. J Mol Endocrinol 2000;24:145–55.

[28] Gejima K, Kawaguchi H, Souda M, Kawashima H, Komokata T, Hamada N, et al. Expression of estrogen receptor-alpha protein in the rat digestive tract. In Vivo 2007;21:487–92.

[29] Campbell-Thompson ML. Estrogen receptor alpha and beta expression in upper gastrointestinal tract with regulation of trefoil factor family 2 mRNA levels in ovariectomized rats. Biochem Biophys Res Commun 1997;240:478–83.

[30] Campbell-Thompson M, Reyher KK, Wilkinson LB. Immunolocalization of estrogen receptor alpha and beta in gastric epithelium and enteric neurons. J Endocrinol 2001;171:65–73.

[31] Isensee J, Meoli L, Zazzu V, Nabzdyk C, Witt H, Soewarto D, et al. Expression pattern of G protein-coupled receptor 30 in LacZ reporter mice. Endocrinology 2009;150:1722–30.

[32] Picotto G, Massheimer V, Boland R. Acute stimulation of intestinal cell calcium influx induced by 17 beta-estradiol via the cAMP messenger system. Mol Cell Endocrinol 1996;119:129–34.

[33] Picotto G, Vazquez G, Boland R. 17beta-oestradiol increases intracellular Ca^{2+} concentration in rat enterocytes. Potential role of phospholipase C-dependent store-operated Ca^{2+} influx. Biochem J 1999;339:71–7.

[34] Choijookhuu N, Hino SI, Oo PS, Batmunkh B, Mohmand NA, Kyaw MT, et al. Ontogenetic changes in the expression of estrogen receptor β in mouse duodenal epithelium. Clin Res Hepatol Gastroenterol 2015;7401:32–7.

[35] Díaz M, Ramírez C, Marin R, Marrero-Alonso J, Gómez T, Alonso R. Acute relaxation of mouse duodenum by estrogens. Evidence for an estrogen receptor-independent modulation of muscle excitability. Eur J Pharmacol 2004;501:161–78.

[36] Uotinen N, Puustinen R, Pasanen S, Manninen T, Kivineva M, Syvälä H, et al. Distribution of progesterone receptor in female mouse tissues. Gen Comp Endocrinol 1999;115:429–41.

[37] Pfaffl MW, Lange IG, Meyer HH. The gastrointestinal tract as target of steroid hormone action: quantification of steroid receptor mRNA expression (AR, ERalpha, ERbeta and PR) in 10 bovine gastrointestinal tract compartments by kinetic RT-PCR. J Steroid Biochem Mol Biol 2003;84:159–66.

[38] Tuo B, Wen G, Wei J, Liu X, Wang X, Zhang Y, et al. Estrogen regulation of duodenal bicarbonate secretion and sex-specific protection of human duodenum. Gastroenterology 2011;141:854–63.

[39] Kawano N, Koji T, Hishikawa Y, Murase K, Murata I, Kohno S. Identification and localization of estrogen receptor alpha- and beta-positive cells in adult male and female mouse intestine at various estrogen levels. Histochem Cell Biol 2004;121:399–405.

[40] González-Montelongo MC, Marín R, Gómez T, Díaz M. Androgens are powerful non-genomic inducers of calcium sensitization in visceral smooth muscle. Steroids 2010;75:533–8.

[41] Pines A, Eckstein N, Dotan I, Ayalon D, Varon D, Barnea O, et al. Effect of estradiol on rat ileum. Gen Pharmacol 1998;31:735–6.

[42] Wada-Hiraike O, Warner M, Gustafsson JA. New developments in oestrogen signalling in colonic epithelium. Biochem Soc Trans 2006;34:1114–16.

[43] Choijookhuu N, Sato Y, Nishino T, Endo D, Hishikawa Y, Koji T. Estrogen-dependent regulation of sodium/hydrogen exchanger-3 (NHE3) expression via estrogen receptor female mouse intestine at various e. Histochem Cell Biol 2012;137:575–87.

[44] Li CP, Ling C, Biancani P, Behar J. Effect of progesterone on colonic motility and fecal output in mice with diarrhea. Neurogastroenterol Motil 2012;24:392.

[45] González-Montelongo MC, Marín R, Gómez T, Marrero-Alonso J, Díaz M. Androgens induce nongenomic stimulation of colonic contractile activity through induction of calcium sensitization and phosphorylation of LC20 and CPI-17. Mol Endocrinol 2010;24:1007–23.

[46] Cheng L, Biancani P, Behar J. Progesterone receptor A mediates VIP inhibition of contraction. Am J Physiol Gastrointest Liver Physiol 2010;298:433–9.

[47] Xiao ZL, Biancani P, Behar J. Effects of progesterone on motility and prostaglandin levels in the distal guinea pig colon. Am J Physiol Gastrointest Liver Physiol 2009;297:886–93.

[48] Campbell-Thompson M, Lynch IJ, Bhardwaj B. Expression of estrogen receptor (ER) subtypes and ERbeta isoforms in colon cancer. Cancer Res 2001;61:632–40.

[49] Konstantinopoulos PA, Kominea A, Vandoros G, Sykiotis GP, Andricopoulos P, Varakis I, et al. Oestrogen receptor beta (ERβ) is abundantly expressed in normal colonic mucosa, but declines in colon adenocarcinoma paralleling the tumour's dedifferentiation. Eur J Cancer 2003;39:1251–8.

[50] Xiao ZL, Pricolo V, Biancani P, Behar J. Role of progesterone signaling in the regulation of G-protein levels in female chronic constipation. Gastroenterology 2005;128:667–75.

[51] Cheng L, Pricolo V, Biancani P, Behar J. Overexpression of progesterone receptor B increases sensitivity of human colon muscle cells to progesterone. Am J Physiol Gastrointest Liver Physiol 2008;295:493–502.

[52] Feng M, Qin J, Wang C, Ye Y, Wang S, Xie D, et al. Estradiol upregulates the expression of oxytocin receptor in colon in rats. Am J Physiol Endocrinol Metab 2009;296:1059–66.

[53] Condliffe SB, Doolan CM, Harvey BJ. 17beta-oestradiol acutely regulates Cl^- secretion in rat distal colonic epithelium. J Physiol 2001;530:47–54.

[54] O'Mahony F, Alzamora R, Betts V, LaPaix F, Carter D, Irnaten M, et al. Female gender-specific inhibition of KCNQ1 channels and chloride secretion by 17beta-estradiol in rat distal colonic crypts. J Biol Chem 2007;282:24563–73.

[55] O'Mahony F, Harvey BJ. Sex and estrous cycle-dependent rapid protein kinase signaling actions of estrogen in distal colonic cells. Steroids 2008;73:889–94.

[56] O'Mahony F, Alzamora R, Chung HL, Thomas W, Harvey BJ. Genomic priming of the antisecretory response to estrogen in rat distal colon throughout the estrous cycle. Mol Endocrinol 2009;23:1885–99.

[57] Doolan CM, Condliffe SB, Harvey BJ. Rapid non-genomic activation of cytosolic cyclic AMP-dependent protein kinase activity and [Ca(2+)](i) by 17beta-oestradiol in female rat distal colon. Br J Pharmacol 2000;129:1375–86.

[58] Doolan CM, Harvey BJ. A Gαs protein-coupled membrane receptor, distinct from the classical oestrogen receptor, transduces rapid effects of oestradiol on $[Ca^{2+}]_i$ in female rat distal colon. Mol Cell Endocrinol 2003;199:87–103.

[59] Mensah-Brown EP, Rizk DE, Patel M, Chandranath SI, Adem A. Effects of ovariectomy and hormone replacement on submucosal collagen and blood vessels of the anal canal of rats. Colorectal Dis 2004;6:481–7.

[60] Oettling G, Franz HB. Mapping of androgen, estrogen and progesterone receptors in the anal continence organ. Eur J Obstet Gynecol Reprod Biol 1998;77:211–16.

[61] Becker JB, Arnold AP, Berkley KJ, Blaustein JD, Eckel LA, Hampson E, et al. Strategies and methods for research on sex differences in brain and behavior. Endocrinology 2005;146:1650–73.

[62] Cunningham M, Gilkeson G. Estrogen receptors in immunity and autoimmunity. Clin Rev Allergy Immunol 2011;40:66–73.

[63] Hinds JP, Stoney B, Wald A. Does gender or the menstrual cycle affect colonic transit? Am J Gastroenterol 1989;84:123–6.

[64] Turnbull GK, Thompson DG, Day S, Martin J, Walker E, Lennard-Jones JE. Relationships between symptoms, menstrual cycle and orocaecal transit in normal and constipated women. Gut 1989;30:30–4.

[65] Bovo P, Paola Brunori M, di Francesco V, Frulloni L, Montesi G, Cavallini G. The menstrual cycle has no effect on gastrointestinal transit time. Evaluation by means of the lactulose H2 breath test. Ital J Gastroenterol 1992;24:449–51.

[66] Degen LP, Phillips SF. Variability of gastrointestinal transit in healthy women and men. Gut 1996;39:299–305.

[67] Müller-Lissner SA, Kamm MA, Scarpignato C, Wald A. Myths and misconceptions about chronic constipation. Am J Gastroenterol 2005;100:232–42.

[68] Wald A, Van Thiel DH, Hoechstetter L, Gavaler JS, Egler KM, Verm R, et al. Gastrointestinal transit: the effect of the menstrual cycle. Gastroenterology 1981;80:1497–500.

[69] Jung HK, Kim DY, Moon IH. Effects of gender and menstrual cycle on colonic transit time in healthy subjects. Korean J Intern Med 2003;18:181–6.

[70] Lampe JW, Fredstrom SB, Slavin JL, Potter JD. Sex differences in colonic function: a randomised trial. Gut 1993;34:531–6.

[71] Meier R, Beglinger C, Dederding JP, Meyer-Wyss B, Fumagalli M, Rowedder A, et al. Influence of age, gender, hormonal status and smoking habits on colonic transit time. Neurogastroenterol Motil 1995;7:235–8.

[72] Everson GT. Gastrointestinal motility in pregnancy. Gastroenterol Clin North Am 1992;21:751–76.

[73] Shah S, Hobbs A, Singh R, Cuevas J, Ignarro LJ, Chaudhuri G. Gastrointestinal motility during pregnancy: role of nitrergic component of NANC nerves. Am J Physiol Regul Integr Comp Physiol 2000;279:R1478–85.

[74] McCrea GL, Miaskowski C, Stotts NA, Macera L, Paul SM, Varma MG. Gender differences in self-reported constipation characteristics, symptoms, and bowel and dietary habits among patients attending a specialty clinic for constipation. Gend Med 2009;6:259–71.

[75] Monstein HJ, Grahn N, Truedsson M, Ohlsson B. Oxytocin and oxytocin-receptor mRNA expression in the human gastrointestinal tract: a polymerase chain reaction study. Regul Pept 2004;119:39–44.

[76] Welch MG, Margolis KG, Li Z, Gershon MD. Oxytocin regulates gastrointestinal motility, inflammation, macromolecular permeability, and mucosal maintenance in mice. Am J Physiol Gastrointest Liver Physiol 2014;307:G848–62.

[77] Welch MG, Tamir H, Gross KJ, Chen J, Anwar M, Gershon MD. Expression and developmental regulation of oxytocin (OT) and oxytocin receptors (OTR) in the enteric nervous system (ENS) and intestinal epithelium. J Comp Neurol 2009;512:256–70.

[78] Lv Y, Feng M, Che T, Sun H, Luo Y, Liu K, et al. CCK mediated the inhibitory effect of oxytocin on the contraction of longitudinal muscle strips of duodenum in male rats. Pflugers Arch 2010;460:1063–71.

[79] Yang X, Xi TF, Li YX, Wang HH, Qin Y, Zhang JP, et al. Oxytocin decreases colonic motility of cold water stressed rats via oxytocin receptors. World J Gastroenterol 2014;20:10886–94.

[80] Bradley CS, Kennedy CM, Turcea AM, Rao SS, Nygaard IE. Constipation in pregnancy: prevalence, symptoms, and risk factors. Obstet Gynecol 2007;110:1351–7.

[81] Cullen G, O'Donoghue D. Constipation and pregnancy. Best Pract Res Clin Gastroenterol 2007;21:807–18.

[82] Turner JR. Intestinal mucosal barrier function in health and disease. Nat Rev Immunol 2009;9:799–809.

[83] Turner JR. Molecular basis of epithelial barrier regulation: from basic mechanisms to clinical application. Am J Pathol 2006;169:1901–9.

[84] Gerbe F, Legraverend C, Jay P. The intestinal epithelium tuft cells: specification and function. Cell Mol Life Sci 2012;69:2907–17.

[85] Saint-Criq V, Rapetti-Mauss R, Yusef YR, Harvey BJ. Estrogen regulation of epithelial ion transport: implications in health and disease. Steroids 2012;77:918–23.

[86] Thiagarajah JR, Verkman AS. Chloride channel-targeted therapy for secretory diarrheas. Curr Opin Pharmacol 2013;13:888–94.

[87] Arjmandi BH, Salih MA, Herbert DC, Sims SH, Kalu DN. Evidence for estrogen receptor-linked calcium transport in the intestine. Bone Miner 1993;21:63–74.

[88] Irnaten M, Blanchard-Gutton N, Harvey BJ. Rapid effects of 17beta-estradiol on epithelial TRPV6 Ca^{2+} channel in human T84 colonic cells. Cell Calcium 2008;44:441–52.

[89] van Abel M, Hoenderop JG, van der Kemp AW, van Leeuwen JP, Bindels RJ. Regulation of the epithelial Ca^{2+} channels in small intestine as studied by quantitative mRNA detection. Am J Physiol Gastrointest Liver Physiol 2003;285:78–85.

[90] Homma H, Hoy E, Xu DZ, Lu Q, Feinman R, Deitch EA. The female intestine is more resistant than the male intestine to gut injury and inflammation when subjected to conditions associated with shock states. Am J Physiol Gastrointest Liver Physiol 2005;288:G466–72.

[91] Sheth SU, Lu Q, Twelker K, Sharpe SM, Qin X, Reino DC, et al. Intestinal mucus layer preservation in female rats attenuates gut injury after trauma-hemorrhagic shock. J Trauma 2010;68:279–88.

[92] Tam A, Wadsworth S, Dorscheid D, Man SF, Sin DD. Estradiol increases mucus synthesis in bronchial epithelial cells. PLoS One 2014;9:e100633.

[93] Pelaseyed T, Bergström JH, Gustafsson JK, Ermund A, Birchenough GM, Schütte A, et al. The mucus and mucins of the goblet cells and enterocytes provide the first defense line of the gastrointestinal tract and interact with the immune system. Immunol Rev 2014;260:8–20.

[94] Diebel ME, Diebel LN, Manke CW, Liberati DM. Estrogen modulates intestinal mucus physiochemical properties and protects against oxidant injury. J Trauma Acute Care Surg 2015;78:94–9.

[95] Ait-Belgnaoui A, Bradesi S, Fioramonti J, Theodorou V, Bueno L. Acute stress-induced hypersensitivity to colonic distension depends upon increase in paracellular permeability: role of myosin light chain kinase. Pain 2005;113:141–7.

[96] Moussa L, Bézirard V, Salvador-Cartier C, Bacquié V, Houdeau E, Théodorou V. A new soy germ fermented ingredient displays estrogenic and protease inhibitor activities able to prevent irritable bowel syndrome-like symptoms in stressed female rats. Clin Nutr 2013;32:51–8.

[97] Looijer-van Langen M, Hotte N, Dieleman LA, Albert E, Mulder C, Madsen KL. Estrogen receptor-mented ingredient displays estrogenic and proteas. Am J Physiol Gastrointest Liver Physiol 2011;300:621–6.

[98] Harris HA, Albert LM, Leathurby Y, Malamas MS, Mewshaw RE, Miller CP, et al. Evaluation of an estrogen receptor-beta agonist in animal models of human disease. Endocrinology 2003;144:4241–9.

[99] Rosenstock SJ, Jørgensen T. Prevalence and incidence of peptic ulcer disease in a Danish County—a prospective cohort study. Gut 1995;36:819–24.

[100] Wu HC, Tuo BG, Wu WM, Gao Y, Xu QQ, Zhao K. Prevalence of peptic ulcer in dyspeptic patients and the influence of age, sex, and *Helicobacter pylori* infection. Dig Dis Sci 2008;53:2650–6.

[101] Kumral ZN, Memi G, Ercan F, Yeğen BC. Estrogen alleviates acetic acid-induced gastric or colonic damage via both ERα- and ERβ-mediated and direct antioxidant mechanisms in rats. Inflammation 2014;37:694–705.

[102] Tuo BG, Wen GR, Seidler U. Phosphatidylinositol 3-kinase is involved in prostaglandin E2-mediated murine duodenal bicarbonate secretion. Am J Physiol Gastrointest Liver Physiol 2007;293:G279–87.

[103] Tuo B, Wen G, Wang X, Xu J, Xie R, Liu X, et al. Estrogen potentiates prostaglandin E_2-stimulated duodenal mucosal HCO_3^- secretion in mice. Am J Physiol Endocrinol Metab 2012;303:111–21.

[104] Hovde Ø, Moum BA. Epidemiology and clinical course of Crohn's disease: results from observational studies. World J Gastroenterol 2012;18:1723–31.

[105] Kane SV, Sable K, Hanauer SB. The menstrual cycle and its effect on inflammatory bowel disease and irritable bowel syndrome: a prevalence study. Am J Gastroenterol 1998;93:1867–72.

[106] Weber AM, Ziegler C, Belinson JL, Mitchinson AR, Widrich T, Faziov V. Gynecologic history of women with inflammatory bowel disease. Obstet Gynecol 1995;86:843–7.

[107] Khalili H, Higuchi LM, Ananthakrishnan AN, Manson JE, Feskanich D, Richter JM, et al. Hormone therapy increases risk of ulcerative colitis but not Crohn's disease. Gastroenterology 2012;143:1199–206.

[108] Kane SV, Reddy D. Hormonal replacement therapy after menopause is protective of disease activity in women with inflammatory bowel disease. Am J Gastroenterol 2008;103:1193–6.

[109] Bernklev T, Jahnsen J, Aadland E, Sauar J, Schulz T, Lygren I, et al. Health-related quality of life in patients with inflammatory bowel disease five years after the initial diagnosis. Scand J Gastroenterol 2004;39:365–73.

[110] Saibeni S, Cortinovis I, Beretta L, Tatarella M, Ferraris L, Rondonotti E, et al. Gender and disease activity influence health-related quality of life in inflammatory bowel diseases. Hepatogastroenterology 2005;52:509–15.

[111] Sankaran-Walters S, Macal M, Grishina I, Nagy L, Goulart L, Coolidge K, et al. Sex differences matter in the gut: effect on mucosal immune activation and inflammation. Biol Sex Differ 2013;4:10.

[112] Goodman WA, Garg RR, Reuter BK, Mattioli B, Rissman EF, Pizarro TT. Loss of estrogen-mediated immunoprotection underlies female gender bias in experimental Crohn's-like ileitis. Mucosal Immunol 2014;7:1255–65.

[113] Chang L, Heitkemper MM. Gender differences in irritable bowel syndrome. Gastroenterology 2002;123:1686–701.

[114] Payne S. Sex, gender, and irritable bowel syndrome: making the connections. Gend Med 2004;1:18–28.

[115] Houghton LA, Lea R, Jackson N, Whorwell PJ. The menstrual cycle affects rectal sensitivity in patients with irritable bowel syndrome but not healthy volunteers. Gut 2002;50:471–4.

[116] Heitkemper MM, Cain KC, Jarrett ME, Burr RL, Hertig V, Bond EF. Symptoms across the menstrual cycle in women with irritable bowel syndrome. Am J Gastroenterol 2003;98:420–30.

[117] Gecse K, Róka R, Ferrier L, Leveque M, Eutamene H, Cartier C, et al. Increased faecal serine protease activity in diarrhoeic IBS patients: a colonic lumenal factor impairing colonic permeability and sensitivity. Gut 2008;57:591–9.

[118] Zhou Q, Zhang B, Verne GN. Intestinal membrane permeability and hypersensitivity in the irritable bowel syndrome. Pain 2009;146:41–6.

[119] Barbara G, Stanghellini V, De Giorgio R, Cremon C, Cottrell GS, Santini D, et al. Activated mast cells in proximity to colonic nerves correlate with abdominal pain in irritable bowel syndrome. Gastroenterology 2004;126:693–702.

[120] Meleine M, Matricon J. Gender-related differences in irritable bowel syndrome: potential mechanisms of sex hormones. World J Gastroenterol 2014;20:6725–43.

[121] Cao DY, Ji Y, Tang B, Traub RJ. Estrogen receptor β activation is antinociceptive in a model of visceral pain in the rat. J Pain 2012;13:685–94.

[122] Bradesi S, Eutamene H, Theodorou V, Fioramonti J, Bueno L. Effect of ovarian hormones on intestinal mast cell reactivity to substance P. Life Sci 2001;68:1047–56.

[123] Duell EJ, Travier N, Lujan-Barroso L, Boutron-Ruault MC, Clavel-Chapelon F, Palli D, et al. Menstrual and reproductive factors, exogenous hormone use, and gastric cancer risk in a cohort of women from the European Prospective Investigation into Cancer and Nutrition. Am J Epidemiol 2010;172:1384–93.

[124] Koo JH, Leong RW. Sex differences in epidemiological, clinical and pathological characteristics of colorectal cancer. J Gastroenterol Hepatol 2010;25:33–42.

[125] Camargo MC, Goto Y, Zabaleta J, Morgan DR, Correa P, Rabkin CS. Sex hormones, hormonal interventions, and gastric cancer risk: a meta-analysis. Cancer Epidemiol Biomarkers Prev 2012;21:20–38.

[126] Barzi A, Lenz AM, Labonte MJ, Lenz HJ. Molecular pathways: estrogen pathway in colorectal cancer. Clin Cancer Res 2013;19:5842–8.

[127] Motylewska E, Stasikowska O, Meleń-Mucha G. The inhibitory effect of diarylpropionitrile, a selective agonist of estrogen receptor beta, on the growth of MC38 colon cancer line. Cancer Lett 2008;276:68–73.

[128] Gan L, He J, Zhang X, Zhang YJ, Yu GZ, Chen Y, et al. Expression profile and prognostic role of sex hormone receptors in gastric cancer. BMC Cancer 2012;12:566.

[129] Foley EF, Jazaeri AA, Shupnik MA, Jazaeri O, Rice LW. Selective loss of estrogen receptor beta in malignant human colon. Cancer Res 2000;60:245–8.

[130] Di Leo A, Barone M, Maiorano E, Tanzi S, Piscitelli D, Marangi S, et al. ER-beta expression in large bowel adenomas: implications in colon carcinogenesis. Dig Liver Dis 2008;40:260–6.

[131] Wilson JM, Coletta PL, Cuthbert RJ, Scott N, MacLennan K, Hawcroft G, et al. Macrophage migration inhibitory factor promotes intestinal tumorigenesis. Gastroenterology 2005;129:1485–503.

[132] Morris KT, Nofchissey RA, Pinchuk IV, Beswick EJ. Chronic macrophage migration inhibitory factor exposure induces mesenchymal epithelial transition and promotes gastric and colon cancers. PLoS One 2014;9:e98656.
[133] Smith A, Contreras C, Ko KH, Chow J, Dong X, Tuo B, et al. Gender-specific protection of estrogen against gastric acid-induced duodenal injury: stimulation of duodenal mucosal bicarbonate secretion. Endocrinology 2008;149:4554–66.
[134] Sartor RB. Mechanisms of disease: pathogenesis of Crohn's disease and ulcerative colitis. Nat Clin Pract Gastroenterol Hepatol 2006;3:390–407.
[135] Mayer L. Evolving paradigms in the pathogenesis of IBD. J Gastroenterol 2010;45:9–16.
[136] Chey WD, Kurlander J, Eswaran S. Irritable bowel syndrome: a clinical review. JAMA 2015;313:949–58.
[137] Garcia Rodriguez LA, Gonzalez-Perez A, Johansson S, Wallander MA. Risk factors for inflammatory bowel disease in the general population. Aliment Pharmacol Ther 2005;22:309–15.
[138] Heitkemper M, Jarrett M. Irritable bowel syndrome: does gender matter? J Psychosom Res 2008;64:583–7.
[139] Lamah M, Scott HJ. Inflammatory bowel disease and pregnancy. Int J Colorectal Dis 2002;17:216–22.
[140] Kosiewicz MM, Dryden GW, Chhabra A, Alard P. Relationship between gut microbiota and development of T cell associated disease. FEBS Lett 2014;588:4195–206.
[141] Verdú EF, Deng Y, Bercik P, Collins SM. Modulatory effects of estrogen in two murine models of experimental colitis. Am J Physiol Gastrointest Liver Physiol 2002;283:G27–36.
[142] Seibel J, Molzberger AF, Hertrampf T, Laudenbach-Leschowski U, Diel P. Oral treatment with genistein reduces the expression of molecular and biochemical markers of inflammation in a rat model of chronic TNBS-induced colitis. Eur J Nutr 2009;48:213–20.
[143] Moussa L, Bézirard V, Salvador-Cartier C, Bacquié V, Lencina C, Lévêque M, et al. A low dose of fermented soy germ alleviates gut barrier injury, hyperalgesia and faecal protease activity in a rat model of inflammatory bowel disease. PLoS One 2012;7:49547.
[144] Ietta F, Bechi N, Romagnoli R, Bhattacharjee J, Realacci M, Di Vito M, et al. 17β-Estradiol modulates the macrophage migration inhibitory factor secretory pathway by regulating ABCA1 expression in human first-trimester placenta. Am J Physiol Endocrinol Metab 2010;298:E411–18.
[145] Gilliver SC, Emmerson E, Bernhagen J, Hardman MJ. MIF: a key player in cutaneous biology and wound healing. Exp Dermatol 2011;20:1–6.
[146] Veillat V, Sengers V, Metz CN, Roger T, Leboeuf M, Mailloux J, et al. Macrophage migration inhibitory factor is involved in a positive feedback loop increasing aromatase expression in endometriosis. Am J Pathol 2012;181:917–27.
[147] Geremia A, Biancheri P, Allan P, Corazza GR, Di Sabatino A. Innate and adaptive immunity in inflammatory bowel disease. Autoimmun Rev 2014;13:3–10.
[148] Papka RE, Storey-Workley M, Shughrue PJ, Merchenthaler I, Collins JJ, Usip S, et al. Estrogen receptor-alpha and beta- immunoreactivity and mRNA in neurons of sensory and autonomic ganglia and spinal cord. Cell Tissue Res 2001;304:193–214.
[149] Dun SL, Brailoiu GC, Gao X, Brailoiu E, Arterburn JB, Prossnitz ER, et al. Expression of estrogen receptor GPR30 in the rat spinal cord and in autonomic and sensory ganglia. J Neurosci Res 2009;87:1610–19.
[150] Acconcia F, Marino M. The effects of 17β-estradiol in cancer are mediated by estrogen receptor signaling at the plasma membrane. Front Physiol 2011;2:30.
[151] Bardin A, Boulle N, Lazennec G, Vignon F, Pujol P. Loss of ERbeta expression as a common step in estrogen-dependent tumor progression. Endocr Relat Cancer 2004;11:537–51.
[152] Heijmans J, Muncan V, Jacobs RJ, de Jonge-Muller ES, Graven L, Biemond I, et al. Intestinal tumorigenesis is not affected by progesterone signaling in rodent models. PLoS One 2011;6:22620.
[153] Schug TT, Janesick A, Blumberg B, Heindel JJ. Endocrine disrupting chemicals and disease susceptibility. J Steroid Biochem Mol Biol 2011;127:204–15.
[154] Rouiller-Fabre V, Guerquin MJ, N'Tumba-Byn T, Muczynski V, Moison D, Tourpin S, et al. Nuclear receptors and endocrine disruptors in fetal and neonatal testes: a gapped landscape. Front Endocrinol 2015;6:58.
[155] Kiyama R, Wada-Kiyama Y. Estrogenic endocrine disruptors: molecular mechanisms of action. Environ Int 2015;83:11–40.
[156] Seibel J, Molzberger AF, Hertrampf T, Laudenbach-Leschowski U, Degen GH, Diel P. In utero and postnatal exposure to a phytoestrogen-enriched diet increases parameters of acute inflammation in a rat model of TNBS-induced colitis. Arch Toxicol 2008;82:941–50.
[157] Ménard S, Guzylack-Piriou L, Lencina C, Leveque M, Naturel M, Sekkal S, et al. Perinatal exposure to a low dose of bisphenol A impaired systemic cellular immune response and predisposes young rats to intestinal parasitic infection. PLoS One 2014;9:112752.
[158] Fulde M, Hornef MW. Maturation of the enteric mucosal innate immune system during the postnatal period. Immunol Rev 2014;260:21–34.
[159] Braniste V, Audebert M, Zalko D, Houdeau E. Bisphenol A in the gut: another break in the wall? In: Bourguignon JP, Jégou B, Kerdelhué B, Toppari J, Christen Y, editors. Multi-system endocrine disruption—research and perspectives in endocrine interactions. Berlin/Heidelberg: Springer-Verlag; 2011. p. 127–44.
[160] Islam MA, Bekele R, Vanden Berg JH, Kuswanti Y, Thapa O, Soltani S, et al. Deconjugation of soy isoflavone glucuronides needed for estrogenic activity. Toxicol In Vitro 2015;29:706–15.
[161] Chen AC, Berhow MA, Tappenden KA, Donovan SM. Genistein inhibits intestinal cell proliferation in piglets. Pediatr Res 2005;57:192–200.
[162] Cotter AA, Jewell C, Cashman KD. The effect of oestrogen and dietary phyto-oestrogens on transepithelial calcium transport in human intestinal-like Caco-2 cells. Br J Nutr 2003;89:755–65.
[163] Yoshino S, Yamaki K, Li X, Sai T, Yanagisawa R, Takano H, et al. Prenatal exposure to bisphenol A up-regulates immune responses, including T helper 1 and T helper 2 responses, in mice. Immunology 2004;112:489–95.
[164] Kuiper GG, Lemmen JG, Carlsson B, Corton JC, Safe SH, van der Saag PT, et al. Interaction of estrogenic chemicals and phytoestrogens with estrogen receptor beta. Endocrinology 1998;139:4252–63.
[165] Shi T, Zhao C, Li Z, Zhang Q, Jin X. Bisphenol a exposure promotes the migration of NCM460 cells via estrogen receptor-mediated integrin β1/MMP-9 pathway. Environ Toxicol 2014;10:1002.
[166] Routledge EJ, White R, Parker MG, Sumpter JP. Differential effects of xenoestrogens on coactivator recruitment by estrogen receptor (ER) alpha and ERbeta. J Biol Chem 2000;275:35986–93.
[167] Swedenborg E, Ruegg J, Makela S, Pongratz I. Endocrine disruptive chemicals: mechanisms of action and involvement in metabolic disorders. J Mol Endocrinol 2009;43:1–10.
[168] Welshons WV, Nagel SC, vom Saal FS. Large effects from small exposures. III. Endocrine mechanisms mediating effects of bisphenol A at levels of human exposure. Endocrinology 2006;147:56–69.

CHAPTER

9

Sex and Gender Differences in Body Composition, Lipid Metabolism, and Glucose Regulation

Kelly Ethun

Department of Pathology and Laboratory Medicine, Emory University School of Medicine, Atlanta, GA, United States

INTRODUCTION

There is substantial evidence that males and females differ in their basic metabolic physiology and in their susceptibility to a variety of clinical pathologies including obesity, insulin resistance, and type II diabetes mellitus [1]. In some instances, these sex-based differences are complex, changing over the lifespan (eg, onset of menopause in women) or as a function of life style (different physical activity levels and/or dietary conditions). A better understanding of sex differences in body composition and energy metabolism should help physicians and patients anticipate these important changes. More importantly, the existence of sex differences in obesity and metabolic disease risk implies that one sex has a specific factor or process that protects them from the disease. If that factor can be enhanced or altered, by direct manipulation of the factor or its downstream pathways, then disease development and/or progression may be attenuated in both sexes. Thus, studying sex differences in energy metabolism may lead to the development of novel therapies [2].

Sex-specific functions of energy metabolism reflect the unique requirements of adult females to sustain gestation and lactation, whereas the energy metabolism of adult males represents a more static state. These differences arise from the gene-dosage effects of sex chromosomes, X and Y, as well as gonadal hormones, including estrogen and progesterone in females and androgens in males [3]. The effects of sex chromosomes on energy metabolism has recently been described using the four core genotypes mouse model which showed that sex chromosome complement independently regulates adiposity, feeding behavior, fatty liver, and glucose homeostasis [3]. The focus of this chapter is sex-specific differences in body fat deposition, lipid metabolism, and glucose homeostasis, and their regulation by sex hormones and adipokines, with an emphasis on studies in humans as well as relevant preclinical animal models.

SEX DIFFERENCES IN BODY FAT DEPOSITION

Body Composition and Metabolic Health Risk

For equivalent body mass index (BMI), women generally have a greater percentage of total body fat and less lean (muscle and bone) mass than men throughout adulthood [4]. Men tend to have central fat distribution, whereas women often have peripheral fat distribution, particularly in the lower body [5–7]. Other common terms to describe central versus peripheral fat distribution, respectively, include android versus gynoid and "apple" shaped versus "pear" shaped phenotypes [6]. Compared to men, premenopausal women also have less visceral white adipose tissue (VAT), and more subcutaneous white adipose tissue (SAT), both in the abdominal and gluteal-femoral regions, as measured by computerized tomography (CT) scan or magnetic resonance imaging (MRI) as shown in Fig. 9.1 [5–7]. For instance, using a highly spectral-spatial technique for fat selective MRI, one study showed that

Sex Differences in Physiology
DOI: http://dx.doi.org/10.1016/B978-0-12-802388-4.00009-4

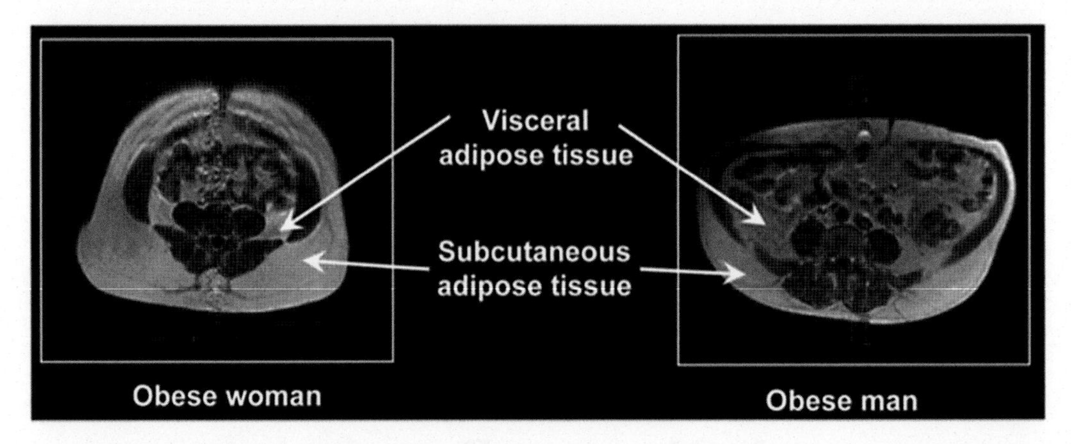

FIGURE 9.1 **Cross-sectional abdominal magnetic resonance gray-level images of an obese woman and an obese man.** Compared to the man, the woman has less visceral adipose tissue, but more subcutaneous adipose tissue. *Source: From Geer EB, Shen W. Gender differences in insulin resistance, body composition, and energy balance. Gend Med 2009;6(Suppl. 1):60–75. Epub 2009/06/12. http://dx.doi.org/10.1016/j.genm.2009.02.002 PubMed PMID: 19318219; PMCID: Pmc2908522.*

female adult subjects had significantly higher fat content in all muscle groups of the lower legs compared to age-matched men [5].

Fat selective MRI and CT are reliable methods to directly and accurately estimate VAT or central obesity among both men and women [8]. However, due to high cost and technical demands associated with these imaging modalities, anthropometric measures, such as waist circumference and waist-to-hip ratio, are commonly used to estimate central obesity in men and women [7]. In contrast, peripheral obesity often found among women is typically measured by hip circumference, thigh circumference, or dual-energy absorptiometry (DEXA) methods.

Although BMI (or general excess weight gain) is a strong predictor of overall mortality in humans [9], central adiposity, particularly increased visceral fat, in both men and women is a significant risk factor for the development of metabolic disorders including insulin resistance, type II diabetes mellitus, and cardiovascular disease [4,10]. In the Diabetes Prevention Program's population of 3234 patients, baseline waist circumference was the strongest predictor of diabetes in both sexes [11]. In addition, central adiposity often found in men is associated with higher risk for metabolic and cardiovascular disorders, compared to women with low VAT and high SAT. Although increased central obesity in both sexes is associated with an increased risk for metabolic syndrome according to the American Heart Association, the cutoff value for increased risk in men is larger than that for women (eg, waist circumference of 102 cm vs 88 cm, respectively) [12], suggesting a possible protective role of estrogen against metabolic syndrome and diabetes in premenopausal women [4]. Furthermore, gluteal-femoral or peripheral fat distribution found more often in women may have protective effects against diabetes and overall mortality [13], suggesting that effective uptake and storage of free fatty acids (FFAs) in lower body SAT may prevent abdominal obesity and associated metabolic disease (see Section "FFA Uptake and Triglyceride Storage in White Adipose Tissue") [14].

Sex Hormones and Fat-Depot Distribution

Estrogens

The regulators of sex differences in fat-depot distribution and their influence on the development of metabolic disorders are not fully understood, but evidence from human and animal studies suggests a predominant role of sex hormones, particularly estrogens and androgens (Fig. 9.2) [15,16]. The preferential deposition of adipose tissue in the periphery (abdominal and gluteal-femoral SAT) among women versus deposition in the viscera among men may be related to the higher level of estrogen in women compared with men [10]. Compared to a premenopausal state, menopause-induced estrogen deficiency in older women is associated with increased visceral adiposity and greater risk for metabolic disease. For example, Kotani et al. performed whole-body CT scans in 87 overweight (BMI $> 25\,kg/m^2$) women ranging in age from 20 to 79 to assess the role of aging on adipose tissue distribution. Postmenopausal women had a 2.6 times larger increase in VAT volume compared with premenopausal women [4]. Other studies show that postmenopausal women who take hormone replacement therapy (HRT), containing synthetic estrogen and progesterone components, have lower waist circumferences and VAT than those who do not take HRT [17,18], further illustrating that the presence of ovarian hormones may help prevent central adiposity in women. In line with human clinical

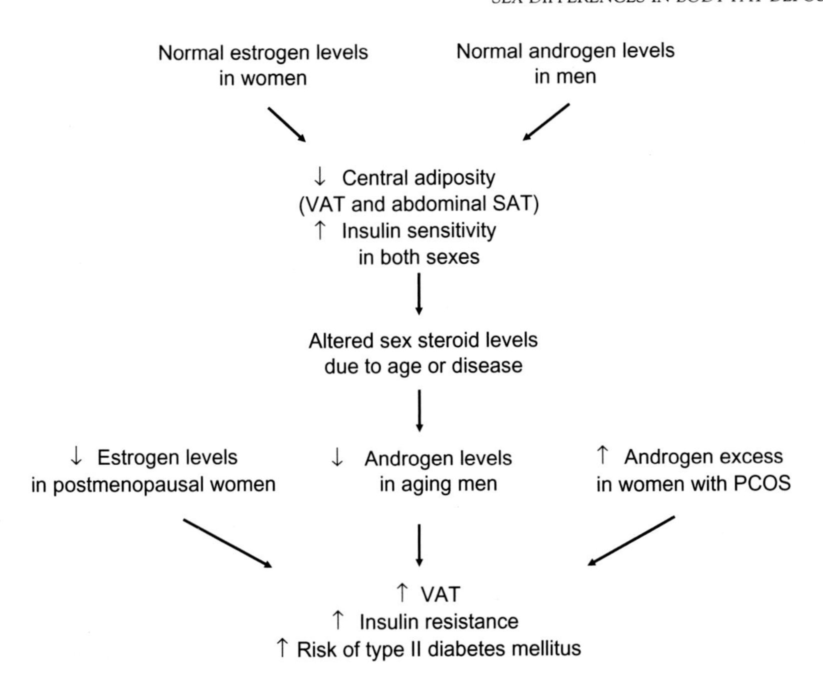

FIGURE 9.2 **Influence of sex steroids (estrogens and androgens) on body fat deposition and metabolic disease risk in men and women.** Normal physiological levels of estrogens and androgens in men and women, respectively, prevent central adiposity and insulin resistance. Aging men and women, however, experience a decline in steroids which results in increased visceral fat deposition, insulin resistance, and risk for type II diabetes mellitus. Furthermore, PCOS, a common cause of hyperandrogenism and anovulation in women, is generally associated with increased visceral fat and insulin resistance.

data, ovariectomized animals have increased visceral adiposity compared to intact animals [19] and estrogen treatment significantly reduces visceral adipose mass and adipocyte size. Together, these data demonstrate that estrogen has protective effects against obesity, in part, by reducing visceral fat accumulation in both humans and animals.

Estrogen may modulate fat accumulation in a depot- and sex-specific manner via differential expression of estrogen receptors (ERs) within white adipose tissue. Both ERα and ERβ mRNA are expressed in mature adipocytes of both women and men, but ERα is the more abundant isoform irrespective of white adipose tissue region [20]. ERβ1 expression (mRNA and protein) is significantly reduced in VAT compared to SAT, whereas the expression of ERβ4 and ERβ5 mRNA isoforms are significantly higher in gluteal-femoral SAT compared to abdominal SAT in both women and men, with higher ERβ mRNA levels in women compared with men [21]. 17β-estradiol in vitro upregulates the expression of both ERα and ERβ mRNA in subcutaneous adipocytes from women, but only ERα mRNA expression is upregulated in subcutaneous and visceral adipocytes from men [20,21].

The use of transgenic mice in which ERα (ERα-/-) and/or ERβ (ERβ-/-) are knocked out has provided valuable insight into the specific roles of each receptor, and a wealth of data suggest that ERα and ERβ may have divergent functions in carbohydrate and lipid metabolism. The precise functions of the two ERs in adipose tissue are not yet clear, but reports of humans with ERα or ERβ polymorphisms together with data obtained from ER-/- mice provide clues regarding the function of each receptor [22]. In male mice, deletion of ERα causes progressive increase in epididymal, perirenal, and inguinal adipose tissue with advancing age. Epididymal and perirenal adipocyte size is increased and adipocyte number is greater in the fat pads. Female ERα-/- mice have increased fat, adipocyte size, number, and cholesterol levels. Both male and female mice lacking ERα are insulin resistant with impaired glucose tolerance [23]. In contrast, absence of ERβ in mice prevents accumulation of triglycerides, preserves regular insulin signaling in liver and skeletal muscle, and improves whole-body insulin sensitivity and glucose tolerance, suggesting a pro-diabetogenic action of ERβ [24]. Use of ERβ agonists in high-fat-diet-fed and ovariectomy-induced obese wild-type mice, however, reduced weight gain, serum cholesterol, glucose, and fat accumulation [25]. Based on these findings, an imbalance between ERα and ERβ (ERα/ERβ ratio) in adipose tissue may have important implications for the development of metabolic diseases. Further mechanistic studies are needed, however, to clarify the intracellular signaling cascades resulting from activation of these ER subtypes and their influence on regional adiposity, adipose tissue substrate metabolism, and the development of metabolic disorders.

Androgens

Androgens also have sex- and depot-specific effects on fat distribution. Normal physiological levels of testosterone and androstenedione in men are associated with decreased total and central adiposity compared to premenopausal women. In a study of 178 young men, waist-to-hip ratio was inversely related with free testosterone [26]. Other studies show that testosterone in men declines with age and is often associated with an increase in VAT [27]. Furthermore, testosterone therapy in aging men decreases visceral fat mass and increases adipose tissue lipolysis by inhibiting lipoprotein lipase (LPL) activity [28].

These sex hormone-induced effects on body fat distribution are important for the metabolic health of men because reduced VAT is associated with improved insulin sensitivity, as demonstrated by a higher glucose disposal rate during hyperinsulinemic-euglycemic (HE) clamps in men treated with testosterone compared to controls [29]. Moreover, androgen receptor knockout mice show higher visceral adiposity and insulin resistance [30], suggesting that androgens may suppress central deposition of white adipose tissue in both human and rodent males. Although hypoandrogenism in men is positively correlated with increased fat accumulation and insulin insensitivity [31], it is currently not known whether these changes represent direct effects of androgen deficiency on target tissues or secondary effects of aging, diet, and/or other lifestyle changes on whole-body metabolism and body composition.

In contrast to males, androgen excess in females is generally associated with increased visceral fat accumulation and insulin resistance. Testosterone and androstenedione levels in women are positively correlated with waist diameter on CT [32] and are independently associated with significantly greater abdominal fat accumulation [33,34]. Obese postmenopausal women treated with testosterone for 9 months developed significantly increased visceral fat compared with women receiving placebo [35]. Furthermore, polycystic ovarian syndrome (PCOS), a common cause of hyperandrogenism and anovulation in women, is generally associated with increased visceral fat and insulin resistance [36].

Adipose Tissue Sex-Hormone Metabolism

Greater lower-body subcutaneous adiposity in women and higher visceral adiposity in men may result not only from the effects of circulating sex hormones, but also from depot-specific adipose tissue sex-hormone metabolism [37]. The gonads and adrenal glands contribute majority of sex hormones to circulation, however, adipose tissue, via enzyme activation and conversion, can contribute up to 50% of circulating testosterone in premenopausal women and 100% of circulating estrogen in postmenopausal women [38]. Adipose tissue-derived aromatase converts androgens to estrogens: androstenedione to estrone and testosterone to estradiol. Adipose tissue 17β-hydroxysteroid dehydrogenase (17β-HSD) converts weaker hormones to stronger hormones: androstenedione to testosterone and estrone to estradiol. SAT expresses relatively more aromatase than 17β-HSD, whereas VAT expresses more 17β-HSD than aromatase. Therefore, higher visceral adiposity in men or women with hyperandrogenemia may be associated with relatively more 17β-HSD expression, resulting in more local androgen biosynthesis [6]. Moreover, knockout mice deficient in aromatase activity to convert androgens to estrogens have increased visceral adiposity and insulin resistance [39], suggesting again that the absence of estrogens promotes visceral adiposity and insulin resistance.

Additional studies are needed to elucidate regional effects of sex-steroid hormones in adipose tissue to further advance our understanding of the contribution of sex hormones to white adipose tissue metabolism. For instance, studies are needed to determine the expression of steroidogenic metabolizing enzymes in relation to ERs in upper- and lower-body white adipose tissue [40]. The impact of sex steroids on body fat distribution, however, is complicated by the metabolism of sex steroids. The liver synthesizes and releases sex hormone-binding globulin (SHBG), which binds circulating estrogen and testosterone. Decreased SHBG levels may be associated with an android body composition [41]. Thus, future studies need to distinguish between total and free circulating sex hormones relative to SHBG levels.

SEX DIFFERENCES IN LIPID METABOLISM

It is important to note that there are two types of fat found in mammals: white and brown adipose tissue. Brown adipose tissue plays a major role in non-shivering thermogenesis in newborn humans and hibernating animals, but its role in human energy metabolism, body weight regulation, and obesity-associated metabolic complications is controversial and has not been elucidated [42]. White adipose tissue, on the other hand, is used to store energy, and thus, has several important functions relevant to human metabolic health.

One of the primary functions of white adipose tissue is the storage and release of triglycerides as a source of energy in the form of FFAs, to be utilized during exercise, fasting, or starvation [40]. The principal determinants of adipose tissue triglyceride metabolism are (1) the rate of triglyceride storage, primarily via lipoprotein lipase (LPL) activity, and (2) the rate of FFA release via lipolysis, which is suppressed by insulin in a postprandial state and stimulated by catecholamines in a fasted state. As such, the upper body (visceral and abdominal subcutaneous) and lower body (gluteal-femoral subcutaneous) adipose tissue depots exhibit key differences

TABLE 9.1 Sex and White Fat Depot-Specific Differences in Lipid Metabolism

Depots		Women				Men			
		Fat mass	FFA release	Insulin resistance	FFA uptake	Fat mass	FFA release	Insulin resistance	FFA uptake
VAT		−	+	+	+	+	++	++	+
SAT	Abdominal	+	++	−	−	−	+	+	+
	Gluteal-femoral	+	− −	−	++	−	−	−	−

in the rates and amounts of FFA release and uptake, and heterogeneity in these factors may be responsible for site- and sex-specific metabolic properties of fat deposition (summarized in Table 9.1).

FFA Uptake and Triglyceride Storage in White Adipose Tissue

The storage of triglycerides in white adipose tissue derived from circulating FFAs depends on the level of expression and activity of LPL in adipocytes, as this enzyme represents the rate-limiting step in the uptake of triglycerides-FFAs from circulation. Many studies have evaluated the depot- and sex-specific regulation of LPL. In early studies of nonobese men, LPL expression and activity were found to be higher in abdominal subcutaneous adipocytes than in adipocytes derived from the gluteal-femoral region. In contrast, LPL expression and activity were shown to be greater in gluteal-femoral fat depot in nonobese women compared to abdominal adipocytes [43]. Other studies show that LPL activity is reduced in VAT versus abdominal SAT-derived adipocytes from obese women, but higher in VAT adipocytes isolated from obese men [44,45]. In the same regard, LPL expression and activity is higher in VAT than in gluteal-femoral SAT in lean and moderately overweight men [45].

Both estrogen and testosterone appear to regulate LPL expression and activity in white adipose tissue in a depot-specific manner [46–49]. Some studies show that testosterone inhibits LPL activity in gluteal-femoral versus abdominal SAT depots of men, possibly contributing to abdominal fat gain observed more commonly in men compared to women [47]; whereas, other investigators have reported that testosterone has no effect on LPL activity in the gluteal-femoral SAT region of men [48]. Estrogen therapy (applied in form of skin patch), on the other hand, decreases LPL activity, but not mRNA expression, in the gluteal-femoral region of women through posttranscriptional modification [46]. Moreover, in vitro studies of subcutaneous abdominal adipocytes isolated from premenopausal women reveal that high concentrations of estradiol significantly reduce LPL expression, while low concentrations of estradiol significantly induces LPL expression in the abdominal SAT depot relative to control [50]. When replicated in vivo, however, the menstrual cycle was found to have no apparent effect on FFA uptake in healthy women for any fat depot, consistent with animal studies including nonhuman primate (NHP) studies [51,52].

Although most FFA uptake in human white adipose tissue is mediated via hydrolysis of circulating triglycerides by LPL, it is now recognized that direct LPL-independent FFA absorption may also contribute to FFA storage in humans [53,54]. Overall, a greater proportion of meal-derived FFAs is stored (primarily by the LPL-independent pathway) in upper-body abdominal SAT versus lower-body gluteal-femoral SAT in both lean and obese men and women, with women storing a greater percentage of dietary FFA in SAT than in men, who tend to store FFAs in VAT [48,51,55]. One point of interest is that the in vivo rate of FFA absorption in these studies was measured per mass of the whole-body fat mass. Thus, level of FFA uptake may be interpreted as greater in SAT due to its larger fat mass and similar or lower in VAT when FFA uptake is determined per cell [40]. Nonetheless, storage of postprandial FFAs per gram of whole-body adipose tissue following a high-fat meal is significantly greater in the gluteal-femoral SAT of women compared to men [56]. In addition, evidence suggests that women with lower-body obesity more efficiently remove meal-derived FFAs from circulation, which correlates with greater postprandial white adipose tissue LPL activity in women compared to men [56]. Women also have higher rates of triglyceride synthesis in their SAT compared to their VAT or any white adipose tissue depot in men [57]. Conversely, studies show that meal-derived FFA absorption in VAT is higher in lean men than women [55]. There are also dissimilarities in the upper-body depots, as postprandial FFA uptake is markedly higher in VAT than in abdominal SAT depot, which in turn is higher than the femoral SAT depot in obese men and women [58]. Interestingly, testosterone therapy induces postprandial FFA uptake preferentially in abdominal SAT and reduces uptake in VAT of middle-aged men [48]. Thus, aged men with low testosterone levels are prone to develop VAT adiposity.

In a postabsorptive state (following an overnight fast), women also exhibit significantly greater direct adipose tissue FFA storage compared to men [53,59]. For instance, Koutsari et al. found that women had higher direct uptake of FFA than men in both abdominal and femoral SAT in a postabsorptive state. Importantly, the storage rates were markedly greater in femoral SAT than abdominal SAT in women, whereas the opposite was observed for men [59]. Shadid et al. compared the postabsorptive direct FFA uptake among lean and obese subjects and found that lean men are more efficient at taking up circulating FFA into the abdominal SAT compared to the femoral SAT depot; while in obese subjects, direct FFA uptake is increased specifically in the femoral SAT of women [53]. Although some studies suggest that variations in LPL-activity may be responsible for the depot-specific differences in the above findings [60], further analyses are needed to understand the underlying mechanisms of FFA storage in upper- and lower-body depots of men and women. Furthermore, sex differences in short- and long-term fatty acid metabolism in both upper- and lower-body depots have not been fully illuminated and thus requires additional investigation.

Overall, direct adipose tissue FFA uptake and storage is significantly greater in women than men in both a postprandial and postabsorptive state, which provides a plausible explanation as to why women have greater total body fat than men [40]. Furthermore, the long-term storage of fatty acids in the lower-body (gluteal-femoral) depot, preferentially in women, may account for the protective properties of this depot from ectopic fat accumulation and metabolic perturbations. For instance, several studies suggest that gluteal-femoral fat distribution found more often in women may be associated with lower risk of diabetes and metabolic syndrome [13]. Nevertheless, the exact regulatory mechanisms of fatty acid metabolism responsible for the beneficial role of the gluteal-femoral depot remain to be elucidated.

Lipolysis and Fatty Acid Mobilization From White Adipose Tissue

Exercise and fasting are potent physiologic lipolytic stimuli. In response to these metabolic stressors, plasma insulin and glucose concentrations decrease and plasma catecholamine, glucagon, and FFA concentrations increase, while fat oxidation is favored over carbohydrate oxidation [61]. Plasma FFAs released from adipose tissue-derived triglycerides via catecholamine-mediated lipolysis are utilized as fuel for the body. FFA release in excess of metabolic demand, however, can lead to elevated plasma FFA availability, which is thought to be responsible for many of the metabolic abnormalities associated with obesity, including hypertriglyceridemia and insulin resistance.

Substantial sex-specific differences in catecholamine-induced lipolysis and FFA mobilization (release/influx into circulation) have been documented in the literature. In general, adipose tissue lipolysis and FFA plasma influx is substantially greater (~40%) in women than in men, even though metabolic health is typically better in women [40]. FFA mobilization rates also return to baseline values more readily in women than in men after removal of the lipolytic stimuli (ie, exercise termination or insulin administration after prolonged fasting). For instance, the increase in glycerol and FFA release rates into plasma during the transition from an overnight fast to a prolonged (22 h) fast are reported to be greater in women than men [62]. Similarly, the relative increase in FFA and glycerol release rates into circulation during moderate-intensity exercise is more pronounced in women than in men [63]. The mechanisms responsible for sex-based differences in exercise- and prolonged fasting-induced lipolytic stimulation of FFA release are not completely understood, but may be due, in part, to increased fat oxidation and more efficient utilization and disposal of FFA in women compared to men (reviewed in [64]).

Growing evidence suggests that there is great heterogeneity in catecholamine-mediated lipolysis between adipose tissue depots of men and women (Table 9.1). For instance, upper-body adipocytes from lean males and females are more responsive to lipolytic adrenergic stimulation, compared to adipocytes from the lower body. Most importantly, these effects are significantly more evident in adipocytes from lean women than lean men [65,66]. In support of these findings, subsequent in vivo studies of lean adults demonstrate that catecholamine-induced lipolysis is significantly greater in the upper-body compared to lower-body SAT depots [67]. Such depot-specific differences in lipolytic adrenergic stimulation are also evident in obese men and women. Both prolonged fasting-induced and exercise-induced lipolytic activity is substantially more pronounced in abdominal than in gluteal-femoral SAT in both men and women, but particularly among overweight women [67,68]. On the contrary, the catecholamine-induced lipolytic response in VAT and gluteal-femoral SAT among obese men is greater than that observed in obese women [67]. Collectively, these studies suggest that body fat distribution greatly influences lipolysis.

Following a meal, insulin is released from pancreatic β-cells and inhibits adipose tissue lipolysis and

FFA release into the circulation [69]. Some studies suggest that the suppression of plasma FFA release following a standard oral glucose load or mixed meal is greater in women than in men [70–72]. This sex difference, however, is likely attributed to men having a larger proportion of insulin-resistant upper-body SAT compared to women [73]. For example, some investigators have found that the upper-body SAT in both obese men and women are more resistant to the antilipolytic effects of insulin compared to lower-body fat depots [70,74,75]; however, FFA release by the upper-body SAT is higher in men than in women, indicating a higher resistance to the antilipolytic effect of meal ingestion (insulin) in the upper-body fat depot in men [76]. Furthermore, VAT in men is more resistant to the antilipolytic effects of insulin in men compared to women [76]. Based on current evidence from insulin infusion studies, FFA suppression is not different between men and women in response to low-dose insulin infusion, which nearly completely suppresses FFA release [73]. Any sex differences in response are likely due to variations in the extent of hyperinsulinemia during an insulin clamp, oral glucose tolerance test (OGTT), or fatty acid composition of meals [73]. The literature currently lacks dose-response studies in both obese and nonobese men and women simultaneously to help disentangle these sex-specific effects.

Sex- and depot-specific differences in lipolytic activity may be due to variations in sex hormones between men and women and/or differential expression of key enzymes involved in lipolysis. A recent study found that premenopausal obese women administered estrogen had increased lipolytic response in abdominal SAT, while estrogen impaired lipolysis in gluteal SAT [77]. These findings suggest that estrogen may be involved in maintaining a gynoid body fat distribution in premenopausal women by inhibiting lipolysis. In addition, some studies show that hormone sensitive lipase (HSL), a key enzyme in the lipolysis pathway, is differentially expressed among fat depots, and its activity is higher in abdominal SAT compared to gluteal SAT [78].

As previously described, circulating FFA derived from triglycerides stored in adipose tissue are an important source of fuel for the body; however, excess FFA plasma availability can lead to various metabolic abnormalities associated with obesity [70] such as ectopic fat accumulation and insulin resistance in liver [79] and muscle [80]. Studies have demonstrated that lipolysis of VAT increases relative to visceral fat mass in both men and women, which results in the release of excess FFAs into the liver via the portal vein. This enhanced influx of FFA impairs hepatic function and promotes insulin resistance [81]. While VAT is considered a major contributor of FFAs to the liver, abdominal SAT is the primary source of excess systemic FFAs to muscle, which mediates peripheral insulin resistance [70]. Together, these findings demonstrate a link between central adiposity (increased VAT and abdominal SAT), depot-specific levels of lipolytic activity, and adverse metabolic consequences of obesity. Specifically, men may be at higher risk for metabolic disorders than women, at least in part, due to central fat distribution and greater FFA release rates from VAT. On the contrary, women often have peripheral fat distribution, particularly in the lower body fat (gluteal-femoral SAT), which exhibits reduced lipolytic activity compared to VAT.

Plasma Lipid Profiles and Hepatic Lipoprotein Kinetics

Sex Differences in Plasma Lipid Profiles and Cardiovascular Health

On average, men have a greater atherogenic plasma lipid profile than premenopausal women. Specifically, men have lower high-density lipoprotein (HDL) and higher low-density lipoprotein (LDL), very low-density lipoprotein (VLDL), total plasma triglycerides, and VLDL-triglycerides concentrations (both during fed and fasting conditions) than age-matched women [60,70,73]. Furthermore, the average size of circulating VLDL is larger in men than women, whereas the average size of LDL is smaller in men [82]. Women also have larger HDL particles than men due to a shift toward more cholesterol-rich HDL [82]. LDL cholesterol is often considered the "bad" cholesterol because it contributes to plaque formation in arteries, leading to the development of coronary heart disease (CHD) and other forms of atherosclerosis. HDL cholesterol is considered "good" cholesterol because it helps remove LDL cholesterol from the arteries. HDL also acts as a scavenger, carrying LDL cholesterol away from the arteries and back to the liver, where it is broken down and passed from the body. Obese subjects compared to lean subjects have an increased risk of cardiovascular disease due to greater fasting plasma concentrations of total triglycerides and greater concentrations of VLDL-triglycerides and LDL particles (VLDL-apolipoprotein B-100) [83,84]. In the same regard, these dissimilarities in plasma lipid profiles account, at least in part, for the increased risk of cardiovascular disease in men compared to women [82].

Hepatic Lipid Kinetics in Lean Men and Women

The mechanisms responsible for the differences in the plasma lipid profile between lean men and women are only partially understood. Interestingly, the lower plasma concentrations of VLDL-triglycerides and LDL particles in lean women compared to lean men are associated with increased (rather than reduced) VLDL-triglycerides and LDL hepatic production and secretion; however, more efficient removal of VLDL-triglycerides and LDL particles from the circulation in lean women compensates for this and results in overall lower steady-state concentrations [85,86]. Conversely, lean women have lower plasma VLDL-apolipoprotein B100 concentrations (fewer VLDL particles) than lean men caused by a reduced hepatic secretion rate of VLDL-apolipoprotein B100 and shorter VLDL-apolipoprotein B100 residence time in circulation when compared to lean men. Consequently, lean women secrete fewer, but more triglyceride-rich VLDL than lean men [86]. Several in vivo human and animal studies indicate that the larger average size of VLDL particles facilitate more efficient clearance of VLDL-triglycerides from circulation by enhancing their susceptibility to hydrolysis by LPL [73]. Sex differences in plasma HDL concentrations are associated with a higher HDL hepatic production rate (without differences in plasma clearance rate) in lean women compared to lean men and both greater HDL apolipoprotein A-II production and plasma removal rate [87]. No information is available regarding sex differences in cholesterol kinetics among lipoprotein fractions in humans; however, data from animals suggest that endogenous sex hormones may play an important role in mediating cholesterol metabolism in a sex-specific manner [88].

Effects of Obesity on Hepatic Lipoprotein Kinetics in Men and Women

It is well-established that obesity is associated with elevated plasma VLDL-triglyceride concentrations and thus a higher risk of CHD in both men and women; however, few studies have examined the contribution of obesity to sex differences on the rate of hepatic lipoprotein secretion and plasma removal. Mittendorfer et al. found that increased adiposity is associated with enhanced hepatic secretion rates of VLDL-triglyceride and VLDL-apolipoprotein B100 in men, but not in women, and with decreased plasma removal rates of VLDL-triglyceride and VLDL-apolipoprotein B100 in women, but in men [89]. The authors also found that visceral fat mass is an important regulator of hepatic VLDL-triglyceride production, presumably through the availability of nonsystemic fatty acids derived from lipolysis of visceral and intrahepatic fat for VLDL-triglyceride synthesis. This finding is consistent with data from other studies demonstrating that the contribution from nonsystemic plasma FFA to total VLDL-triglyceride secretion is greater in obese compared with lean men and women, suggesting that a significant proportion of VLDL-triglyceride produced by the liver in obese individuals is derived from FFA available from lipolysis of visceral and intrahepatic fat and de novo hepatic lipogenesis [73]. Together, these findings demonstrate that sex and obesity have independent effects on the regulation of VLDL-triglyceride production by the liver.

Sex Hormones and Cardiovascular Health

Despite common belief, females do not have a more favorable cardiovascular lipid profile than men solely due to higher estrogens levels. In fact, findings from several studies examining the normal physiological changes in the hormonal milieu (ie, due to menopause or throughout the menstrual cycle) and the effects of exogenous sex steroid administration, demonstrate that sex hormones (estrogens, androgens, and progesterone) have little to no impact on plasma lipid homeostasis (recently reviewed in [90]). Furthermore, parenterally administered estrogens (intravenous or intramuscular routes) have either no effect or very modest beneficial effects, whereas orally administrated estrogens increase plasma triglyceride concentrations, consistent with observed sex differences in lipid metabolism and likely results from the hepatic "first-pass effect." Similarly, progesterone and androgens mimic only in part the differences in plasma lipid profiles between men and women.

The lack of randomized, placebo-controlled, dose-response studies and obvious difficulties in reproducing sex hormonal milieu experimentally (eg, the normal fluctuations of estradiol and progesterone during the menstrual cycle or the circadian rhythm of testosterone) limits interpretation of available human data [90]. NHPs, specifically ovariectomized macaque species, however, have been used extensively to study the effects of endogenous and exogenous hormones on serum lipid profiles and cardiovascular risk in postmenopausal women [91]. For instance, following long-term consumption of a high-fat diet, several reports demonstrate reduced coronary and peripheral atherosclerosis among ovariectomized female macaques treated with estrogen alone and estrogen-progesterone combination therapy compared to controls. Consistent with human clinical data, these NHP reports also showed that exogenous sex steroids have little impact on serum lipid profiles, consistent with human clinical data [91]. Due to numerous similarities in both male and female reproductive physiology with humans, NHP species may be useful in sex-comparative studies [92].

Other factors besides sex hormones likely contribute to sex differences in lipid metabolism. For instance, insulin and adipokines are important regulators of lipid metabolism; however, little is known regarding the possible sex differences in the regulation of lipid metabolism by insulin and various adipokines [93].

Furthermore, it is unlikely that differences in body composition between men and women are solely responsible for this phenomenon because sex differences in plasma lipid profiles persist even when men and women are matched for percentage of body fat [94]. Nonetheless, as previously described, increased adiposity appears to affect lipid kinetics differently in men and women [89], which may, in part, explain why men are at higher risk for cardiovascular disease than women.

Adipokines

In addition to its ability to store triglycerides during energy excess and release fatty acids to meet systemic energy needs, white adipose tissue is also an important endocrine organ that produces an array of signaling molecules, termed adipokines. Adipokines mediate normal local and whole-body metabolism via auto/paracrine and endocrine mechanisms. For instance, well-known adipokines, leptin and adiponectin, communicate with other organs including brain, liver, muscle, the immune system, and adipose tissue itself to regulate appetite and maintain energy homeostasis. Leptin and adiponectin, however, also regulate inflammatory responses and their levels of expression are altered in obesity. Leptin is considered a pro-inflammatory adipokine and fat accumulation upregulates its expression [95]. In contrast, adiponectin has anti-inflammatory properties and its expression is reduced with obesity [96]. Thus, the pro-inflammatory state induced by obesity can partly be attributed to an upregulation and downregulation of pro- and anti-inflammatory adipokines, respectively. Notably, the dysregulation of adipokines in obesity have been implicated in peripheral tissue insulin resistance, type II diabetes, and cardiovascular disease, at least, in part, due to an imbalance in the expression of pro- and anti-inflammatory adipokines [97]. Because sex steroids and fat depot heavily influence adipokine expression, sex differences in metabolic disorder risk may partly be attributed to variances in adipokine production [98].

Leptin

Leptin is a key hormone in the central regulation of metabolism and transfers a catabolic signal to the brain to inhibit food intake and increase energy expenditure. Leptin is produced by white adipose tissue in direct proportion to fat content and crosses the blood–brain barrier to interact with leptin receptors in the hypothalamus [99]. Circulating leptin concentrations have been reported to be up to four times higher in women compared to men of similar age [32]. Similarly, several cross-sectional studies have shown significant sex differences in circulating leptin levels during puberty with higher values in girls independent of BMI [100,101]. The cause of higher leptin levels in females is not clearly understood, but these observations suggest a strong sex steroid effect. Serum testosterone levels are inversely related to leptin levels in men [102] and testosterone treatment decreases leptin levels and improves insulin sensitivity in hypogonadal men [103]. Conversely, estrogen therapy increases leptin levels, suggesting that estrogen has a stimulatory effect on leptin concentrations in females [104].

Higher leptin concentrations in premenopausal women compared with men may also reflect higher adiposity, particularly higher levels of SAT, found in women [98]. Some studies have also shown that leptin concentrations are positively correlated with central obesity and fat cell size in women but not men [32]. Notably, higher leptin levels in women are associated with larger adipocytes in SAT but not VAT area on CT [32], which is in agreement with other studies reporting that leptin is predominantly expressed by SAT as opposed to VAT, particularly in women [105,106]. Furthermore, in vivo human studies demonstrate that secreted leptin levels are higher in femoral SAT than abdominal SAT [107], which is more commonly found in obese women compared to men.

Adiponectin

Adiponectin is derived solely from adipose tissue and is reduced in obesity and states of insulin resistance. Adiponectin regulates glucose metabolism by reducing gluconeogenesis in the liver [108,109] and improves insulin sensitivity in the liver and muscle by increasing FFA oxidation [108]. Similar to leptin, adiponectin levels have been reported to be higher in women than in men [110] as well as in adolescent girls compared to boys (boys 17% lower) [111]. For example, one cross-sectional study of >1000 participants reported that median adiponectin levels were significantly higher in women than in men, even after adjusting for BMI [112]. The same study also showed that reduced levels of adiponectin were more closely associated with diabetes mellitus in women than in men, though both sexes demonstrated an association between low adiponectin levels and other risk factors of the metabolic syndrome including abdominal adiposity, which was significantly associated in both men and women.

Current evidence demonstrates no relationship between adiponectin and estrogen levels, as several studies have reported no difference in plasma adiponectin levels in pre- and postmenopausal women or in postmenopausal women receiving HRT [112]. In addition, ovariectomy in mice does not alter plasma adiponectin levels. On the contrary, androgens have suppressive effects on plasma adiponectin levels and reduced levels of adiponectin may be associated with increased insulin resistance [113]. For example, decreases in testosterone levels with castration are associated with high levels of plasma adiponectin and improved insulin sensitivity [113]. Testosterone treatment in castrated mice also reduces plasma adiponectin levels. Furthermore, in vitro studies report that testosterone reduces adiponectin secretion in adipocytes cells. Adiponectin plasma levels are also positively associated with femoral and gluteal SAT and negatively correlated with VAT, particularly in men [114]. Together, these findings indicate that lower adiponectin levels in men than in women may be due to the inhibitory effect of androgens on adiponectin levels, higher visceral adiposity, or lower insulin sensitivity in men [113].

Other Adipokines

During the past decade, several novel adipokines have been identified that are highly regulated by obesity and may contribute to metabolic dysfunction. Their role in human metabolism and disease has recently been reviewed [40]; however, important findings are summarized here.

Chemerin, also known as retinoic acid receptor responder 2 or tazarotene-induced gene 2, is a recently identified adipokine that may modulate adipogenesis, inflammation, and glucose homeostasis [115]. Chemerin is synthesized by adipocytes [116] and adipocyte hypertrophy increases chemerin secretion [117]. Knockout mice deficient in chemerin have impaired insulin sensitivity and reduced adipose tissue interleukin-6 (IL-6) and tumor necrosis factor-alpha (TNF-α) expression [118]. Obese humans with metabolic syndrome have increased levels of circulating and gluteal SAT-secreted chemerin [119]. In addition, high circulating levels of chemerin are positively associated with VAT adiposity [120] and low circulating concentrations of estradiol [121]. Furthermore, a recent study investigating chemerin expression in paired abdominal SAT and VAT biopsies from obese humans showed higher chemerin expression in VAT than in abdominal SAT in men; however, the inverse was observed in women. Interestingly, women with PCOS in the same study showed a similar pattern to men, suggesting a role of androgens in sex differences regarding chemerin expression in adipose tissue [122]. Moreover, Alfadda and colleagues found a significant negative relationship between chemerin mRNA expression in abdominal SAT and plasma chemerin levels, which was associated with obesity, but not with insulin resistance [123]. Current evidence suggests that the molecular actions of chemerin that mediate adipose tissue metabolism and glucose homeostasis are fat depot-specific. Thus, further analyses of chemerin transduction pathways in the specific fat depots are needed to determine these distinct mechanisms.

Lipocalin 2 (also known as neutrophil gelatinase-associated lipocalin and 24p3) transports lipophilic substances such as retinoids, steroids, and arachidonic acid. High circulating levels of lipocalin 2 are associated with adiposity [124]; however, current evidence is unclear whether or not lipocalin 2 improves or impairs insulin sensitivity [40]. Similar to chemerin, men have higher expression of lipocalin 2 in VAT than in abdominal SAT, whereas the opposite is observed for women [122]. Additional studies are warranted to determine how these differences confer adipose tissue depot-specific function and whole-body glucose homeostasis.

Glypican-4 (Gpc4) is newly identified adipokines, synthesized predominantly by adipocytes. In women, serum Gpc4 levels increase proportionally with increases in BMI. While in men, serum Gpc4 levels increase from a lean to overweight phenotype, but in obese men, Gpc4 levels decrease to those observed in lean men. The distinct mechanism of this finding is not known, but maybe related to enhanced insulin resistance [125].

Omentin is another novel adipokines primarily synthetized by stromavascular cells. Omentin gene expression is higher in VAT than in abdominal SAT with no significant sex differences [122], but sex still appears to be a significant predictor of circulating omentin levels [121]. For instance, men have higher circulating levels of omentin compared to women and testosterone levels are inversely related to omentin levels [121]. Other studies have shown that omentin has anti-inflammatory properties and serum levels are reduced in adults with obesity and type II diabetes [40]. Based on this evidence, omentin may help reduce inflammation and improve insulin sensitivity; however, the clinical relevance of these findings needs to be explored in further studies.

It is critical to distinguish whether expression of these novel adipokines are compensatory reactions to increased metabolic distress, that is, adipocyte hypertrophy, in an effort to undergo normal adipose tissue remodeling, or if they have a pathogenic role leading to metabolic dysfunction [40]. Select adipokines may exert dual pro- or anti-inflammatory or insulin-resistant or insulin-sensitive effects, relative to normal adipocyte remodeling patterns in VAT and SAT depots. For instance, adipokines that decrease insulin resistance and show pro-inflammatory actions, including chemerin and lipocalin 2, appear to be produced more in VAT than SAT depots and more in men than in women. This may result from their secretion by hypertrophic adipocytes and macrophages in an inflammatory state during white adipose tissue remodeling, which occurs more frequently in men. While adipokines with insulin sensitizing and/or anti-inflammatory properties, such as Gpc4 and omentin, appear to increase mostly in the VAT (more in women than men), in an attempt to ameliorate or resolve local inflammation, and increase much less or decrease in abdominal SAT (less in women than men) [40]. Nevertheless, sex steroids and fat content regulate the expression of these novel adipokines; therefore, further investigations may provide insight into their contribution to metabolic dysfunction.

SEX DIFFERENCES IN GLUCOSE METABOLISM AND INSULIN ACTION

Secreted by the β cells of pancreatic islets, insulin is the primary regulator of plasma glucose concentration. In a basal state, it reduces endogenous glucose production (>90% from the liver) and stimulates peripheral glucose uptake. Whole-body insulin sensitivity reflects the combined sensitivity of liver, adipose tissue, and skeletal muscle to insulin. Despite having a higher percentage of body fat, the prevalence of type II diabetes mellitus and insulin resistance is lower in women than in men [32]. As shown in Table 9.2, these sex differences are explained by higher whole-body insulin sensitivity and greater serum glucose clearance rates in women versus men [126]. Human data are consistent with rodent studies, which further demonstrate greater insulin sensitivity and lower resistance among female rodents compared to male rodents fed a high-fat diet [127].

Table 9.2 also displays sex-based dissimilarities in glucose metabolism during exercise. Further discussion of how men and women differently utilize glucose during exercise is beyond the scope of this chapter. Nevertheless, the reader should understand that during exercise, women with higher estradiol concentrations, compared to men, oxidize significantly more lipids and less carbohydrates, deplete less muscle glycogen, and exhibit reduced hepatic gluconeogenesis [128,129]. The potential mechanisms responsible for these sex-specific metabolic responses to exercise include lower sympathetic nerve activity and greater type I and type II muscle fiber density in women [130]. Additional information regarding sex-specific differences in skeletal muscle metabolism can be found in recent reviews [130–132] and Chapter 10 of this book. The remainder of this chapter will primarily focus on skeletal muscle (as well as whole-body) insulin sensitivity and glucose homeostasis.

Whole Body and Skeletal Muscle Insulin Sensitivity

Several techniques are commonly used in research to study whole-body insulin sensitivity including the HE clamp technique and whole-body glucose tolerance tests [133]. Although labor intensive and technically challenging, the HE clamp technique is considered the gold standard for the direct measurement of insulin resistance. As part of this technique, plasma insulin levels are maintained at a hyperphysiological level by a continuous infusion of insulin, whereas the plasma glucose concentration is maintained at a normal physiological level by a variable glucose infusion rate (GIR). Once a steady state is achieved, the GIR equals glucose uptake by all tissues in the body and is therefore a direct measure of whole-body insulin sensitivity. Glucose tolerance tests including the OGTT and intravenous glucose tolerance test (IVGTT) are used in both research and clinically to evaluate the body's capability to metabolize glucose and ability of the liver to store excess glucose as glycogen. Insulin sensitivity can also be measured directly in the skeletal muscle using forearm or leg arterio-venous (A-V) balance technique and stable isotope tracers measured by positron emission tomography (PET). This technique employs flexible catheters to collect blood samples from the femoral artery and vein to determine the net uptake and release of glucose based on the A-V concentration difference.

Numerous studies have evaluated the impact of sex on insulin sensitivity using HE clamps and whole-body glucose tolerance tests. For instance, in a large cross-sectional study, including 1800 subjects, the prevalence of impaired fasting glucose was higher in men than in women (17% vs 13%) [134]. When the influence of sex was further assessed on glucose regulation in a cohort of approximately 8000 Swedish men and

TABLE 9.2 Sex Differences in Glucose Metabolism

	Basal State					Exercise-Induced		
	Diabetes risk	Insulin resistance	Insulin sensitivity	Glucose clearance	Insulin secretion	Carbohydrate utilization	Hepatic gluconeogenesis	Muscle glycogen depletion
Women	−	−	+	+	+	−	−	−
Men	+	+	−	−	+	+	+	+

women by an OGTT, the prevalence of impaired fasting glucose and type II diabetes mellitus was twice as high in men compared to women [135]. Additionally, in a Danish population study of 380 men and women matched for age and BMI, the glucose clearance rate during an IVGTT was 15% higher in women compared to men [136]. Collectively, these findings indicate that glucose effectiveness and clearance rates are higher in women than in men.

To obtain valid conclusions, insulin sensitivity should also be measured in the skeletal muscle directly. Additionally, confounding variables such as adiposity, fat distribution, hormonal status, and aerobic fitness level might complicate interpretations; thus, it is crucial to have matched men and women in regard to body composition, maximal oxygen uptake (VO_2-peak) per lean body mass (LBM), training status, and menstrual cyclicity when determining sex differences in insulin sensitivity [131]. To investigate possible sex differences in insulin-stimulated glucose clearance (insulin action and peripheral glucose disposal) directly in skeletal muscle, HE studies have expressed the GIR relative to the size of LBM, which gives a rough estimate of glucose clearance by skeletal muscle. Other studies have applied the forearm or leg A-V balance technique as previously described. Using the forearm A-V balance technique in healthy premenopausal women after an oral glucose load, Paula and colleagues showed that glucose uptake related to forearm muscle mass was 37% higher in women compared to men [137]. Subsequently, Nuutila et al. observed a 47% greater rate of glucose uptake in muscles of women compared to men, measured by PET scanning during a HE clamp [138]. Together, this evidence demonstrates that women have greater glucose uptake in their skeletal muscle when stimulated by physiological insulin concentrations.

The mechanisms responsible for the sex differences in insulin sensitivity are poorly understood. One study found no differences in insulin secretion between young men and women [139] whereas others have reported that young women have higher first-phase insulin secretion than age-matched men [140]. Sex differences in insulin signaling are tissue-specific and much of the work, thus far, has been conducted in rodents. For example, VAT from female mice is more responsive to low physiological levels of insulin and show greater increases in Akt and extracellular signal-related kinase phosphorylation and lipogenesis, whereas male adipocytes show activation only at higher insulin concentrations [141]. Moreover, greater rates of insulin-induced glucose uptake in female skeletal muscle and adipocytes correlate with higher expression levels of muscle mRNAs encoding glucose transporter-4 and metabolic enzymes [142].

Ovarian Hormones, Menstrual Cycle, and Insulin Sensitivity

A growing body of evidence emphasizes the importance of ovarian steroidal hormones in the regulation of glucose metabolism and insulin sensitivity. The use of oral contraceptives among premenopausal women, consisting of ethinyl estradiol and synthetic progesterone, is associated with lower insulin sensitivity and reduced circulating levels of estradiol [143], suggesting that estradiol may protect females against insulin resistance. Consistent with this idea, studies in estrogen-deficient aromatase-knockout and ovariectomized wild-type mice show that estrogen replacement therapy can protect female mice against hepatic steatosis and improve mitochondrial β-oxidation and insulin sensitivity [141,144,145]. Estrogen has also been found to protect against hyperglycemia in rodent models of diabetes, by decreasing hepatic gluconeogenesis and enhancing glucose transport into the muscle [146,147].

Longitudinal studies of adolescents show that sex differences in whole-body insulin sensitivity do not arise until after puberty, further underscoring the role of sex hormones in normal insulin action. Moran and colleagues performed repeated HE clamps on a large cohort of adolescents between the ages of 11 and 19 years, demonstrating a divergent pattern of insulin sensitivity across development. When expressed per kilogram of LBM, insulin sensitivity was reduced in adolescent males during puberty, whereas it was increased in adolescent females and became significantly higher in women than men by 19 years of age [148]. These findings may be attributed to the relative lower levels of estradiol in adolescent males compared to females.

Menopause and the cessation of ovarian hormone production in aging women are also associated with an increased risk of insulin resistance and type II diabetes mellitus, and subsequent estrogen replacement therapy decreases this risk [149]. Additional longitudinal HE clamp studies among peri- and postmenopausal women, however, are needed to determine whether estradiol per se, or other sex hormones, influences the development of insulin resistance across the menopausal transition. Other confounding variables such as age and changes in physical activity should be considered in such studies as well [131].

Interestingly, the complete lack of estrogen synthesis or activity in men is associated with insulin resistance. Specifically, men with a mutation in the aromatase gene, which catalyzes the final step in the biosynthesis of estradiol from androgens, leads to a diabetic phenotype [150] and a mutation of the ERα gene in men is associated with insulin resistance [151].

Thus, the relevance of estradiol for glucose homeostasis appears to extend to men.

The molecular mechanisms responsible for estrogen-mediated improvement in insulin sensitivity are tissue and pathway-specific. The effects of estrogen on various signal transduction and metabolic pathways in adipocytes, myocytes, and hepatocytes have thoroughly been described in a recent review [152]. Briefly, data from humans and animals suggest that estrogens may have beneficial effects on insulin sensitivity through a number of possible mechanisms: (1) direct effects on glucose homeostasis and substrate metabolism, (2) effects on oxidative stress and pro-inflammatory cytokines, or (3) involvement in adipose tissue metabolism and body composition, as previously described in this chapter.

Glucose Homeostasis and Menstrual Cycle

Data from rodent studies indicate that glucose metabolism is influenced by changes in estradiol and progesterone during the estrous cycle, when energy demand is high. Specifically, estradiol promotes insulin sensitivity, whereas progesterone promotes insulin resistance by blocking the positive effect of estradiol on glucose uptake and enhancing the activity of the β-oxidation pathway in skeletal muscle during exercise [153]. Evidence from human studies, however, suggest that whole-body insulin sensitivity is not subject to major changes during the menstrual cycle, despite large fluctuations in estradiol and progesterone across the follicular and luteal phases of the menstrual cycle, respectively. For instance, some studies show no change in whole-body metabolic rate and respiratory exchange ratio at rest [154] or basal plasma glucose and insulin [155] when the follicular phase (high estrogen-to-progesterone ratio) and luteal phase (high progesterone-to-estrogen ratio) are compared. In addition, when insulin sensitivity was evaluated during the menstrual cycle in young premenopausal women by an IVGTT, no differences in insulin sensitivity were observed between the luteal and mid-follicular phases [156]. Other studies have, however, found a small reduction in insulin sensitivity during the luteal phase, as assessed by homeostatic HOMA-IR or IVGTT [157,158]. Similarly, some studies show that glucose metabolism and insulin action during exercise is affected by the menstrual cycle in which glucose appearance and disappearance rates are higher during the follicular phase compared to the luteal phase [159].

Estrogen, Oxidative Stress, and Inflammatory Cytokines

Estradiol also has antioxidant properties, and confers increased resistance to oxidative stress through estrogen-induced expression of genes encoding the antioxidant enzymes superoxide dismutase and glutathione peroxidase [160]. Consequently, data from animal studies have found that mitochondria from females produce fewer reactive oxygen species than those from males [161]. Furthermore, estrogens exert protective effects on pancreatic β-cell function by reducing islet amyloid formation in ovariectomized mouse models of type II diabetes [162] and prevent streptozotocin (STZ)-induced β-cell apoptosis in both sexes [163]. These findings support the hypothesis that estrogen is protective against the development of insulin resistance and type II diabetes mellitus.

Estrogen also has been associated with anti-inflammatory properties [164]. One study showed that circulating level of TNF-α were sevenfold higher in ovariectomized rats compared with intact subjects [165]. Human studies have also found that menopause is associated with increased cytokine levels, including TNF-α, IL-1, and IL-6; these cytokine levels are substantially lower in women receiving HRT [166]. Given the connection between increased pro-inflammatory cytokines and obesity-related complications such as insulin resistance [167], the association between estrogen and decreased cytokines may play a role in sex differences in insulin resistance.

Androgens and Insulin Sensitivity

As previously described, androgen deficiency in men is associated with insulin resistance and obesity and subsequent testosterone replacement therapy improves insulin sensitivity and promotes fat mass loss. In contrast, androgen excess in women with PCOS is associated with insulin resistance and obesity [31]. The molecular mechanism for these sex-specific effects is not known, but current evidence indicates that testosterone has direct effects on skeletal muscle and pancreatic islet function. For instance, testosterone enhances the commitment of pluripotent mesenchymal stem cells to myogenic lineage [168], which may improve energy expenditure and reduce fat mass accumulation, contributing to improved insulin sensitivity in hormone-supplemented men. PGC1α (Peroxisome proliferator-activated receptor gamma coactivator 1-alpha) is an important regulator of mitochondrial biogenesis and function and reduced levels of this transcription factor have been found in type II diabetic patients [169]. A human study, however, has reported that testosterone increases the expression of PGC1α in skeletal muscle, stimulating mitochondrial biogenesis, substrate oxidation, and muscle insulin sensitivity [170].

Data from rodent studies have also shown that testosterone has a direct effect on pancreatic islet function by inducing insulin gene expression and release into circulation. This hyperinsulinemia then results in the development of impaired peripheral and adipocyte insulin sensitivity in female rats, even after transient hyperandrogenemia [171]. Other rodent studies report that testosterone exerts sex-specific protective effects against STZ-induced β-cell apoptosis [172] in which testosterone administration to normally STZ-resistant female rodents makes their β-cells more susceptible to apoptosis [173], thus illustrating the importance of androgen/estrogen ratio in β-cell survival in female rodents. Further investigations are needed, however, to determine whether these findings in rodents translate to humans, especially under different diet conditions.

It is also currently not clear whether the relationship between insulin resistance and hypogonadism represents direct effects of hormonal imbalances on target tissues or secondary effects of aging or changes in lifestyle on body composition and whole-body metabolism. Due to their close phylogenetic relation to humans, NHP represent a translational animal model for such metabolic studies. One of the primary advantages of using NHP is the ability to control variables that may impact whole-body metabolism such as diet, physical activity, and prior medical history. Similar to findings in humans, female rhesus macaque have significantly higher levels of insulin-induced glucose disposal than males [174] and estrogens can enhance glucose homeostasis in female macaques [175]. On the contrary, adult female macaques administered exogenous testosterone develop insulin resistance; however, this effect is only apparent when the animals are fed a high-fat diet [52]. Moreover, prenatal exposure to excess androgens leads to impairment in insulin secretion and action in rhesus macaque mothers and their female offspring [176]. These findings are consistent with the notion that prenatal exposure of females to androgens may result in metabolic reprogramming toward a male-like phenotype [176].

CONCLUSION

In summary, women have more body fat and less muscle mass than men. Women also store more lipids in lower-body fat, particularly in the abdominal and gluteal-femoral regions, than men; whereas men store more lipid in VAT than SAT and oxidize dietary FFAs more readily than women. These differences reflect the unique energy requirements of each sex. The female body has evolved to endure gestation and lactation, conditions with high-energy demands. Higher insulin sensitivity and lower muscle mass observed in women is more beneficial for energy storage of glucose and less beneficial for its utilization. In addition, more efficient insulin inhibition of lipolysis results in higher triglyceride storage in female white adipose tissue compared to men. These differences are primarily mediated by sex hormones including estrogen and progesterone in females and androgens in males. Estrogens and androgens exert beneficial metabolic effects by reducing body fat accumulation and improving insulin sensitivity in females and males, respectively. Sex steroids also heavily influence adipokine expression from various fat depots. For instance, androgens inhibit adiponectin secretion from VAT adipocytes and reduced plasma levels of adiponectin are associated with increased insulin resistance. Thus, the risk of insulin resistance and type II diabetes mellitus associated with obesity may be higher in men compared to women due to the inhibitory effect of androgens on adiponectin levels, higher visceral adiposity, and lower insulin sensitivity in men.

ABBREVIATIONS

17β-HSD 17β-hydroxysteroid dehydrogenase
BMI Body mass index
CT Computerized tomography
DEXA Dual-energy absorptiometry
ERs Estrogen receptors
FFAs Free fatty acids
GIR Glucose infusion rate
HE Hyperinsulinemic-euglycemic clamp
HRT Hormone replacement therapy
IL-1 Interleukin-1
IL-6 Interleukin-6
IVGTT Intravenous glucose tolerance test
LBM Lean body mass
LPL Lipoprotein lipase
MRI Magnetic resonance imaging
NHP Nonhuman primates
OGTT Oral glucose tolerance test
PCOS Polycystic ovarian syndrome
PET Positron emission tomography
PGC1α Peroxisome proliferator-activated receptor gamma coactivator 1-alpha
SAT Subcutaneous white adipose tissue
SHBG Sex hormone-binding globulin
STZ Streptozotocin
TNF-α Tumor necrosis factor-alpha
VAT Visceral white adipose tissue

References

[1] Lovejoy JC, Sainsbury A, Stock Conference Working G. Sex differences in obesity and the regulation of energy homeostasis. Obes Rev 2009;10(2):154–67. Available from: http://dx.doi.org/10.1111/j.1467-789X.2008.00529.x. PubMed PMID: 19021872.

[2] Arnold AP. Promoting the understanding of sex differences to enhance equity and excellence in biomedical science. Biol Sex Differ 2010;1(1):1. Available from: http://dx.doi.org/10.1186/2042-6410-1-1. PubMed PMID: 21208467; PMCID: 3010102.

[3] Link JC, Chen X, Arnold AP, Reue K. Metabolic impact of sex chromosomes. Adipocyte 2013;2(2):74–9. Available from: http://dx.doi.org/10.4161/adip.23320. PubMed PMID: 23805402; PMCID: 3661109.

[4] Kotani K, Tokunaga K, Fujioka S, Kobatake T, Keno Y, Yoshida S, et al. Sexual dimorphism of age-related changes in whole-body fat distribution in the obese. Int J Obes Relat Metab Disord 1994;18(4):207–12. PubMed PMID: 8044194.

[5] Machann J, Bachmann OP, Brechtel K, Dahl DB, Wietek B, Klumpp B, et al. Lipid content in the musculature of the lower leg assessed by fat selective MRI: intra- and interindividual differences and correlation with anthropometric and metabolic data. J Magn Reson Imaging 2003;17(3):350–7. Epub 2003/02/21. Available from: http://dx.doi.org/10.1002/jmri.10255. PubMed PMID: 12594726.

[6] Geer EB, Shen W. Gender differences in insulin resistance, body composition, and energy balance. Gend Med 2009;6 (Suppl. 1):60–75. Epub 2009/06/12. Available from: http://dx.doi.org/10.1016/j.genm.2009.02.002. PubMed PMID: 19318219; PMCID: Pmc2908522.

[7] Bray GA, Jablonski KA, Fujimoto WY, Barrett-Connor E, Haffner S, Hanson RL, et al. Relation of central adiposity and body mass index to the development of diabetes in the Diabetes Prevention Program. Am J Clin Nutr 2008;87 (5):1212–8. PubMed PMID: 18469241; PMCID: 2517222.

[8] Shen W, Wang Z, Punyanita M, Lei J, Sinav A, Kral JG, et al. Adipose tissue quantification by imaging methods: a proposed classification. Obes Res 2003;11(1):5–16. Available from: http://dx.doi.org/10.1038/oby.2003.3. PubMed PMID: 12529479; PMCID: 1894646.

[9] Prospective Studies C, Whitlock G, Lewington S, Sherliker P, Clarke R, Emberson J, et al. Body-mass index and cause-specific mortality in 900 000 adults: collaborative analyses of 57 prospective studies. Lancet 2009;373(9669):1083–96. Available from: http://dx.doi.org/10.1016/S0140-6736(09)60318-4. PubMed PMID: 19299006; PMCID: 2662372.

[10] Kissebah AH, Krakower GR. Regional adiposity and morbidity. Physiol Rev 1994;74(4):761–811. PubMed PMID: 7938225.

[11] Diabetes Prevention Program Research Group. Relationship of body size and shape to the development of diabetes in the diabetes prevention program. Obesity (Silver Spring) 2006;14 (11):2107–17. Available from: http://dx.doi.org/10.1038/oby.2006.246. PubMed PMID: 17135629; PMCID: 2373982.

[12] Grundy SM, Cleeman JI, Daniels SR, Donato KA, Eckel RH, Franklin BA, et al. Diagnosis and management of the metabolic syndrome: an American Heart Association/National Heart, Lung, and Blood Institute scientific statement. Curr Opin Cardiol 2006;21(1):1–6. PubMed PMID: 16355022.

[13] Pischon T, Boeing H, Hoffmann K, Bergmann M, Schulze MB, Overvad K, et al. General and abdominal adiposity and risk of death in Europe. N Engl J Med 2008;359(20):2105–20. Available from: http://dx.doi.org/10.1056/NEJMoa0801891. PubMed PMID: 19005195.

[14] Hernandez TL, Kittelson JM, Law CK, Ketch LL, Stob NR, Lindstrom RC, et al. Fat redistribution following suction lipectomy: defense of body fat and patterns of restoration. Obesity (Silver Spring) 2011;19(7):1388–95. Available from: http://dx.doi.org/10.1038/oby.2011.64. PubMed PMID: 21475140.

[15] Shi H, Clegg DJ. Sex differences in the regulation of body weight. Physiol Behav 2009;97(2):199–204. Available from: http://dx.doi.org/10.1016/j.physbeh.2009.02.017. PubMed PMID: 19250944.

[16] Stevens J, Katz EG, Huxley RR. Associations between gender, age and waist circumference. Eur J Clin Nutr 2010;64(1):6–15. Available from: http://dx.doi.org/10.1038/ejcn.2009.101. PubMed PMID: 19738633.

[17] Espeland MA, Stefanick ML, Kritz-Silverstein D, Fineberg SE, Waclawiw MA, James MK, et al. Effect of postmenopausal hormone therapy on body weight and waist and hip girths. Postmenopausal Estrogen-Progestin Interventions Study Investigators. J Clin Endocrinol Metab 1997;82(5):1549–56. Available from: http://dx.doi.org/10.1210/jcem.82.5.3925. PubMed PMID: 9141548.

[18] Munoz J, Derstine A, Gower BA. Fat distribution and insulin sensitivity in postmenopausal women: influence of hormone replacement. Obes Res 2002;10(6):424–31. Available from: http://dx.doi.org/10.1038/oby.2002.59. PubMed PMID: 12055317.

[19] D'Eon TM, Souza SC, Aronovitz M, Obin MS, Fried SK, Greenberg AS. Estrogen regulation of adiposity and fuel partitioning. Evidence of genomic and non-genomic regulation of lipogenic and oxidative pathways. J Biol Chem 2005;280 (43):35983–91. Available from: http://dx.doi.org/10.1074/jbc.M507339200. PubMed PMID: 16109719.

[20] Dieudonne MN, Leneveu MC, Giudicelli Y, Pecquery R. Evidence for functional estrogen receptors alpha and beta in human adipose cells: regional specificities and regulation by estrogens. Am J Physiol Cell Physiol 2004;286(3):C655–61. Available from: http://dx.doi.org/10.1152/ajpcell.00321.2003. PubMed PMID: 14761887.

[21] Pedersen SB, Bruun JM, Hube F, Kristensen K, Hauner H, Richelsen B. Demonstration of estrogen receptor subtypes alpha and beta in human adipose tissue: influences of adipose cell differentiation and fat depot localization. Mol Cell Endocrinol 2001;182(1):27–37. PubMed PMID: 11500236.

[22] Barros RP, Gustafsson JA. Estrogen receptors and the metabolic network. Cell Metab 2011;14(3):289–99. Available from: http://dx.doi.org/10.1016/j.cmet.2011.08.005. PubMed PMID: 21907136.

[23] Heine PA, Taylor JA, Iwamoto GA, Lubahn DB, Cooke PS. Increased adipose tissue in male and female estrogen receptor-alpha knockout mice. Proc Natl Acad Sci USA 2000;97 (23):12729–34. Available from: http://dx.doi.org/10.1073/pnas.97.23.12729. PubMed PMID: 11070086; PMCID: PMC18832.

[24] Foryst-Ludwig A, Clemenz M, Hohmann S, Hartge M, Sprang C, Frost N, et al. Metabolic actions of estrogen receptor beta (ERbeta) are mediated by a negative cross-talk with PPARgamma. PLoS Genet 2008;4(6):e1000108. Available from: http://dx.doi.org/10.1371/journal.pgen.1000108. PubMed PMID: 18584035; PMCID: PMC2432036.

[25] Yepuru M, Eswaraka J, Kearbey JD, Barrett CM, Raghow S, Veverka KA, et al. Estrogen receptor-{beta}-selective ligands alleviate high-fat diet- and ovariectomy-induced obesity in mice. J Biol Chem 2010;285(41):31292–303. Available from: http://dx.doi.org/10.1074/jbc.M110.147850. PubMed PMID: 20657011; PMCID: PMC2951204.

[26] Haffner SM, Valdez RA, Stern MP, Katz MS. Obesity, body fat distribution and sex hormones in men. Int J Obes Relat Metab Disord 1993;17(11):643–9. PubMed PMID: 8281222

[27] Derby CA, Zilber S, Brambilla D, Morales KH, McKinlay JB. Body mass index, waist circumference and waist to hip ratio and change in sex steroid hormones: the Massachusetts Male Ageing Study. Clin Endocrinol (Oxf) 2006;65(1):125–31. Available from: http://dx.doi.org/10.1111/j.1365-2265.2006.02560.x. PubMed PMID: 16817831.

[28] Marin P, Holmang S, Jonsson L, Sjostrom L, Kvist H, Holm G, et al. The effects of testosterone treatment on body composition and metabolism in middle-aged obese men. Int J Obes Relat Metab Disord 1992;16(12):991–7. PubMed PMID: 1335979.

[29] Marin P, Holmang S, Gustafsson C, Jonsson L, Kvist H, Elander A, et al. Androgen treatment of abdominally obese men. Obes Res 1993;1(4):245–51. PubMed PMID: 16350577.

[30] Fan W, Yanase T, Nomura M, Okabe T, Goto K, Sato T, et al. Androgen receptor null male mice develop late-onset obesity caused by decreased energy expenditure and lipolytic activity but show normal insulin sensitivity with high adiponectin secretion. Diabetes 2005;54(4):1000–8. PubMed PMID: 15793238.

[31] Ding EL, Song Y, Malik VS, Liu S. Sex differences of endogenous sex hormones and risk of type 2 diabetes: a systematic review and meta-analysis. JAMA 2006;295(11):1288–99. Available from: http://dx.doi.org/10.1001/jama.295.11.1288. PubMed PMID: 16537739.

[32] Garaulet M, Perex-Llamas F, Fuente T, Zamora S, Tebar FJ. Anthropometric, computed tomography and fat cell data in an obese population: relationship with insulin, leptin, tumor necrosis factor-alpha, sex hormone-binding globulin and sex hormones. Eur J Endocrinol 2000;143(5):657–66. PubMed PMID: 11078990.

[33] Evans DJ, Hoffmann RG, Kalkhoff RK, Kissebah AH. Relationship of androgenic activity to body fat topography, fat cell morphology, and metabolic aberrations in premenopausal women. J Clin Endocrinol Metab 1983;57(2):304–10. Available from: http://dx.doi.org/10.1210/jcem-57-2-304. PubMed PMID: 6345569.

[34] Haffner SM, Katz MS, Dunn JF. Increased upper body and overall adiposity is associated with decreased sex hormone binding globulin in postmenopausal women. Int J Obes 1991;15 (7):471–8. PubMed PMID: 1894424.

[35] Lovejoy JC, Bray GA, Bourgeois MO, Macchiavelli R, Rood JC, Greeson C, et al. Exogenous androgens influence body composition and regional body fat distribution in obese postmenopausal women–a clinical research center study. J Clin Endocrinol Metab 1996;81(6):2198–203. Available from: http://dx.doi.org/10.1210/jcem.81.6.8964851. PubMed PMID: 8964851.

[36] Dunaif A. Hyperandrogenic anovulation (PCOS): a unique disorder of insulin action associated with an increased risk of non-insulin-dependent diabetes mellitus. Am J Med 1995;98 (1A):33S–39SS. PubMed PMID: 7825639.

[37] Kershaw EE, Flier JS. Adipose tissue as an endocrine organ. J Clin Endocrinol Metab 2004;89(6):2548–56. Available from: http://dx.doi.org/10.1210/jc.2004-0395. PubMed PMID: 15181022.

[38] Belanger C, Luu-The V, Dupont P, Tchernof A. Adipose tissue intracrinology: potential importance of local androgen/estrogen metabolism in the regulation of adiposity. Horm Metab Res 2002;34(11–12):737–45. Available from: http://dx.doi.org/10.1055/s-2002-38265. PubMed PMID: 12660892.

[39] Jones ME, Thorburn AW, Britt KL, Hewitt KN, Wreford NG, Proietto J, et al. Aromatase-deficient (ArKO) mice have a phenotype of increased adiposity. Proc Natl Acad Sci USA 2000;97(23):12735–40. Available from: http://dx.doi.org/10.1073/pnas.97.23.12735. PubMed PMID: 11070087; PMCID: 18833.

[40] White UA, Tchoukalova YD. Sex dimorphism and depot differences in adipose tissue function. Biochim Biophys Acta 2014;1842 (3):377–92. Epub 2013/05/21. Available from: http://dx.doi.org/10.1016/j.bbadis.2013.05.006. PubMed PMID: 23684841; PMCID: Pmc3926193.

[41] Hautanen A. Synthesis and regulation of sex hormone-binding globulin in obesity. Int J Obes Relat Metab Disord 2000;24 (Suppl. 2):S64–70. PubMed PMID: 10997612.

[42] Ravussin E, Galgani JE. The implication of brown adipose tissue for humans. Annu Rev Nutr 2011;31:33–47. Available from: http://dx.doi.org/10.1146/annurev-nutr-072610-145209. PubMed PMID: 21548774; PMCID: PMC4404503.

[43] Arner P, Lithell H, Wahrenberg H, Bronnegard M. Expression of lipoprotein lipase in different human subcutaneous adipose tissue regions. J Lipid Res 1991;32(3):423–9. PubMed PMID: 2066672.

[44] Fried SK, Kral JG. Sex differences in regional distribution of fat cell size and lipoprotein lipase activity in morbidly obese patients. Int J Obes 1987;11(2):129–40. PubMed PMID: 3610466.

[45] Boivin A, Brochu G, Marceau S, Marceau P, Hould FS, Tchernof A. Regional differences in adipose tissue metabolism in obese men. Metabolism 2007;56(4):533–40. Available from: http://dx.doi.org/10.1016/j.metabol.2006.11.015. PubMed PMID: 17379013.

[46] Price TM, O'Brien SN, Welter BH, George R, Anandjiwala J, Kilgore M. Estrogen regulation of adipose tissue lipoprotein lipase–possible mechanism of body fat distribution. Am J Obstet Gynecol 1998;178(1 Pt 1):101–7. PubMed PMID: 9465811.

[47] Ramirez ME, McMurry MP, Wiebke GA, Felten KJ, Ren K, Meikle AW, et al. Evidence for sex steroid inhibition of lipoprotein lipase in men: comparison of abdominal and femoral adipose tissue. Metabolism 1997;46(2):179–85. PubMed PMID: 9030826.

[48] Marin P, Oden B, Bjorntorp P. Assimilation and mobilization of triglycerides in subcutaneous abdominal and femoral adipose tissue in vivo in men: effects of androgens. J Clin Endocrinol Metab 1995;80(1):239–43. Available from: http://dx.doi.org/10.1210/jcem.80.1.7829619. PubMed PMID: 7829619.

[49] Blouin K, Nadeau M, Perreault M, Veilleux A, Drolet R, Marceau P, et al. Effects of androgens on adipocyte differentiation and adipose tissue explant metabolism in men and women. Clin Endocrinol (Oxf) 2010;72(2):176–88. Available from: http://dx.doi.org/10.1111/j.1365-2265.2009.03645.x. PubMed PMID: 19500113.

[50] Palin SL, McTernan PG, Anderson LA, Sturdee DW, Barnett AH, Kumar S. 17Beta-estradiol and anti-estrogen ICI:compound 182,780 regulate expression of lipoprotein lipase and hormone-sensitive lipase in isolated subcutaneous abdominal adipocytes. Metabolism 2003;52(4):383–8. Available from: http://dx.doi.org/10.1053/meta.2003.50088. PubMed PMID: 12701046.

[51] Uranga AP, Levine J, Jensen M. Isotope tracer measures of meal fatty acid metabolism: reproducibility and effects of the menstrual cycle. Am J Physiol Endocrinol Metab 2005;288(3):E547–55. Available from: http://dx.doi.org/10.1152/ajpendo.00340.2004. PubMed PMID: 15507534.

[52] Varlamov O, Chu MP, McGee WK, Cameron JL, O'Rourke RW, Meyer KA, et al. Ovarian cycle-specific regulation of adipose tissue lipid storage by testosterone in female nonhuman primates. Endocrinology 2013;154(11):4126–35. Available from: http://dx.doi.org/10.1210/en.2013-1428. PubMed PMID: 24008344; PMCID: 3800767.

[53] Shadid S, Koutsari C, Jensen MD. Direct free fatty acid uptake into human adipocytes in vivo: relation to body fat distribution. Diabetes 2007;56(5):1369–75. Available from: http://dx.doi.org/10.2337/db06-1680. PubMed PMID: 17287467.

[54] Bickerton AS, Roberts R, Fielding BA, Hodson L, Blaak EE, Wagenmakers AJ, et al. Preferential uptake of dietary fatty acids in adipose tissue and muscle in the postprandial period. Diabetes 2007;56(1):168–76. Available from: http://dx.doi.org/10.2337/db06-0822. PubMed PMID: 17192479.

[55] Romanski SA, Nelson RM, Jensen MD. Meal fatty acid uptake in adipose tissue: gender effects in nonobese humans. Am J Physiol Endocrinol Metab 2000;279(2):E455–62. PubMed PMID: 10913047.

[56] Votruba SB, Jensen MD. Sex-specific differences in leg fat uptake are revealed with a high-fat meal. Am J Physiol Endocrinol Metab 2006;291(5):E1115–23. Available from: http://dx.doi.org/10.1152/ajpendo.00196.2006. PubMed PMID: 16803856.

[57] Edens NK, Fried SK, Kral JG, Hirsch J, Leibel RL. In vitro lipid synthesis in human adipose tissue from three abdominal sites. Am J Physiol 1993;265(3 Pt 1):E374–9. PubMed PMID: 8214046.

[58] Jensen MD, Sarr MG, Dumesic DA, Southorn PA, Levine JA. Regional uptake of meal fatty acids in humans. Am J Physiol Endocrinol Metab 2003;285(6):E1282–8. Available from: http://dx.doi.org/10.1152/ajpendo.00220.2003. PubMed PMID: 12915396.

[59] Koutsari C, Ali AH, Mundi MS, Jensen MD. Storage of circulating free fatty acid in adipose tissue of postabsorptive humans: quantitative measures and implications for body fat distribution. Diabetes 2011;60(8):2032–40. Available from: http://dx.doi.org/10.2337/db11-0154. PubMed PMID: 21659500; PMCID: 3142075.

[60] Nguyen TT, Hernandez Mijares A, Johnson CM, Jensen MD. Postprandial leg and splanchnic fatty acid metabolism in nonobese men and women. Am J Physiol 1996;271(6 Pt 1):E965–72. PubMed PMID: 8997213.

[61] Bergman BC, Cornier MA, Horton TJ, Bessesen DH. Effects of fasting on insulin action and glucose kinetics in lean and obese men and women. Am J Physiol Endocrinol Metab 2007;293(4):E1103–11. Available from: http://dx.doi.org/10.1152/ajpendo.00613.2006. PubMed PMID: 17684102.

[62] Mittendorfer B, Horowitz JF, Klein S. Gender differences in lipid and glucose kinetics during short-term fasting. Am J Physiol Endocrinol Metab 2001;281(6):E1333–9. PubMed PMID: 11701450.

[63] Moro C, Pillard F, de Glisezinski I, Crampes F, Thalamas C, Harant I, et al. Sex differences in lipolysis-regulating mechanisms in overweight subjects: effect of exercise intensity. Obesity (Silver Spring) 2007;15(9):2245–55. Available from: http://dx.doi.org/10.1038/oby.2007.267. PubMed PMID: 17890493.

[64] Karpe F, Dickmann JR, Frayn KN. Fatty acids, obesity, and insulin resistance: time for a reevaluation. Diabetes 2011;60(10):2441–9. Available from: http://dx.doi.org/10.2337/db11-0425. PubMed PMID: 21948998; PMCID: 3178283.

[65] Wahrenberg H, Lonnqvist F, Arner P. Mechanisms underlying regional differences in lipolysis in human adipose tissue. J Clin Invest 1989;84(2):458–67. Available from: http://dx.doi.org/10.1172/JCI114187. PubMed PMID: 2503539; PMCID: 548904.

[66] Fried SK, Leibel RL, Edens NK, Kral JG. Lipolysis in intraabdominal adipose tissues of obese women and men. Obes Res 1993;1(6):443–8. PubMed PMID: 16353332.

[67] Jensen MD, Cryer PE, Johnson CM, Murray MJ. Effects of epinephrine on regional free fatty acid and energy metabolism in men and women. Am J Physiol 1996;270(2 Pt 1):E259–64. PubMed PMID: 8779947.

[68] Arner P, Kriegholm E, Engfeldt P, Bolinder J. Adrenergic regulation of lipolysis in situ at rest and during exercise. J Clin Invest 1990;85(3):893–8. Available from: http://dx.doi.org/10.1172/JCI114516. PubMed PMID: 2312732; PMCID: 296507.

[69] Abbasi F, McLaughlin T, Lamendola C, Reaven GM. The relationship between glucose disposal in response to physiological hyperinsulinemia and basal glucose and free fatty acid concentrations in healthy volunteers. J Clin Endocrinol Metab 2000;85(3):1251–4. Available from: http://dx.doi.org/10.1210/jcem.85.3.6450. PubMed PMID: 10720071.

[70] Jensen MD. Gender differences in regional fatty acid metabolism before and after meal ingestion. J Clin Invest 1995;96(5):2297–303. Available from: http://dx.doi.org/10.1172/JCI118285. PubMed PMID: 7593616; PMCID: 185880.

[71] McKeigue PM, Laws A, Chen YD, Marmot MG, Reaven GM. Relation of plasma triglyceride and apoB levels to insulin-mediated suppression of nonesterified fatty acids. Possible explanation for sex differences in lipoprotein pattern. Arterioscler Thromb 1993;13(8):1187–92. PubMed PMID: 8343493.

[72] Sumner AE, Kushner H, Tulenko TN, Falkner B, Marsh JB. The relationship in African-Americans of sex differences in insulin-mediated suppression of nonesterified fatty acids to sex differences in fasting triglyceride levels. Metabolism 1997;46(4):400–5. PubMed PMID: 9109843.

[73] Magkos F, Mittendorfer B. Gender differences in lipid metabolism and the effect of obesity. Obstet Gynecol Clin North Am 2009;36(2):245–65. vii. Epub 2009/06/09. Available from: http://dx.doi.org/10.1016/j.ogc.2009.03.001. PubMed PMID: 19501312.

[74] Dowling HJ, Fried SK, Pi-Sunyer FX. Insulin resistance in adipocytes of obese women: effects of body fat distribution and race. Metabolism 1995;44(8):987–95. PubMed PMID: 7637656.

[75] Jensen MD, Haymond MW, Rizza RA, Cryer PE, Miles JM. Influence of body fat distribution on free fatty acid metabolism in obesity. J Clin Invest 1989;83(4):1168–73. Available from: http://dx.doi.org/10.1172/JCI113997. PubMed PMID: 2649512; PMCID: 303803.

[76] Blaak E. Gender differences in fat metabolism. Curr Opin Clin Nutr Metab Care 2001;4(6):499–502. Epub 2001/11/14. PubMed PMID: 11706283.

[77] Gavin KM, Cooper EE, Raymer DK, Hickner RC. Estradiol effects on subcutaneous adipose tissue lipolysis in premenopausal women are adipose tissue depot specific and treatment dependent. Am J Physiol Endocrinol Metab 2013;304(11):E1167–74. Available from: http://dx.doi.org/10.1152/ajpendo.00023.2013. PubMed PMID: 23531620.

[78] Tan GD, Goossens GH, Humphreys SM, Vidal H, Karpe F. Upper and lower body adipose tissue function: a direct comparison of fat mobilization in humans. Obes Res 2004;12(1):114–18. Available from: http://dx.doi.org/10.1038/oby.2004.15. PubMed PMID: 14742849.

[79] Saloranta C, Franssila-Kallunki A, Ekstrand A, Taskinen MR, Groop L. Modulation of hepatic glucose production by non-esterified fatty acids in type 2 (non-insulin-dependent) diabetes mellitus. Diabetologia 1991;34(6):409–15. PubMed PMID: 1884899.

[80] Boden G, Chen X. Effects of fat on glucose uptake and utilization in patients with non-insulin-dependent diabetes. J Clin Invest 1995;96(3):1261–8. Available from: http://dx.doi.org/10.1172/JCI118160. PubMed PMID: 7657800; PMCID: 185747.

[81] Nielsen S, Guo Z, Johnson CM, Hensrud DD, Jensen MD. Splanchnic lipolysis in human obesity. J Clin Invest 2004;113(11):1582–8. Available from: http://dx.doi.org/10.1172/JCI21047. PubMed PMID: 15173884; PMCID: 419492.

[82] Freedman DS, Otvos JD, Jeyarajah EJ, Shalaurova I, Cupples LA, Parise H, et al. Sex and age differences in lipoprotein subclasses measured by nuclear magnetic resonance spectroscopy: the Framingham Study. Clin Chem 2004;50(7):1189–200. Available from: http://dx.doi.org/10.1373/clinchem.2004.032763. PubMed PMID: 15107310.

[83] Chan DC, Barrett HP, Watts GF. Dyslipidemia in visceral obesity: mechanisms, implications, and therapy. Am J Cardiovasc Drugs 2004;4(4):227–46. PubMed PMID: 15285698.

[84] Goff Jr DC, D'Agostino Jr RB, Haffner SM, Otvos JD. Insulin resistance and adiposity influence lipoprotein size and subclass concentrations. Results from the Insulin Resistance Atherosclerosis Study. Metabolism 2005;54(2):264–70. Available from: http://dx.doi.org/10.1016/j.metabol.2004.09.002. PubMed PMID: 15690322.

[85] Matthan NR, Jalbert SM, Barrett PH, Dolnikowski GG, Schaefer EJ, Lichtenstein AH. Gender-specific differences in the kinetics of nonfasting TRL, IDL, and LDL apolipoprotein B-100 in men and premenopausal women. Arterioscler Thromb Vasc Biol 2008;28(10):1838–43. Available from: http://dx.doi.org/10.1161/ATVBAHA.108.163931. PubMed PMID: 18658047; PMCID: 2872098.

[86] Magkos F, Patterson BW, Mohammed BS, Klein S, Mittendorfer B. Women produce fewer but triglyceride-richer very low-density lipoproteins than men. J Clin Endocrinol Metab 2007;92(4):1311–18. Available from: http://dx.doi.org/10.1210/jc.2006-2215. PubMed PMID: 17264179.

[87] Schaefer EJ, Zech LA, Jenkins LL, Bronzert TJ, Rubalcaba EA, Lindgren FT, et al. Human apolipoprotein A-I and A-II metabolism. J Lipid Res 1982;23(6):850–62. PubMed PMID: 6813411.

[88] Hewitt KN, Boon WC, Murata Y, Jones ME, Simpson ER. The aromatase knockout mouse presents with a sexually dimorphic disruptito cholesterol homeostasis. Endocrinology 2003;144(9):3895–903. Available from: http://dx.doi.org/10.1210/en.2003-0244. PubMed PMID: 12933663.

[89] Mittendorfer B, Patterson BW, Klein S. Effect of sex and obesity on basal VLDL-triacylglycerol kinetics. Am J Clin Nutr 2003;77(3):573–9. PubMed PMID: 12600845.

[90] Wang X, Magkos F, Mittendorfer B. Sex differences in lipid and lipoprotein metabolism: it's not just about sex hormones. J Clin Endocrinol Metab 2011;96(4):885–93. Available from: http://dx.doi.org/10.1210/jc.2010-2061. PubMed PMID: 21474685; PMCID: 3070248.

[91] Clarkson TB, Mehaffey MH. Coronary heart disease of females: lessons learned from nonhuman primates. Am J Primatol 2009;71(9):785–93. Available from: http://dx.doi.org/10.1002/ajp.20693. PubMed PMID: 19382155.

[92] Kaplan JR, Adams MR, Clarkson TB, Manuck SB, Shively CA, Williams JK. Psychosocial factors, sex differences, and atherosclerosis: lessons from animal models. Psychosom Med 1996;58(6):598–611. PubMed PMID: 8948008.

[93] Magkos F, Wang X, Mittendorfer B. Metabolic actions of insulin in men and women. Nutrition 2010;26(7–8):686–93. Available from: http://dx.doi.org/10.1016/j.nut.2009.10.013. PubMed PMID: 20392600; PMCID: 2893237.

[94] Magkos F, Mohammed BS, Mittendorfer B. Effect of obesity on the plasma lipoprotein subclass profile in normoglycemic and normolipidemic men and women. Int J Obes 2008;32(11):1655–64. Available from: http://dx.doi.org/10.1038/ijo.2008.164. PubMed PMID: 18779822; PMCID: 2584161.

[95] Iikuni N, Lam QL, Lu L, Matarese G, La Cava A. Leptin and Inflammation. Curr Immunol Rev 2008;4(2):70–9. Available from: http://dx.doi.org/10.2174/157339508784325046. PubMed PMID: 20198122; PMCID: 2829991.

[96] Arita Y, Kihara S, Ouchi N, Takahashi M, Maeda K, Miyagawa J, et al. Paradoxical decrease of an adipose-specific protein, adiponectin, in obesity. Biochem Biophys Res Commun 1999;257(1):79–83. PubMed PMID: 10092513.

[97] Bluher M. Are there still healthy obese patients? Curr Opin Endocrinol Diabetes Obes 2012;19(5):341–6. Available from: http://dx.doi.org/10.1097/MED.0b013e328357f0a3. PubMed PMID: 22895358.

[98] Manolopoulos KN, Karpe F, Frayn KN. Gluteofemoral body fat as a determinant of metabolic health. Int J Obes 2010;34(6):949–59. Available from: http://dx.doi.org/10.1038/ijo.2009.286. PubMed PMID: 20065965.

[99] Elmquist JK, Elias CF, Saper CB. From lesions to leptin: hypothalamic control of food intake and body weight. Neuron 1999;22(2):221–32. PubMed PMID: 10069329.

[100] Clayton PE, Gill MS, Hall CM, Tillmann V, Whatmore AJ, Price DA. Serum leptin through childhood and adolescence. Clin Endocrinol (Oxf) 1997;46(6):727–33. PubMed PMID: 9274704.

[101] Koester-Weber T, Valtuena J, Breidenassel C, Beghin L, Plada M, Moreno S, et al. Reference values for leptin, cortisol, insulin and glucose, among European adolescents and their association with adiposity: the Helena study. Nutr Hosp 2014;30(n05):1181–90. Epub 2014/11/05. Available from: http://dx.doi.org/10.3305/nh.2014.30.5.7982. PubMed PMID: 25365025.

[102] Luukkaa V, Pesonen U, Huhtaniemi I, Lehtonen A, Tilvis R, Tuomilehto J, et al. Inverse correlation between serum testosterone and leptin in men. J Clin Endocrinol Metab 1998;83(9):3243–6. Available from: http://dx.doi.org/10.1210/jcem.83.9.5134. PubMed PMID: 9745436.

[103] Simon D, Charles MA, Lahlou N, Nahoul K, Oppert JM, Gouault-Heilmann M, et al. Androgen therapy improves insulin sensitivity and decreases leptin level in healthy adult men with low plasma total testosterone: a 3-month randomized placebo-controlled trial. Diabetes care 2001;24(12):2149–51. PubMed PMID: 11723098.

[104] Demerath EW, Towne B, Wisemandle W, Blangero J, Chumlea WC, Siervogel RM. Serum leptin concentration, body composition, and gonadal hormones during puberty. Int J Obes Relat Metab Disord 1999;23(7):678–85. PubMed PMID: 10454100.

[105] Hube F, Lietz U, Igel M, Jensen PB, Tornqvist H, Joost HG, et al. Difference in leptin mRNA levels between omental and subcutaneous abdominal adipose tissue from obese humans. Horm Metab Res 1996;28(12):690–3. Available from: http://dx.doi.org/10.1055/s-2007-979879. PubMed PMID: 9013743.

[106] Montague CT, Prins JB, Sanders L, Zhang J, Sewter CP, Digby J, et al. Depot-related gene expression in human subcutaneous and omental adipocytes. Diabetes 1998;47(9):1384–91. PubMed PMID: 9726225.

[107] Nielsen NB, Hojbjerre L, Sonne MP, Alibegovic AC, Vaag A, Dela F, et al. Interstitial concentrations of adipokines in subcutaneous abdominal and femoral adipose tissue. Regul Pept 2009;155(1–3):39–45. Available from: http://dx.doi.org/10.1016/j.regpep.2009.04.010. PubMed PMID: 19376162.

[108] Yamauchi T, Kamon J, Waki H, Terauchi Y, Kubota N, Hara K, et al. The fat-derived hormone adiponectin reverses insulin resistance associated with both lipoatrophy and obesity. Nat Med 2001;7(8):941–6. Available from: http://dx.doi.org/10.1038/90984. PubMed PMID: 11479627.

[109] Batterham RL, Cowley MA, Small CJ, Herzog H, Cohen MA, Dakin CL, et al. Gut hormone PYY(3-36) physiologically inhibits food intake. Nature 2002;418(6898):650–4. Available from: http://dx.doi.org/10.1038/nature02666. PubMed PMID: 12167864.

[110] Cnop M, Havel PJ, Utzschneider KM, Carr DB, Sinha MK, Boyko EJ, et al. Relationship of adiponectin to body fat distribution, insulin sensitivity and plasma lipoproteins: evidence for independent roles of age and sex. Diabetologia 2003;46(4):459–69. Available from: http://dx.doi.org/10.1007/s00125-003-1074-z. PubMed PMID: 12687327.

[111] Punthakee Z, Delvin EE, O'Loughlin J, Paradis G, Levy E, Platt RW, et al. Adiponectin, adiposity, and insulin resistance in children and adolescents. J Clin Endocrinol Metab 2006;91(6):2119–25. Epub 2006/03/16. Available from: http://dx.doi.org/10.1210/jc.2005-2346. PubMed PMID: 16537675.

[112] Salas-Salvado J, Granada M, Bullo M, Corominas A, Casas P, Foz M. Plasma adiponectin distribution in a Mediterranean population and its association with cardiovascular risk factors and metabolic syndrome. Metabolism 2007;56(11):1486–92. Available from: http://dx.doi.org/10.1016/j.metabol.2007.06.014. PubMed PMID: 17950098.

[113] Nishizawa H, Shimomura I, Kishida K, Maeda N, Kuriyama H, Nagaretani H, et al. Androgens decrease plasma adiponectin, an insulin-sensitizing adipocyte-derived protein. Diabetes 2002;51(9):2734–41. PubMed PMID: 12196466.

[114] Fisher FM, McTernan PG, Valsamakis G, Chetty R, Harte AL, Anwar AJ, et al. Differences in adiponectin protein expression: effect of fat depots and type 2 diabetic status. Horm Metab Res 2002;34(11–12):650–4. Available from: http://dx.doi.org/10.1055/s-2002-38246. PubMed PMID: 12660876.

[115] Roh SG, Song SH, Choi KC, Katoh K, Wittamer V, Parmentier M, et al. Chemerin–a new adipokine that modulates adipogenesis via its own receptor. Biochem Biophys Res Commun 2007;362(4):1013–18. Available from: http://dx.doi.org/10.1016/j.bbrc.2007.08.104. PubMed PMID: 17767914.

[116] Goralski KB, McCarthy TC, Hanniman EA, Zabel BA, Butcher EC, Parlee SD, et al. Chemerin, a novel adipokine that regulates adipogenesis and adipocyte metabolism. J Biol Chem 2007;282 (38):28175–88. Available from: http://dx.doi.org/10.1074/jbc.M700793200. PubMed PMID: 17635925.

[117] Bauer S, Wanninger J, Schmidhofer S, Weigert J, Neumeier M, Dorn C, et al. Sterol regulatory element-binding protein 2 (SREBP2) activation after excess triglyceride storage induces chemerin in hypertrophic adipocytes. Endocrinology 2011;152 (1):26–35. Available from: http://dx.doi.org/10.1210/en.2010-1157. PubMed PMID: 21084441.

[118] Ernst MC, Haidl ID, Zuniga LA, Dranse HJ, Rourke JL, Zabel BA, et al. Disruption of the chemokine-like receptor-1 (CMKLR1) gene is associated with reduced adiposity and glucose intolerance. Endocrinology 2012;153(2):672–82. Available from: http://dx.doi.org/10.1210/en.2011-1490. PubMed PMID: 22186410; PMCID: 3275396.

[119] Jialal I, Devaraj S, Kaur H, Adams-Huet B, Bremer AA. Increased chemerin and decreased omentin-1 in both adipose tissue and plasma in nascent metabolic syndrome. J Clin Endocrinol Metab 2013;98(3):E514–17. Available from: http://dx.doi.org/10.1210/jc.2012-3673. PubMed PMID: 23303213.

[120] Shin HY, Lee DC, Chu SH, Jeon JY, Lee MK, Im JA, et al. Chemerin levels are positively correlated with abdominal visceral fat accumulation. Clin Endocrinol (Oxf) 2012;77(1):47–50. Available from: http://dx.doi.org/10.1111/j.1365-2265.2011.04217.x. PubMed PMID: 21895733.

[121] Luque-Ramirez M, Martinez-Garcia MA, Montes-Nieto R, Fernandez-Duran E, Insenser M, Alpanes M, et al. Sexual dimorphism in adipose tissue function as evidenced by circulating adipokine concentrations in the fasting state and after an oral glucose challenge. Hum Reprod 2013;28(7):1908–18. Available from: http://dx.doi.org/10.1093/humrep/det097. PubMed PMID: 23559188.

[122] Martinez-Garcia MA, Montes-Nieto R, Fernandez-Duran E, Insenser M, Luque-Ramirez M, Escobar-Morreale HF. Evidence for masculinization of adipokine gene expression in visceral and subcutaneous adipose tissue of obese women with polycystic ovary syndrome (PCOS). J Clin Endocrinol Metab 2013;98(2):E388–96. Available from: http://dx.doi.org/10.1210/jc.2012-3414. PubMed PMID: 23337724.

[123] Alfadda AA, Sallam RM, Chishti MA, Moustafa AS, Fatma S, Alomaim WS, et al. Differential patterns of serum concentration and adipose tissue expression of chemerin in obesity: adipose depot specificity and gender dimorphism. Mol Cells 2012;33(6):591–6. Available from: http://dx.doi.org/10.1007/s10059-012-0012-7. PubMed PMID: 22544171; PMCID: 3887762.

[124] Moreno-Navarrete JM, Manco M, Ibanez J, Garcia-Fuentes E, Ortega F, Gorostiaga E, et al. Metabolic endotoxemia and saturated fat contribute to circulating NGAL concentrations in subjects with insulin resistance. Int J Obes 2010;34(2):240–9. Available from: http://dx.doi.org/10.1038/ijo.2009.242. PubMed PMID: 19949414.

[125] Ussar S, Bezy O, Bluher M, Kahn CR. Glypican-4 enhances insulin signaling via interaction with the insulin receptor and serves as a novel adipokine. Diabetes 2012;61(9):2289–98. Available from: http://dx.doi.org/10.2337/db11-1395. PubMed PMID: 22751693; PMCID: 3425403.

[126] Frias JP, Macaraeg GB, Ofrecio J, Yu JG, Olefsky JM, Kruszynska YT. Decreased susceptibility to fatty acid-induced peripheral tissue insulin resistance in women. Diabetes 2001;50(6):1344–50. PubMed PMID: 11375335.

[127] Corsetti JP, Sparks JD, Peterson RG, Smith RL, Sparks CE. Effect of dietary fat on the development of non-insulin dependent diabetes mellitus in obese Zucker diabetic fatty male and female rats. Atherosclerosis 2000;148(2):231–41. PubMed PMID: 10657558.

[128] Carter S, McKenzie S, Mourtzakis M, Mahoney DJ, Tarnopolsky MA. Short-term 17beta-estradiol decreases glucose R(a) but not whole body metabolism during endurance exercise. J Appl Physiol 2001;90(1):139–46. PubMed PMID: 11133904.

[129] Carter SL, Rennie C, Tarnopolsky MA. Substrate utilization during endurance exercise in men and women after endurance training. Am J Physiol Endocrinol Metab 2001;280(6): E898–907. PubMed PMID: 11350771.

[130] Tarnopolsky MA. Sex differences in exercise metabolism and the role of 17-beta estradiol. Med Sci Sports Exerc 2008;40 (4):648–54. Available from: http://dx.doi.org/10.1249/MSS.0b013e31816212ff. PubMed PMID: 18317381.

[131] Lundsgaard AM, Kiens B. Gender differences in skeletal muscle substrate metabolism - molecular mechanisms and insulin sensitivity. Front Endocrinol 2014;5:195. Available from: http://dx.doi.org/10.3389/fendo.2014.00195. PubMed PMID: 25431568; PMCID: 4230199.

[132] Burd NA, Tang JE, Moore DR, Phillips SM. Exercise training and protein metabolism: influences of contraction, protein intake, and sex-based differences. J Appl Physiol 2009;106(5):1692–701. Available from: http://dx.doi.org/10.1152/japplphysiol.91351.2008. PubMed PMID: 19036897.

[133] Muniyappa R, Lee S, Chen H, Quon MJ. Current approaches for assessing insulin sensitivity and resistance in vivo: advantages, limitations, and appropriate usage. Am J Physiol Endocrinol Metab 2008;294(1):E15–26. Epub 2007/10/25. Available from: http://dx.doi.org/10.1152/ajpendo.00645.2007. PubMed PMID: 17957034.

[134] Munguia-Miranda C, Sanchez-Barrera RG, Tuz K, Alonso-Garcia AL, Cruz M. [Impaired fasting glucose detection in blood donors population]. Rev Med Inst Mex Seguro Soc 2009;47(1):17–24. PubMed PMID: 19624959.

[135] Kuhl J, Hilding A, Ostenson CG, Grill V, Efendic S, Bavenholm P. Characterisation of subjects with early abnormalities of glucose tolerance in the Stockholm Diabetes Prevention Programme: the impact of sex and type 2 diabetes heredity. Diabetologia 2005;48(1):35–40. Available from: http://dx.doi.org/10.1007/s00125-004-1614-1. PubMed PMID: 15619073.

[136] Clausen JO, Borch-Johnsen K, Ibsen H, Bergman RN, Hougaard P, Winther K, et al. Insulin sensitivity index, acute insulin response, and glucose effectiveness in a population-based sample of 380 young healthy Caucasians. Analysis of the impact of gender, body fat, physical fitness, and life-style factors. J Clin Invest 1996;98(5):1195–209. Available from: http://dx.doi.org/10.1172/JCI118903. PubMed PMID: 8787683; PMCID: 507542.

[137] Paula FJ, Pimenta WP, Saad MJ, Paccola GM, Piccinato CE, Foss MC. Sex-related differences in peripheral glucose metabolism in normal subjects. Diabet Metab 1990;16(3):234–9. PubMed PMID: 2210019.

[138] Nuutila P, Knuuti MJ, Maki M, Laine H, Ruotsalainen U, Teras M, et al. Gender and insulin sensitivity in the heart and in skeletal muscles. Studies using positron emission tomography. Diabetes 1995;44(1):31–6. PubMed PMID: 7813811.

[139] Basu R, Dalla Man C, Campioni M, Basu A, Klee G, Toffolo G, et al. Effects of age and sex on postprandial glucose metabolism: differences in glucose turnover, insulin secretion, insulin action, and hepatic insulin extraction. Diabetes 2006;55 (7):2001–14. Available from: http://dx.doi.org/10.2337/db05-1692. PubMed PMID: 16804069.

[140] Flanagan DE, Holt RI, Owens PC, Cockington RJ, Moore VM, Robinson JS, et al. Gender differences in the insulin-like growth factor axis response to a glucose load. Acta Physiol 2006;187 (3):371–8. Available from: http://dx.doi.org/10.1111/j.1748-1716.2006.01581.x. PubMed PMID: 16776662.

[141] Macotela Y, Boucher J, Tran TT, Kahn CR. Sex and depot differences in adipocyte insulin sensitivity and glucose metabolism. Diabetes 2009;58(4):803–12. Available from: http://dx.doi.org/10.2337/db08-1054. PubMed PMID: 19136652; PMCID: 2661589

[142] Rune A, Salehzadeh F, Szekeres F, Kuhn I, Osler ME, Al-Khalili L. Evidence against a sexual dimorphism in glucose and fatty acid metabolism in skeletal muscle cultures from age-matched men and post-menopausal women. Acta Physiol 2009;197 (3):207–15. Available from: http://dx.doi.org/10.1111/j.1748-1716.2009.02010.x. PubMed PMID: 19508405.

[143] Perseghin G, Scifo P, Pagliato E, Battezzati A, Benedini S, Soldini L, et al. Gender factors affect fatty acids-induced insulin resistance in nonobese humans: effects of oral steroidal contraception. J Clin Endocrinol Metab 2001;86(7):3188–96. Available from: http://dx.doi.org/10.1210/jcem.86.7.7666. PubMed PMID: 11443187.

[144] Nemoto Y, Toda K, Ono M, Fujikawa-Adachi K, Saibara T, Onishi S, et al. Altered expression of fatty acid-metabolizing enzymes in aromatase-deficient mice. J Clin Invest 2000;105 (12):1819–25. Available from: http://dx.doi.org/10.1172/JCI9575. PubMed PMID: 10862797; PMCID: 378513.

[145] Camporez JP, Jornayvaz FR, Lee HY, Kanda S, Guigni BA, Kahn M, et al. Cellular mechanism by which estradiol protects female ovariectomized mice from high-fat diet-induced hepatic and muscle insulin resistance. Endocrinology 2013;154 (3):1021–8. Available from: http://dx.doi.org/10.1210/en.2012-1989. PubMed PMID: 23364948; PMCID: 3578999.

[146] Louet JF, LeMay C, Mauvais-Jarvis F. Antidiabetic actions of estrogen: insight from human and genetic mouse models. Curr Atheroscler Rep 2004;6(3):180–5. PubMed PMID: 15068742.

[147] Rincon J, Holmang A, Wahlstrom EO, Lonnroth P, Bjorntorp P, Zierath JR, et al. Mechanisms behind insulin resistance in rat skeletal muscle after oophorectomy and additional testosterone treatment. Diabetes 1996;45(5):615–21. PubMed PMID: 8621012.

[148] Moran A, Jacobs Jr DR, Steinberger J, Steffen LM, Pankow JS, Hong CP, et al. Changes in insulin resistance and cardiovascular risk during adolescence: establishment of differential risk in males and females. Circulation 2008;117(18):2361–8. Available from: http://dx.doi.org/10.1161/CIRCULATIONAHA.107.704569. PubMed PMID: 18427135.

[149] Polotsky HN, Polotsky AJ. Metabolic implications of menopause. Semin Reprod Med 2010;28(5):426–34. Available from: http://dx.doi.org/10.1055/s-0030-1262902. PubMed PMID: 20865657.

[150] Jones ME, Boon WC, Proietto J, Simpson ER. Of mice and men: the evolving phenotype of aromatase deficiency. Trends Endocrinol Metab 2006;17(2):55–64. Available from: http://dx.doi.org/10.1016/j.tem.2006.01.004. PubMed PMID: 16480891.

[151] Smith EP, Boyd J, Frank GR, Takahashi H, Cohen RM, Specker B, et al. Estrogen resistance caused by a mutation in the estrogen-receptor gene in a man. N Engl J Med 1994;331 (16):1056–61. Available from: http://dx.doi.org/10.1056/NEJM199410203311604. PubMed PMID: 8090165.

[152] Jelenik T, Roden M. How estrogens prevent from lipid-induced insulin resistance. Endocrinology 2013;154(3):989–92. Available from: http://dx.doi.org/10.1210/en.2013-1112. PubMed PMID: 23429711.

[153] Hatta H, Atomi Y, Shinohara S, Yamamoto Y, Yamada S. The effects of ovarian hormones on glucose and fatty acid oxidation during exercise in female ovariectomized rats. Horm Metab Res 1988;20(10):609–11. Available from: http://dx.doi.org/10.1055/s-2007-1010897. PubMed PMID: 3220443.

[154] Piers LS, Diggavi SN, Rijskamp J, van Raaij JM, Shetty PS, Hautvast JG. Resting metabolic rate and thermic effect of a meal in the follicular and luteal phases of the menstrual cycle in well-nourished Indian women. Am J Clin Nutr 1995;61 (2):296–302. PubMed PMID: 7840066.

[155] Horton TJ, Miller EK, Glueck D, Tench K. No effect of menstrual cycle phase on glucose kinetics and fuel oxidation during moderate-intensity exercise. Am J Physiol Endocrinol Metab 2002;282(4):E752–62. Available from: http://dx.doi.org/10.1152/ajpendo.00238.2001. PubMed PMID: 11882494; PMCID: 3124251.

[156] Bingley CA, Gitau R, Lovegrove JA. Impact of menstrual cycle phase on insulin sensitivity measures and fasting lipids. Horm Metab Res 2008;40(12):901–6. Available from: http://dx.doi.org/10.1055/s-0028-1082081. PubMed PMID: 18726830.

[157] Yeung EH, Zhang C, Mumford SL, Ye A, Trevisan M, Chen L, et al. Longitudinal study of insulin resistance and sex hormones over the menstrual cycle: the BioCycle Study. J Clin Endocrinol Metab 2010;95(12):5435–42. Available from: http://dx.doi.org/10.1210/jc.2010-0702. PubMed PMID: 20843950; PMCID: 2999972.

[158] Valdes CT, Elkind-Hirsch KE. Intravenous glucose tolerance test-derived insulin sensitivity changes during the menstrual cycle. J Clin Endocrinol Metab 1991;72(3):642–6. Available from: http://dx.doi.org/10.1210/jcem-72-3-642. PubMed PMID: 1997519.

[159] Campbell SE, Angus DJ, Febbraio MA. Glucose kinetics and exercise performance during phases of the menstrual cycle: effect of glucose ingestion. Am J Physiol Endocrinol Metab 2001;281(4):E817–25. PubMed PMID: 11551860.

[160] Baba T, Shimizu T, Suzuki Y, Ogawara M, Isono K, Koseki H, et al. Estrogen, insulin, and dietary signals cooperatively regulate longevity signals to enhance resistance to oxidative stress in mice. J Biol Chem 2005;280(16):16417–26. Available from: http://dx.doi.org/10.1074/jbc.M500924200. PubMed PMID: 15713666.

[161] Vina J, Borras C, Gambini J, Sastre J, Pallardo FV. Why females live longer than males: control of longevity by sex hormones. Sci Aging Knowledge Environ 2005;2005(23):pe17. Available from: http://dx.doi.org/10.1126/sageke.2005.23.pe17. PubMed PMID: 15944465.

[162] Kahn SE, Andrikopoulos S, Verchere CB, Wang F, Hull RL, Vidal J. Oophorectomy promotes islet amyloid formation in a transgenic mouse model of type II diabetes. Diabetologia 2000;43 (10):1309–12. Available from: http://dx.doi.org/10.1007/s001250051527. PubMed PMID: 11079750.

[163] Le May C, Chu K, Hu M, Ortega CS, Simpson ER, Korach KS, et al. Estrogens protect pancreatic beta-cells from apoptosis and prevent insulin-deficient diabetes mellitus in mice. Proc Natl Acad Sci USA 2006;103(24):9232–7. Available from: http://dx.doi.org/10.1073/pnas.0602956103. PubMed PMID: 16754860; PMCID: 1482595.

[164] Dantas AP, Sandberg K. Estrogen regulation of tumor necrosis factor-alpha: a missing link between menopause and cardiovascular risk in women? Hypertension 2005;46(1):21–2. Available from: http://dx.doi.org/10.1161/01.HYP.0000169038.67923.b0. PubMed PMID: 15911737.

[165] Arenas IA, Armstrong SJ, Xu Y, Davidge ST. Chronic tumor necrosis factor-alpha inhibition enhances NO modulation of vascular function in estrogen-deficient rats. Hypertension 2005;46(1):76–81. Available from: http://dx.doi.org/10.1161/01.HYP.0000168925.98963.ef. PubMed PMID: 15911738.

[166] Pfeilschifter J, Koditz R, Pfohl M, Schatz H. Changes in proinflammatory cytokine activity after menopause. Endocr Rev 2002;23(1):90–119. Available from: http://dx.doi.org/10.1210/edrv.23.1.0456. PubMed PMID: 11844745.

[167] Steinberg GR. Inflammation in obesity is the common link between defects in fatty acid metabolism and insulin resistance. Cell cycle 2007;6(8):888–94. PubMed PMID: 17438370.

[168] Singh R, Artaza JN, Taylor WE, Gonzalez-Cadavid NF, Bhasin S. Androgens stimulate myogenic differentiation and inhibit adipogenesis in C3H 10T1/2 pluripotent cells through an androgen receptor-mediated pathway. Endocrinology 2003;144(11):5081–8. Available from: http://dx.doi.org/10.1210/en.2003-0741. PubMed PMID: 12960001.

[169] Mootha VK, Lindgren CM, Eriksson KF, Subramanian A, Sihag S, Lehar J, et al. PGC-1alpha-responsive genes involved in oxidative phosphorylation are coordinately downregulated in human diabetes. Nat Genet 2003;34(3):267–73. Available from: http://dx.doi.org/10.1038/ng1180. PubMed PMID: 12808457.

[170] Pitteloud N, Mootha VK, Dwyer AA, Hardin M, Lee H, Eriksson KF, et al. Relationship between testosterone levels, insulin sensitivity, and mitochondrial function in men. Diabetes Care 2005;28(7):1636–42. PubMed PMID: 15983313.

[171] Morimoto S, Fernandez-Mejia C, Romero-Navarro G, Morales-Peza N, Diaz-Sanchez V. Testosterone effect on insulin content, messenger ribonucleic acid levels, promoter activity, and secretion in the rat. Endocrinology 2001;142(4):1442–7. Available from: http://dx.doi.org/10.1210/endo.142.4.8069. PubMed PMID: 11250923.

[172] Morimoto S, Mendoza-Rodriguez CA, Hiriart M, Larrieta ME, Vital P, Cerbon MA. Protective effect of testosterone on early apoptotic damage induced by streptozotocin in rat pancreas. J Endocrinol 2005;187(2):217–24. Available from: http://dx.doi.org/10.1677/joe.1.06357. PubMed PMID: 16293769.

[173] Paik SG, Michelis MA, Kim YT, Shin S. Induction of insulin-dependent diabetes by streptozotocin. Inhibition by estrogens and potentiation by androgens. Diabetes 1982;31(8 Pt 1):724–9. PubMed PMID: 6219020.

[174] Ramsey JJ, Laatsch JL, Kemnitz JW. Age and gender differences in body composition, energy expenditure, and glucoregulation of adult rhesus monkeys. J Med Primatol 2000;29(1):11–9. PubMed PMID: 10870670.

[175] Kemnitz JW, Gibber JR, Lindsay KA, Eisele SG. Effects of ovarian hormones on eating behaviors, body weight, and glucoregulation in rhesus monkeys. Horm Behav 1989;23(2):235–50. PubMed PMID: 2663699.

[176] Abbott DH, Bruns CR, Barnett DK, Dunaif A, Goodfriend TL, Dumesic DA, et al. Experimentally induced gestational androgen excess disrupts glucoregulation in rhesus monkey dams and their female offspring. Am J Physiol Endocrinol Metab 2010;299(5):E741–51. Available from: http://dx.doi.org/10.1152/ajpendo.00058.2010. PubMed PMID: 20682841; PMCID: 2980359.

CHAPTER

10

Sex Hormone Influenced Differences in Skeletal Muscle Responses to Aging and Exercise

Marybeth Brown[1] *and Peter Tiidus*[2]

[1]Department of Physical Therapy, School of Health Professions, University of Missouri, Columbia, MO, United States
[2]Department of Kinesiology, Brock University, St. Catharines, ON, Canada

INTRODUCTION

It is abundantly evident that both men and women lose muscle mass and strength with advancing age. Males tend to lose muscle and strength beginning in the late third or early fourth decade although this modest loss does not appear to be related to a decline in testosterone (T2) as circulating T2 has been reported to decline about 1% per year beginning at approximately the age of 40 years [1]. After the age of 40, there is a strong correlation between muscle mass loss and the decline in T2 [2]. However, earlier hormone decline has been reported to occur in ~20% of aging men [3]. Women lose approximately 10–15% of their muscle mass between the ages of 25 and the onset of menopause, but thereafter the decline accelerates to an average of 2% per year during the menopause and beyond [1]. Although the average age of menopause is 51 years, there can be as much as a 20-year difference in age of onset in menopausal symptoms (NIH). The average lifespan in the United States has increased to 78.7 years and thus, men and women now spend a significant portion of their lives in a reduced sex hormone state.

Although this chapter is focused on skeletal muscle, brief mention of the fact that estrogen (E2) and T2 are important to multiple tissues is warranted. E2 effects are mediated through two primary receptors, alpha and beta, with evidence for a third receptor, a G-protein receptor [4]. E2 receptors have been found on nearly all tissues examined including bone, brain, liver, reproductive organs, skin, blood vessels, and heart [5]. Additionally, E2 receptors have been found in the cytoplasm of some cells, and on mitochondria [6] and endoplasmic reticulum, which has implications for a variety of functions outside the scope of this chapter. T2 effects are mediated primarily through androgen receptors (ARs), which too, are represented in multiple tissues including skin, bone, and testes. The focus in this chapter on skeletal muscle ignores potential multiple other system effects that are also important, systems that may impact directly or indirectly on muscle quality and quantity.

E2 LOSS AND AGING MUSCLE

Cross-sectional studies reveal a strong association between the decline in E2 and muscle mass loss in aging females [7–10]. Muscle mass begins an accelerated decline, an average of 2% per year, coincident to the reduction in E2 levels at menopause. A potent anabolic effect of E2 also has been demonstrated in young women in their early 20s who are given gonadotropin-releasing hormone to reduce endogenous E2 release. A loss of lean mass with a concomitant increase in fat mass occurs with startling rapidity [11]. Within approximately 1 week there is a more than a 2 kg loss of lean mass, which is most likely muscle, with a concomitant increase in fat mass such that body weight does not change. Thus, there appears to be an anabolic effect of E2 between adulthood and menopause. As will be noted later, when E2 is given back to young and menopausal women who are E2 deficient there is an increase in muscle mass and more importantly muscle force production.

Of late, there has been recognition that the reduction in lean muscle mass is only part of the story: muscle force production is affected to an even greater extent

Sex Differences in Physiology
DOI: http://dx.doi.org/10.1016/B978-0-12-802388-4.00010-0

with the loss of E2 [12–14]. In 1993 Phillips et al. published an important study of strength in women on hormone replacement therapy (HRT) and women who were not [15]. What made this particular study distinctive was the fact that force production of the adductor pollicis muscle was normalized to its cross-sectional area. Findings revealed that women receiving hormone therapy had comparable force/unit muscle mass over a several decade span whereas women not taking HRT had a greater loss in force than muscle cross-sectional area. Their data also suggested that by the age of 70 years the force preservation advantage of HRT users had essentially disappeared. These data have not been replicated and it is unclear if the benefits of HRT are limited to women who are younger than 70 years.

The disproportionate gap between muscle mass and force loss for women with the menopause was recently illustrated by Charlier et al. who studied 578 women between the ages of 18 and 78 years [10]. Each subject had their total lean body muscle mass calculated following bioelectrical impedance and height assessment. Isometric, isotonic, and isokinetic force production of the quadriceps was measured in each subject and expressed as relative to skeletal muscle mass. Women were divided into three groups based on their relative muscle mass value compared to normative young controls. The number of women that fell one or two standard deviations below the normative reference group increased from 18% for women in their 40s to 27% for women in their 50s to 57% for women in their 60s. The declines in strength were greater than the loss in mass but proportionate and also showed an accelerated loss with menopause, more so in isokinetic and isotonic values than isometric [10].

One aspect of aging in current times that is not clearly identified as a potential contributor to muscle compromise is sarcopenic obesity. More and more women are becoming overweight with current estimates at 40% of adults being obese (Centers for Disease Control). Fat is an E2-producing tissue and the extent to which this E2 depot may influence aging is unknown. Also unclear is whether the increase in abdominal fat is a contributor to intramuscular fat in older women who are obese. The extent to which intramuscular fat is a major contributor to the higher drop in specific muscle tension with aging is unclear. Case reports suggest this may be the case but parsing out the contribution of intramuscular fat, age-related decline, inactivity, and hormone loss remains to be done [16,17]. Additionally, it has been recently demonstrated that muscle myosin phosphorylation may be compromised with E2 loss. Myosin phosphorylation occurs in response to muscle contraction and leads to potentiation of actin-myosin cross bridge formation and force development [18]. This very recent study demonstrated that ovariectomy in a rodent model will attenuate hind limb muscle tetanic force development along with a loss in myosin phosphorylation while a return of E2 will normalize both. Hence, multiple mechanisms related to E2 effects on muscle force development may influence muscle force loss in postmenopausal women.

One of the most compelling studies to demonstrate that force production following loss of ovarian hormone is reduced was done by Moran et al. [13] who removed the ovaries of mature mice and waited for 60 days. Subsequently, they studied single fibers from the soleus and extensor digitorum longus (EDL) and determined that force production was reduced approximately 18% in both soleus and EDL compared to controls. They also determined that protein content (total and myosin) was unchanged but that the fraction of strong-binding myosin was lower with ovariectomy. Proportion of strongly bound myosin explained the force reduction in EDL, which was the only muscle they could study under the conditions imposed by electron spin microscopy. Although not replicated, these findings strongly indicate that ovarian hormones (later determined to be E2) affect muscle force at the single fiber, specifically at the actin-myosin level. Later study of single fibers revealed similar findings, a significantly higher specific force in both type I (27%) and type II (23%) single fibers in women taking HRT compared to nonusers [37].

In each of the cited studies E2 content was determined on the basis of blood serum values. Muscle, however, is an intracrine organ and produces its own intramuscular E2, T2, and dehydroepiandrosterone (DHEA). The potential role of intramuscular E2 was not known until recently when Pöllänen and colleagues [19] determined muscle properties and physical performance capabilities in postmenopausal women who were and were not taking HRT. This study was unique in that subjects were monozygotic twin pairs, one of whom was taking HRT and had been for an average of 7 years. Findings indicated that women taking HRT had higher systemic E2 but intramuscular E2 was the same for the twin pairs. Interestingly, intramuscular E2, T2, and DHEA were associated with knee extension strength and power, explaining 36–39% of the variance in knee extension strength, and 37–45% of the variance in vertical jump height (power). These investigators further determined that human muscle cells (myoblasts —> myotubes) harvested from the monozygotic twin pairs were capable of synthesizing E2 and its receptors [19].

When premenopausal women are compared to postmenopausal women, striking differences in physical performance capacity emerge. Premenopausal women have faster walking speeds, greater fat-free mass, more skeletal muscle mass, higher strength values, and significantly more muscle power [19–21]. Evidence also

TABLE 10.1 Fundamental Age-Related Changes in Skeletal Muscle

Skeletal muscle atrophy	Reduced numbers of fibers
	Reduced cross-sectional area, particularly in type II fibers
Muscle structural alterations affecting force production	Increased adipose tissue, inter- and intramuscular infiltration
	Increased fibrotic tissue
Satellite cell changes	Fewer cells with aging
	Delayed or reduced response to injury
	Diminished hypertrophy response to weight-training
Single muscle fiber changes	Diminished force production (force/cross-sectional area), possibly at the cross-bridge
	Altered myosin protein content
Excitation-coupling	Diminished release of calcium at the sarcoplasmic reticulum, fewer dihyropyradine and ryanodine receptors
	Anatomical changes at the synapse
Fiber type changes	Type-grouping
	Fewer type II fibers—transformation?
Increased collagen cross-linking	Stiffer connective tissues
	Slower movement time, decreased range of motion

Many of the events listed above are associated, in part, with loss of sex hormones and/or, are modified following administration of HRT.

suggests that HRT given at the time of menopause can delay the loss of muscle mass, strength, and power, and maintain a better body composition [19]. Premenopausal women in the study by Pöllänen et al. averaged 32 years, postmenopausal women 58 years but knee extension strength in the premenopausal women was more than 30% greater than strength in the older women suggesting a strength decline for women approaching the menopause was in excess of 1%/year. A 2009 meta-analysis of studies in which E2 was given to postmenopausal women suggests a 5% greater strength benefit, which is likely a modest estimate [22]. Given the probability of women living three to four decades beyond menopause, findings are compelling as over the age of 80 years more than half of all women require physical assistance, having lost their independence [1,23].

One additional aspect of the aging skeletal muscle that warrants mention is its reduced ability to regenerate [24–27]. Older muscle is more susceptible to injury for a variety of reasons (Table 10.1) and failure to fully repair a muscle that is already undergoing decline in mass and force may be a contributor to the ultimate loss of independence in many women. The ability of muscle to repair is dependent upon resident stem cells (satellite cells). With aging and loss of E2 fewer stem cells are present and fewer satellite cells are recruited to repair muscle injury (LaColla et al. [28], brooks, Vasconsuelo et al. [29]). There is evidence that the signals that activate satellite cells are delayed or lacking, which has direct consequences for the extent and speed to which the injury will repair. There is also evidence that satellite cells in aging muscle are susceptible to spontaneous apoptosis [112]. Type I fibers have 4–5 × the number of resident satellite cells than type II fibers and it is the fast type II fibers that appear to be preferentially lost with age. The work of Tiidus et al. has demonstrated that E2 attenuates the inflammatory portion of the injury cycle, which determines satellite cell recruitment [25–27,30]. The potential role of E2 (and T2 in men) to foster muscle repair in an aged organism is poorly understood. In a young organism it is apparent that E2 will attenuate the extent of injury to skeletal muscle.

E2 and Exercise

Exercise of any sort has positive benefits. Weight-lifting in particular is an anabolic stimulus resulting in greater muscle mass and strength if the load is sufficiently heavy. Theoretically, it is possible that lifting exercise combined with HRT has additive effects, which was the primary question addressed by Taaffe et al. [14] in a year-long study of postmenopausal women. Eighty women were randomly assigned to one of four groups: control ($n = 20$), HRT alone, exercise alone, and HRT + exercise. HRT was E2 and progesterone combined. Exercise consisted of bounding over obstacles, jumping, hopping, skipping, and traditional weight-lifting activities for the upper extremities. Two days of exercise were supervised but four days of activity at home were encouraged. After 1 year, women were reevaluated with muscle mass as the primary outcome measure, assessed by CT. Secondary outcomes were cross-sectional area of the quadriceps and hamstrings, jump height, running speed, and knee extension strength. Quadriceps muscle mass was significantly higher in the HRT, Exercise, and Exercise plus HRT groups compared with controls. For hamstrings, the Exercise + HRT and HRT alone groups had higher muscle mass than controls. Mass in the exercise/HRT groups was higher primarily because the intervention attenuated the 1–2% muscle mass loss that occurred in the nonexercise controls. Findings suggest that HRT may be as anabolic as the exercise given in this study with no clear distinction between the two. However, women exercised an average of once/week and it may be that the plyometric exercise given was insufficient to actually increase thigh muscle size. More study is needed of menopausal women who undergo traditional strength-training in conjunction with HRT or E2 supplementation to better define expected outcomes and to determine the underlying mechanisms of the augmented strength gain.

One additional study of postmenopausal women between the ages of 60 and 72 years suggests that HRT may not be effective in later years (well beyond the menopause) to augment muscle mass [31]. Women who had previously been inactive were put on a center-based program of stretching/balance activities for 3 months followed by 3 months of traditional strength and finally 3 months of aerobic training. Twenty-two of the forty-two subjects were on HRT for the entire 9 months. Both groups gained significant amounts of upper and lower extremity strength (eg, bench press, leg press, knee extension) with no differences between groups. Average strength gains were 16% for the women who were not on HRT and 17% for the women who were. It is unclear whether women were too far postmenopause to derive muscular benefit from HRT or if the weight-training component of the study was of sufficient duration to optimize muscle mass and strength differences with and without hormone therapy.

In support of this hypothesis a recent rodent study noted that when E2 replacement was delayed by several months in ovariectomized rats (equivalent to starting HRT several years after menopausal onset in human females), the previously reported augmentation benefits of E2 on exercise-induced muscle satellite cell activation was lost [25]. Since postexercise augmentation of satellite cell number is indicative of the anabolic response, the beneficial effects of E2 on muscle hypertrophy may be compromised if E2 replacement or HRT is not started proximal to onset of menopause [25].

Several studies have shown that older women do not gain muscle mass and strength to the same degree as young women [32,33]. There was speculation that perhaps postmenopausal women were unable to undergo protein synthesis to the same degree as young women although evidence of support of this speculation is lacking. Unclear as well was whether satellite cell activation in the older women in this study was diminished and whether low E2 values played a role.

Although, compared to premenopausal women, postmenopausal women have a blunted strength gain in response to strength-training exercise, it should be emphasized that older women are still capable of gaining muscle mass and strength. Maximizing strength is imperative for women at all ages, but in particular, in the later years when functional independence may become jeopardized. Thus far findings suggest that to derive the greatest amount of benefit from an exercise stimulus, women who are premenopausal or in early menopause and on HRT will derive the greatest benefit.

Indirect E2 Effects on Aging Skeletal Muscle in Females

Physical inactivity is a major contributor to the decline in skeletal muscle mass and strength in older women, and for many women, is associated with a decline in functional capability, and loss of independence [113]. The decline in E2 has been recognized as one potent contributor to physical inactivity. Rodent studies are particularly noteworthy as the reduction in spontaneous daily movement following the removal of ovarian hormones by ovariectomy or OVX is highly significant, averaging about 80%. Moreover, the onset of reduced activity occurs almost immediately following the removal of ovarian hormones (Fig. 10.1). That the ovarian hormone responsible for the inactivity effects is E2 has been demonstrated by several investigators [34–37]. When E2 is given back to OVX rats or mice, spontaneous wheel running distance is restored to pre-OVX levels. An example of the potent effect of E2 is provided in Fig. 10.1 for mice that were 4 months of age at the beginning of the study (unpublished data).

Hormone Treatment Effects in Aging Women

The greatest number of insights on the effects of E2-based therapy in postmenopausal women has come

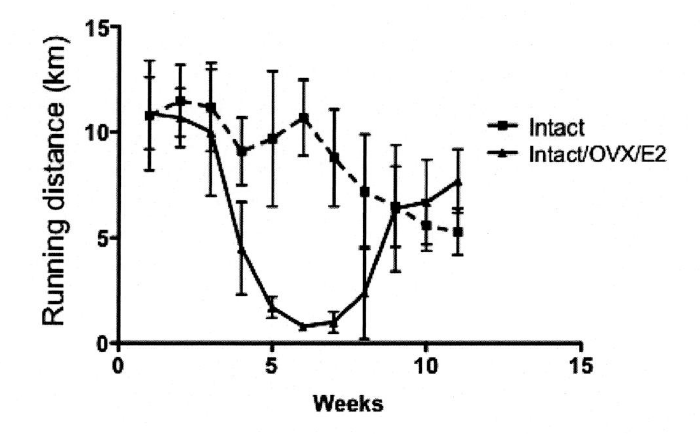

FIGURE 10.1 Intact mice (n=8) were followed for 11 weeks. Running distance was recorded daily. Weekly averages are provided. The second group (n=8) was intact for 3 weeks and then OVX, and followed for 4 weeks in the OVX state. Subsequently, E2 was given daily to OVX animals and spontaneous running distances returned to pre-OVX levels.

from the twin study, conducted in Finland [19,38,39]. Thirteen postmenopausal monozygotic twin pairs were recruited, one of whom was on HRT and the other was not. Women taking HRT had been on their hormone regimen an average of 7 years. Sisters taking HRT had significantly more lean muscle mass, greater muscle strength (particularly power), and a higher functional profile than their twin, suggesting that women supplemented during the perimenopausal phase of life maintain their lean mass and strength, at least into their 60s. Whether the benefit of HRT extends beyond that point in time is unclear. Women who are in their 70s and 80s do not appear to gain lean mass with HRT which has led to the "timing hypothesis," meaning that the benefits of HRT for skeletal muscle are lost on women beyond the menopause.

Mechanisms of E2 or HRT Supplementation

Since the original review on this topic was written several years ago [30] considerable work has been done to elucidate the mechanisms of E2 effects on skeletal muscle. Pathways have been elaborated, the role of satellite cells examined and genetic changes are beginning to be understood [40–49]. In some instances, investigators parsed out the independent E2 and progesterone effects [50], for example, determined the fractional rate of myofibrillar protein synthesis rate using the stable isotope technique in 10 women who had undergone hysterectomy and/or oophorectomy and had been on E2 or HRT for an average of 16 years. Comparisons were made to 10 age-matched postmenopausal women who were not taking any supplementation. At rest, myofibrillar protein synthesis rate was lower in HRT users than in controls. One knee-extension exercise bout enhanced myofibrillar protein synthesis rate only in HRT users. Also observed were higher androstenedione concentrations in the controls compared to women taking HRT. Androstenedione has been touted as a muscle-building substance but evidence to support this is limited [51]. Nonetheless, investigators could not exclude the possibility that androstenedione differences could have influenced the results. Findings do indicate that resistance exercise combined with HRT seems to have a counteracting effect on the reduced myofibrillar fractional synthesis rate in hysterectomized/oophorectomized women.

The Finnish twin study revealed that women taking HRT had significantly more IGF-1 expression (40% more mechano growth factor or IGF-1Ec) and higher expression of IGF-1 receptor than co-twin controls not taking HRT [21] and was associated with higher thigh mass. The IGF-1/PI3K/AKT pathway is key to activating protein synthesis and inhibiting muscle degradation. Thus, more IGF-1 expression suggests that HRT is enhancing the net balance of muscle protein turnover. Microarray analysis of tissues from a subset of the women in this study revealed that HRT was associated with a higher expression of 22 genes related to energy metabolism, organization of the cytoskeleton, cell–environment interactions, and responses to nutrition [65]. Higher gene expression explained 18–26% of the variation in muscle composition and power (19–20%). Dieli-Conwright et al. [52] reported greater resting levels in expression of quadriceps muscle mRNA of proanabolic markers, such as MyoD, myogenin, Myf5, and greater suppression of proteolytic markers, such as FOXO3A and MURF-1, as well as the negative growth regulator, myostatin, in 50- to 57-year-old women taking HRT relative to nonusers. These differences in muscle anabolic signaling between HRT users and nonusers were further accentuated following a single bout of resistance exercise.

This enhancement of a proanabolic environment in muscle at rest and following eccentric exercise in postmenopausal women using HRT may be an important factor in helping to maintain muscle mass and strength in menopausal women, as well as in enhancing the anabolic effects of resistance training. The mechanisms by which HRT augments this anabolic signaling in skeletal muscle has yet to be fully determined.

The ability of E2 and HRT to diminish systemic and muscle inflammatory factors at rest and following exercise may also directly and indirectly affect muscle mass and exercise-induced muscle damage. It has been consistently demonstrated that E2 replacement in ovariectomized rodents will result in diminished postexercise neutrophil infiltration into skeletal muscle [53,114,115]. Earlier studies also clearly demonstrated that E2 diminished membrane disruption in isolated muscles as determined by loss of creatine kinase and that this was likely due to its effects on membrane stability, membrane fluidity, and antioxidant actions [54,55]. Reduced postexercise muscle creatine kinase loss has also been observed in young adult human females relative to males, with this effect being attributed to higher E2 levels in females [56,57]. Reduced postexercise disruption of muscle membranes by E2 may directly or indirectly influence postexercise inflammatory responses and leukocyte infiltration into skeletal muscle [58]. E2 administration in sedentary ovariectomized rats has also been reported to upregulate constitutive heat shock protein 72 (Hsp72) in muscle to levels that are induced by acute exercise [116]. Because Hsps can provide protection against exercise-induced muscle damage-age, the ability of E2 to upregulate constitutive Hsp72 expression in skeletal muscle may also afford muscle increased protection and, hence, limit inflammatory responses and exercise-induced damage.

Satellite cells are believed to be an important contributor to skeletal muscle homeostasis. Reportedly, with aging, some muscles, particularly those of the limbs, lose satellite cells, which may be a contributor to age-related muscle mass decline (sarcopenia) [59]. Recent study in murine muscle revealed that during adulthood (12–20 months) satellite cells continually contribute to muscle myofibers. The satellite cell contribution remained comparable between 12- and 20-month animals in the EDL, TA, Plantaris, and soleus but gastrocnemius showed a 22% decline in satellite cell number. In spite of comparable satellite cell contribution for most limb muscles between 12 and 20 months, all muscles studied lost myofiber area suggesting possible independent events. The sex of the animals used in this study was not specified thus, potential male/female age-related differences are unknown.

The possible relationship between E2 and satellite cell function has been studied to some extent. When female rats undergo a downhill running exercise protocol for 90 min to induce muscle damage, satellite cells are recruited for repair in animals that have a normal E2 complement. For rats that are E2 deficient satellite cell recruitment is diminished and likely delayed. Enns et al. [26] and Tiidus [58] determined that E2 stimulates satellite cell activation 72 h following the downhill running protocol, which occurs through alpha E2 receptors. Subsequently the PI3K/AKT pathway is activated and as also noted in human studies, it may be that E2 activation of this pathway is one of a number of important factors in its anabolic influence on muscle anabolic responses. Recently, this research group further determined that if E2 supplementation is delayed in ovariectomized rats, there is a much diminished satellite cell response compared to rats that are supplemented 2 weeks postovariectomy. It should be noted however, that postexercise satellite cell numbers were significantly increased in both the soleus and EDL muscles, regardless of the presence or absence of E2. Hormone replacement augmented satellite cell numbers, which has implications for aging postmenopausal women who sustain skeletal muscle damage [58]. It has been hypothesized that part of age-related sarcopenia is the accumulation of repeated bouts of muscle damage [1].

E2, HRT, and Recovery from Muscle Atrophy and Damage

The preponderance of human studies demonstrates a positive E2 or HRT effect on muscle mass. Most animal-based studies too have demonstrated positive effects of E2 on muscle mass [35,36], with, for example, consistent anabolic effects seen in cattle [117]. Recent findings have also demonstrated a positive influence of E2 or HRT on muscle mass regain following muscle atrophy or injury in rodents. Following hind-limb unweighting-induced muscle atrophy in rats, several studies have reported that ovariectomy will inhibit and E2 replacement will enhance muscle mass and strength recovery [60,61]. Data further suggest that a physical activity-based rehabilitation program following hind limb unweighting-induced muscle atrophy in rats will not optimally restore muscle mass in ovariectomized rats without E2 replacement [62]. An additional randomized control study by this research group further confirmed the development of increased skeletal muscle and lean body mass in postmenopausal females aged 50–57 years given HRT over 1 year relative to placebo-supplemented controls, who had no increases over this time period [31]. These findings combined with previous studies demonstrating greater retention in lean body mass and muscle with HRT in postmenopausal females [63] and greater gains in exercise-induced muscle mass and strength [14,64] highlight the potential importance and effectiveness of HRT in maintaining a resistance training-induced accretion of muscle mass in aging females. Table 10.2 summarizes the results of several relevant studies reflecting a diversity of results relative to HRT and E2 influence on muscle mass in postmenopausal females.

Contraindications to E2 in Aging Women

Even though a clear association exists between E2 supplementation and muscle mass and strength in perimenopausal women, it should be recalled that the Women's Health Initiative or WHI was terminated prematurely because of a higher incidence of stroke among women who were taking HRT. The E2 only arm of the study was halted prematurely 2 years later for higher incidence of stroke and deep vein thrombosis. Lost in the furor at the time was the fact that women in both trials had substantially less colorectal cancer and fewer hip fractures.

Following trial cessation, additional analyses were done on the data that revealed that women between the ages of 50 and 59 who were taking hormone had less risk of breast cancer and less risk of heart disease than women not taking E2 or HRT. The risk/benefit profiles of hormone therapy users changed with each successive decade with less return for women in their 70s compared to women in their 50s and 60s. There is evidence as well that women with preexisting heart disease did not benefit from taking hormones (https://www.whi.org). Thus, there are benefits of E2 and HRT at all ages (eg, bone, colorectal health) but also risks, which may include heightened susceptibility

TABLE 10.2 Summary of Hormone Replacement Therapy Effects, Including 17-β Estradiol, on Parameters Related to Skeletal Muscle Mass in Postmenopausal Women

Study group	Subjects	Intervention/dose	Outcomes
Ronkainen et al. [65]	13 postmenopausal twin pairs discordant for HRT	E2 only (n=5) 1–2 mg	Women on E2 had greater thigh muscle CSA (6%), more power, faster walk speed
		E2 daily for an average of 7 years	Women on HRT had 13% more relative muscle mass
		E2 + progesterone (n=6) 1–2 mg E2 daily	
		Tibilone (n=4) 2.5 mg daily	
Sorenson et al. [63]	16 postmenopausal women 55±3 years	Cross-over design	Placebo group lost lean body mass (0.996 ± 1.58 kg)
		Half of subjects on HRT	
		4 mg E2 for 22 days, 1mg E2 for 6 days, followed by 1 mg noethisterone acetate for 10 days	HRT group gained significant lean body mass (0.347 ± 0.858 kg)
		12 week washout period between placebo and HRT phases	
Qaisar et al. [39]	13 postmenopausal twin pairs discordant for HRT (see Ronkainen et al. [65])	See Ronkainen et al. [65]	Vastus lateralis single
			Fiber CSA not different between pairs but specific tension of type I and type II single fibers was significantly greater
Taaffe et al. [14]	51 postmenopausal women 50–57 years	HRT n=14, 1–2 mg E2	All 3 treatment groups significantly increased lean body mass
		Resistance exercise alone, n=12	Quadriceps and posterior thigh CSA significantly increased in HRT and HRT + Exercise groups
		HRT + exercise (n=10) 1 mg norethisterone	
		Controls n=15	
Widrick et al. [66]	17 postmenopausal women 45–54 years	8 women on HRT	Type I and type II CSA not different between groups
		9 women no HRT	
		Quadriceps biopsy-single fiber study	
Skeleton et al. [67]	102 women 5–15 years postmenopause	50 women on HRT 6–12 mos.	No differences in muscle thickness between groups
		CSA of adductor pollicis	
Sipilä et al. [20,64,68]	80 postmenopausal women 50–57 years	20 controls	Quadriceps CSA increased significantly in HRT and HRT plus exercise groups
		20 HRT	Lower leg CSA greater in exercise + HRT than exercise group alone
		20 Exercise	
		20 HRT + Exercise	
		Resistance exercise	
		Upper extremities, plyometric exercise lower extremities	

to stroke, deep vein thrombosis, breast cancer, and myocardial infarction. Benefits appear to be greater at younger ages (perimenopausal women) than later in life, supporting the "timing hypothesis" [69].

Alternatives to HRT

Given potential detrimental effects of E2 or HRT on peri- and postmenopausal women it raises the question of whether some of the beneficial effects of hormone therapy could be gained using alternatives (phytoestrogens, SERMS) such as genistein, tamoxifen or Raloxifene [70,71]. Velders et al. [72] ovariectomized female rats and subsequently supplemented them with E2, genistein, or vehicle. Rats receiving genistein and E2 had higher myosin heavy chain (MHC)-I expression compared to controls. E2 and genistein also increased MHC IIb expression when intense treadmill exercise was combined with treatment. Findings suggest that MHC-IIb adaptation to treadmill exercise was hormone dependent which has implications for older women participating in exercise programs. Phytoestrogens may exert additional benefit as they have a high affinity for the alpha E2 receptor [73] and they have been found to reduce inflammation [74]. In one of the few studies performed [75] randomized postmenopausal women to isoflavone or placebo groups for 24 weeks. Women taking 70 mg/day isoflavone had a 3% increase in fat-free mass and a 2.2% increase in leg fat-free mass suggesting these women may have more strength. Conversely, Moeller et al. [76] randomized postmenopausal women to an isoflavone-rich soy protein group (40 mg/d) or whey protein and found no differences in lean mass between groups. Soy protein ingestion in conjunction with strength training did not result in strength increases over and above those achieved by strengthening alone [77]. Ovariectomized female rats show more CK leakage due to the lack of circulating E2. Treatment with both E2- and tamoxifen resulted in a 60% reduction of the CK leakage, suggesting that tamoxifen like E2, reduces contraction-induced muscle damage in the rat [78]. Our lab has found no benefit of Tamoxifen on either muscle mass or strength (unpublished). Tibilone is the only compound to date that has significantly increased lean body mass, knee extension strength, and decreased total body fat content but it is not yet FDA approved in the United States [79]. Thus, there is some evidence for alterative replacement benefits but it is not overwhelming.

SKELETAL MUSCLE AGING IN MEN

Men are endowed with approximately 35% more muscle mass and strength than women. Because men start with so much more muscle, the decline in muscle mass and strength with age is not as likely to have the severe penalty as the declines in mass and force in women. T2 effects on skeletal muscle are dramatic as evidenced by the advent of puberty in boys. T2 directly influences muscle protein synthesis and the muscle increase in pubescent boys is the consequence of hormonal action. Although a bit player, E2 also has a role in determining the amount of muscle mass in males but does not appear to have an impact on contractile force [68,80]. E2 is aromatized from T2 and both hormones determine total muscle mass at maturity.

T2 is produced by Leydig cells in the testes, and its effects are mediated primarily through ARs. Skeletal muscle is well endowed with AR as are satellite cells, the primary muscle stem cell residing beneath the basal lamina of individual skeletal muscle fibers [28]. ARs have also been found on pluripotent stem cells which may play a role in muscle hypertrophy and/or the repair of muscle following injury but evidence for either of these possibilities is lacking in humans [81,82]. Once T2 interacts with the AR in skeletal muscle it is translocated to the nucleus to induce protein synthesis. Importantly, T2 also maintains or increases muscle mass by inhibiting the ubiquitin-proteasome catabolic system [28].

Because muscle mass in aging men begins to decline before T2 shows a notable downturn it has been questioned whether lifestyle is a contributor to the advent of early muscle loss. Fewer than 25% of men currently meet recommended physical activity minimums (ACSM) and the increase in male obesity is alarming. Because adipose tissue is an endocrine organ and secretes E2, it will be of interest to see future studies of aging in obese men. Will older obese men have higher than normal E2 values than men who are not obese? Will T2 values be higher or lower than normal? Will muscle loss be attenuated or accelerated under altered hormonal conditions? Confounding these questions however, will be physical activity or, more likely, the lack of activity in overweight men.

T2 and Aging

As noted, T2 is a key determinant of muscle mass at all ages primarily through its effects on protein metabolism [83,84]. Approximately 20% of men older than 60 years and 50% of men older than 80 years have low T2 values. Hormone concentration begins to decline in the mid-30s by approximately 1–3% per year. The sensitivity of skeletal muscle to T2 also diminishes with age [3]. Exercise, whether aerobic or resistance training, increases T2 levels and it is probable that lifestyle, specifically activity level, factors into the magnitude of T2 decline [85,86]. Although the focus of this chapter is on muscle aging, it is important to bear in mind that T2 plays a significant role in glucose and lipid homeostasis, which has a direct effect on muscle performance. Muscle mass is a direct determinant of bone mass and both are influenced by T2.

Hormone Treatment of Aging Men

T2 is a key determinant of total muscle mass at all ages. Suppression of T2 levels in men results in a reduction in fat-free mass (primarily muscle) and fractional muscle protein synthesis [87–90]. Because androgens have direct anabolic effects on skeletal muscle it was logical to begin studies of T2 replacement in older men who were frail and/or hypogonadal. The early studies of T2-replacement of hypogonadal men older than 60 years demonstrated that T2 increased muscle mass and strength in a dose-response fashion, that fat mass decreased, but functional capability in frail older men improved to a modest extent or not at all [88]. These physical changes were accompanied by an increase in muscle protein net balance with a concomitant decrease in muscle protein breakdown [91]. However, studies incorporating supra-physiological doses of T2 were not without secondary side effects, including polycythemia, elevated PSA values, edema, elevated LDL with a concomitant decrease in HDL, skin issues, sleep apnea, and pain. Nevertheless, most studies, those that bring T2 levels to eugonadal levels, have not shown that many deleterious events associated with T2 supplementation and almost all show some increases in muscle mass [84,87,92–95,118]. There is little evidence to support an improvement in physical capacity or function even with gains in muscle mass, suggesting a possible gap between quantity and quality of muscle.

T2 replacement in older men resulted in an average gain of 1.7 kg in fat-free mass, most of which is likely muscle [3]. In an average 165 lb man, 1.7 kg of muscle represents approximately 5% of his total mass. A concomitant strength gain in conjunction with the increase in mass has not been a consistent feature of T2 supplementation studies but if strength also increased by 5%, the end resultant would be a 4–5-year delay in frailty or loss of independence. T2 apparently increases the size of both type I and type II fibers which necessitates an increase in myonuclei and satellite cells [96–98].

T2 and Exercise Effects

Weight-training exercise results in a significant increase in muscle mass and strength. In reality the improvements in muscle mass and strength with exercise exceed the gains made by T2 supplementation alone. For example, Villanueva et al. [99] resistance trained a group of older men for 12 weeks and showed a 100% increase in leg strength. Bammann et al. [33] also demonstrated marked improvements in strength following resistance exercise in older men and changes in functional capacity. Studies combining strength training and T2 hormone therapy in older men are few but suggest that T2 may slightly augment strength or have no effect [84,88,100]. Of note recently is that studies in which T2 was delivered transdermally do not show an increase in strength [101,102]. This finding contrasts with T2 injection investigations that do show significant increases in strength.

Some older men are incapable or unwilling to exercise. Findings for intramuscular injections of T2 suggest that strength can be augmented to a modest extent in older men. Under these circumstances T2 injections are likely warranted, particularly if loss of independence is imminent.

Mechanisms of T2 Effects

The loss of T2 with age is hypothesized to be associated with the reduced number of satellite cells. Fewer satellite cells is likely responsible for the attenuated response to muscle damage in older men [90,103–107]. The blunted response is mediated by IL-6 which has been determined to be a positive regulator of satellite cell proliferation. IL-6 levels increase briefly with injury which is associated with the proliferation of satellite cells for repair. With age, IL-6 is chronically elevated which may be one reason for the blunted repair response to muscle injury [106] and for the reduced efficacy of the protein synthesis pathways, notably IGF-1 [92].

Indirect Effects of T2 on Aging Skeletal Muscle in Males

There is an association between inactivity and the loss of skeletal muscle mass and strength that occurs with age, in women and men. Rodent studies in male rats demonstrate that removal of T2 through orchiectomy results in a reduction in spontaneous wheel running activity. It should be noted however that male rats cover approximately one-fourth the nightly distance of females and thus, the decline in running distance is less in absolute as well as relative terms. These rodent data suggest that the T2-related reduction in physical activity may not have the impact on the aging muscular system in males as the reduction in E2 in females. When T2 is given back to orchiectomized rats, running distances equilibrate to baseline.

One potential consequence of an inactivity-associated loss of muscle mass and strength is a reduction of the amount of force imposed on the skeletal system. The greatest stimulus for bone accretion comes from the forces that muscle imposes on the skeleton. Potentially, for males and females, the loss of sex hormones with aging can have a negative effect on bone quantity and quality, directly via hormone action and indirectly through inactivity.

Adverse Events Associated with T2 Treatment

A recent meta-analysis of 12 studies of men with T2 administration within physiological limits revealed the risk of potential deleterious side effects was small [89,95]. In a recent study of 308 men over the age of 60 years, differences between untreated and treated groups were negligible for intima-media thickness and coronary artery calcium score [108]. Interestingly, there were no differences between groups in quality of life. Hematocrit, HDL, and PSA values were significantly higher in men on hormone therapy. However, additional study of 150,000 men did not reveal an increase in prostate cancer with hormone use [119]. In contrast to findings by Sattler et al. [89], Xu et al. [109] conducted a meta-analysis in which they examined cardiovascular adverse events in 27 randomized controlled trials and found the odds ratio of a cardiovascular adverse event was 1.54. Borst et al. [101] and Yarrow et al. [102] found that the method of T2 supplementation made a difference in adverse cardiovascular events with the use of oral drugs producing a higher risk than transdermal patches. It should be noted however that low T2 is associated with poor cardiovascular health including heart failure, artery disease, stroke, and mortality (shores). Several investigators have noted that recovery of the pituitary/hormonal axis was blunted or markedly diminished following the cessation of hormone therapy, a potential concern in aging, hypogonadal men.

Thus, as noted with E2, hormone supplementation of older men carries some risk. Of interest is that approximately half of the men given the option to continue T2 following the duration of study, chose to quit taking T2 [89].

Alternatives to T2

Theoretically, selective AR modulators have the potential to positively influence skeletal muscle and bone in aging males but to date, there is little indication that an effective product has been produced or determined to increase muscle mass and strength without side effects. Trenbolone has been found to elevate expression of anabolic genes in ORX rats but evidence for effectiveness in humans is lacking [104]. Attention has turned in recent years to DHEA, a naturally occurring hormone that also decreases with age, and there are several clinical studies indicating that hormone supplementation to youthful levels has minimal effects. In one randomized controlled trial of older men (and women), DHEA or placebo was given for 10 months. During the final 4 months a subset of subjects undertook resistance training. DHEA alone did not increase strength or thigh muscle volume. In subjects who weight-trained, strength and muscle mass were significantly augmented in the DHEA supplementation group [110]. In an earlier study of men between the ages of 60 and 80 years, DHEA was given for 1 year. Muscle strength and muscle cross-sectional area did not change [51]. A recent meta-analysis of 25 randomized controlled trials revealed that DHEA supplementation was associated with a reduction in fat mass, but no other variables. The effect on body composition was due entirely to increases in endogenous T2 and E2. Thus, the effectiveness, if any, of DHEA was due to its conversion to bioactive metabolites [111].

SUMMARY

While some of the age-related decline in skeletal muscle has been elucidated, there is still a great deal of work required to determine the best ways to counteract sarcopenia and prevent loss of independence in both men and women. Theoretically promising, hormone supplementation is a new frontier and there is much work that needs to be done to better understand hormone effects on the fundamental process of muscle aging. The interaction of hormones or their analogs with the existing hormonal axis needs to be further explored. The role of physical activity/inactivity and hormone expression and loss is another avenue for exploration. Finally, the effects of obesity on the entire aging process are underappreciated.

References

[1] Frontera WR, Hughes VA, Fielding RA, et al. Aging of skeletal muscle: a 12-yr long study. J Appl Physiol 2000;88:1321–6.
[2] Vermeulen A, Goemaere S, Kaufman JM. Testosterone, body composition and aging: a review. J Endocrinol Invest 1999;22 (Suppl. 5):110–16.
[3] Bhasin S. Testosterone supplementation for aging-associated sarcopenia. J Gerontol Med Sci 2003;11:1002–8.
[4] Baltgalvis KA, Greising SM, Warren GL, Lowe DA. Estrogen regulates estrogen receptors and antioxidant gene expression in mouse skeletal muscle. PLOS one 2010;5:1–11.
[5] Foryst-Ludwig A, Kintscher U. Metabolic impact of estrogen signaling through ERalpha and ERbeta. J Steroid Biochem Mol Biol 2010;122:74–81.

[6] Cavalcanti-de-Albuquerque JPA, Salvador IC, Martins EL, et al. Role of estrogen on skeletal muscle mitochondrial function in ovariectomized rats: a time course study in different fiber types. J Appl Physiol 2014;116:779–89.
[7] Miljkovic N, Lim J-Y, Miljkovic I, Frontera WR. Aging of skeletal muscle fibers. Ann Rehabil Med 2015;39:155–62.
[8] Greeves JP, Cable NT, Reilly T, Kingsland C. Changes in muscle strength in women following the menopause: a longitudinal assessment of the efficacy of hormone replacement therapy. Clin Sci 1999;97:79–84.
[9] Sirola J, Rikkonen T. Muscle performance after the menopause. J Br Menopause Soc 2005;11:45–50.
[10] Charlier R, Mertens E, Lefevre J, Thomis M. Muscle mass and muscle function over the adult life span: a cross-sectional study in Flemish adults. Arch Gerontol Geriatr 2015;61:161–7.
[11] Yamasaki H, Douchi T, Yamamoto S, Oki T, Kuwahata R, Nagata Y. Body fat distribution and body composition during GnRH agonist therapy. Obstet Gynecol 2001;97:338–42.
[12] Moran AL, Nelson SA, Landisch RM, et al. Estradiol replacement reverses ovariectomy-induced muscle contractile and myosin dysfunction in mature female mice. J Appl Physiol 2007;102:1387–93.
[13] Moran AL, Warren GL, Lowe DA. Removal of ovarian hormones from mature mice detrimentally affects muscle contractile function and myosin structural distribution. J Appl Physiol 2006;100:548–59.
[14] Taaffe ER, Sipilä S, Cheng S, et al. The effect of hormone replacement therapy and/or exercise on skeletal muscle attenuation in postmenopausal women: a year long intervention. Clin Physiol Funct Imaging 2005;25:297–304.
[15] Phillips SK, Rook KM, Siddle NC, Bruce SA, Woledge RC. Muscle weakness in women occurs at an earlier age than in men, but strength is preserved by hormone replacement therapy. Clin Sci (Lond) 1993;84:95–8.
[16] Figueroa A, Going SB, Milliken LA, et al. Effects of exercise training and hormone replacement on lean and fat mass in postmenopausal women. J Gerontol A Biol Sci Med Sci 2004;58L266–70L.
[17] Tuttle LJ, Sinacore DR, Mueller MJ. Intermuscular adipose tissue is muscle specific and associated with poor functional performance. J Aging Res 2012;2012:172957. Available from: http://dx.doi.org/10.1155/2012/172957. Epub 2012 May 14.
[18] Vandenboom R. Modulation of skeletal muscle contraction by myosin phosphorylation. Compr Physiol 2016 (in press).
[19] Pöllänen E, Kangas R, Horttanainen M, et al. Intramuscular sex steroids are associated with skeletal muscle strength and power in women with different hormonal status. Aging Cell 2015;14:236–48.
[20] Sipilä S, Finni T, Kovanen V. Estrogen influences on neuromuscular function in postmenopausal women. Calcif Tissue Int 2015;96:222–33.
[21] Pöllänen E, Ronkainen PHA, Horttanainen M, et al. Effects of combined hormone replacement therapy or its effective agents on the IGF-1 pathway in skeletal muscle. Growth Horm IGF Res 2010;20:372–9.
[22] Greising SM, Baltgalvis KA, Lowe DA, Warren GL. Hormone therapy and skeletal muscle strength: a meta-analysis. J Gerontol Med Sci 2009;64A:1071–81.
[23] Bea JW, Zhao Q, Cauley JA, et al. Effect of hormone therapy on lean body mass, falls and fractures: six-year results from the Women's Health Initiative Hormone trials. Menopause 2011;18:44–52.
[24] Kahlert S, Grohé C, Karas RH, et al. Effects of estrogen on skeletal myoblast growth. Biochem Biophys Res Commun 1997;232:373–8.
[25] Mangan G, Iqbal S, Hubbard A, Hamilton V, Bombardier E, Tiidus PM. Delay in post-ovariectomy estrogen replacement negates estrogen-induced augmentation of post-exercise muscle satellite cell proliferation. Can J Physiol Pharmacol 2015;93:945–51.
[26] Enns DL, Iqbal S, Tiidus PM. Oestrogen receptors mediate oestrogen-induced increases in post-exercise rat skeletal muscle satellite cells. Acta Physiol 2008;194:81–93.
[27] Bombardier E, Vigna C, Bloemberg D, et al. The role of estrogen receptor-α in estrogen-mediated regulation of basal and exercise-induced Hsp70 and Hsp27 expression in rat soleus. Can J Physiol Pharmacol 2013;91:823–9.
[28] LaColla A, Pronsato L, Milanesi L, Basconsuelo A. 17-ß estradiol and testosterone in sarcopenia: role of satellite cells. Ageing Res Rev 2015;24:166–77.
[29] Vasconsuelo A, Milanesi L, Boland R. 17-ß estradiol abrogates apoptosis in murine skeletal muscle cells through estrogen receptors: role of the PI3K/Akt pathway. J Endocrinol 2008; 196L385–94.
[30] Tiidus PM, Lowe DA, Brown M. Estrogen replacement and skeletal muscle: mechanisms and population health. J Appl Physiol 2013;115:569–78.
[31] Brown M, Birge SJ, Kohrt WM. Hormone replacement therapy does not augment gains in muscle strength or fat-free mass in response to weight-bearing exercise. J Gerontol A Biol Sci Med Sci 1997;52:B166–70.
[32] Kosek DJ, Kim JS, Petrella JK, Cross JM, Bamman MM. Efficacy of 3 days/wk resistance training on myofiber hypertrophy and myogenic mechanisms in young vs. older adults. J Appl Physiol 2006;101:531–44.
[33] Bamman MM, Ragan RC, Kim JS, et al. Myogenic protein expression before and after resistance loading in 26- and 64-yr-old men and women. J Appl Physiol 2004;97:1329–37.
[34] Meeuwsen IB, Samson MM, Verhaar HJ. Evaluation of the applicability of HRT as a preservative of muscle strength in women. Maturitas 2000;36:49–61.
[35] Greising SM, Carey RS, Blackford JE, et al. Estradiol treatment, physical activity, and muscle function in ovarian-senescent mice. Exp Gerontol 2011;46:685–93.
[36] Greising SM, Baltgavis A, Kosir AM, et al. Estradiol's beneficial effect on murine muscle function is independent of muscle activity. J Appl Physiol 2011;110:109–15.
[37] Gorzek JF, Hendrickson KC, Forstner JP, et al. Estradiol and tamoxifen reverse ovariectomy-induced physical inactivity in mice. Med Sci Sports Exerc 2007;39:248–56.
[38] Olivieri F, Ahtianen M, Lazzarini R, et al. Hormone replacement therapy enhances IGF-1 signaling in skeletal muscle by diminishing miR-182 and miR-223 expressions: a study on postmenopausal monozygotic twin pairs. Aging Cell 2014;13:850–61.
[39] Qaisar R, Renaud G, Hedstrom Y, et al. Hormone replacement therapy improves contractile function and myonuclear organization of single muscle fibers from postmenopausal monozygotic female twin pairs. J Physiol 2013;591:2333–44.
[40] Wiik A, Hellsten Y, Berthelson P, et al. Activation of estrogen response elements is mediated both via estrogen and muscle contractions in rat skeletal muscle myotubes. Am J Physiol Cell Physiol 2008;296:C215–20.
[41] Hatae J, Takami N, Lin H, Hona A, Inoue R. 17-ß estradiol-induced enhancement of estrogen receptor biosynthesis via MAPK pathway in mouse skeletal muscle myoblasts. J Physiol Sci 2009;59:181–90.
[42] Dieli-Conwright CM, Spektor TM, Rice JC, Sattler FR, Schroeder ET. Influence of hormone replacement therapy on eccentric exercise induced myogenic expression in postmenopausal women. J Appl Physiol 2009;107:1381–8.

[43] Ronda AC, Vasconsuelo A, Boland R. 17ß-estradiol protects mitochondrial functions through extracellular-signal-regulataed kinase in C2C12 muscle cells. Cell Physiol Biochem 2013;32:1011023.

[44] Wiik A, Hellstein Y, Berthelson P, et al. Activation of estrogen response elements is mediated by both estrogen and muscle contractions in rat skeletal muscle myotubes. Am J Physiol Cell Physiol 2009;296:C215–20.

[45] Wend K, Wend P, Krum SA. Tissue-specific effects of loss of estrogen during menopause and aging. Front Endocrinol 2012;3:19. Available from: http://dx.doi.org/10.3389/fendo.2012.00019.

[46] Smith GI, Yoshino J, Reeds DN, et al. Testosterone and progesterone, but not estradiol, stimulate muscle protein synthesis in postmenopausal women. Endocr Res 2014;99:256–65.

[47] Toivonen MHM, Pöllänen E, Ahtianen M, et al. OGT and OGA expression in postmenopausal skeletal muscle associates with hormone replacement therapy and muscle cross-sectional area. Exp Gerontol 2013;48:1501–4.

[48] Wiik A, Glenmark B, Ekman M, et al. Oestrogen receptor ß is expressed in adult human skeletal muscle both at the mRNA and protein level. Acta Physiol Scand 2003;179:381–7.

[49] Wiik A, Ekman M, Johansson O, et al. Expression of both oestrogen receptor alpha and beta in human skeletal muscle tissue. Histochem Cell Biol 2009;131:181–9.

[50] Hansen M, Kjaer M. Influence of sex and estrogen on musculotendinous protein turnover at rest and after exercise. Exerc Sport Sci Rev 2014;42:183–94.

[51] Percheron G, Hogrel JY, Denot-Ledunois S, et al. Effect of a 1-year oral administration of dehydroepiandrosterone to 60- to 80-year-old individuals on muscle function and cross-sectional area: a double-blind placebo-controlled trial. Arch Intern Med 2003;163:720–7.

[52] Dieli-Conwright CM, Spektor TM, Sattler FR, Schroeder ET. Hormone replacement therapy and messenger RNA expression of estrogen receptor coregulators after exercise in postmenopausal women. Med Sci Sports Exerc 2010;42:422–9.

[53] Tiidus PM, Holden D, Bombadier F, et al. Estrogen effect on post-exercise skeletal muscle neutrophil infiltration and calpain activity. Can J Physiol Pharmacol 2001;79:400–6.

[54] Bar PR, Amelink GJ. Protection against muscle damage exerted by oestrogen: hormonal or antioxidant action? Biochem Soc Trans 1997;25:50–4.

[55] Whiting KP, Restall CJ, Brain PF. Steroid hormone-induced effects on membrane fluidity and their potential roles in non-genomic mechanisms. Life Sci 2000;67:743–57.

[56] Wolf MR, Fragalia MS, Volek JS, et al. Sex differences in creating kinase after acute heavy resistance exercise on circulating granulocyte estradiol receptors. Eur J Appl Physiol 2012;112:3335–40.

[57] Stupka N, Tarnapolsky MA, Yardley NJ, Phillips SM. Cellular adaptations to repeated eccentric exercise-induced muscle damage. J Appl Physiol 2001;91:1669–78.

[58] Tiidus PM. Influence of estrogen and gender on muscle damage, inflammation and repair. Exerc Sport Sci Rev 2003;31:40–4.

[59] Keefe AC, Lawson JA, Flygare SD, et al. Muscle stem cells contribute to myotubes in sedentary adult mice. Nat Commun 2015;215:7087–93.

[60] McClung JM, Davis JM, Wilson MA, et al. Estrogen status and skeletal muscle recovery from disuse atrophy. J Appl Physiol 2006;100:2012–23.

[61] Sitnick M, Foley AM, Brown M, Spangenburg EE. Ovariectomy prevents the recovery of atrophied gastrocnemius muscle mass. J Appl Physiol 2006;100:286–93.

[62] Brown M, Ferreira JA, Foley AM, Hemmann KM. A rehabilitation program to remediate skeletal muscle atrophy in an estrogen-deficient organism may be ineffective. Eur J Appl Physiol 2012;112:91–104.

[63] Sorensen MB, Rosenfalck AM, Hojgaard L, Ottensen B. Obesity and sarcopenia after menopause are reversed by sex hormone replacement therapy. Obes Res 2001;9:622–6.

[64] Sipilä S, Taaffe DR, Cheng S, et al. Effects of hormone replacement therapy and high impact physical exercise on skeletal muscle in post-menopausal women: a randomized placebo controlled trial. Clin Sci 2001;101:147–57.

[65] Ronkainen PHA, Kovanen V, Alen M, et al. Postmenopausal hormone replacement therapy modifies skeletal muscle composition and function; a study with monozygotic twin pairs. J Appl Physiol 2009;107:25–33.

[66] Widrick JJ, Maddalozzo GF, Lewis D, et al. Morphological and functional characteristics of skeletal muscle fibers from hormone-replaced and nonreplaced postmenopausal women. J Gerontol A Biol Sci Med Sci 2003;131:3–10.

[67] Skeleton DA, Phillips SK, Bruce A, et al. Hormone replacement therapy increases isometric muscle strength of adductor pollicis in post-menopausal women. Clin Sci 1999;96:357–64.

[68] Sipilä S, Narici M, Kjaer M, et al. Sex hormones and skeletal muscle weakness. Biogerontology 2013;14:231–45.

[69] LaCroix AZ, Chlebowski RT, Manson JE, et al. Health outcomes after stopping conjugated equine estrogens among postmenopausal women with prior hysterectomy. JAMA 2001;305:1305–14.

[70] Somjen D, Katzburg S, Knoll E, et al. DT56a (Femarelle): a natural selective estrogen receptor modulator (SERM). Steroid Biochem Mol Biol 2007;104:252–8.

[71] Chillbeck PD, Cornish SM. Effect of estrogenic compounds (estrogen or phytoestrogens) combined with exercise on bone and muscle mass in older individuals. Appl Physiol Nutr Metab 2008;33:200–12.

[72] Velders M, Solzbacher M, Schleipen B, et al. Estradiol and genistein antagonize the ovariectomy effects on skeletal muscle myosin heavy chain expression via ER-β mediated pathways. J Steroid Biochem Mol Biol 2010;120:53–9.

[73] Hertrampf T, Seibel J, Laudenbach U, et al. Analysis of the effects of oestrogen receptor α (ERα)- and ERβ-selective ligands given in combination to ovariectomized rats. Br J Pharmacol 2008;153:1432–7.

[74] Dieli-Conwright CM, Spektor TM, Rice JC, Schroeder ET. Oestradiol and SERM treatments influence oestrogen receptor coregulator gene expression in human skeletal muscle cells. Acta Physiol 2009;197:187–96.

[75] Aubertin-Leheudre M, Lord C, Khalil A, Dionne IJ. Six months of isoflavone supplement increases fat-free mass in obese-sarcopenia women: a randomized double-blind controlled trial. Eur J Clin Nutr 2007;61:1442–4.

[76] Moeller LE, Peterson CT, Hanson KB, et al. Isoflavone-rich soy protein prevents loss of hip lean mass but does not prevent the shift in regional fat distribution in postmenopausal women. Menopause 2003;10:322–31.

[77] Maesta N, Nahas EA, Nahas-Neto J, et al. Effects of soy protein and resistance exercise on body composition and blood lipids in postmenopausal women. Maturitas 2007;56:350–8.

[78] Koot RW, Amelink GJ, Blankenstein MA, Bär PR. Tamoxifen and oestrogen both protect the rat muscle against physiological damage. J Steroid Biochem Mol Biol 1991;40:689–95.

[79] Jacobsen DE, Samson MM, Kezic S, Verhaar HJJ. Postmenopausal HRT and tibilone in relation to muscle strength and body composition. Maturitas 2007;58:7–18.

[80] Brown M, Ning J, Ferreira JA, Bogener JL, Lubahn DB. Estrogen receptor-alpha and -beta and aromatase knockout effects on lower limb muscle mass and contractile function in female mice. Am J Physiol Endocrinol Metab 2009;296: E854–61.
[81] Horstman AM, Dillon EL, Urban RJ, Sheffield-Moore M. The role of androgens and estrogens on healthy aging and longevity. J Gerontol: Biol Sci Med Sci 2012;67:1140–52.
[82] Svensson J, Movérare-Skrtic S, Windahl S, Swanson C, Sjögren K. Stimulation of both estrogen and androgen receptors maintains skeletal muscle mass in gonadectomized male mice but mainly via different pathways. J Mol Endocrinol 2010;45:45–57.
[83] Basaulto-Alarcon C, Varela D, Duran J, Maas R, Estrada M. Sarcopenia and androgens: a link between pathology and treatment. Front Endocrinol 2014;5:1–11.
[84] Srinvas-Shankar U, Roberts SA, Connolly MJ, et al. Effects of testosterone on muscle strength, physical function, body composition, and quality of life in intermediate-frail and frail elderly men: a randomized, double-blind, placebo-controlled trial. J Clin Endocrinol Metab 2010;95:639–50.
[85] Marcell TJ, Harman SM, Urban RJ, et al. Comparison of GH, IGF-1 and testosterone with mRNA receptors and myostatin in skeletal muscle in older men. Am J Physiol Endocrinol 2001;281:1159–64.
[86] Mudali S, Dobs AS. Effects of testosterone on body composition of the aging male. Mech Ageing Dev 2004;125:297–304.
[87] Urban RJ. Growth hormone and testosterone: anabolic effects on muscle. Horm Res Paediatr 2011;76:81–3.
[88] Bhasin S, Calof OM, Storer TW, et al. Drug insight: testosterone and selective androgen receptor modulators as anabolic therapies for chronic illness and aging. Nat Clin Pract Endocrinol Metab 2006;2:146–59.
[89] Sattler FR, Bhasin S, He Y, et al. Durability of the effects of testosterone and growth hormone supplementation on older community-dwelling older men: the HORMA trial. Clin Endocrinol 2011;75:103–11.
[90] Bhasin S, Taylor WE, Singh R, et al. The mechanisms of androgen effects on body composition: mesenchymal pluripotent cell as the target of androgen action. J Gerontol 2003;58A:1103–10.
[91] Ferrando AA, Sheffield-Moore M, Yeckel CW, et al. Testosterone administration to older men improves muscle function: molecular and physiological mechanisms. Am J Physiol Endocrinol Metab 2002;282:E601–7.
[92] Urban RJ, Dillon EL, Choudhary S, et al. Translational studies in older men using testosterone to treat sarcopenia. Trans Am Clin Climatol Assoc 2014;125:27–44.
[93] Hajjar RR, Kaiser FE, Morley JE. Outcomes of long-term testosterone replacement in older hypogondal males: a retrospective analysis. J Clin Endocrinol Metab 1997;82:3793–6.
[94] Rhoden EL, Morgentaler A. Risks of testosterone-replacement therapy and recommendations for monitoring. N Engl J Med 2004;350:482–92.
[95] Fernández-Balsells MM, Murad MH, Lane M, et al. Adverse effects of testosterone therapy in adult men: a systematic review and meta-analysis. J Clin Endocrinol Metab 2010;95:2560–75.
[96] Basaulto-Alarcón C, Jorquera G, Altamirano F, et al. Testosterone signals through mTOR and androgen receptor to induce muscle hypertrophy. Med Sci Sports Exerc 2013;45:1712–20.
[97] Velders M, Diel P. How sex hormones promote skeletal muscle regeneration. Sports Med 2013;43:1089–100.
[98] Sinha I, Sinha-Hakim AP, Wagers AJ, Sinha-Hikim I. Testosterone is essential for skeletal muscle growth in aged mice in a heterochronic parabiosis model. Cell Tissue Res 2014;357:815–21.
[99] Villanueva MG, Lane CJ, Schroeder ET. Short rest interval lengths between sets optimally enhance body composition and performance with 8 weeks of strength resistance training in older men. Eur J Appl Physiol 2015;115:295–308.
[100] Brill KT, Weltman AL, Gentili A, et al. Single and combined effects of growth hormone and testosterone administration on measures of body composition, physical performance, mood, sexual function, bone turnover, and muscle gene expression in healthy older men. J Clin Endocrinol Metab 2002;87:5649–57.
[101] Borst SE, Yarrow JF, Conover CF, et al. Musculoskeletal and prostate effects of combined testosterone and finasteride administration in older hypogonadal men: a randomized, controlled trial. Am J Physiol Endocrinol Metab 2014;306:E433–42.
[102] Yarrow JF, McCoy SC, Borst SE. Intracrine and myotrophic roles of 5α-reductase and androgens: a review. Med Sci Sports Exerc 2012;44:818–26.
[103] Kadi F. Cellular and molecular mechanisms responsible for the action of testosterone on human skeletal muscle. A basis for illegal performance enhancement. Br J Pharmacol 2008;154:522–8.
[104] Ye F, McCoy SC, Ross HH, et al. Transcriptional regulation of myotrophic actions by testosterone and trenbolone on androgen-responsive muscle. Steroids 2014;87:59–66.
[105] Nielsen S, Hvid T, Kelly M, et al. Muscle specific miRNAs are induced by testosterone and independently upregulated by age. Front Physiol 2014;23:394–9.
[106] Serra C, Tangherlini F, Rudy S, et al. Testosterone improves the regeneration of old and young mouse skeletal muscle. J Gerontol A Biol Sci Med Sci 2013;68:17–26.
[107] Harjola V-P, Jänkälä H, Härkönen M. The effect of androgen status on skeletal muscle myosin heavy chain mRNA and protein levels in rats recovering from immobilition. Eur J Appl Physiol 2000;83:427–33.
[108] Snyder PJ, Ellenberg SS, Cunningham GR, et al. The testosterone trials: seven coordinated trials of testosterone treatment in elderly men. Clin Trials 2014;11:362–75.
[109] Xu L, Freeman G, Cowling BJ, Schooling CM. Testosterone therapy and cardiovascular events among men: a systematic review and meta-analysis of placebo-controlled randomized trials. BMC Med 2013;11:108–15.
[110] Villareal DT, Holloszy JO. DHEA enhances effects of weight training on muscle mass and strength in elderly men and women. Am J Physiol Endocrinol Metab 2006;291:1003–8.
[111] Corona G, Tastrelli G, Giagulli VA, et al. Dehydroepiandrosterone supplementation in elderly men: a meta-analysis study of place-controlled trials. J Clin Endocrinol Metab 2013;98:3614–26.
[112] Galuzzo P, Rastellli C, Bulzomi P, Acconcia F, Pallottini V, Marino M. 17-β-Estradiol regulates the first steps of skeletal muscle cell differentation via ERα-mediated signals. Am J Physiol Cell Physiol 2009;297:C1249–62.
[113] Stenholm S, Shardell M, Bandinelli S, Guralnik JM, Ferrucci L. Editor's choice: physiological factors contributing to mobility loss over 9 years of follow-up—results from the InCHIANTI study. J Gerontol A Biol Sci Med Sci 2015;70(5):591–7. Available from: http://dx.doi.org/10.1093/gerona/glv004.
[114] Enns DL, Tiidus PM. The influence of estrogen on skeletal muscle. Sports Med 2012;40:41–58.
[115] Iqbal S, Thomas A, Bunyan K, Tiidus PM. Progesterone and estrogen influence postexercise leukocyte infiltration in overiectomized female rats. Appl Physiol Nutr Metab 2008;33:1207–12. Available from: http://dx.doi.org/10.1139/H08-108.

[116] Bombardier E, Vigna C, Iqbal S, Tiidus PM, Tupling AR. Effects of ovarian sex hormones and downhill running on fiber-type-specific HSP70 expression in rat soleus. J Appl Physiol 2009;106(6):2009–15. Available from: http://dx.doi.org/10.1152/japplphysiol.91573.2008.

[117] Sauerwein H, Meyer HH. Androgen and estrogen receptors in bovine skeletal muscle: relation to steroid-induced allometric muscle growth. J Animal Sci 1989;67(1):206–12.

[118] Bhasin S, Calof OM, Storer TW, et al. Drug insight: testosterone and selective androgen receptor modulators as anabolic therapies for chronic illness and aging. Nat Clin Pract 2006;2(3):146–59.

[119] Kaplan AL, Lenis AT, Shah A, Rajfer J, Hu JC. Testosterone replacement therapy in men with prostate cancer: a time-varying analysis. J Sex Med 2015;12(2):374–80.

CHAPTER

11

Strategies and Approaches for Studying Sex Differences in Physiology

Margaret M. McCarthy

Department of Pharmacology, University of Maryland School of Medicine, Baltimore, MD, United States

INTRODUCTION

The discipline of physiology is no stranger to sex differences given that reproductive biology is by definition differences in the physiology of males and females driven by the gonads characteristic of each sex. Indeed the physiology of reproduction is so vast that the fourth edition of a two-volume tome bearing that name has 52 chapters and 114 authors [1]. But this emphasis on studying both sexes in great depth does not transfer to other aspects of physiology. An analysis of articles published in 2009 in the *Journal of Physiology* and three American J. Physiology Journals (renal & cardiovascular, GI and liver) revealed an almost 4:1 ratio of male to female subjects (animal only) and over 20% not reporting the sex at all (Fig. 11.1). Only 12.5% of physiology studies reported using animals of both sexes, and of those, only 30% actually analyzed the data by sex [2]. At the same time, equal sex representation was far greater in clinical studies involving humans, presumably due to the US National Institutes of Health Revitalization Act of 1993 which mandated inclusion of women and minorities. So how did this happen?

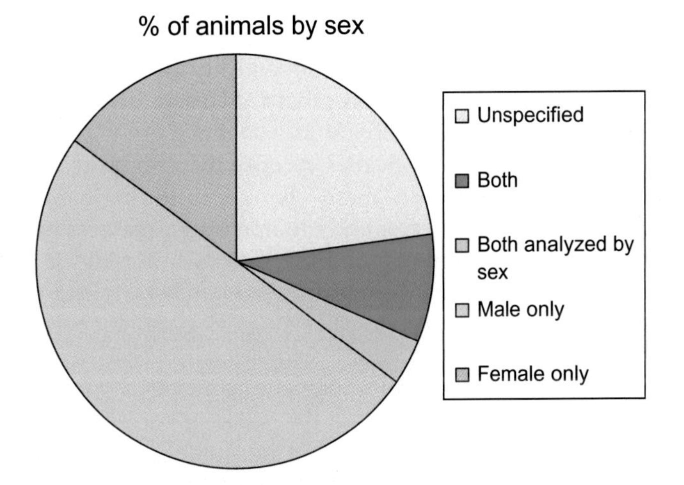

FIGURE 11.1 **Animal representation by sex in physiology journals.** The reporting and analyses of animals by sex were assessed in the *Journal of Physiology* and three American physiology journals (renal, GI and liver, cardiovascular) for the year 2009. *Source: Redrawn from data presented in Beery AK, Zucker I. Sex bias in neuroscience and biomedical research. Neurosci Biobehav Rev 2010;35:565–72.*

HISTORICAL BARRIERS TO THE STUDY OF BOTH SEXES

As scientists our goal is to control variables that we believe are not relevant to the phenomenon of study in order to manipulate those we believe are important and thereby further our understanding. Thus if sex is not important, it is best to not let it vary and just focus on one sex. This would be an example of well-intentioned benign neglect. Conversely one might argue that the dominant barrier to incorporating sex as a variable outside the context of reproduction is the profound importance of sex. As the center of gravity in the study of reproductive physiology, the determination of sex became defined by its own weightiness as the source of all things. But like the Big Bang, once it was over, other more sophisticated processes shaped the galaxies and stars therein. Thus once sex was determined, other physiological processes proceed independently. But we now know that the echo of the Big Bang continues to

Sex Differences in Physiology
DOI: http://dx.doi.org/10.1016/B978-0-12-802388-4.00011-2

impact all aspects of the known universe, and so too does sex impact much of physiology.

If we begin with fundamentals, a female is defined as the member of the species that produces large and relatively infrequent gametes that are then "fertilized" by the small, motile, and abundant gametes of the males. In aquatic species the fertilization might occur outside the body, but the general relationship still holds. As we move into terrestrial species the eggs are usually fertilized internally and eventually, when we get to true mammals, the embryo is nourished internally and after being summarily pushed into the world is fed exclusively by the female who possesses the requisite mammary glands. Female mammals thus make a greater investment, from a physiological stand point, than males in their offspring. This higher investment has presumably led to the evolution of "female choice" by which females only mate when there is a suitable mate and she is in the proper physiological state (with some primates including humans, excepted). Moreover, most females are not sexually receptive all the time, instead, they "cycle" or opportunistically become available depending on conditions. In many species sexual receptivity is closely tied to seasonal changes, but in laboratory rodents there is a 4–5-day recurring cycle. Sexual receptivity is a powerful force as it encompasses many changes including behavior, lining of the uterus, body temperature, metabolism, appetite, sleep patterns, and the list goes on. And it is this variability that most alarms researchers trying to study organ systems or physiological parameters not involved in reproduction. Females are best to be avoided as they introduce a level of unwanted variability, the thinking goes. But is the thinking correct? For a long time it was considered an immutable truth, with grant and manuscript reviewers insisting on control for the stage of the reproductive cycle in any female subjects. In retrospect this may have done more harm than good, as the fallacy of this approach has recently been illustrated empirically. Analyses of 293 articles published between 2009 and 2012 that included female mice at random stages of the estrus cycle as well as males allowed for assessment of the coefficient of variance generated from close to 10,000 measurements of 30 broad categories. The outcome was no greater variance in females compared to males on any of the traits measured, and in some cases greater variation in males [3]. A smaller more focused study on pain thresholds found similar results, no greater impact of estrous cycle on variability in females than that observed in males [4]. The lack of higher variability in females has been recently extended to include gene expression changes using microarray databases in both human and animal models [5]. Interestingly, housing conditions, either grouped or individual, substantially increased variation in both sexes but with variation in many endpoints for males exceeding that of group-housed females and larger than that accounted for in females by the estrous cycle [3]. Interestingly the authors argue that to reduce variability animals should be singly housed, a move that would increase research costs far more than including females in study design, not to mention that social isolation is a major stressor for most rodents.

WHY STUDY SEX DIFFERENCES?

Once past the argument that including females would more than double the number of subjects required, some scientists maintain they should not be required to include both sexes because sex differences are not that important, no different say than differences caused by age or body mass. Why give special attention to sex? The answer is that sex is a singular variable that for the most part does not change, and it categorizes two halves of every population. The importance of sex may vary across the life span, but once a male, always a male. It is arguably the easiest biological trait to identify (with perhaps the exception of some monomorphic species with no external genitalia such as some birds). Moreover, as noted above, the impact of sex begins at the very earliest stages of development and, as will be discussed in other portions of this chapter, includes chromosome complement, hormones developmentally and in adulthood, and behavior and experience. All of these aspects of sex have been sculpted by evolutionary forces to optimize fitness within that sex, and these forces have the potential to vary substantially given the different investments and reproductive strategies of males and females. Thus studying any physiological parameter in one sex is essentially studying only half the population until proven otherwise. That this is a bad thing should be self-evident but is made tangibly clear by the fact that 8 out of 10 prescription drugs withdrawn from the US market during a 3-year period was due to unanticipated negative consequences in women (reviewed in [6]). We will never know, but it is plausible to assume that if the preclinical research had paid closer attention to sex as a variable, negative outcomes could have been avoided.

Ultimately, as researchers become more comfortable with including both sexes in their experimental design and tracking the results accordingly, they may discover the heuristic value inherent in comparing and contrasting males and females. Novel mechanisms related to pain [7], cell death [8], synaptic physiology [9], and synapse formation [10] are just a few examples of biological processes that would have remained undiscovered if not for the active comparison of males and females. Thus even if a particular endpoint is the same in males and females, by incorporating both sexes in the experimental design researchers can continue their investigation but remain alert to the potential that some aspects of the signaling pathway, response metric, gene expression etc. may diverge in males and females. This does not mean that the group size of all experiments needs to be doubled and everything essentially repeated in females. There are several reasonable approaches. One is to pick a particular relevant endpoint and conduct a pilot study with sufficient numbers of males and females to detect even a modest sex difference. Alternatively, increasing the sample size by a few animals but splitting the numbers evenly between males and females in multiple experiments would allow for broader detection of more robust sex differences without being overly burdensome. If sex differences are detected, the investigator now has a new tool for interrogating their system. If there is no evidence of a sex difference, then the work is broadly relevant.

WHERE DO SEX DIFFERENCES COME FROM?

Appropriate attention to sex as a variable is enhanced by an appreciation of the various ways in which males and females come to differ. Although sex of the gonads is determined in some species by temperature or is facultative in response to social and environmental variables (some reptiles and fish), in other species the sex of the gonad depends on chromosome complement, which in mammals and Drosophila is XX and XY and in birds is ZW and ZZ for females and males respectively. Combinations of the XY and ZW system are found in various species and some combine genetic and temperature sex determination [11]. In mammals there are two aspects of chromosome complement that are sexually dimorphic. The first is that only in female cells is there inactivation of one X chromosome, so this is a sex-specific phenomenon that likely impacts phenotypic differences is ways not fully appreciated [12]. The second is the presence of the *Sry* gene of the Y chromosome that directs the embryonic gonad toward development as a testis. If the *Sry* gene is absent, as in females, or mutated, the gonad will by default develop into an ovary [13,14]. This process occurs early in development, at 11 days postconception in the mouse in which gestation lasts approximately 3 weeks [15]. Steroid production by the developing testis also begins embryonically and drives the formation of the male reproductive tract and genitalia. Later in gestation and immediately after birth there is a surge in androgen production by the fetal testis and this impacts on the developing brain to organize masculine brain characteristics. This sensitive period for brain sexual differentiation extends into the postnatal period in rats and mice but is considered more or less complete in utero in nonhuman primates and humans (reviewed in [16]). The relative importance of androgens versus their metabolized byproducts, estrogens, also varies across species and physiological and behavioral endpoints. The infantile and juvenile period, including an extended childhood in humans, is characterized by a lack of circulating gonadal steroids but this time is not without sex differences as many parameters are set by the differences in perinatal hormone exposure and may or may not be impacted by genes found on the sex chromosomes [17]. One vitally important parameter that is organized by early hormone exposure is the timing and nature of puberty which begins earlier in females and is characterized by the establishment of reproductive cyclicity versus males which begin continuous production of sperm in response to elevated androgens. Some secondary sex characteristics of both males and females are also permanently established at this time while others are dependent upon the continued presence of circulating gonadal steroids in adulthood [18–21].

Sex chromosome genes and gonadal hormones are biological variables that determine or modulate sex differences, and among these differences are behaviors. The act of engaging in certain behaviors can in turn impact back on the physiology of an individual, particularly in terms of neural physiology. But there is also the impact of the behavior of other individuals toward each other based on perceptions of sex or gender. Gender is defined as both a person's and society's perception of one's sex, so this term is used exclusively for humans. In humans gender impacts every aspect of life, including one's "identity" and it is difficult for us to know just how much this impacts on sex differences in physiology and behavior because biological and social factors are confounded in humans. Even in animals we can document that as early as the first few days of life individuals are treated differently based on sex. In rodents, the dam engages in more anogenital grooming of her male offspring than her female pups and this impacts the development of select motor neurons [22], social behavior [23], and expression of particular genes in the brain [24]. Newborn rat pups emit ultrasonic vocalizations when separated from the nest and the frequency and characteristics of these calls differ in males and females. More importantly, the dam is paying attention and preferentially retrieves those pups with male calling patterns (even if they are female but induced to call like males by genetic manipulation) [25]. These naturally occurring differences in animal behavior may be further confounded by animal husbandry practices in our animal care facilities. Male mice are notoriously intolerant of each other and so as adults are often housed alone, whereas females are usually group-housed in order to cut down on cage space. When male mice are housed together there is considerable discord.

Male rats are more congenial but when housed in a group still establish a dominance hierarchy so that within one cage males may differ considerably in their stress level. As noted above, the variance induced by group housing is as great or greater than that induced by sex. These are just a few examples of the way the external world can impact the sexes differently and highlights the importance of considering all aspects of an animal's existence. Taking this to the extreme, it was recently determined that even the microbiota of the gut can differ in male and female mice [26]. The importance of all of these variables; hormones, genetics, behavior, and experience, and how to control them will be discussed in detail later in this chapter.

WHAT IS THE NATURE OF SEX DIFFERENCES?

The last foundational principle to consider is what are we really talking about when it comes to sex differences. While this may seem obvious, it quickly is not when one begins to contemplate the matter in some detail. Foremost is the issue of nomenclature. The expressions "sexually dimorphic" or "sex dimorphism" are used liberally and frequently as descriptors of even the smallest of sex differences. This might seem trivial but the terms "di-" and "morph" mean "two-" and "forms," suggesting that males and females are of entirely different forms. In other words it is an exaggeration. Rarely are endpoints in males and females of entirely different forms and when they are, it usually is associated with reproduction (ie, gonads, genitalia, etc.). More often what people are really referring to is a sex "difference," meaning an endpoint that lies along a continuum but the mean value differs in populations of males and females, with overlap of the two sexes. This could include physiological parameters such as body weight, circulating insulin, heart rate, bone density, height, and so on. Sometimes the sex difference is large, as in height in humans, and other times it is very small and requires a large number of subjects to detect. How many subjects is a function of variability, if the endpoint is highly variable or the fidelity of the measurement technique is low, more subjects will be required. Conversely, if the difference in mean is large and the variability is low, many fewer subjects may be required to reliably detect a sex difference. A power analysis, which takes these factors into account, is often used to calculate the number of subjects to be used to achieve statistically significant differences between groups. However, it is important to distinguish between the power calculation for detecting an effect of a manipulation versus differences in the effect of a manipulation on males versus females (a treatment by sex interaction). These two things are often conflated and may unnecessarily deter an investigator from including males and females when in fact they could use only a few more animals than the original number of subjects as proposed by the original power calculation but include both sexes and detect even a modest sex difference, provided the variance is not too high.

After an effect of sex is detected it is equally important to consider the same variables that went into the power calculation, magnitude of the difference and variability in response, both of which are incorporated under the term of effect size, also referred to as "Cohen's *d*" and calculated as the mean difference divided by the pooled standard deviation for both groups. In other words, the difference between groups is scaled in standard deviation units. General guidelines for meaningfulness of the *d* is small ($d = 0.2$), medium ($d = 0.5$), and large ($d = 0.8$) effects but these are of course guidelines and should be treated as such. More importantly, though, is for there to be some kind of metric for the magnitude and impact of a sex difference instead of treating any difference that is detectable by a t-test and $p < 0.05$ as a sexual dimorphism. Some argue that we should dispense with the term sex "difference" all together and substitute "sex effect" [27], which is actually quite appropriate as really what we are studying is the effect of sex, not the difference of sex.

But sometimes there are genuine sexual dimorphisms outside the context of reproduction. For instance the pattern of growth hormone secretion from the anterior pituitary is markedly different in males and females, being pulsatile in one sex and more cyclical in the other [28,29]. Given the centrality of growth hormone to so many aspects of growth, development, and adult physiology, this sex difference has profound implications despite the observation that on average, growth hormone levels do not differ much in males and females. Protecting the salience of this biological phenomenon requires using the proper terminology here and elsewhere.

Lastly, sex differences can derive in unexpected ways. Sometimes the two sexes might solve the same problem but using different approaches. This is illustrated in different spatial learning strategies [30] and neural circuits regulating pair-bonding [31] and parenting behavior [32]. The reason for the divergence in paths to the same end is not always clear but in some cases may be the indirect result of the reproductive system specific to each sex. For instance in the case of parenting behavior, in mammalian brains the female has a neural circuit for nurturing that is closely tied to lactation, but males lack such a system. In one species of vole, males have evolved a distinct circuit that serves this purpose and they behave as respectable "dads." In a closely related vole species lacking this specialized circuit, males show no parental tendencies [33]. Thus both species exhibit a sex difference in the neural circuits of parenting, but in one species the behavior is the same in males and females, meaning they both exhibit parental behavior, but in the other species behavior toward offspring is distinctly different, in fact dimorphic. This phenomenon of a convergence on the same behavior via different routes has been labeled "compensation" as it refers to the male compensating for his lack of a neural circuit for parenting with the creation of a new one [34]. It has also been referred to as "convergence" since the two sexes are converging on the same endpoint and this can occur independent of the need to compensate for reproductively imposed restrictions [35].

The contrast to "convergence" is "divergence," a process whereby the two sexes are at the same point under baseline conditions but they diverge in response to a challenge or stressor. Learning is a good example of divergence, with males improving under stressful conditions and female performance degrading, a process paralleled by opposing changes in hippocampal synapses [36]. Convergence and divergence can occur in response to other variables such as aging, with sex differences either appearing or disappearing at various phases. Given the dynamic and often hidden nature of these forms of sex differences they can be among the most challenging to either control or pursue experimentally but may also be among the most important biologically.

All of these aspects of sex differences in physiology will be important to bear in mind as we enter a new era of research in which major funding agencies require attention to the central biological variable of sex. The Canadian Institutes of Health Research requires applicants to respond to questions regarding whether they have included male and female subjects and if not, why not. This simple act has significantly increased the representation of both sexes in preclinical research funded by that agency. In an even bolder step, the National Institutes of Health of the US recently released guidelines regarding the incorporation of sex as a biological variable (http://grants.nih.gov/grants/guide/notice-files/NOT-OD-15-102.html), and notes that the appropriate adherence to these guidelines will be a review criterion.

The following four mini-chapters will provide practical considerations and specific advice regarding the variable of sex in physiological research.

The four areas of focus are:

1. Sex chromosome effects
2. Organizational effects of hormones
3. Activational effects of hormones
4. Special considerations for studying sex within humans

The importance of considering sex in the study of physiology is coupled to the essential need for appropriate experimental design, techniques, and analyses. The following four sections were composed by experts who have systematically assessed the variable of sex on multiple levels. These sections serve as a starting point to facilitate the successful consideration of sex as a variable by researchers beginning to appreciate the implications of sex for their specific research and the larger biological context in which it resides.

CHAPTER

11.1

How to Study Sex Differences Caused by "Sex Chromosome Effects" on Tissues

Arthur P. Arnold

Department of Integrative Biology & Physiology, UCLA, Los Angeles, CA, United States

Sex differences in traits are caused by two broad categories of mechanisms. In species in which the sex of the gonads is determined by genetic factors (instead of environmental factors such as temperature [1]), the sex chromosomes of males and females differ in the number or types of genes, which in turn act during the animal's life span to cause sex differences in traits. We refer to these mechanisms as "ontogenetic," because they are considered to affect the ontogeny of each male and female. Contrasted with this category are sex differences that result from population-level forces, because some environmental or biological factors influence a disproportionate number of individuals of one sex rather than the other, shifting the mean trait value for that sex and creating an average sex difference. An example of a population-level sex difference is the process dubbed "Mother's Curse" [2,3]. Because the mitochondrial genome is inherited only through the female lineage, and is never passed from males to their progeny, it is possible for mitochondrial gene variants to be selected that are advantageous for females but disadvantageous for males, placing a genetic load (disease, mortality) on males more than females. In this section we focus on "ontogenetic" processes only, and discuss experimental strategies to categorize and discover specific factors that cause sex differences in mammals.

Our conceptual framework is that all sex differences stem from the genetic imbalance of the X and Y chromosomes, which are the only factors that are currently known to differ in male and females at the beginning of the individual's life, in the zygote [4]. Thus, all sex differences have to be traceable back to an unavoidable and essential sex difference in X or Y genes. Although sexually dimorphic development is controlled in large part by autosomal genes (because these genes represent about 95% of the genome and therefore are the main constituents of gene pathways that control any trait), the autosomal genes are generally thought to be on an average equivalent in the two sexes. Accordingly, differences between the sexes cannot originate with autosomal genes, but must be established by the differential action of X and Y genes. Nevertheless, modifications in autosomal genes would be expected to disrupt or alter sexual differentiation because the autosomal genes are downstream from the primary sex-biasing effects of X or Y genes.

The X and Y chromosomes have evolved from an ancient autosomal chromosomal pair [5]. The Y chromosome, which in most mammalian species carries the dominant testis-determining gene *Sry*, has lost most of its DNA in the course of evolution, so that it is small, and contains relatively few different genes [6] (less than 20 protein-coding genes in the mouse). Some Y genes are present in multiple (sometimes many) copies, increasing the total tally of Y genes if each variant within amplified gene families is counted separately. In contrast, the X chromosome is large and has many more distinct single-copy genes [7] (about 1000 coding genes in the mouse). The two chromosomes share an identical region, the pseudoautosomal region (PAR), which is the region of pairing and crossing over that occurs during the first meiotic division of spermatogenesis. The PAR genes are the same on the X and Y chromosomes, so all XX and XY individuals have two copies of the PAR. Thus, the PAR genes are not thought to be the origin of sex differences.

Genetic sex differences are thought to stem from three kinds of "imbalance" in XX versus XY cells [4]:

1. XY cells have non-PAR Y genes, which might make them different than XX cells.
2. XX cells have two copies of each non-PAR X gene, which might make them different from XY cells, which have only one copy. Because X inactivation silences one of the two X chromosomes in each nongermline cell of XX individuals, X genes are normally expressed from one X chromosome in XX cells, similar to the expression from the single X chromosome of XY cells. X inactivation is a great leveler, which eliminates most sex differences in

expression of X genes that would otherwise have occurred because of the sex difference in the number of copies of X genes [8]. However, a minority (perhaps 3% in mice and 15% in humans) [9,10] of X genes escapes inactivation and these genes are expressed from both the active and inactive X chromosomes, which can result in constitutively greater expression of those genes in XX cells than XY cells. The "X escapees" therefore represent one source of inherent sexual imbalance in XX versus XY cells, caused by the imbalance in the number of X chromosomes.

3. The third general category of imbalance results from the sex difference in parental imprint on X genes. XY individuals inherit their X chromosome from the mother, whereas XX individuals inherit an X chromosome from each parent. Thus, the effects of a paternal X imprint are experienced only by XX individuals. The process of random X inactivation means that tissues of XX individuals are mosaics, with some cells inactivating the maternal X and experiencing the effects of the paternal X chromosome, and other cells inactivating the paternal X chromosome and experiencing the effects of the maternal X chromosome. Thus, the XX female's tissues will tend to average out the effects of the two parental imprints on X genes, but XY males will show only the effects of the maternal imprint. Parent-of-origin (imprinting) effects on autosomes are equally felt by XX and XY offspring, and thus are not thought to be the primary cause of sex differences in traits.

We seek to establish a comprehensive list of all of the basic mechanisms that can lead to sex differences, to provide a framework for conducting experiments to test various possibilities to explain any observed sex difference in mammals. The "top three" list discussed in the last paragraph may not be exhaustive. Some evidence suggests the existence of a fourth category, based on studies of *Drosophila* and mammalian cells [11–14]. In mammalian cells, the presence of a large inactivated X chromosome, in XX but not XY cells, might sequester molecular factors that make DNA heterochromatic and inactive, so that such factors are less available to shut down expression of some regions of autosomes [13,14]. Good experimental strategies exist to test for the "top three" list, as discussed in the following.

The most important gene for sexual differentiation is *Sry*, encoded by the Y chromosome. This gene is expressed in males just before differentiation of the embryonic gonadal ridge, and triggers a molecular cascade that results in development of testes [15]. In the absence of *Sry*, even when the rest of the Y chromosome is present, autosomal or X chromosome genes cause differentiation of ovaries. The sexual differentiation of the gonads sets up lifelong sex differences in the secretion of gonadal hormones, which act to cause sex differences in many tissues. These hormonal effects can be permanent "organizational" effects, or reversible "activational" effects [16,17]. Indeed, until the late 20th century, all sex differences in traits were thought to be proximal effects of gonadal sex steroid hormones.

If our goal is to test for X and Y gene effects that cause sex differences in specific traits, outside of the well-established gonadal hormonal mechanisms, the simplest approach is to manipulate the number and type of sex chromosomes to determine how they influence the traits. For example, if we wish to determine if Y genes act within the brain to make the XY brain different than XX in mice, we might give a Y chromosome to XX female mice, or remove it from XY males, measuring the effects on phenotype. The problem is that giving a female a Y chromosome also makes her a gonadal male. The effects of non-*Sry* Y genes acting throughout the body are always confounded with the effects of testicular secretions. How does one separate the two? One approach is to remove *Sry* from the Y chromosome, so that one can compare groups of mice that differ in the presence of non-*Sry* Y genes, but which have the same type of gonad. This approach is used in the Four Core Genotypes (FCG) Mouse Model.

THE FOUR CORE GENOTYPES MOUSE MODEL

FCG mice have a "Y minus" chromosome, which is deleted for *Sry* [18–23]. XY^- mice therefore have ovaries and are called females (XYF). Some FCG mice also have a *Sry* transgene inserted onto chromosome 3 [22]. Thus, XY^- (*Sry* +) mice (a.k.a. XYM) have testes and are fertile males. Because the *Sry* transgene segregates independently of sex chromosome complement, XYM mice make four kinds of sperm, with either an X or Y^- chromosome, each with or without *Sry* in four combinations. These sperm produce the four core genotypes: XXF, XXM, XYF, XYM (Fig. 11.2, Table 11.1). The progeny represent a 2 X 2 design comparing the effects of two factors, sex chromosome complement (XX vs XY^-) and *Sry* (present (gonadal male) versus absent (gonadal female)). To compare traits measured in FCG mice, an appropriate statistical test is the two-way ANOVA, which measures if there are main effects of sex chromosome complement or gonadal sex, or interactions of the two.

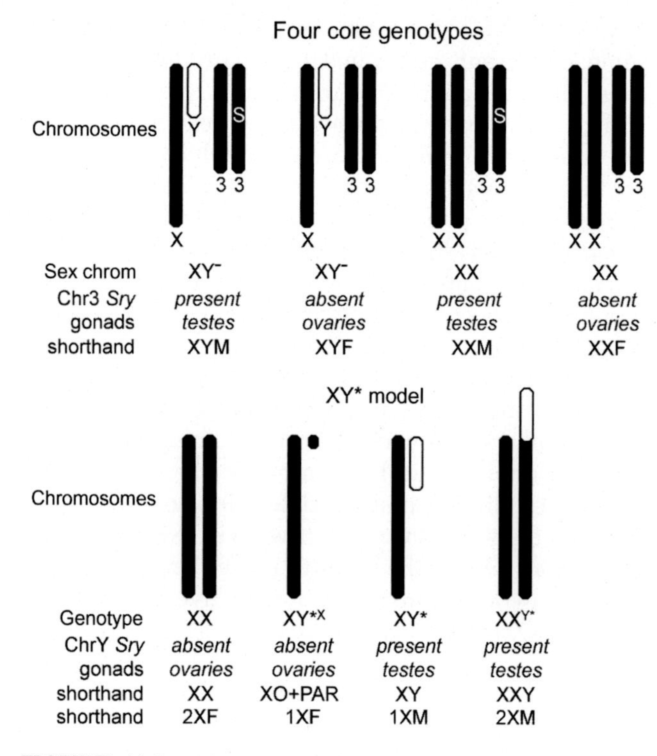

FIGURE 11.2 Schematic diagram showing genetic and phenotypic differences in groups of the Four Core Genotypes (FCG) and XY* model. More detailed discussion of these models is elsewhere for FCG [19,20,22,33] and XY* [21,30,31,33].

The primary goal of the FCG model is to ask if the normal male (XY) and female (XX) complement of sex chromosomes act outside of the gonads to cause sex differences in phenotypes of any tissues. For this purpose, we compare traits in XX and XY mice with the same type of gonad (XXF vs XYF; XXM vs XYM) to see if there are group differences. A second goal is to determine if the phenotype is affected by the type of gonadal secretions, testicular versus ovarian. If we find a difference in XXM and XXF, or in XXM versus XXF, we conclude that the difference is likely caused by the different gonadal secretions (Fig. 11.3). (The gonadal sex of the mouse in this model is confounded with the presence/absence of the *Sry* transgene, however, so the FCG model does not separate the effects of *Sry* on the gonads, resulting in male–female differences in gonadal secretions, from its effects on other tissues.)

These comparisons come with some important caveats. One critical question is whether XX and XY mice with the same type of gonad have identical levels of gonadal hormones. The short answer is that such differences have not been demonstrated, but that we can never know the answer to that question under all possible conditions. In some traits that are sexually differentiated by the action of gonadal hormones, XXM FCG mice are quite similar to wild-type (WT) XY

TABLE 11.1 The Table Shows Group Comparisons in the FCG and XY* Models, Possible Group Differences that can be Observed, and Interpretations of Those Outcomes

Groups	Group comparisons	Tests effect of	Trait outcomes	Interpretation	Notes
FCG MODEL					
XXF, XYF, XXM, XYM	XXF vs XYF XXM vs XYM	XX vs XY	XX > XY or XY > XX	Sex chromosome complement influences the trait	If possible, compare groups that have the same level of gonadal hormones
	XXF vs XXM XYF vs XYM	*Sry*	F > M or F < M	Group differences in trait are probably caused by group differences in gonadal hormones	Group differences in traits could also be caused by group differences in the effects of *Sry* acting outside of the gonads
XY* MODEL					
1XF, 2XF, 1XM, 2XM	1XF vs 1XM 2XF vs 2XM	Y chrom number (1 vs 0)	1XF > 1XM or 1XF < 1XM 2XF > 2XM or 2XF < 2XM	Group differences in trait are caused by the presence/absence of the Y chromosome	If the FCG experiment shows that a sex difference is caused by sex chromosome effects but not by gonadal hormone effects, the finding of a Y chromosome effect here implies that the Y chromosome's effect is not via induction of testicular development, but because of action of Y chromosome genes outside of the gonad
	1XF vs 2XF 1XM vs 2XM	X chrom number (2 vs 1)	1XF > 2XF or 1XF < 2XF 1XM > 2XM or 1XM < 2XM	Group differences in the trait are caused by the number of X chromosomes	If a sex chromosome effect on the trait has been detected using FCG, this outcome indicates that the sex chromosome effect is caused by the number of X chromosomes, at least in part

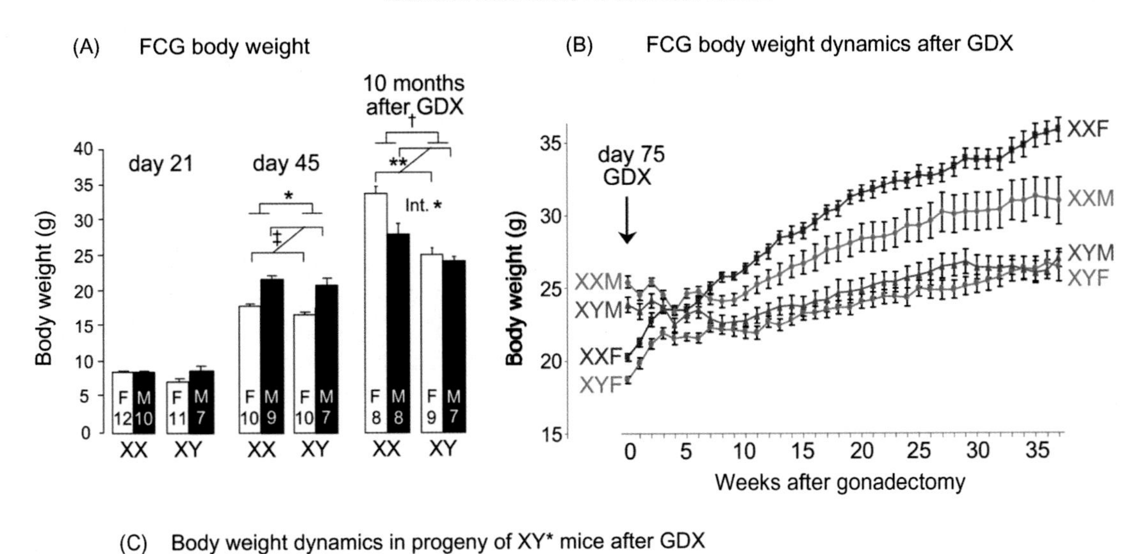

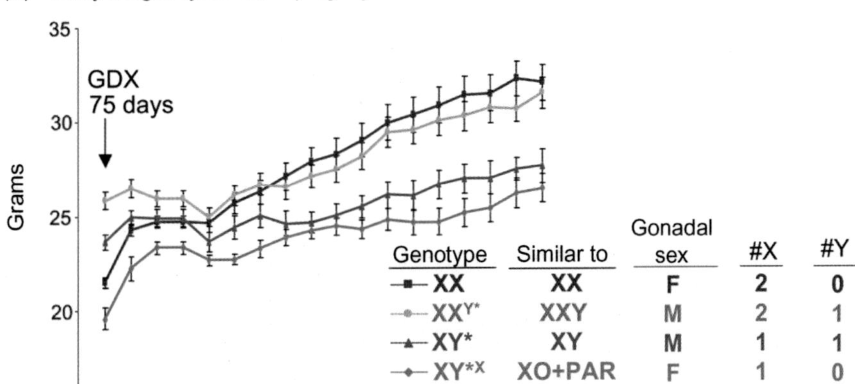

FIGURE 11.3 Sex chromosome and gonadal hormonal control of sex differences in body weight, discovered using the Four Core Genotypes and XY* models. (A) At age 21 days, FCG mice did not differ in body weight. Twenty-four days later (day 45), mice with testes had greater body weight than mice with ovaries (‡$p < 0.000001$), and mice with XX sex chromosomes weighed slightly more than XY (*$p < 0.05$). Ten months after gonadectomy (GDX) at 75 days of age, mice with XX sex chromosomes were much heavier than mice with XY sex chromosomes (†$p < 0.0001$), and females were heavier than males (**$p < 0.01$). There was a significant interaction of the effects of sex chromosome complement and gonadal sex (Int, *$p < 0.05$). (B) Body weight in gonadally intact FCG mice at day 75 (week 0) and after GDX at day 75. At day 75 in gonad-intact mice when all sex-biasing factors were operating, mice with testes were heavier than mice with ovaries, and mice with XX sex chromosomes were heavier than mice with XY sex chromosomes. The sex difference caused by activational effects of gonadal secretions disappeared in the first month after GDX, after which XX mice gained more weight than XY mice. The difference between XX and XY groups was much greater long after gonads were removed, compared to when gonads were present. (C) The sex chromosome effect on body weight is confirmed in the XY* model and found to be caused by the number of X chromosomes. Mice were again weighed at day 75 gonad-intact and then GDX. After GDX, mice with two X chromosomes gained weight more than mice with one X chromosome ($p < 0.000001$). The presence or absence of the Y chromosome had little effect. *Source: From Chen X, McClusky R, Chen J, Beaven SW, Tontonoz P, Arnold AP, et al. The number of X chromosomes causes sex differences in adiposity in mice. PLoS Genet 2012;8:e1002709.*

males, and XYF are similar to WT XX females, suggesting the lack of major differences in gonadal hormones between XX and XY mice of the same sex [24]. Several studies have measured levels of gonadal hormones in adult XX and XY FCG mice of the same sex, but no differences have been reported [25–28]. Moreover, there is evidence against the idea that XX and XY mice of the same gonadal sex have major differences in the levels of androgens prenatally, because a bioassay for prenatal androgens (anogenital distance) shows no difference between XXM versus XYM, or between XXF versus XYF [22]. To date, therefore, the levels of gonadal hormones are found to be roughly similar in XX and XY mice. However, XX and XY adult males are reported to differ in the levels of gonadotrophins [27], and XXF but not XYF mice are fertile in some genetic backgrounds, so they may well not have identical hormone levels. Thus, we do not assume that gonadal hormone levels are identical under all conditions in XX and XY FCG mice of the same sex.

Many studies using FCG mice circumvent the possibility of group differences in the level of gonadal hormones by comparing groups that have been gonadectomized with or without equal treatment with gonadal hormones in adulthood [19,29]. This method eliminates any group differences in circulating (activational) gonadal hormone levels, so that XX versus XY differences in traits cannot be explained by actions of gonadal hormones at the time of measurement of the trait. An illustrative example comes from the study of sex differences in body weight in FCG mice [29] (Fig. 11.3). Adult male mice weigh more than female mice. Gonadectomy of adult mice eliminates the sex difference in body weight, supporting the idea that the sex difference was primarily the result of ongoing effects of testicular and ovarian secretions, which are activational effects that are reversed by removal of the gonads. However, by about 2 months after GDX of FCG mice, the XX mice have gained more weight than XY mice, no matter whether the mice previously had testes or ovaries. The XX > XY difference becomes more extreme at later ages after GDX. Because the XX > XY difference occurs within both sexes (ie, in mice with either testes or ovaries), the sex chromosome effect is unlikely to be the result of a simple difference in gonadal secretions in the XX and XY groups prior to GDX. Moreover, the XX versus XY difference emerges after gonadal hormones have been removed, so they cannot be caused by gonadal secretions after GDX. An advantage of the FCG model is that it tests for sex chromosome effects in two different gonadal hormonal conditions (XXF vs XYF, XXM vs XYM), which helps to establish the generality of the sex chromosome effect. A sex chromosome effect on a trait detected with FCG mice means that the difference is caused either by the presence or absence of the Y chromosome, or the different effects of one versus two X chromosomes. The XY* model helps to distinguish between these two possibilities.

THE XY* MODEL

The XY* model is the progeny of XY* males. The Y* chromosome has a modified PAR, which can recombine abnormally with the X chromosome during spermatogenesis, producing mice with unusual numbers of types of sex chromosomes. The official names for these genotypes are described elsewhere [30–33], but the genotypes are very similar to XX, XO + PAR, XY, and XXY (Figs. 11.2 and 11.3). These groups can also be named by the number of X chromosomes and their gonadal sex: 2XF, 1XF, 1XM, 2XM, respectively. In outbred but not inbred genetic backgrounds, XO mice are also produced. The XO + PAR genotype has one X chromosome plus a small second sex chromosome that contains mostly a PAR (Fig. 11.2). Comparing mice without Y chromosomes (XO + PAR or XX) to those with a Y chromosome (XY or XXY), one can detect whether presence versus absence of the Y chromosome has an effect on the trait of interest (Table 11.1). Comparing mice with one X chromosome (XO + PAR or XY) to those with two X chromosomes (XX, XXY) yields information about the effects of one versus two X chromosomes.

The XY* model has the advantage that it can confirm the sex chromosome effects found using the FCG model. For example, in the body weight example discussed above, GDX XX mice weigh more than GDX XY mice, a result that is confirmed in the XY* model which shows that body weight is greater in mice with two X chromosomes (XX, XXY) than in mice with one X chromosome (XO + PAR, XY), and the presence or absence of the Y chromosome does not have any significant effect. Thus, the XY* model confirms the sex chromosome effect in an entirely different model in mice that are not deleted for the *Sry* region of the Y chromosome, and that do not have any transgenes. The combination of the two models increases confidence in the sex chromosome effect, and localizes it to the X chromosome.

The XY* model is useful by itself for testing X and Y chromosome effects [23,34], without any prior test for sex chromosome effects using the FCG model. There are advantages in testing for sex chromosome effects using both models. One benefit is the value of independent replication of sex chromosome effects in two models. A second advantage is that the FCG model can identify sex differences that are caused by sex chromosome complement, but that are not affected by *Sry* or type of gonad. Under those conditions, finding a Y chromosome effect in the XY* model means that the Y effect is unlikely to be caused by its effects on the gonads, but is likely an effect of Y genes acting outside of the gonad (Table 11.1).

Although the FCG and XY* models represent the easiest and most often used approaches for detecting sex chromosome effects outside of the gonads, other methods are available (discussed by [21,35]). These include SF1 knockout mice, which allows comparison of XX and XY mice that never developed gonads [36–38]. Also useful are Y chromosome consomic strains, which comprise groups of mice with the same genetic background but with different Y chromosomes [39,40].

FINDING THE GENES

If a sex chromosome effect is localized to the X or Y chromosome using XY* mice, the next step is to identify the specific genes that could account for the effect.

Here, the best approach is to compare mice that are genetically identical, except that they have a difference in expression of only one candidate gene, which mimics the inherent XX versus XY difference in expression of that gene. For Y genes, for example, there are two basic approaches. The first is to knock out individual Y genes, comparing XY mice that have or lack a specific gene, which tests if each Y gene is necessary to produce the Y chromosome effect on the trait. The second is to add a transgene encoding an individual Y gene to an XO female mouse (vs no transgene control), to determine if the individual Y gene is sufficient to mimic the effect of a whole Y chromosome on the trait.

To find an X "escapee" gene that causes sex differences because the gene is inherently expressed higher in XX than XY, again there are two approaches. The first is to compare XX mice that are identical except that one group are normal WT females with two expressed copies of the X gene, and XX females in which one of the copies of the genes is knocked out. This experiment tests if two versus one copy of the gene mimics the effects of two versus one copy of the whole X chromosome, detected in XY* mice. The second approach for X escapees is to add a copy of the gene as a transgene to XY males or XO females, to determine if the second copy of that one gene is sufficient to cause the effect of a whole second X chromosome.

To find X genes that cause a sex chromosome effect because of differences in parental imprint, one can compare XO mice that are identical except that they have different parental imprints on X genes [41]. To date, there are no published reports of these kinds of manipulations in mice that identify specific genes that cause sex chromosome-induced sex differences in traits.

THE FUTURE OF SEX CHROMOSOME EFFECTS

Once specific X or Y genes are found that cause sex differences, the sex chromosome effects will need to be integrated with the parallel-acting effects of gonadal hormones [4]. For instance, in the example of body weight discussed here, both the testes and ovaries secrete hormones during adulthood, which have a large effect to cause sex differences in body weight. The effect of X chromosome number can be seen both when gonads are present and when they are absent, but the effect is much larger after gonadectomy. Thus, the two factors appear to reduce the effects of each other [29,42,43]. The mechanisms of this interaction are completely unexplored. For example, do the X chromosome gene(s) regulate the same cellular mechanisms as gonadal hormones, or do they oppose each other by acting on different cellular mechanisms that both impact the emergent phenotype of body weight? That will be a fascinating question to answer in a variety of sexually dimorphic systems that show both hormonal and sex chromosome effects.

References

[1] Kohno S, Parrott BB, Yatsu R, Miyagawa S, Moore BC, Iguchi T, et al. Gonadal differentiation in reptiles exhibiting environmental sex determination. Sex Dev 2014;8(5):208–26.

[2] Innocenti P, Morrow EH, Dowling DK. Experimental evidence supports a sex-specific selective sieve in mitochondrial genome evolution. Science 2011;332(6031):845–8.

[3] Wolff JN, Gemmell NJ. Mitochondria, maternal inheritance, and asymmetric fitness: why males die younger. BioEssays 2012;35(2):93–9.

[4] Arnold AP. The end of gonad-centric sex determination in mammals. Trends Genet 2011;28:55–61.

[5] Lahn BT, Page DC. Four evolutionary strata on the human X chromosome. Science 1999;286:964–7.

[6] Bachtrog D. Y-chromosome evolution: emerging insights into processes of Y-chromosome degeneration. Nat Rev Genet 2013;14(2):113–24.

[7] Graves JAM. Sex chromosome specialization and degeneration in mammals. Cell 2006;124(5):901–14.

[8] Itoh Y, Melamed E, Yang X, Kampf K, Wang S, Yehya N, et al. Dosage compensation is less effective in birds than in mammals. J Biol 2007;6:2.

[9] Berletch JB, Yang F, Disteche CM. Escape from X inactivation in mice and humans. Genome Biol 2010;11(6):213.

[10] Yang F, Babak T, Shendure J, Disteche CM. Global survey of escape from X inactivation by RNA-sequencing in mouse. Genome Res 2010;20(5):614–22.

[11] Zhou J, Sackton TB, Martinsen L, Lemos B, Eickbush TH, Hartl DL. Y chromosome mediates ribosomal DNA silencing and modulates the chromatin state in Drosophila. Proc Natl Acad Sci USA 2012;109(25):9941–6.

[12] Lemos B, Araripe LO, Hartl DL. Polymorphic Y chromosomes harbor cryptic variation with manifold functional consequences. Science 2008;319(5859):91–3.

[13] Wijchers PJ, Yandim C, Panousopoulou E, Ahmad M, Harker N, Saveliev A, et al. Sexual dimorphism in mammalian autosomal gene regulation is determined not only by Sry but by sex chromosome complement as well. Dev Cell 2010;19(3):477–84.

[14] Wijchers PJ, Festenstein RJ. Epigenetic regulation of autosomal gene expression by sex chromosomes. Trends Genet 2011;27:132–40.

[15] Kim Y, Capel B. Balancing the bipotential gonad between alternative organ fates: a new perspective on an old problem. Dev Dyn 2006;235(9):2292–300.

[16] Arnold AP. The organizational-activational hypothesis as the foundation for a unified theory of sexual differentiation of all mammalian tissues. Horm Behav 2009;55(5):570–8.

[17] Phoenix CH, Goy RW, Gerall AA, Young WC. Organizing action of prenatally administered testosterone propionate on the tissues mediating mating behavior in the female guinea pig. Endocrinology 1959;65:369–82.

[18] Lovell-Badge R, Robertson E. XY female mice resulting from a heritable mutation in the primary testis-determining gene, Tdy. Development 1990;109:635–46.

[19] De Vries GJ, Rissman EF, Simerly RB, Yang LY, Scordalakes EM, Auger CJ, et al. A model system for study of sex

chromosome effects on sexually dimorphic neural and behavioral traits. J Neurosci 2002;22(20):9005–14.
[20] Arnold AP, Chen X. What does the "four core genotypes" mouse model tell us about sex differences in the brain and other tissues? Front Neuroendocrinol 2009;30(1):1–9.
[21] Arnold AP. Conceptual frameworks and mouse models for studying sex differences in physiology and disease: why compensation changes the game. Exp Neurol 2014;259:2–9.
[22] Itoh Y, Mackie R, Kampf K, Domadia S, Brown JD, O'Neill R, et al. Four core genotypes mouse model: localization of the Sry transgene and bioassay for testicular hormone levels. BMC Res Notes 2015;8:69.
[23] Cox KH, Bonthuis PJ, Rissman EF. Mouse model systems to study sex chromosome genes and behavior: relevance to humans. Front Neuroendocrinol 2014;35(4):405–19.
[24] De Vries GJ, Simerly RB. Anatomy, development, and function of sexually dimorphic neural circuits in the mammalian brain. In: Pfaff DW, Arnold AP, Etgen AM, Fahrbach SE, Rubin RT, editors. Hormones, brain, and behavior. San Diego: Academic Press; 2002. p. 137–91.
[25] Gatewood JD, Wills A, Shetty S, Xu J, Arnold AP, Burgoyne PS, et al. Sex chromosome complement and gonadal sex influence aggressive and parental behaviors in mice. J Neurosci 2006;26(8):2335–42.
[26] Sasidhar MV, Itoh N, Gold SM, Lawson GW, Voskuhl RR. The XX sex chromosome complement in mice is associated with increased spontaneous lupus compared with XY. Ann Rheum Dis 2012;71:1418–22.
[27] Corre C, Friedel M, Vousden DA, Metcalf A, Spring S, Qiu LR, et al. Separate effects of sex hormones and sex chromosomes on brain structure and function revealed by high-resolution magnetic resonance imaging and spatial navigation assessment of the Four Core Genotype mouse model. Brain Struct Funct 2014; [epub Dec 2]
[28] Holaskova I, Franko J, Goodman RL, Arnold AP, Schafer R. The XX sex chromosome complement is required in male and female mice for enhancement of immunity induced by exposure to 3,4-Dichloropropionanilide. Am J Reprod Immunol Microbiol 2015;74(2):136–47.
[29] Chen X, McClusky R, Chen J, Beaven SW, Tontonoz P, Arnold AP, et al. The number of X chromosomes causes sex differences in adiposity in mice. PLoS Genet 2012;8:e1002709.
[30] Eicher EM, Hale DW, Hunt PA, Lee BK, Tucker PK, King TR, et al. The mouse Y* chromosome involves a complex rearrangement, including interstitial positioning of the pseudoautosomal region. Cytogenet Cell Genet 1991;57:221–30.
[31] Burgoyne PS, Mahadevaiah SK, Perry J, Palmer SJ, Ashworth A. The Y* rearrangement in mice: new insights into a perplexing PAR. Cytogenet Cell Genet 1998;80(1–4):37–40.
[32] Chen X, Watkins R, Delot E, Reliene R, Schiestl RH, Burgoyne PS, et al. Sex difference in neural tube defects in p53-null mice is caused by differences in the complement of X not Y genes. Dev Neurobiol 2008;68(2):265–73.
[33] Chen X, McClusky R, Itoh Y, Reue K, Arnold AP. X and Y chromosome complement influence adiposity and metabolism in mice. Endocrinology 2013;154:1092–104.
[34] Bonthuis PJ, Cox KH, Rissman EF. X-chromosome dosage affects male sexual behavior. Horm Behav 2012;64:565–72.
[35] Arnold AP. Mouse models for evaluating sex chromosome effects that cause sex differences in non-gonadal tissues. J Neuroendocrinol 2009;21(4):377–86.
[36] Budefeld T, Grgurevic N, Tobet SA, Majdic G. Sex differences in brain developing in the presence or absence of gonads. Dev Neurobiol 2008;68(7):981–95.
[37] Grgurevic N, Budefeld T, Spanic T, Tobet SA, Majdic G. Evidence that sex chromosome genes affect sexual differentiation of female sexual behavior. Horm Behav 2012;61(5):719–24.
[38] Majdic G, Tobet S. Cooperation of sex chromosomal genes and endocrine influences for hypothalamic sexual differentiation. Front Neuroendocrinol 2011;32(2):137–45.
[39] Case LK, Wall EH, Dragon JA, Saligrama N, Krementsov DN, Moussawi M, et al. The Y chromosome as a regulatory element shaping immune cell transcriptomes and susceptibility to autoimmune disease. Genome Res 2013;23(9):1474–85.
[40] Spach KM, Blake M, Bunn JY, McElvany B, Noubade R, Blankenhorn EP, et al. Cutting edge: the Y chromosome controls the age-dependent experimental allergic encephalomyelitis sexual dimorphism in SJL/J mice. J Immunol 2009;182(4):1789–93.
[41] Davies W, Isles A, Smith R, Karunadasa D, Burrmann D, Humby T, et al. Xlr3b is a new imprinted candidate for X-linked parent-of-origin effects on cognitive function in mice. Nat Genet 2005;37(6):625–9.
[42] Arnold AP, Chen X, Link JC, Itoh Y, Reue K. Cell-autonomous sex determination outside of the gonad. Dev Dyn 2013;242(4):371–9.
[43] Chen X, Wang L, Loh D, Colwell C, Tache Y, Reue K, et al. Sex differences in diurnal rhythms of food intake in mice caused by gonadal hormones and complement of sex chromosomes. Horm Behav 2015;75:55–63.

CHAPTER

11.2

Organizational Influences of the Gonadal Steroid Hormones: Lessons Learned Through the Hypothalamic-Pituitary-Adrenal (HPA) Axis

Victor Viau and Leyla Innala

Laboratory of Neuroendocrine Function, Department of Cellular and Physiological Sciences, University of British Columbia, Vancouver, BC, Canada

INTRODUCTION

The HPA axis represents one of many defense systems available for meeting the demands of threats to homeostasis or stress. Circadian or stressor-induced activation of the HPA axis stimulates the release of the glucocorticoid steroid hormones from the adrenals, cortisol in humans and corticosterone (CORT) in the rodent. The principal physiological function of glucocorticoids is to promote the mobilization and redistribution of energy stores necessary for the preservation of homeostatic mechanisms and survival. As several types of mood, metabolic, immune, and cardiovascular disorders are associated with sustained activation of the HPA axis and/or increased exposure to CORT in circulation [1], this neuroendocrine axis must be tightly regulated. Sex differences in HPA output responses under challenging situations in humans, as well as in animal models of stress, continue to point to critical roles for the gonadal steroid hormones in adulthood to actively shape the magnitude of the HPA axis response, in addition to immune, autonomic, behavioral, and other neuroendocrine response systems. Thus, sex differences in stress and coping styles could provide the basis for widespread sex disparities in the prevalence, onset, and symptomatology of several types of illnesses.

The original organizational–activational hypothesis of gonadal hormone-dependent sex differences [2] has rightfully inspired an incredible amount of research dedicated to understanding how the steroid milieu during early development can organize or permanently alter the function of central mediators of reproductive, social, and sexual behavior. Since the 1960s, it has also long been known and/or predicted that manipulations of either the adrenal, gonadal, or thyroid systems in newborn rats can irreversibly alter the functioning of each other, in addition to behaviors comprising emotional reactions [3]. Several lines of research on phenotypic, connectional, and functional grounds continue to underscore the existence of multiple networks and mechanistic targets for hormonally mediated sexual differentiation of the brain in the context of reproductive physiology, but with far less attention paid to the neurobiology of stress. Here we will provide evidence from our lab and those of others to indicate that gonadal steroid hormone signaling over the perinatal period is capable of organizing the HPA axis response to stress. Because several disease states in humans are associated with major changes in reproductive endocrine status, we posit that early changes in gonadal hormone release and exposure may also come to shape individual differences in stress responses and predisposition to disease.

Brain regions implicated as gonadal sensitive targets for the development of sexually dimorphic circuits (and those relevant to reproduction) overlap extensively with several cell groups associated with HPA axis control, including within the limbic system (hippocampus, amygdala), cortex, thalamus, and hypothalamus, as well as within several midline forebrain and hypothalamic relays, including the septum and various subnuclei of the bed nucleus of the stria terminalis [4]. That this overlap reflects such a widespread and potential influence of the gonadal hormones to organize the HPA axis warrants some caution. But by all accounts, proper stress responding and adaptation require a huge amount of substrate dedicated to the processing of experiential, sensory, cognitive, and emotional information, in addition to mounting appropriate neuroendocrine, autonomic, and behavioral responses [5]. Moreover, gonadal hormone secretion and signaling are disrupted by homeostatic threat, but can also redirect the strength of influence of stress on the body. Changes in gonadal status that mediate or that are subject to organizational influences, therefore, are no less agents of homeostasis as they are for reproductive fitness.

By and large, the preoptic area has received significant attention, and deservedly so, based on the remarkable potency of the gonadal hormones to organize its form and function with respect to reproductive physiology and behavior. Several provocative studies already underscore that there is a complex of gonadal hormone dependent mechanism guiding the sexual differentiation of the preoptic nuclei, to include genetic and epigenetic regulation of cell death, neurogenesis, differentiation, and migration, and for these to shoulder the organization of cell volume and morphology, and ultimately synaptic strength, communication, and plasticity [6–9]. It is reasonable to assume that some combinations of these mechanisms likely explain the origins of neural sex differences in other parts of the brain, yet to be considered for the organization of other adaptive systems. Furthermore, neuroanatomical and physiological evidence suggests that the preoptic nuclei are poised to register changes in gonadal status to the paraventricular nucleus of the hypothalamus (PVH), the pivotal driver of the HPA axis. In this context, we provide a framework here for understanding how the gonadal hormones might act in the developing brain to reach out to the PVH and its extended circuitries to organize adaptive responses to stress, as well as to provide a template for assessing the organizational influences of sex steroids on other systems of interest.

SEX DIFFERENCES IN THE HPA AXIS

The HPA axis involves the sequential release of a chain of hormones from the hypothalamus and anterior pituitary, ultimately regulating the de novo synthesis and release of the glucocorticoid steroid hormones from the adrenals. The HPA axis is governed by the PVH, which receives multimodal input from several brain regions capable of relaying limbic-related (autonomic, emotional), homeostatic (blood-borne), and somatosensory information [10]. Thus, depending on the nature of the stimulus and whether it represents a threat to homeostasis, either real or perceived, the PVH is positioned to integrate and respond to a variety of challenges. The neuroendocrine arm of the PVH is contained within a specific group of neurons located in the medial parvocellular, dorsal part of the nucleus (Fig. 11.4). As with several other hypothalamic neuroendocrine systems, these neurosecretory neurons secrete peptide stores from the median eminence into the pituitary portal system to stimulate the anterior pituitary, and foremost among these include corticotropin-releasing hormone (CRH) and arginine vasopressin (AVP). In humans and rodents, AVP alone is a weak secretagogue of adrenocorticotropin (ACTH), but synergizes with CRH on the release of ACTH from anterior pituitary corticotropes [11,12], which then stimulates the synthesis and release of glucocorticoids from the adrenal gland. The amplitude of circadian and stress-induced HPA activity is limited, at least in part, by the negative-feedback effects of glucocorticoids, mediated at multiple target sites, including the hippocampus, cortex, hypothalamus, and anterior pituitary [13]. Thus, the magnitude of the HPA axis

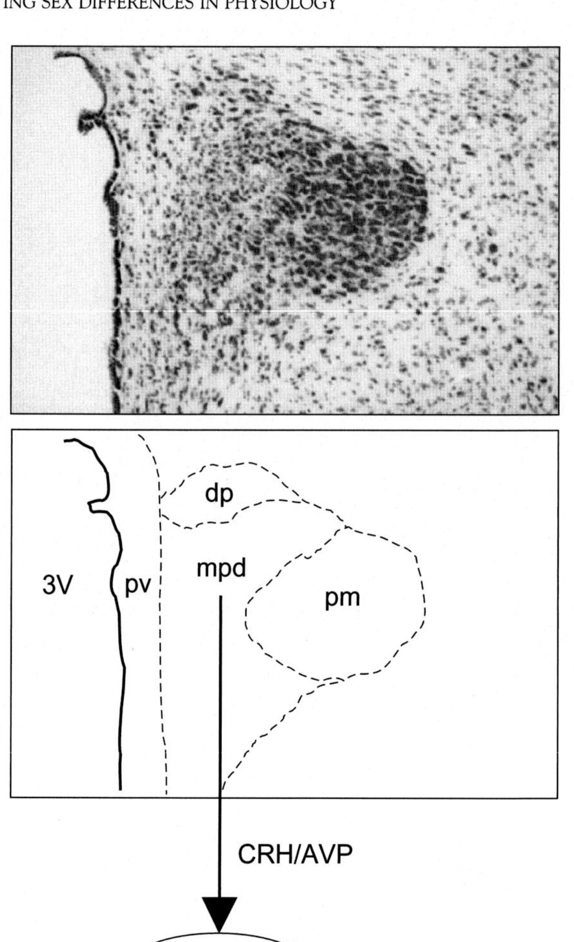

FIGURE 11.4 Cells occupying the medial parvocellular, dorsal (mpd) part of the paraventricular nucleus of the hypothalamus (PVH) represent the neuroendocrine arm of the HPA axis. Corticotropin-releasing hormone (CRH) and arginine vasopressin (AVP) stimulate the synthesis and release of adrenocorticotropin (ACTH) from anterior pituitary corticotrope cells, which in turn stimulates the synthesis and release of corticosterone (CORT) from the adrenals. AVP may also gain access to this system from cells occupying the posterior magnocellular (pm) part of the nucleus. pv, periventricular; dp, dorsal parvocellular; 3V, third ventricles. Further parcellation of different cell types in the PVH is described elsewhere [10].

response represents the net of stimulatory drive and glucocorticoid mediated negative feedback inhibition of PVH motor neuron and/or pituitary responses.

In rodents, females secrete higher levels of CRH than males, and higher levels of CORT in response to various stressors. The gonadal hormones are at least partly responsible for these sex differences, as castrated adult males show increases in ACTH/CORT responses, reversible with testosterone replacement or with dihydrotestosterone, a potent nonaromatizable androgen receptor agonist, whereas ovariectomized adult females show decreases in ACTH/CORT responses, reversible with estradiol replacement [14]. Consistent with a central site of action, the inhibitory effect of testosterone on HPA output in the rat is recapitulated by decreases in AVP, whereas increments in systemic levels of estradiol associate with increases in CRH expression within neuroendocrine PVH neurons directed at the median eminence [15]. In humans without psychiatric illness, the sex difference in stress HPA axis function is not so apparent on the surface [16–19]. Thus, men often show similar if not higher levels of cortisol than women in response to various acute challenges. However, this does not discount an underlying influence for the gonadal hormones to regulate the HPA axis in humans. Estradiol treatment enhances HPA responses to psychosocial stress in men, whereas testosterone replacement during hypogonadism can suppress the adrenal cortisol response to ACTH. Hypothalamic CRH content is higher in women compared to men, and CRH stimulation induces higher cortisol responses in women. As revealed under glucocorticoid synthesis blockade, there is an increased drive to the HPA axis during the follicular phase than during the luteal phase of the menstrual cycle when estrogen secretion is increased and decreased, respectively. Taken together, the findings discussed here underscore an intrinsic influence of the gonadal hormones on the HPA axis in humans, similar to findings in animal studies.

ORGANIZATIONAL–ACTIVATIONAL EFFECTS ON THE HPA AXIS

As recently reviewed [20], there is evidence to suggest that sex differences and gonadal hormone influences in the HPA axis follow both organizational–activational and aromatization hypotheses, as proposed for the sexual differentiation of reproductive physiology and behavior. As described above, and reviewed elsewhere [15], manipulations of the gonadal axis by castration and testosterone replacement indicate that the principal modulatory or activational effect of androgens in adult male rodents is to decrease the magnitude of the HPA axis response, at least under acute stress conditions. In contrast, manipulations of the gonadal hormones over the perinatal period in males, either in the form of castration, androgen receptor antagonism, or by blocking the conversion/aromatization of testosterone to estradiol, are often met by permanent effects on the HPA axis that are not reversed with adult testosterone replacement [21,22]. In these frameworks, early androgen and/or estradiol exposure is responsible for driving a masculine phenotype, whereas the prevention of brain exposure to circulating estradiol in females, as a result of alpha-fetoprotein sequestration in the bloodstream, represents a default mechanism for the expression of a feminine phenotype.

In so far as the terms masculinization and defeminization in males (and by default, feminization in females) are used as conceptual frameworks for describing the neural capacity of sexual behavior, we hesitate to employ these in describing the early roles of the gonadal hormones in shaping other systems of interest, including the HPA axis. As evident in both humans and rodents, males do not always show lower glucocorticoid responses to stress, whereas the relative strength or direction of the sex difference in ACTH and adrenal responses can vary as a function of age, social and reproductive status, as well as in response to different variables of stress (eg, duration, frequency, chronicity, in addition to experiential and predictive factors). Thus, unlike the sexual dimorphic endpoints of reproductive behavior and physiology that may be absent or more or less prevalent in one sex over the other, sex differences in stress HPA axis output and other response systems might not always exist along a functional continuum [23,24] potentially involving different signal transduction pathways and substrate. Thus, while administering estradiol to newborn female rats can decrease CORT responses and CRH expression levels to those resembling adult males [25–28], for example, there remains little conceptual imperative for describing these treatment effects as "masculinizing." Nonetheless, these findings and those of several others provide numerous indications that the gonadal hormones can have enduring influences on intrinsic and extrinsic markers of the HPA axis.

Male rats experience a surge in testosterone that occurs on days 18–19 of gestation [29], and again during the first few hours of birth [30]. Both of these periods are critical, as increases in the magnitude of the HPA response to stress in adulthood follow the administration of the androgen receptor antagonist flutamide restricted to either the prenatal or postnatal period [31–33]. In contrast to the effects of flutamide administered through the perinatal period to increase CRH and AVP mRNA levels in the PVH of adult animals, there appears to be no effect of postnatal castration, with or without testosterone replacement, to alter steady state levels of these transcripts in the PVH [34]. However, there is an enduring influence of postnatal castration to increase the stress-induced activation of PVH neurons, reversible with neonatal, but not with adult testosterone replacement. Thus, there are likely multiple, but functionally distinct periods of influence over which testosterone and its metabolites are capable of organizing and operating on different aspects of the HPA axis [34,35].

Newborn females receiving injections of estradiol or testosterone show reduced stress HPA responses in adulthood [25,31,32]. The effect of large doses of exogenous estradiol in females to overcome alpha-fetoprotein can be interpreted as one that mimics the actions of endogenous estradiol in males. That they respond in a similar manner to estrogen and testosterone provides every indication that newborn males and females possess functional androgen and estrogen receptors in the brain, in addition to aromatase [36–38]. The untoward effects of excess androgen and estradiol exposure, however, do not argue against a normal endogenous role for the gonadal steroids in the developing female brain. Indeed, blocking estradiol synthesis and the estrogen receptor in newborn rodents can disrupt female, as well as male sex behavior [39]. Estradiol and testosterone have been detected in discrete regions of the prenatal and postnatal male and female brain, whose concentrations vary separately and independently between brain regions, as well as from precursors derived from the gonads and adrenals, underscoring a mechanism of central steroidogenesis [40,41]. As far as we can tell, no study to date has systematically explored in female rodents how naturally occurring variations in endogenous steroids (either centrally or peripherally derived), may come to alter sexual differentiation in general, let alone the organization of the HPA axis, other CNS pathways, or other organ systems.

Despite the similar neonatal imprinting of estrogen and testosterone to alter basal and stress-related HPA output in males and females, at least one underlying difference remains. Whereas the effects of neonatal estradiol exposure to decrease ACTH/CORT responses in females may be restored with adult estradiol replacement, most neonatal manipulations in males (ie, castration, androgen receptor, or aromatase blockade) are not reversible with adult testosterone replacement. In this case, the organizational influence in females appears to be secondary to a decrease in estradiol synthesis, rather than a change in neural circuits related to the HPA axis. Whether this continues to hold true in adult females bearing estrogen synthesis blockers or receptor antagonists at birth, and in the context of different types of stress and reproductive states, remains to be seen. The weight of evidence in males is clear enough to suggest that organizational gonadal hormone influences on ACTH/CORT responses are mediated, at least in part, by alterations in gonadal steroid sensitive projections to the HPA axis.

MAPPING GONADAL HORMONE SENSITIVE CIRCUITS IN THE BRAIN

In order to approach the central nervous system bases for the organizational influences of the gonadal hormones in general, there remains precedence for studying sex differences in the size, distribution, and functional connectivity of the gonadal steroid hormone receptors. Importantly, previous findings indicate that neonatal gonadal hormone exposure alters the uptake and metabolism of androgen and estrogen in the adult brain, in addition to the expression and binding capacity of androgen and estrogen receptors [42,43]. Here we describe our efforts in mapping the distribution and connectivity of androgen and estrogen sensitive neurons with respect to the PVH and its extended circuitries, as identical strategies may also be applied to any candidate nucleus and afferents of interest. We then revisit the connectivity of the medial preoptic nucleus to underscore that gonadal hormone sensitive cells in this region are only one synapse removed from the HPA axis, in addition to other mediators of homeostasis.

The Paraventricular Nucleus of the Hypothalamus

Organizational influences on HPA output responses are often matched by alterations in ACTH synthesis and release that follow changes in CRH and AVP expression in the PVH. These findings would suggest a local site of influence, if only because androgen and estrogen receptors are found in the PVH. Based on our connectivity studies, however, androgen receptors are not localized to the hypophysiotropic zone of the PVH where CRH expressing cells are amassed, but almost exclusively to cell groups identified as spinal cord projecting [44]. The distribution of the principal estrogen receptor in the PVH, the estrogen receptor-beta isoform, is as restrictive as the profile of the androgen receptor with respect to preautonomic neurons, as well as to those projecting to various sensory relays in the medulla. These findings underscore a role for the gonadal steroids to regulate directly the long descending arm of the PVH to influence sensory and autonomic function, but to regulate HPA output they must operate indirectly or upstream from the PVH [45].

In pursuit of these candidate mediators of gonadal status to the HPA axis, we have begun to characterize androgen receptor staining within cells identified as projecting to the PVH in adult male rats bearing retrograde tracer injections aimed at the caudal, CRH expressing part of the nucleus [46]. The connectional data predicts that androgens can act on a very large assortment of multimodal inputs to the PVH, including within the midbrain, pons, and medulla to regulate the processing of different types of sensory information, in addition to various midline structures in the forebrain that are in a position to integrate ascending sensory and descending limbic-related information to the HPA axis (Fig. 11.5). Various cell groups in the forebrain showing the largest numbers of cells identified as androgen receptor expressing and projecting to the PVH region, overlap with those already implicated as gonadal steroid sensitive targets for the development of sexually dimorphic circuits. These would include, but are not limited to, the ventromedial hypothalamus, bed nuclei of the stria terminalis and preoptic area, including the anteroventral periventricular and medial preoptic nuclei.

As we attempt to follow the organizational–activational hypothesis as a foundation for the development of individual variations and sex differences in the HPA axis, the large assortment of candidate afferent mediators of gonadal status on the PVH described just above, likely still underestimates the substrates and mechanisms of sex steroid regulation of the stress response. For example, the expression of androgen and estrogen receptors, as well as the morphology, size, connectivity and/or function of the adult hippocampus, cortex, and amygdala are all altered by neonatal estradiol and testosterone exposure. Despite the potent influences of these limbic structures on ACTH/CORT release, none of these project directly to the PVH. Any basal or stress-induced influence of the limbic system on the HPA axis requires an obligate complex of projections to and through various intermediaries [47,48]. Once again, the bed nucleus of the stria terminalis and medial preoptic area stand out; based on their capacities for relaying and distributing information flow between the limbic system and the PVH, and sensitivity to both organizational and activational influences of the gonadal steroids.

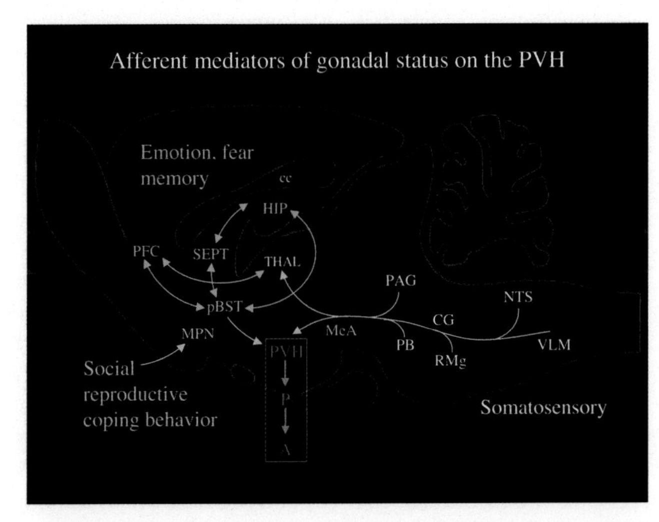

FIGURE 11.5 Schematic summarizing the largest of cell groups identified as projecting to the PVH region and displaying androgen receptor (AR) immunoreactivity. Candidate afferent mediators of gonadal status on PVH function are represented within several forebrain and medial hypothalamic cell groups, posterior division of the bed nucleus of the stria terminalis (pBST), medial preoptic nucleus (MPN), thalamus (THAL), and within various hindbrain sensory and reticular cell groups, including the peracqueductal gray (PAG), parabrachial nucleus (PB), central gray (CG), raphe magnus (RMg), nucleus of the solitary tract (NTS), and ventrolateral medulla (VLM). Brain regions in green represent AR-rich structures providing indirect input to the PVH, including the prefrontal cortex (PFC), lateral septum (LS), hippocampus (HIP), and the medial nucleus of the amygdala (MeA). Additional discussion of gonadal sensitive afferents to the PVH and its extended circuitries is described elsewhere [46].

The Medial Preoptic Nucleus Revisited

To date, no previous study has established directly the relevant substrate and cellular origins by which the gonadal sex steroids may come to permanently alter stress HPA axis function. Beyond representing an important nodal point for integrating sensory and gonadal steroid influences on reproductive physiology and behavior, the medial preoptic area and its intrinsic nuclei, including the anteroventral periventricular nucleus and the cell dense medial preoptic nucleus, the sexually dimorphic nucleus of the preoptic area [49], extensively project to several regions of the forebrain and brainstem to regulate a variety of behaviors and physiological mechanisms related to homeostasis [50–52]. In this regard, it should not be so surprising that emotional or psychogenic stimuli recruit such a multitude of highly interconnected cell groups in the limbic forebrain, including the lateral septum, amygdala, bed nucleus, hypothalamus, hippocampus, and frontal cortex [53–55]. What deserves emphasis here is that the medial preoptic nucleus targets all of these cell groups, clearly placing it among several central homeostatic response systems.

Based on our previous connectivity and functional experiments [46,56,57], the medial preoptic nucleus is in a position to register changes in circulating testosterone, as well as poised to pass this information onto the HPA axis directly to decrease the biosynthetic capacity and stress-induced recruitment of PVH neurons and ACTH/CORT responses. The medial preoptic nucleus also projects to other stress responding nuclei, including the medial amygdala and posterior bed nuclei. Both of these show a high degree of sexual dimorphism, are extremely sensitive to neonatal imprinting and activational effects of testosterone and estradiol, and the strength of their projections to various hypothalamic nuclei renders them no less important in organizational influences on reproductive behavior and physiology [58,59]. However, AVP-containing pathways from the medial amygdala and posterior bed nuclei also target multiple limbic-related structures in the forebrain and hindbrain to include the septum and lateral habenula, the central gray, dorsal raphe nucleus, and locus coeruleus [15,60–63], capable of effecting a broad array of behaviors associated with emotional and coping responses (Fig. 11.6).

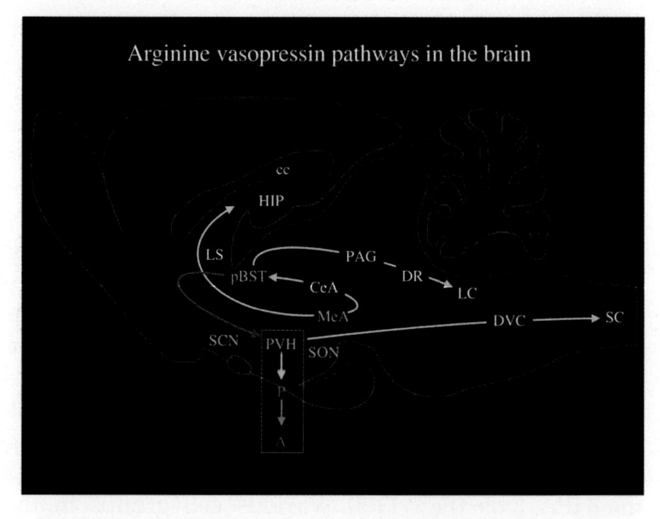

FIGURE 11.6 Schematic summarizing arginine vasopressin (AVP) containing pathways to illustrate the depth to which central AVP could act to coordinate a broad array of neuroendocrine, emotional, and behavioral responses to stress. Cell groups in green identify major sites of AVP synthesis, including the suprachiasmatic (SCN), supraoptic (SON), and paraventricular nuclei (PVH) of the hypothalamus, and the medial amygdala (MeA) and posterior bed nuclei (pBST). Given the dependence of this central AVP system to shifts in gonadal status and connectivity to the HPA axis, we posit that the inhibitory effects of testosterone on PVH neuroendocrine neurons (red arrow-line) are coupled to upstream effects on these extrahypothalamic sources of AVP. Unlike their hypothalamic counterparts, AVP producing cell groups in the MeA and pBST are also extremely and uniquely sensitive to variations in gonadal status in adulthood, as well as during early development. Cell groups targeted by AVP containing pathways (white): CG, central gray; DVC, dorsal vagal complex; Dr, dorsal raphe nucleus; LC, locus coeruleus, LS, lateral septum; SC, spinal cord. Further details of the functional connectivity of these AVP containing cell groups are described elsewhere [15,64].

In adult males, castration causes a testosterone-reversible decrease in AVP expression in the medial amygdala and posterior bed nuclei in adults. Whereas postnatal castration or prenatal flutamide treatment produces the same result to decrease AVP expression in these regions, they are not reversible with adult testosterone replacement [22,34]. This effect of neonatal castration to desensitize AVP responses in the medial amygdala and bed nuclei to testosterone is met by a decrease in androgen receptors in these same regions, as well as by larger HPA output responses to stress. These findings imply that sources of extrahypothalamic AVP, unlike their neuroendocrine partner in the PVH, provide an inhibitory influence on the HPA axis. Thus, we propose that neonatal testosterone organizes the adult neuroendocrine response by priming relevant androgen sensitive, AVP containing pathways to the HPA axis [64,65]. Based on previous connectivity and lesion experiments, both the medial amygdala and bed nuclei of the stria terminalis are in a position to provide inhibitory input to the PVH directly and/or its perinuclear zone [66,67], as well as to modulate several sources of descending limbic inputs, including those from the prefrontal cortex, lateral septum, and ventral subiculum. Our androgen implant and antagonist studies would also implicate a role for androgen receptors in the medial preoptic area to concurrently modulate neuropeptide expression and/or patterns of neuronal activation in the PVH, medial amygdala, and lateral septum in the face of changing levels of testosterone in circulation [68]. Because the medial preoptic nucleus projects extensively and directly to the medial amygdala and bed nuclei of the stria terminalis, it is realistically only one synapse removed from coordinating a vast number of brain regions dedicated to the processing of experiential, sensory, cognitive, and emotional information, taking the organizational influences of the gonadal steroid hormones into the realm of stress, learning, and memory. Indeed, prenatal flutamide treatment has been shown to block stress-induced learning [69], as well as the dependency of adult males on the central vasopressin 1A receptor to express social recognition memories [70]. Moreover, postnatal antagonism of aromatase or the androgen receptor also blocks the capacity of adult males to show stress HPA axis habituation [71], a form of non-associative learning that also requires the activation of central vasopressin 1A receptors [72,73].

STRATEGIES FOR RESEARCH ON GONADAL STEROID ORGANIZATIONAL INFLUENCES ON THE HPA AXIS

There are many ways and means for tackling directly the nature by which sex and individual differences in stress HPA axis function may be explained by variations in androgen and/or estradiol exposure

TABLE 11.2 Strategies for Discerning the Organizational Basis for Gonadal Steroid Influences and Sex Differences in the HPA Axis in Male and Female Rodents Manipulated During the Postnatal Period

ORGANIZATIONAL, INDEPENDENT VARIABLES (POSTNATAL WEEKS 1–3)

Gonadectomy (GDX) ± gonadal steroid replacement:

- GDX
- GDX + testosterone (or estradiol)
- GDX + testosterone (or estradiol) + flutamide
- GDX + testosterone + ATD
- GDX + DHT

Gonadal-intact ± gonadal steroid receptor antagonist

- Flutamide
- ICI182, 780 (ER antagonist)
- ATD
- ATD + flutamide

ACTIVATIONAL, INDEPENDENT VARIABLES (ADULT ANIMALS >60 DAYS OLD)

Steroid replacement (GDX neonates):

- Testosterone
- Estradiol
- DHT

GDX ± steroid replacement (gonadal-intact neonates):

- GDX
- GDX + testosterone (or estradiol)
- GDX + DHT

DEPENDENT VARIABLES (ADULT ANIMALS >60 DAYS OLD)

HPA axis

- Circadian or stress-related (ACTH, CORT)
- Adrenal (steroidogenesis)
- Pituitary (POMC expression)
- PVH (CRH, AVP mRNA, and protein)
- Activation (eg, Fos expression PVH)
- Corticosterone binding globulin
- Glucocorticoid receptor expression

PVH extended circuitries

- Connectivity and strength (tract tracing)
- Phenotype (neurotransmitter, neuropeptide)
- Stress sensitivity (Fos expression)
- Gonadal sensitivity (AR and ER distribution and expression)

HPG axis

- Gonadal status (testosterone, estradiol, progesterone)
- Sexually dimorphic nucleus (morphology)

during development (Table 11.2). Several comprehensive guidelines and experimental approaches for differentiating the organizational influences of different types of gonadal steroids can be found elsewhere [24,74]. To facilitate pursuit of this type of research in the rodent, here we emphasize some additional concepts for entering into longitudinal studies of gonadal-adrenal interactions, including choice, route, and timing of neonatal drug delivery, control and monitoring of maternal behavior and nutritional load, as well as experimental designs for distinguishing organizational and activational effects.

Timing of Gonadal Steroid Hormone and Drug Delivery

As discussed above, there are likely multiple and functionally distinct periods of influence over which the gonadal steroids are capable of operating on and organizing the HPA axis. If the size and cytoarchitecture of the medial preoptic nucleus provide any indication of the period for the sexual differentiation of the rodent brain, the onset and termination of this gonadal steroid-sensitive period would appear to occur anywhere between day 18 of gestation and postnatal day 5, where hormone modulatory effects abruptly end [75,76]. However, the ontogeny of projections from the preoptic area, including those from the anteroventral periventricular nucleus would suggest that these pathways continue to develop through the first and third postnatal weeks depending on the target innervated [77]. Moreover, the effects of neonatal testosterone to organize extrahypothalamic AVP would also appear to extend into the second postnatal week [78]. Taken together, these findings provide some welcome relief in so far as manipulating the gonadal steroid milieu during the postnatal period in the rodent that need not be so restricted to the first few days of life. As discussed above, several previous studies make clear that aromatase or androgen receptor blockade during the prenatal period has demonstrable effects on the HPA axis, at least in males. However, the extent to which these results speak to direct effects on the developing fetal brain versus maternal physiology is equivocal. Based on the weight of evidence to suggest important influences of the gonadal steroids on the HPA axis during the postnatal period, manipulations over this interval remain an effective and practical starting point.

Permanent versus Temporary Hormone Removal

Another consideration is whether to gonadectomize newborn male and female rodents (± relevant types of steroid replacement) or to superimpose different regimens of gonadal steroid receptor agonists/antagonists in gonadal-intact animals (depending on the HPA axis element or system of interest). Both of these general strategies are valid and allow for different types of questions to be asked. For example, removing the testes/ovaries during different postnatal intervals provides a means for discerning distinct periods of steroid influence. Pharmacological approaches can likewise provide the same information, albeit with one important caveat. For example, flutamide treatment in the neonatal male rodent designed to test the endogenous role for the androgen receptor, is considered a transient form of castration. Unlike neonatal gonadectomy, therefore, this allows the gonadal hormones free to roam later in life during adulthood and puberty, the latter representing an additional period for the sexual differentiation of the brain [79] and HPA axis [80–82]. Thus, whereas neonatal gonadectomy and sex steroid replacement studies continue to inform and are more flexible in terms of testing steroid receptor antagonists against a background of different types of gonadal hormone replacement, any effect observed in this case may not be so easily interpreted as one that reflects an organizational influence restricted to the postnatal period.

Soluble esterified steroids (eg, estradiol benzoate, testosterone propionate) with half-lives ~24 h provide some advantage in terms of limiting the number of drug administrations. However, failure to see reliable organizational changes in this scenario may be no less a product of neonatal stress and disruptions to maternal care, or as indicated above, potential requirements for gonadal-adrenal interactions that extend beyond the first week of life. To offset these potential confounds, subcutaneous Silastic implants containing gonadal steroids or other agents of interest can be surgically introduced during the postnatal period. We have employed this delivery method in our previous studies, whereby implant surgeries can be performed shortly after birth, thereafter litters remain intact until capsules are removed on day 21 day of weaning [71].

Controlling for Confounding Effects on Maternal Care

While this route of drug delivery effectively limits nest disturbance, assessments of mother–pup interactions must still be considered, as variations in maternal care have a pronounced influence to alter neuroendocrine responses to stress, at least in male offspring [83]. Several additional controls are also required to decrease between-litter effects in maternal care and nutritional load, including limiting the number of experimental animals per litter, minimizing the number of cage cleanings, and employing litters of equal size and sex ratios. In addition to using animals bearing blank or vehicle-filled capsules at birth, subsets of unhandled, nonexperimental animals may also serve as supplementary controls for surgery and handling effects. Finally, each study should use only one experimental animal per-litter for any given test as a single measure, to empirically control for within-litter effects [84].

Assessing Adult Hormone Concentrations

As described above, tests of steroid removal, hormone replacement, metabolic blockade, or steroid receptor antagonism best reveal the underlying nature by which the gonadal steroids activate sex differences in HPA axis function. By the same token, assessing the protracted effects of the neonatal gonadal hormones to alter HPA output in adults cannot stand alone without measuring circulating gonadal steroids, testing equivalent levels of adult gonadal hormone replacement between treatment groups, and/or comparing neonatal treated animals against those manipulated as adults. As an example, whereas we previously found an enduring influence of neonatal androgen receptor antagonism or aromatase blockade to abolish the capacity of adult male rats to show repeated restraint stress-induced declines in corticosterone responses, we also showed that this process of stress habituation remains intact in animals castrated as adults [71]. Thus, without this comparator we could not conclude that this change in the HPA axis was hormonally organized. Nevertheless, the nature by which these neonatal treatments reflect a change in brain circuits mediating stress habitation and/or those differentially activated by the gonadal steroid hormones remains to be seen.

Sex Differences in Response to Hormone Removal and Replacement

While there is evidence to indicate that neonatal estrogen or androgen treatment can organize the HPA axis response in female rodents, only one study to date has compared the effects of neonatal ovariectomy with or without adult gonadal steroid replacement to alter HPA axis function [25]. In this study, there was only a very subtle effect of neonatal ovariectomy to decrease the normal stimulatory effect of adult estrogen/progesterone replacement on HPA output responses to acute restraint stress exposure. As argued elsewhere [24], the neonatal gonadal steroids may exert similar influences on HPA-related endpoints in adult males and females that may be met by different neural underpinnings. Thus, whereas neonatal ovariectomy appears to have only a minor influence on the HPA axis, this should still not dissuade pursuit of possible changes in intrinsic markers of HPA axis function (eg, CRH expression, adrenal steroidogenesis) or those associated with changes in the activity of structures supplying stress-related input to the PVH (eg, afferent connectivity, steroid receptor expression, Fos responses).

Finally, marked cyclic variations in stress neuroendocrine responses in adult females likely deter studies employing both male and female subjects. Changes in vaginal cytology can be used as a surrogate marker of estrous cycle phase that requires 4–5 groups and large subsets of animals. Alternatively, assessing variations in plasma estradiol concentrations can be used for detecting possible treatment-related shifts in HPA function attributed to individual changes in ovarian status, requiring relatively smaller numbers of test animals. It is also important to note that males also show dynamic variations in testosterone under basal conditions, as well as in response to stress. Thus, changes in gonadal status is no less an important dependent variable of interest in males as it is for females.

CONCLUSIONS AND FUTURE DIRECTIONS

Determining where and how neonatal gonadal steroids may come to alter the expression of their cognate receptors in adults remains an important area of investigation. Possible targets and mechanisms of organizational influences involve changes in steroid metabolism that redirect cellular responses to different types of soluble steroid and membrane-bound receptors, in addition to transcription factors and receptor co-repressors/co-activators to alter steroid receptor signaling, transactivation, and specificity of gene activation [85]. The extent to which the neonatal hormones contribute to changes in the density of afferent projections to the PVH, in addition to their containment of gonadal steroid receptors, also remains worthy of pursuit [9,86–89]. The delayed ontogeny observed in the HPA axis, including changes in the density of relevant projections to the PVH, adds further to the possibility that the gonadal steroids act during different periods of development to operate on distinct elements of physiological functions.

There is a lack of studies that explore directly the endogenous roles for the neonatal gonadal hormones in females. Nonetheless, several studies continue to underscore the pervasive intimacy by which the gonadal and adrenal systems interact in both males and females on neuroendocrine responses during puberty, adulthood, and aging, as well as during disease. Thus, we predict that early changes in gonadal hormone signaling are critical for any perinatal environmental event, specifically to alter stress HPA axis function, but likely to affect other systems. When designing experiments to study sex differences, it is vital to recognize the enduring impact of the gonadal steroid hormones during the perinatal period on physiological function during adulthood. This chapter outlined methods with which to manipulate the gonadal steroid hormones during development so as to assess the impact of these organizational influences on physiology and disease.

Acknowledgments

Support for some of the work presented came from the Canadian Institutes of Health Research grants MOP-136856 and MOP-136840 and the Natural Sciences and Engineering Research Council of Canada grant RGPIN-2014-05714 (VV).

References

[1] Viau V. Functional cross-talk between the hypothalamic-pituitary-gonadal and -adrenal axes. J Neuroendocrinol 2002; 14:506–13.
[2] Phoenix CH, Goy RW, Gerall AA, Young WC. Organizing action of prenatally administered testosterone propionate on the tissues mediating mating behavior in the female guinea pig. Endocrinology 1959;65:369–82.
[3] Levine S, Mullins Jr RF. Hormonal influences on brain organization in infant rats. Science (New York, NY) 1966;152:1585–92.
[4] Simerly RB. Wired for reproduction: organization and development of sexually dimorphic circuits in the mammalian forebrain. Annu Rev Neurosci 2002;25:507–36.
[5] Radley J, Morilak D, Viau V, Campeau S. Chronic stress and brain plasticity: mechanisms underlying adaptive and maladaptive changes and implications for stress-related CNS disorders. Neurosci Biobehav Rev 2015;58:79–91.
[6] McCarthy MM, Arnold AP. Reframing sexual differentiation of the brain. Nat Neurosci 2011;14:677–83.
[7] Lenz KM, Nugent BM, McCarthy MM. Sexual differentiation of the rodent brain: dogma and beyond. Front Neurosci 2012;6:1–13.
[8] Nugent BM, Wright CL, Shetty AC, Hodes GE, Lenz KM, Mahurkar A, et al. Brain feminization requires active repression of masculinization via DNA methylation. Nat Neurosci 2015;18:690–7.
[9] McCarthy MM, Pickett LA, VanRyzin JW, Kight KE. Surprising origins of sex differences in the brain. Horm Behav 2015; 76:3–10.
[10] Viau V, Sawchenko PE. Hypophysiotropic neurons of the paraventricular nucleus respond in spatially, temporally, and phenotypically differentiated manners to acute vs. repeated restraint stress: rapid publication. J Comp Neurol 2002; 445:293–307.
[11] DeBold CR, Sheldon WR, DeCherney GS, Jackson RV, Alexander AN, Vale W, et al. Arginine vasopressin potentiates adrenocorticotropin release induced by ovine corticotropin-releasing factor. J Clin Invest 1984;73:533–8.
[12] Antoni FA. Hypothalamic control of adrenocorticotropin secretion: advances since the discovery of 41-residue corticotropin-releasing factor. Endocr Rev 1986;7:351–78.
[13] Derijk RH, de Kloet ER. Corticosteroid receptor polymorphisms: determinants of vulnerability and resilience. Eur J Pharmacol 2008;583:303–11.
[14] Goel N, Workman JL, Lee TT, Innala L, Viau V. Sex differences in the HPA axis. Compr Physiol 2014;4:1121–55.
[15] Williamson M, Bingham B, Viau V. Central organization of androgen-sensitive pathways to the hypothalamic-pituitary-adrenal axis: implications for individual differences in responses to homeostatic threat and predisposition to disease. Prog Neuropsychopharmacol Biol Psychiatry 2005;29:1239–48.
[16] Toufexis D, Rivarola MA, Lara H, Viau V. Stress and the reproductive axis. J Neuroendocrinol 2014;26:573–86.
[17] Young EA. Sex differences and the HPA axis: implications for psychiatric disease. J Gend Specif Med 1998;1:21–7.
[18] Young EA, Altemus M. Puberty, ovarian steroids, and stress. Ann N Y Acad Sci 2004;1021:124–33.

[19] Young EA, Korszun A. The hypothalamic-pituitary-gonadal axis in mood disorders. Endocrinol Metab Clin North Am 2002;31:63–78.
[20] Walker C-D, McCormick CM. Development of the stress axis: maternal and environmental influences. In: Pfaff DW, Arnold AP, Etgen AM, Fahrbach SE, Rubin RT, editors. Hormones, brain, and behavior. 2nd ed., San Diego: Academic Press; 2009. p. 1931–74.
[21] McCormick CM, Furey BF, Child M, Sawyer MJ, Donohue SM. Neonatal sex hormones have "organizational" effects on the hypothalamic-pituitary-adrenal axis of male rats. Brain Res Develop Brain Res 1998;105:295–307.
[22] McCormick CM, Mahoney E. Persistent effects of prenatal, neonatal, or adult treatment with flutamide on the hypothalamic-pituitary-adrenal stress response of adult male rats. Horm Behav 1999;35:90–101.
[23] Viau V, Bingham B, Davis J, Lee P, Wong M. Gender and puberty interact on the stress-induced activation of parvocellular neurosecretory neurons and corticotropin-releasing hormone messenger ribonucleic acid expression in the rat. Endocrinology 2005;146:137–46.
[24] McCarthy MM, Arnold AP, Ball GF, Blaustein JD, De Vries GJ. Sex differences in the brain: the not so inconvenient truth. J Neurosci 2012;32:2241–7.
[25] McCormick CM. Effect of neonatal ovariectomy and estradiol treatment on corticosterone release in response to stress in the adult female rat. Stress (Amsterdam, Netherlands) 2011;14:82–7.
[26] Patchev VK, Hayashi S, Orikasa C, Almeida OF. Implications of estrogen-dependent brain organization for gender differences in hypothalamo-pituitary-adrenal regulation. FASEB J 1995;9:419–23.
[27] Patchev VK, Hayashi S, Orikasa C, Almeida OF. Ontogeny of gender-specific responsiveness to stress and glucocorticoids in the rat and its determination by the neonatal gonadal steroid environment. Stress (Amsterdam, Netherlands) 1999;3:41–54.
[28] Patchev AV, Gotz F, Rohde W. Differential role of estrogen receptor isoforms in sex-specific brain organization. FASEB J 2004;18:1568–70.
[29] Weisz J, Ward IL. Plasma testosterone and progesterone titers of pregnant rats, their male and female fetuses, and neonatal offspring. Endocrinology 1980;106:306–16.
[30] Baum MJ, Brand T, Ooms M, Vreeburg JT, Slob AK. Immediate postnatal rise in whole body androgen content in male rats: correlation with increased testicular content and reduced body clearance of testosterone. Biol Psychol 1988;38:980–6.
[31] Seale JV, Wood SA, Atkinson HC, Lightman SL, Harbuz MS. Organizational role for testosterone and estrogen on adult hypothalamic-pituitary-adrenal axis activity in the male rat. Endocrinology 2005;146:1973–82.
[32] Seale JV, Wood SA, Atkinson HC, Harbuz MS, Lightman SL. Postnatal masculinization alters the HPA axis phenotype in the adult female rat. J Physiol 2005;563:265–74.
[33] Evuarherhe O, Leggett JD, Waite EJ, Kershaw YM, Atkinson HC, Lightman SL. Organizational role for pubertal androgens on adult hypothalamic-pituitary-adrenal sensitivity to testosterone in the male rat. J Physiol 2009;587:2977–85.
[34] Bingham B, Viau V. Neonatal gonadectomy and adult testosterone replacement suggest an involvement of limbic arginine vasopressin and androgen receptors in the organization of the hypothalamic-pituitary-adrenal axis. Endocrinology 2008;149:3581–91.
[35] Bingham B, Wang NX, Innala L, Viau V. Postnatal aromatase blockade increases c-fos mRNA responses to acute restraint stress in adult male rats. Endocrinology 2012;153:1603–8.
[36] MacLusky NJ, Lieberburg I, McEwen BS. The development of estrogen receptor systems in the rat brain: perinatal development. Brain Res 1979;178:129–42.
[37] Lieberburg I, MacLusky N, McEwen BS. Androgen receptors in the perinatal rat brain. Brain Res 1980;196:125–38.
[38] Shinoda K, Nagano M, Osawa Y. Neuronal aromatase expression in preoptic, strial, and amygdaloid regions during late prenatal and early postnatal development in the rat. J Comp Neurol 1994;343:113–29.
[39] Lenz KM, McCarthy MM. Organized for sex—steroid hormones and the developing hypothalamus. Eur J Neurosci 2010;32:2096–104.
[40] Amateau SK, Alt JJ, Stamps CL, McCarthy MM. Brain estradiol content in newborn rats: sex differences, regional heterogeneity, and possible de novo synthesis by the female telencephalon. Endocrinology 2004;145:2906–17.
[41] Konkle AT, McCarthy MM. Developmental time course of estradiol, testosterone, and dihydrotestosterone levels in discrete regions of male and female rat brain. Endocrinology 2011;152:223–35.
[42] McEwen BS, Pfaff DW. Factors influencing sex hormone uptake by rat brain regions. I. Effects of neonatal treatment, hypophysectomy, and competing steroid on estradiol uptake. Brain Res 1970;21:1–16.
[43] McEwen BS, Pfaff DW, Zigmond RE. Factors influencing sex hormone uptake by rat brain regions. II. Effects of neonatal treatment and hypophysectomy on testosterone uptake. Brain Res 1970;21:17–28.
[44] Bingham B, Williamson M, Viau V. Androgen and estrogen receptor-beta distribution within spinal-projecting and neurosecretory neurons in the paraventricular nucleus of the male rat. J Comp Neurol 2006;499:911–23.
[45] Lund TD, Hinds LR, Handa RJ. The androgen 5alpha-dihydrotestosterone and its metabolite 5alpha-androstan-3beta, 17beta-diol inhibit the hypothalamo-pituitary-adrenal response to stress by acting through estrogen receptor beta-expressing neurons in the hypothalamus. J Neurosci 2006;26:1448–56.
[46] Williamson M, Viau V. Androgen receptor expressing neurons that project to the paraventricular nucleus of the hypothalamus in the male rat. J Comp Neurol 2007;503:717–40.
[47] Herman JP, Cullinan WE. Neurocircuitry of stress: central control of the hypothalamo-pituitary-adrenocortical axis. Trends Neurosci 1997;20:78–84.
[48] Herman JP, Figueiredo H, Mueller NK, Ulrich-Lai Y, Ostrander MM, Choi DC, et al. Central mechanisms of stress integration: hierarchical circuitry controlling hypothalamo-pituitary-adrenocortical responsiveness. Front Neuroendocrinol 2003;24:151–80.
[49] Gorski RA, Harlan RE, Jacobson CD, Shryne JE, Southam AM. Evidence for the existence of a sexually dimorphic nucleus in the preoptic area of the rat. J Comp Neurol 1980;193:529–39.
[50] Simerly RB, Swanson LW, Gorski RA. Demonstration of a sexual dimorphism in the distribution of serotonin-immunoreactive fibers in the medial preoptic nucleus of the rat. J Comp Neurol 1984;225:151–66.
[51] Simerly RB, Swanson LW. The organization of neural inputs to the medial preoptic nucleus of the rat. J Comp Neurol 1986;246:312–42.
[52] Simerly RB, Swanson LW. Projections of the medial preoptic nucleus: a Phaseolus vulgaris leucoagglutinin anterograde tract-tracing study in the rat. J Comp Neurol 1988;270:209–42.
[53] Cullinan WE, Helmreich DL, Watson SJ. Fos expression in forebrain afferents to the hypothalamic paraventricular nucleus following swim stress. J Comp Neurol 1996;368:88–99.

[54] Dayas CV, Buller KM, Crane JW, Xu Y, Day TA. Stressor categorization: acute physical and psychological stressors elicit distinctive recruitment patterns in the amygdala and in medullary noradrenergic cell groups. Eur J Neurosci 2001;14:1143–52.
[55] Radley JJ, Gosselink KL, Sawchenko PE. A discrete GABAergic relay mediates medial prefrontal cortical inhibition of the neuroendocrine stress response. J Neurosci 2009;29:7330–40.
[56] Viau V, Meaney MJ. The inhibitory effect of testosterone on hypothalamic-pituitary-adrenal responses to stress is mediated by the medial preoptic area. J Neurosci 1996;16:1866–76.
[57] Williamson M, Viau V. Selective contributions of the medial preoptic nucleus to testosterone-dependent regulation of the paraventricular nucleus of the hypothalamus and the HPA axis. Am J Physiol Regul Integr Comp Physiol 2008;295:R1020–1030.
[58] de Vries GJ, Miller MA. Anatomy and function of extrahypothalamic vasopressin systems in the brain. Prog Brain Res 1998;119:3–20.
[59] de Vries GJ. Sex differences in vasopressin and oxytocin innervation of the brain. Prog Brain Res 2008;170:17–27.
[60] Buijs RM, Van Eden CG. The integration of stress by the hypothalamus, amygdala and prefrontal cortex: balance between the autonomic nervous system and the neuroendocrine system. Prog Brain Res 2000;126:117–32.
[61] Kalsbeek A, Palm IF, Buijs RM. Central vasopressin systems and steroid hormones. Prog Brain Res 2002;139:57–73.
[62] Liebsch G, Wotjak CT, Landgraf R, Engelmann M. Septal vasopressin modulates anxiety-related behaviour in rats. Neurosci Lett 1996;217:101–4.
[63] Landgraf R, Neumann ID. Vasopressin and oxytocin release within the brain: a dynamic concept of multiple and variable modes of neuropeptide communication. Front Neuroendocrinol 2004;25:150–76.
[64] Ring RH. The central vasopressinergic system: examining the opportunities for psychiatric drug development. Curr Pharm Des 2005;11:205–25.
[65] Gray M, Bingham B, Viau V. A comparison of two repeated restraint stress paradigms on hypothalamic-pituitary-adrenal axis habituation, gonadal status and central neuropeptide expression in adult male rats. J Neuroendocrinol 2010; 22:92–101.
[66] Dayas CV, Buller KM, Day TA. Neuroendocrine responses to an emotional stressor: evidence for involvement of the medial but not the central amygdala. Eur J Neurosci 1999;11:2312–22.
[67] Choi DC, Furay AR, Evanson NK, Ostrander MM, Ulrich-Lai YM, Herman JP. Bed nucleus of the stria terminalis subregions differentially regulate hypothalamic-pituitary-adrenal axis activity: implications for the integration of limbic inputs. J Neurosci 2007;27:2025–34.
[68] Williamson M, Bingham B, Gray M, Innala L, Viau V. The medial preoptic nucleus integrates the central influences of testosterone on the paraventricular nucleus of the hypothalamus and its extended circuitries. J Neurosci 2010;30:11762–70.
[69] Shors TJ, Miesegaes G. Testosterone in utero and at birth dictates how stressful experience will affect learning in adulthood. Proc Natl Acad Sci USA 2002;99:13955–60.
[70] Axelson JF, Smith M, Duarte M. Prenatal flutamide treatment eliminates the adult male rat's dependency upon vasopressin when forming social-olfactory memories. Horm Behav 1999;36:109–18.
[71] Bingham B, Gray M, Sun T, Viau V. Postnatal blockade of androgen receptors or aromatase impair the expression of stress hypothalamic-pituitary-adrenal axis habituation in adult male rats. Psychoneuroendocrinology 2011;36:249–57.
[72] Gray M, Innala L, Viau V. Central vasopressin V1A receptor blockade impedes hypothalamic-pituitary-adrenal habituation to repeated restraint stress exposure in adult male rats. Neuropsychopharmacology 2012;37:2712–19.
[73] Gray M, Innala L, Viau V. Central vasopressin V1A receptor blockade alters patterns of cellular activation and prevents glucocorticoid habituation to repeated restraint stress exposure. Int J Neuropsychopharmacol 2014;17:2005–15.
[74] Becker JB, Arnold AP, Berkley KJ, Blaustein JD, Eckel LA, Hampson E, et al. Strategies and methods for research on sex differences in brain and behavior. Endocrinology 2005;146:1650–73.
[75] Rhees RW, Shryne JE, Gorski RA. Onset of the hormone-sensitive perinatal period for sexual differentiation of the sexually dimorphic nucleus of the preoptic area in female rats. J Neurobiol 1990;21:781–6.
[76] Rhees RW, Shryne JE, Gorski RA. Termination of the hormone-sensitive period for differentiation of the sexually dimorphic nucleus of the preoptic area in male and female rats. Brain Res Dev Brain Res 1990;52:17–23.
[77] Polston EK, Simerly RB. Ontogeny of the projections from the anteroventral periventricular nucleus of the hypothalamus in the female rat. J Comp Neurol 2006;495:122–32.
[78] De Vries G. Sex differences in neurotransmitters systems; vasopressin as an example handbook of neurochemistry and molecular neurobiology. Springer; 2007. p. 487–512.
[79] Schulz KM, Molenda-Figueira HA, Sisk CL. Back to the future: the organizational-activational hypothesis adapted to puberty and adolescence. Horm Behav 2009;55:597–604.
[80] Gomez F, Manalo S, Dallman MF. Androgen-sensitive changes in regulation of restraint-induced adrenocorticotropin secretion between early and late puberty in male rats. Endocrinology 2004;145:59–70.
[81] McCormick CM, Mathews IZ. HPA function in adolescence: role of sex hormones in its regulation and the enduring consequences of exposure to stressors. Pharmacol Biochem Behav 2007;86:220–33.
[82] McCormick CM, Mathews IZ. Adolescent development, hypothalamic-pituitary-adrenal function, and programming of adult learning and memory. Prog Neuropsychopharmacol Biol Psychiatry 2010;34:756–65.
[83] Meaney MJ. Maternal care, gene expression, and the transmission of individual differences in stress reactivity across generations. Annu Rev Neurosci 2001;24:1161–92.
[84] Abbey H, Howard E. Statistical procedure in developmental studies on species with multiple offspring. Dev Psychobiol 1973;6:329–35.
[85] Nugent BM, Schwarz JM, McCarthy MM. Hormonally mediated epigenetic changes to steroid receptors in the developing brain: implications for sexual differentiation. Horm Behav 2011;59:338–44.
[86] Gu G, Cornea A, Simerly RB. Sexual differentiation of projections from the principal nucleus of the bed nuclei of the stria terminalis. J Comp Neurol 2003;460:542–62.
[87] Gu GB, Simerly RB. Projections of the sexually dimorphic anteroventral periventricular nucleus in the female rat. J Comp Neurol 1997;384:142–64.
[88] Hutton LA, Gu G, Simerly RB. Development of a sexually dimorphic projection from the bed nuclei of the stria terminalis to the anteroventral periventricular nucleus in the rat. J Neurosci 1998;18:3003–13.
[89] Ibanez MA, Gu G, Simerly RB. Target-dependent sexual differentiation of a limbic-hypothalamic neural pathway. J Neurosci 2001;21:5652–9.

CHAPTER

11.3

Strategies and Approaches for the Study of Activational Influences of Gonadal Hormones on Sex Differences in Physiology and Behavior

Jill B. Becker

Patricia Y. Gurin Collegiate Professor of Psychology, Molecular & Behavioral Neuroscience Institute, University of Michigan, Ann Arbor, MI, United States

Many sex differences, but not all (see Tables 11.2, 11.4 and 11.6), will be influenced by circulating hormones from the ovaries and/or testes. Effects of gonadal hormones that directly influence a particular trait in the postpubertal animal are referred to as *activational influences*. If it has been determined that there are sex differences in a particular trait in gonad-intact males and females, and one wants to investigate the role of gonadal hormones in the trait, there are a number of approaches one can use. These approaches will be described in this section.

THE GONADAL HORMONES

The ovaries and the testes are the two primary sources of activational influences of hormones on sex differences. The ovaries tend to get most of the attention when it comes to discussions of sex differences, but testicular influences on peripheral systems and the brain can also have significant effects that may result in sex differences and should not be disregarded [1–5].

As described in Table 11.3, the primary active hormone secreted by the testes is testosterone. Testosterone can act directly on androgen receptors (AR) in the target cells, or be converted to one of its active metabolites, estradiol or dihydrotestosterone (DHT). The enzymes to metabolize testosterone to these active metabolites are found primarily within the target cells. For example, in the brain, aromatase converts testosterone to estradiol, which then acts on estradiol receptors (ER). In the skin of the scrotum and in muscle and other sites, 5α-reductase converts testosterone to DHT which is more potent at AR than is testosterone [6].

The female reproductive cycle varies by species in the length of the various phases of ovarian function, whether there is menstruation, and whether there is a spontaneous luteal phase in the cycle (Table 11.3). In humans and some other primates, the reproductive cycle is referred to as a menstrual cycle because the transition from one cycle to the next is marked by the sloughing off of the lining of the uterus when estrogen and progesterone decline at the end of the luteal phase. In other species, the period around ovulation is marked by the female being reproductively receptive to the male, or in estrus.

The human menstrual cycle and the estrous cycle of the rat, mouse, or hamster are not completely analogous (for a more complete discussion, see Ref. [7]). The human menstrual cycle consists of three phases: follicular, periovulatory, and luteal. During the 10–12-day follicular phase, estradiol is secreted from the ovary as the follicle develops, with concentrations of estradiol increasing throughout the phase. The periovulatory phase is 2–4 days long, and consists of a rapid increase in estradiol that then induces a surge of luteinizing hormone (LH) that stimulates the follicle to ovulate. The remnant of the follicle is called the corpus luteum and it releases the hormones of the luteal phase that lasts 10–12 days. During the luteal phase relatively high concentrations of both estradiol and progesterone are released to prepare the uterus for implantation of a fertilized embryo. Menstruation occurs at the conclusion of the luteal phase with the decrease in estradiol and progesterone, and overlaps with the beginning of the follicular phase (Table 11.3).

Small rodents such as the female rat, mouse, or hamster have an estrous cycle that is highly efficient—as soon as ovulation occurs the cycle begins again. This is because the corpus luteum does not become active unless mating occurs and so they don't have a spontaneous luteal phase. The estrous cycle consists of a 3–4-day follicular phase (the first day is called metestrus followed by diestrus). This is rapidly followed by the periovulatory period (proestrus) when estradiol peaks and followed by the release of LH to induce ovulation and an increase in progesterone. Behavioral receptivity occurs coincident with

TABLE 11.3 Gonadal Hormones in Males and Females: The Active Metabolites and Receptors

		Females			
		Menstrual cycle: Estrous cycle of rats, mice, and hamsters	Follicular Meta-estrus/ diestrus	Periovulatory Proestrus/estrus	Luteal No luteal phase
	Males				
Hormones produced by adult gonads	Testosterone	Estradiol Progesterone	*Estradiol* is secreted from the ovary as the follicle develops	A rapid increase in *estradiol* triggers the LH-surge which induces ovulation. In the estrous cycle *progesterone* is also released by LH	Release of relatively high concentrations of both *estradiol* and *progesterone* from the corpus luteum (remains of follicle after ovulation)
Receptors for hormones	Androgen receptors	Estradiol receptors (ERα & ERβ)—slow acting: hours to days mERα, mERβ, GPR30—fast acting seconds to minutes			
Active metabolites				Dihydrotestosterone (DHT) Estradiol (E2)	Estrone Estriol
Where metabolites are produced	In target cells (eg, brain or secondary sex tissues)	Liver			
Receptors for metabolites	DHT: androgen receptors E2: estradiol receptors (ERα & ERβ)	Estrone and Estriol both act at all Estradiol receptors (ERα & ERβ, mERα, mERβ, and GPR30). Estradiol is more potent than either metabolite at both receptors. Estriol may be an antagonists at GPR30 [18,19]			

ovulation during estrus. By the onset of behavioral estrus, estradiol and progesterone levels have fallen. The estrous cycle terminates with ovulation after the brief increases in estradiol and progesterone that last less than 12 h. The estrous cycle then begins again with the next follicular phase (Table 11.3), unless the female becomes pregnant. This means that it is not possible to find analogous events during the estrous cycle that mimic the luteal or premenstrual phases in women because rats and mice do not have these phases. Nevertheless, it is possible to investigate many effects of ovarian hormones on a trait in female rodents.

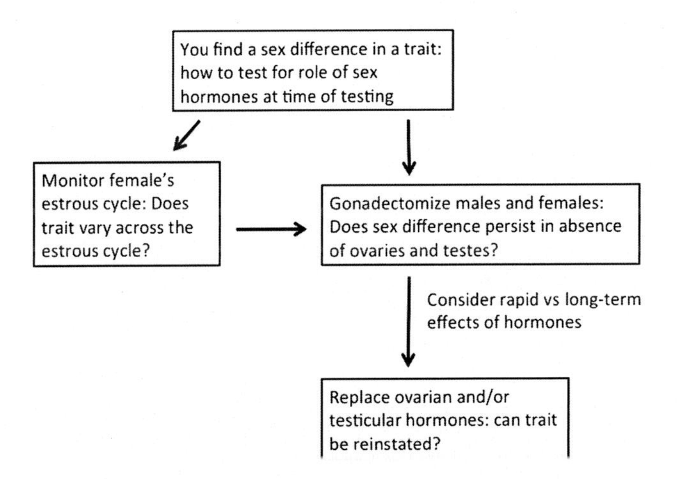

FIGURE 11.7 Flow chart for investigating activational effects of hormones in basic animal research.

YOU FIND A SEX DIFFERENCE IN A TRAIT—WHAT NOW?

If you find that there is a sex difference in a trait that you are studying in a small rodent, and you want to determine which hormones are affecting the trait you have two options (Fig. 11.7). You can monitor the estrous cycle of the females in your study and determine whether the trait is affected by ovarian hormones or you can remove the hormones from the males and/or the females and then selectively replace hormones. We will briefly review both options.

Monitoring the Estrous Cycle

Methods for monitoring the estrous cycle have been described previously in considerable detail and the reader is referred to these sources for additional information [7,8]. The key elements to keep in mind are: (1) it is necessary to obtain vaginal smears or swabs at the same time every day to get reliable information and (2) without multiple days of smears the readings are not accurate as there are some days of the cycle can look like others. For example, vaginal cytology on the morning of meta-estrus frequently looks like proestrus. Additionally, the presence of cornified cells that characterized estrus can persist for multiple days in some females, so unless the female is also behaviorally receptive, cornified cells in the absence of other data (behavioral or 5–7 additional days of vaginal cytology) are not sufficient to conclude a female is in estrus.

If you are studying both males and females and you are repeatedly testing the same females on all days of the cycle, remember that you also need to repeatedly test the male animals in your study to control for repeated testing of the females. If independent groups of animals are used for the different days of the cycle, then only one group of males is needed, but it is a good practice to obtain data from males and females at the same time in order to control for any extraneous environmental influences (ie, fluctuations in animal care facilities).

On the other hand, if you find a sex difference and you want to test both males and females, but you are not specifically interested in the role of gonadal hormones, another approach is to test the females across the cycle and then choose one phase for studying the female phenotype. For example, in studies of sex differences in obesity one might choose to look at females during diestrus as it is known that estradiol decreases feeding behavior, so eating behavior will be most stable during diestrus.

Gonadectomy and Hormone Replacement

Alternatively, one can remove the sources of hormones by castration (CAST) of males or ovariectomy (OVX) of females and determine whether the sex difference in the trait persists. If a trait is due solely to developmental sex differences caused by chromosomes or perinatal hormones the sex difference will still be there. If the sex difference is dependent on activational effects of hormones it will no longer be evident. It is usually a good idea to wait 10–14 days after gonadectomy before testing to allow the hormones to dissipate from the system.

Once a possible activational effect of gonadal hormones has been established, there are many ways to replace hormones and demonstrate the causal relationship. One can replace hormones chronically with pellet implants (commercially available), silastic capsules, or osmotic minipumps, or repeated systemic injections (s.c., i.p., i.v.). For studies of rapid and acute effects of gonadal hormones systemic injections (s.c., i.p., i.v.) or intracranial administrations are most frequently used [7]. The relative merit of each method depends on the question being posed and it is important to consider the caveat of endogenous hormone removal via gonadectomy on the basal expression of hormone receptors and the implications for data interpretation [2].

Then there is the issue of whether to give the free hormone or the esterified hormone (eg, 17ß-estradiol vs estradiol benzoate, testosterone vs testosterone propionate). The esterified hormone must be metabolized by the liver prior to hormone availability to act at hormone receptors, so the rise in serum hormone values and the subsequent decrease is more gradual than for the free hormone. In spite of the fact that the rate of delivery to target organs is modified by the esterification, estradiol benzoate and testosterone propionate are the most frequently used forms of these hormones for systemic administration, even when rapid effects are being studied. Progesterone is given only in the unmodified form [7].

When examining the effects of testosterone, one must also consider whether the effects are mediated by one of its active metabolites (Table 11.3). Administering testosterone will produce effects that could be mediated by testosterone, estradiol, or DHT. One option is to deliver DHT alone or estradiol alone and see if the effects can be produced independent of testosterone. If only a partial response is found in response to either or both treatments, in other words if the complete effect of testosterone is not mimicked, then it is a good idea to also test a group that receives both estradiol and DHT to determine whether the effects of testosterone are mediated by both pathways (Table 11.4).

Another strategy is to inhibit the production of the metabolites. The conversion of testosterone to estradiol can be prevented with an aromatase inhibitor, such as fadrozole. The conversion to DHT can be blocked with an inhibitor of 5α-reductase, such as finasteride. Each of these treatments has its advantages and disadvantages. For example, 5α-reductase also converts progesterone to allopregnanalone so effects of finasteride may not be due solely to blocking DHT synthesis [9].

Following ovariectomy the experimenter can replace estradiol, progesterone, or both. If you wait for the endogenous hormone to dissipate after ovariectomy, there will also be a reduction in the number of ERs, because steroid hormone receptors are induced by their target ligand. In some species, such as the mouse, it is necessary to repeatedly prime with estradiol treatment before a complete response to estradiol is seen (for a discussion see Ref. [7]).

Considering estradiol first, its effects may be mediated by a number of different receptors which can also be investigated (Table 11.5). The long-term effects of estradiol are mediated by estradiol receptor-alpha (ERα) and/or estradiol receptor-beta (ERß). Both of these ERs are classical intracellular steroid receptors. There are also rapid effects of estradiol mediated by the classical ERα and ERß coupled via caveolin to metabotropic receptors (referred to as mERα and mERß) [10–12], as well as the novel G-protein coupled

TABLE 11.4 Hormone Treatment for Testosterone Replacement Post-Castration or Manipulations of Effects Mediated by Testosterone

Hormone to replace	Agonist	Antagonist	Other manipulation
Testosterone (T) acts at androgen receptors (AR) and can be converted to estradiol to act at (ER)	*Testosterone*	*Flutamide*[a] (acts in brain and periphery) *Bicalutamide*[a] (peripherally active only)	
	Testosterone propionate		
	Estradiol (E2) mimics effect at ER (see also Table 11.5)	See Table 11.5	*Fadrozole* inhibits E2 synthesis from T
	Dihydrotestosterone (DHT) (metabolite Active at AR)		*Finasteride*[a] inhibits DHT synthesis from T
	Cl-4AS-1[a]		

[a]*Available from Tocris Bioscience, Ellisville, MO.*

TABLE 11.5 Hormone Treatment for Estradiol Replacement Post-Ovariectomy or Manipulations of Effects Mediated by Specific Estradiol Receptors

Hormone to replace or site of action	Agonist	Antagonist
Estradiol (E2)	*Estradiol*	*ICI182, 780*[a] Best antagonist at ERα and ERß and mER but ICI182, 780 does not cross the blood–brain barrier
	Estradiol benzoate (EB) There are other esterified compounds, but EB is most common	*Tamoxifen*[a] (antagonist/inverse agonist depending on ER)
		Raloxifen[a] (selective ER antagonist)
Estradiol Receptor alpha (ERα)	1,3,5-*tris* (4-hydroxyphenyl)-4-propyl-1H-pyrazole (*PPT*)[a]	There are compounds that are reported to be selective antagonists, but none has been used widely
Estradiol Receptor beta (ERß)	2, 3-*bis* (4-hydroxyphenyl) propionitrile (*DPN*)[a]	
	4-[4,4-Difluoro-1-(2-fluorophenyl) cyclohexyl] phenol (*AC 186*)[a]	
GPR30	*G1*[a]	*G-15*[a]
	Tamoxifen[a]	*G-36*[a]

[a]*Available from Tocris Bioscience, Ellisville, MO.*

membrane ER, GPR30 [13–15]. It is possible to selectively activate the ER receptor subtypes, although so far there are no compounds that selectively activate mERα and mERß without also activating intracellular ERα and ERß (and vice versa). While some selective antagonists for ERα and ERß have been reported (and are available commercially), none have been used extensively, and whether they can be used successfully in vivo is not well established. Agonists and antagonists for GPR30 are available (Table 11.5).

Even though the proteins for ERα and ERß are apparently the same throughout the body, the cells in which the receptors are expressed can affect the ability of ligands to activate the receptor. Thus, it has been possible to target ER in breast and uterine cells with novel molecules that act as ER antagonists in these cells, while simultaneously activating ER in bone and other tissues. These compounds are referred to as selective estrogen receptor modulators (SERMs) [16]. For some questions related to the activational effects of estradiol the SERMs can be quite valuable; a few are listed in Table 11.5, but this is an active area of drug development for breast cancer treatment, so more may become available over time.

If one wants to investigate the role of progesterone in a trait, it can be difficult to dissociate from the effects of estradiol, since synthesis of the classical progesterone receptors (PR) is induced by estradiol. Therefore, many effects of progesterone are not seen unless an animal is previously treated with a dose of estradiol. For example, sexual behavior is seen 4–6 h after progesterone treatment only in rats or hamsters treated 48 h earlier with a dose of estradiol [7]. Progesterone is usually used in the unesterified form (Table 11.6). There are ligands

TABLE 11.6 Hormone Treatment for Progesterone Replacement Post-Ovariectomy or Manipulations of Effects Mediated by Progesterone Receptors

Hormone to replace or site of action	Agonist	Antagonist	Other actions
Progesterone Note: progesterone receptor (PR) synthesis is induced by estradiol) treatment over a 48 h time period. For full effect of progesterone at PR prior treatment with estradiol is required	*Progesterone*	*Mifepristone*[a] Also known as RU-486 antagonist at progesterone and glucocorticoid receptors	*Progesterone* is converted by 5α reductase to allopreganolone which is active at the GABA-A receptor where it is a positive allosteric modulator
	Levonorgesterol[a]		

[a]*Available from Tocris Bioscience, Ellisville, MO.*

developed for contraception that have progesterone agonist and antagonist effects, but many are cross-reactive with glucocorticoid and AR. Progesterone also has rapid effects mediated by actions at the membrane as progesterone [17], although selective ligands have not yet been developed for these G-protein coupled receptors. Additionally, progesterone is converted by 5α-reductase to allopreganolone which is active at the GABA-A receptor where it is a positive allosteric modulator. If there are rapid effects of progesterone, without estradiol pretreatment, the mechanism is likely to be through these membrane PR.

Next, one must consider how long after hormone administration to begin testing and how to give the hormones. Rapid effects of steroid hormones can be seen within seconds at the cellular level and within minutes at the whole animal level [10–12]. The effects of the classical steroid receptors can take many days, as is the case for reproductive behavior. Many people begin by mimicking the hormones of the reproductive cycle of the species they are studying, and this is a successful strategy for demonstrating that the trait under investigation is indeed modulated by the gonadal hormones. Hormones are dissolved in peanut oil and injected s.c. at the nape of the neck. This strategy does not demonstrate which hormones are necessary or sufficient, nor does it identify where the hormones are producing the effect. Subsequent studies can determine how and where the hormones are acting by varying the times between treatment and testing to determine the class of receptor mediating the effect, as well as by using standard physiological techniques appropriate for the system under investigation to determine site of action (see also [7]).

CONCLUSIONS

The role of activational effects of hormones has evolved from being the study of reproduction to the roles for gonadal hormones in all aspects of animal physiology and behavior. As we think about how males and females differ, we need to understand how the reproductive hormones act in the body and how that can differentially impact males and females. The guidelines described briefly here are general principles for the investigator, but knowledge of the specific system under investigation should be used to guide the experimental design.

References

[1] Chen JR, Wang TJ, Lim SH, Wang YJ, Tseng GF. Testosterone modulation of dendritic spines of somatosensory cortical pyramidal neurons. Brain Struct Funct 2013;218(6):1407–17.

[2] Viau V, Meaney MJ. Testosterone-dependent variations in plasma and intrapituitary corticosteroid binding globulin and stress hypothalamic-pituitary-adrenal activity in the male rat. J Endocrinol 2004;181(2):223–31.

[3] Fiber JM, Swann JM. Testosterone differentially influences sex-specific pheromone-stimulated fos expression in limbic regions of Syrian hamsters. Horm Behav 1996;30(4):455–73.

[4] Lewis C, Dluzen DE. Testosterone enhances dopamine depletion by methamphetamine in male, but not female, mice. Neurosci Lett 2008;448(1):130–3.

[5] Shemisa K, Kunnathur V, Liu B, Salvaterra TJ, Dluzen DE. Testosterone modulation of striatal dopamine output in orchidectomized mice. Synapse 2006;60(5):347–53.

[6] Wilkinson M, Brown R. An introduction to neuroendocrinology. 2nd ed. Cambridge, UK: Cambridge University Press; 2015.

[7] Becker JB, Arnold AP, Berkley KJ, Blaustein JD, Eckel LA, Hampson E, et al. Strategies and methods for research on sex differences in brain and behavior. Endocrinology 2005;146 (4):1650–73.

[8] Becker JB, Berkley K, Geary N, Hampson E, Herman JP, Young EA. Sex differences in the brain: from genes to behavior. Oxford, UK: Oxford University Press; 2008.

[9] Anker JJ, Holtz NA, Zlebnik N, Carroll ME. Effects of allopregnanolone on the reinstatement of cocaine-seeking behavior in male and female rats. Psychopharmacology (Berl) 2009;203(1):63–72.

[10] Srivastava DP, Waters EM, Mermelstein PG, Kramar EA, Shors TJ, Liu F. Rapid estrogen signaling in the brain: implications for the fine-tuning of neuronal circuitry. J Neurosci 2011;31 (45):16056–63.

[11] Grove-Strawser D, Boulware MI, Mermelstein PG. Membrane estrogen receptors activate the metabotropic glutamate receptors mglur5 and mglur3 to bidirectionally regulate creb phosphorylation in female rat striatal neurons. Neuroscience 2010;170(4):1045–55.

[12] Micevych PE, Mermelstein PG. Membrane estrogen receptors acting through metabotropic glutamate receptors: an emerging mechanism of estrogen action in brain. Mol Neurobiol 2008;38(1):66–77.
[13] Filardo EJ, Thomas P. Gpr30: a seven-transmembrane-spanning estrogen receptor that triggers egf release. Trends Endocrinol Metab 2005;16(8):362–7.
[14] Small KM, Nag S, Mokha SS. Activation of membrane receptors attenuates opioid receptor-like1 receptor-mediated antinociception via an erk-dependent non-genomic mechanism. Neuroscience 2013;255:177–90.
[15] Thomas P, Pang Y, Filardo EJ, Dong J. Identity of an estrogen membrane receptor coupled to a g protein in human breast cancer cells. Endocrinology 2005;146(2):624–32.
[16] Cyr M, Calon F, Morissette M, Di Paolo T. Estrogenic modulation of brain activity: implications for schizophrenia and Parkinson's disease. J Psychiatry Neurosci 2002;27(1):12–27.
[17] Thomas P. Characteristics of membrane progestin receptor alpha (mpra) and progesterone membrane receptor component 1 (pgmrc1) and their roles in mediating rapid progestin actions. Front Neuroendocrinol 2008;29:292–312.
[18] Katzenellenbogen BS. Biology and receptor interactions of estriol and estriol derivatives in vitro and in vivo. J Steroid Biochem 1984;20(4B):1033–7.
[19] Lappano R, Rosano C, De Marco P, De Francesco EM, Pezzi V, Maggiolini M. Estriol acts as a GPR30 antagonist in estrogen receptor-negative breast cancer cells. Mol Cell Endocrinol 2010;320(1–2):162–70.

C H A P T E R

11.4

Human Methodologies in the Study of Sex Differences

Emily J. Bartley and Margarete Ribeiro-Dasilva

University of Florida, College of Dentistry, Pain Research and Intervention Center of Excellence, Gainesville, FL, United States

INTRODUCTION

An appreciation for the influence of sex differences in health has burgeoned over recent years. Through most of this text we have taken great care to focus on sex differences with isolation of the concept of gender from that of sex. As indicated earlier, gender is an additional level of distinction that is reserved for humans. Consistent with previous chapters in this text, we will refer to *sex differences* in terms of physiological differences among males and females. In addition, because we are specifically discussing humans in this section, we will also refer to *gender* as a social construct that is influenced not only by biology, but also shaped through experience and the environment. The increased focus in sex differences is dramatic when both sex and gender are taken into account, as demonstrated by the 40-fold increase in the amount of sex and gender publications over the past five decades (Fig. 11.8A). This effect is even greater when health and disease are considered (Fig. 11.8B). It is widely acknowledged that sex differences impact a vast array of health-related outcomes including the onset, course, and severity of disease, as well as treatment response and mortality [1]. Indeed, while men are at increased risk for major, life-threatening illnesses (eg, cardiovascular disease, lung cancer) and have higher rates of mortality, women often report higher levels of psychiatric disease and have a greater susceptibility to chronic health conditions (eg, persistent pain, diabetes). Evidence also supports the differential manifestation of health-related symptoms among men and women [2,3]. Although sex hormones have been implicated as a significant contributor to sexually dimorphic differences in health and functioning, it is likely that sex-specific variance is also due to genetic, environmental, lifestyle, psychological, and sociocultural factors [1]. While the previous chapters review the emerging literature on sex differences across the spectrum of health and disease, the purpose of the current chapter is to provide an overview of standard human methodologies by which to examine sex differences. Specifically, we will discuss sex hormone and neuroimaging methods that are commonly utilized in the assessment of sex differences, and provide a brief review of the literature examining psychosocial methods of gender role measurement. We will conclude with a brief commentary on the implications of sex-based differences in health and disease and discuss future directions for research.

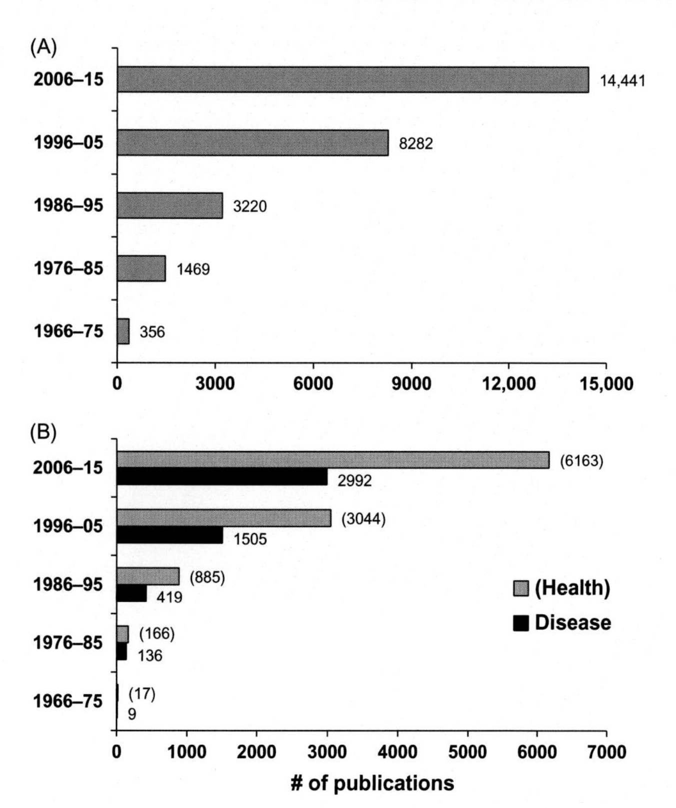

FIGURE 11.8 **Sex and gender citations.** This figure presents results from a PubMed search reflecting citations dating over the past 50 years (1966–2015). The terms ("sex differences" or "gender differences"), ("sex differences" or "gender differences" and "health"), and ("sex differences" or "gender differences" and "disease") were entered, yielding a total of 27,768 papers that have been published on sex and gender differences (A) since 1966, and 10,275 and 5061 citations on sex/gender differences in areas of health and disease (B), respectively. *Note:* PubMed search conducted on Nov. 9, 2015.

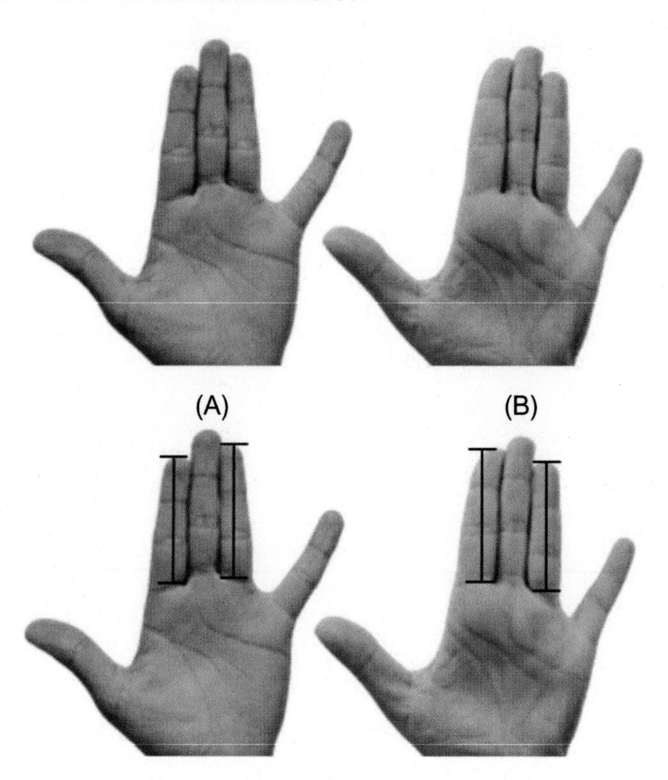

FIGURE 11.9 **Finger length ratio.** The figure presents an example of different finger lengths. In panel A (figures oriented on the left), the second finger (2D) is shorter than the fourth finger (4D), indicating more prenatal testosterone exposure (high 2D:4D ratio). In panel B (figures oriented on the right), the second finger is longer than the fourth finger, reflecting less prenatal testosterone exposure (low 2D:4D ratio). *Note:* Both panels represent female hands.

PRENATAL HORMONE EXPOSURE

Hormones can exert influences starting as early as in-utero, with prenatal hormones being a determinant and/or a cofactor in the development of adult physical, cognitive, and behavioral characteristics [4]. The production of testosterone, and subsequent conversion into dihydrotestosterone, occurs around weeks 6–12 of pregnancy and this exposure is key in the production of male gonads [5] (see chapter: Chromosomal and Endocrinological Origins of Sex). During the prenatal period, testosterone organizes tissues so that they respond differently to gonadal hormones in adulthood [6]. Unlike assessment of organizational effects of prenatal hormones in preclinical models (see mini-chapter: Strategies and Approaches for the Study of Activational Influences of Gonadal Hormones on Sex Differences in Physiology and Behavior) the assessment of organization effects of hormones in humans and their impact on sex differences across health-related conditions can be challenging, as ethically researchers cannot measure hormone exposure or alter hormones in a developing fetus. Instead, human investigation relies on indirect methods and observation of naturally occurring abnormalities as markers of hormone exposure [7,8]. Below, we briefly address methodologies of sex hormone assessment related to prenatal hormone exposure in humans.

Finger Length

The second-to-fourth digit ratio (2D:4D) [8,9] is frequently used as a marker of prenatal androgen exposure (See Fig. 11.9) similar to the use of aniogenital distance in rodents (see mini-chapter: Organizational Influences of the Gonadal Steroid Hormones: Lessons Learned Through the Hypothalamic-Pituitary-Adrenal (HPA) Axis). 2D:4D is established in-utero by the 13th or 14th week postconception [10,11], and it demonstrates substantial stability over the lifetime [12,13]. While a lower 2D:4D indicates a higher level of prenatal testosterone exposure, a higher ratio reflects a lower exposure to testosterone (ie, higher estrogen exposure) [10,14–16]. These effects are sexually dimorphic, as females tend to have a higher 2D:4D ratio than males [17]. Measurement of finger length ratio is typically made by scanning the hands and drawing a line from the top of each finger to the middle of the lowest crease of the finger. To avoid bias, at least two people, double-blinded, independently measure the lengths of the index (second digit: 2D) and ring fingers (fourth digit: 4D). The digit ratios are computed by dividing the (mean) length of the index finger by the (mean) length of the ring finger for each hand. Studies indicate that differences in 2D:4D ratio are associated with levels of aggression, occupational interests, major psychiatric illness, and the incidence of pain [13,15,16,18].

Amniotic Fluid and Umbilical Cord Blood

Prenatal hormone exposure can also be measured through amniotic fluid and umbilical cord blood. Hormone levels measured from amniotic fluids are very precise and reflect the exposure during fetal formation. Further, umbilical cord blood can be easily collected from uncomplicated pregnancies following delivery. However, a disadvantage of these procedures is that sampling amniotic fluid may place the fetus at risk and can only be obtained from women undergoing an amniocentesis due to prenatal diagnostic screening at 12–16 weeks of pregnancy [19]. Moreover, the umbilical cord sampling only reflects fetal sex steroid concentrations at late gestation. Nevertheless, studies have demonstrated consistent sex differences in sex steroid concentrations in both amniotic fluid and umbilical cord blood [19–21], suggesting that these approaches can be used to examine the relationship between early-life sex steroid exposure and sex differences in development.

ADULT HORMONE EXPOSURE

Sex hormones, including estrogen, progesterone, and testosterone, exert both organizational and activational effects throughout the life span. While all three are active in men and women, absolute blood levels of these hormones as well as their temporal patterns differ across sex. In general, male hormone levels remain relatively steady after puberty and undergo only minor levels of fluctuation compared to females. The most noticeable change for men occurs in the reduction of testosterone, which gradually decreases with age. In contrast, women exhibit more dynamic changes in sex hormones—varying within and across the menstrual cycle, pregnancy, and after menopause. During the reproductive years, the first half of the female menstrual cycle (ie, follicular phase) is characterized by relatively low levels of estrogen and progesterone, with estrogen gradually increasing and reaching a peak just prior to ovulation. During the second half of the cycle (ie, luteal phase), both estrogen and progesterone heighten midway (although the increase is more modest for estrogen), and then steadily decrease over the remainder of the luteal phase until the onset of menstruation. Variation in testosterone across the menstrual cycle is less dramatic. While testosterone levels remain much lower than estrogen and progesterone, there is a small peak in testosterone that occurs around ovulation [22,23]. A significant body of evidence supports the role that gonadal hormones have on physical and emotional health in women, with symptom exacerbation in some clinical conditions (eg, irritable bowel syndrome, interstitial cystitis, tension- and migraine-type headache, temporomandibular disorder, depression, and anxiety) occurring during periods of greater hormone fluctuation (ie, perimenstrually). These effects are generally attenuated in women using oral contraceptives as the hormonal milieu is more regulated and less prone to fluctuations [24–27].

Differences in circulating gonadal hormone concentrations have been studied in terms of sex differences across a number of conditions including health, mood, cognition, and psychopathology [28–30]. Increasing evidence suggests that estrogen can have similar, as well as opposing effects in men and women, which is likely due to underlying sexual dimorphisms that occur in some brain processes but not others [31]. For instance, varying levels of estrogen can lead to a "pro-inflammatory" state and increase the risk for certain medical conditions; however, estrogen can also serve as a protective factor against inflammation [32–35]. Several methodologies have been developed to accurately measure and understand how circulating hormones influence health, as well as understand how receptors and gene expression play a role in health-related sex differences.

Sex Hormone Measurement

Serum and plasma-based measurements are considered to be the "gold-standard" for the assessment of sex hormones; however, the use of saliva as a diagnostic tool has also been found to be a reliable and valid indicator of sex hormone levels [25,29,30,36–38]. Some advantages include its cost-effectiveness, ease in sampling, and noninvasiveness [37,39–41]. Moreover, salivary levels of sex hormones represent the biologically active fraction of the hormones present in circulation [29,30], and thus serve as a measurement of the amount of hormones available to affect tissue receptors. Given this, it has been suggested that salivary-based measurements may provide a more accurate assessment of sex hormone effects [37,39,42]. An alternative method for measuring sex hormones (namely estradiol) is through the blood spot technique using a finger stick [29]. Although more invasive than saliva, it is less intrusive than venous collection, has good reliability, and demonstrates strong correlations with circulating blood levels of estradiol. Despite the disadvantages associated with venipuncture (ie, painfulness, time-consuming, requires certified personnel for sample collection), this method is still considered to be a more reliable and accurate method for measuring circulating hormones.

Hormone Receptor Expression

Hormones exert their effects by binding to their respective receptors. While a majority of studies have focused on the circulating amount of hormones, studies have also examined hormone receptor expression in the investigation of sex differences across various medical conditions. Receptor binding studies have demonstrated that early in postnatal life, estrogen binds not only to reproductive tissue, but also to areas of the brain such as the hippocampus and cortex in a unique way for each sex [43–45]. For ethical reasons, studies investigating hormone receptor expression in the brain have only been performed in animals, and to our knowledge, there has been only one report on sex differences in estrogen receptor expression using cadaver tissue to examine spinal trigeminal nuclei [46]. However, peripheral blood cell studies also show differences in the expression of hormone receptors [47], and the methodologies employed in a majority of these cases use real-time polymerase chain reaction (PCR) to measure the mRNA expression of different hormone receptors [48,49] (discussed further in chapter: Sex Differences in Neuroanatomy and Neurophysiology: Implications for Brain Function, Behavior, and Neurological Disease).

Sex-Specific Gene Expression

Sex differences in the development and incidence of neurodegenerative and neuropsychiatric diseases may also be related to patterns of gene expression that differ among men and women [50]. Through the use of gene sequencing and gene expression profiling, several studies have examined the activational effects of hormones on genes in terms of explaining phenotypic differences among the sexes. For instance, in a microarray study by van Nas and colleagues [51], the expression of approximately 2600 genes (using mouse liver tissue as a model) was sex-biased when gonads were intact, and some genes exhibited tissue-specific patterns of expression. Conversely, after gonadectomy only a small portion of genes ($N = 12$) were sex-specific. These findings suggest that sexually dimorphic differences in genes are influenced by sex hormones in adulthood, an outcome that could lead to sex-specific phenotypic differences in various disease states [52,53]. Sex differences in gene expression can also be generated by sex chromosome influences (see mini-chapter: Organizational Influences of the Gonadal Steroid Hormones: Lessons Learned Through the Hypothalamic-Pituitary-Adrenal (HPA) Axis), but this level of assessment is not feasible in the human.

Sex Hormone Manipulation

Sex hormone fluctuation can impact the risk of certain medical conditions and lead to differential expression among men and women (eg, pain, depression, irritable bowel syndrome). For this reason, researchers have investigated sex hormone manipulation (eg, oral contraceptives, surgical menopause, antiandrogenic therapy, hormone therapy) in an attempt to better understand hormonal influence on sex differences. For example, animal studies have found that ovariectomized female animals demonstrate a similar drug response to males, while normal females present with opposite responses when compared to both male and ovariectomized females [54]. In humans, hormone therapy is known to increase the prevalence of various conditions such as back pain [55], temporomandibular disorder (TMD) [56], carpal tunnel syndrome [57], and cervical cancer [58]. A study of transsexuals undergoing hormonal treatment revealed that approximately one-third of the male-to-female subjects undergoing estradiol/antiandrogen treatment developed chronic pain, whereas approximately half of the female-to-male subjects treated with testosterone reported a significant improvement in pain (ie, headache) that was present before the initiation of treatment [59]. Exogenous hormone treatment has also been associated with a reduction of clinical pain [60,61], and lowered risk for various medical conditions such as coronary artery disease [62] and ovarian cancer [63]. Overall, these studies highlight the influence that sex hormones have on health and symptom manifestation in men and women, and the need for the assessment of pharmaceuticals in both sexes.

BRAIN IMAGING METHODOLOGIES

Brain imaging has become an inherently valuable tool in examining the structure (anatomy) and function (physiology) of brain systems. With the proliferation of neuroimaging techniques over recent years, a growing number of studies have highlighted sex differences in the brain (for excellent reviews, see [64,65]) including variation among men and women in brain anatomy [66–68], structural and functional connectivity [69–71], synthesis of neurotransmitters [72,73], and responsivity to evoked stimuli [74,75]. While a thorough discussion of sex differences in neuroanatomy and neurophysiology is provided in chapter "Sex Differences in Neuroanatomy and Neurophysiology: Implications for Brain Function, Behavior, and Neurological Disease," the following section will focus on common methodologies by which to assess structural and functional brain processes.

Structural Imaging

One of the earliest methods for imaging the human brain was computed tomography (CT), a technique producing two-dimensional x-ray images of scanned tissue. However, with the advent of more sophisticated technologies over the years, magnetic resonance imaging (MRI) has become integral toward examining anatomical differences in brain structure. Structural neuroimaging provides a map of the neuroanatomy of the brain by producing high-contrast computerized images of tissue structure and volume through the use of a strong magnetic field and radio waves, and the body's own hydrogen molecules (eg, in water). One of the benefits of structural MRI has been its use clinically, as it can characterize patterns of brain maturation and loss, and has become an important diagnostic tool for numerous disease pathologies [76].

Techniques such as voxel-based morphometry (VBM) and diffusion tensor imaging (DTI) characterize regional gray and white matter differences in structural magnetic resonance images, respectively. VBM uses statistical parametric mapping, a voxel-based method to identify group-related differences in brain structure. Prior to analysis, individual brain images of patients are registered to a group template (ie, spatial normalization) so that each image is in the same stereotactic space and has the same shape, size, and overall pattern of structure. Images are segmented into different tissue classes (ie, white matter vs gray matter vs cerebrospinal fluid) and then undergo a smoothing process where each individual voxel represents a weighted mean of its own and surrounding values. This method allows for more normally distributed data and increases the sensitivity to detect very small differences in gray matter. Statistical analyses are typically formed using t-tests (usually by a voxel-by-voxel basis) to identify regions where gray matter concentration varies across groups [77–80].

DTI extends conventional neuroimaging methods by measuring the characterization of white matter, and more specifically, water mobility (termed Brownian motion) through white matter tracts [81–85]. The direction, magnitude, and orientation of water diffusion through tissues are mapped onto a three-dimensional space to provide an output of white matter connectivity and tissue microstructure. An asset of this methodology is that it can detect abnormalities in white matter that are not visible with traditional structural imaging methods [82,86]. However, it is highly sensitive to patient motion, eddy currents, and scanner noise, which when corrected can result in lower image resolution [87].

Functional Imaging

Functional magnetic resonance imaging (fMRI) and positron emission tomography (PET) measure regional blood flow changes as an indicator (albeit indirect) of neuronal activity, thus making these methods popular in the measurement of brain activation under task-related performance. fMRI generally uses a technique called blood-oxygenation-level dependent (BOLD) contrast imaging, which is a contrast agent that reflects the interaction between oxygenated and deoxygenated blood and captures increases in blood flow in areas of increased neuronal activity—termed, the hemodynamic response. Conversely, PET requires the injection of a rapidly decaying isotope (usually by water or glucose) into the bloodstream. The isotopes are delivered to the brain and accumulate in areas of increased cerebral activity (ie, areas of increased blood flow). After, PET isolates concentrations of the radioisotope which are subsequently displayed through a three-dimensional image depicting functional processing. Similar to PET, single-photon emission computed tomography (SPECT) is another form of emission tomography. However, it differs from PET in that it detects single photons rather than dual photon emissions, and uses a low-level radioisotope with a longer half-life. While PET technology produces less acoustic noise than fMRI and is not heavily impacted by patient movement, this method is limited by its high cost and has lower spatial and temporal resolution than fMRI. Further, the radioactive component of the procedure limits its use among specific patient populations [88–90].

Compared to other functional neuroimaging methods, electroencephalography (EEG) and magnetoencephalography (MEG) are direct indicators of neuronal activity and measure electrical potentials and magnetic fields produced from real-time measurement of brain activity. Neuronal activity is captured via EEG electrodes affixed to the scalp or through a highly sensitive magnetometer called a superconducting quantum interference device (SQUID). Both of these methodologies provide a measurement of localized and diffuse brain activity and are relatively noninvasive. However, identification of the source signal is a common challenge associated with EEG and MEG (ie, inverse problem); therefore, the use of localization algorithms is often necessary for detecting the source [91]. There are also distinctions between EEG and MEG that are important to note. For instance, the spatial resolution for MEG is superior when compared to EEG, and can provide measurement with more accuracy and within a millimeter precision. Further, the magnetic fields are less distorted with MEG than the electrical fields measured by EEG. Despite this, EEG is generally superior in terms of having lower cost, can be combined with fMRI to provide both high temporal and spatial resolution data, and can measure more neurons with a greater sensitivity in measurement [87,92,93]. Given these differences, most research paradigms incorporate EEG and MEG simultaneously to capitalize on their independent strengths.

PSYCHOSOCIAL METHODOLOGIES

In addition to biological determinants, there is also compelling evidence that psychosocial mechanisms account for differences between men and women in health and functioning. For instance, men and women tend to cope differently with stress, with women being more emotionally focused in nature and using adaptive coping strategies including positive self-talk and seeking social support from family and friends. In contrast, men are found to use more problem-focused methods for stress coping, with a tendency to engage in maladaptive strategies such as substance abuse and behavioral avoidance [94–99]. It is believed that these differences originate and are shaped developmentally through early-life learning experiences and encouragement of gender role stereotypes (eg, masculinity and femininity). However, it is important to note that gender-related behaviors are influenced by prenatal and neonatal hormone exposure [100], but are also modified across culture and sociocultural expectations of culturally bound behavior. Although a number of tools exist whereby to assess psychosocial differences among men and women, we will limit our discussion to methodologies examining sociocultural gender role stereotypes and focus on the most prominent and commonly used measures.

The BEM Sex Role Inventory (BRSI) [101] and the Personal Attributes Questionnaire (PAQ) [102] are two widely implemented measures that examine traits associated with masculinity and femininity. The BSRI consists of 60 personality characteristics in which participants rate the extent to which they encompass each trait. Items are equally distributed across 20 masculine (eg, self-sufficient, ambitious, independent), 20 feminine (eg, affectionate, flatterable, sympathetic), and 20 gender neutral (eg, tactful, unsystematic, happy) characteristics. Based upon dominance in a particular gender role, four classifications are derived: masculine, feminine, androgynous (scoring high on both masculine and feminine-oriented items), and undifferentiated (scoring below the normed mean on masculine and feminine items). Similarly, the PAQ measures levels of "instrumentality" and "expressivity" across masculinity and femininity, respectively. The questionnaire consists of 24 bipolar-rated items (eg, "not at all aggressive" to "very aggressive"; "not at all emotional" to "very emotional") with respondents rating the degree to which they comprise each characteristic. Overall, these two measures are empirically sound and demonstrate good psychometric properties; however, they have been criticized as being outdated due to temporal changes in gender roles over the past 40 years [103]. Despite this, they are still widely used and have been implemented in several studies exploring gender roles across various cultures and ages [104,105]. Further, they have demonstrated strong clinical utility in terms of understanding the effects of gender role on sex-based differences in health.

Evidence suggests that gender roles impact the perception of stressful events, and strong commitment to a particular gender role stereotype can generate unhealthy and maladaptive stress. In an attempt to examine this, Eisler and colleagues developed the Masculine [106] and Feminine [107] Gender Role Stress Scales (MGRS; FGRS). The MGRS and FGRS consist of 40 and 39 items, respectively, and assess the degree to which cognitive, behavioral, and environmental events are construed as stressful according to contemporary gender role perceptions. Items reflect stressors that could be deemed as particularly stress-inducing for each specific gender. The MGRS (scales: physical inadequacy, emotional expressiveness, subordination to women, intellectual inferiority, performance failure) and FGRS (scales: fear of unemotional relationships, fear of physical unattractiveness, fear of victimization, fear of behaving assertively, fear of not being nurturant) both comprise five subscales. Higher levels of feminine gender role stress have been associated with greater affective distress including eating disorders, depression, and somatization [107], while higher levels of masculine-related stress are linked to higher levels of anger, anxiety, cardiovascular and pulmonary disease, alcoholism, and drug abuse [108].

Several studies have reported on sex differences in both clinical and experimental pain, with women generally experiencing greater pain sensitivity including lower thresholds and tolerances, higher pain ratings, greater pain facilitation, and attenuated pain inhibition [97,109–111]. Further, a number of studies have found higher prevalence of both acute and chronic pain in women, as well as greater widespread pain [97,112]. While physiological factors have been reported in the pathogenesis of sex differences in pain, psychosocial factors have also been implicated. For example, social learning expectations about pain and gender role cues [113] are associated with pain report [114], and manipulating gender roles across men and women can influence pain responses [115]. Created specifically for the examination of pain attributions, the Gender Role Expectations of Pain (GREP) is a 12-item questionnaire that measures gender stereotype expectations of pain sensitivity, pain endurance, and willingness to report pain [114,116]. Using a set of visual analog scales, respondents estimate their individual response to pain compared to the typical man or woman, but also provide stereotypical perceptions of pain responsivity of the typical man or woman. In general, men are typically perceived as less willing to report pain, while women are viewed as being more pain sensitive and experiencing less pain endurance, an effect generally endorsed by both sexes [115]. Studies using the GREP have found that scores significantly predict sensitivity to experimental pain [116,117], mediate sex differences in cold pain sensitivity [118], and even vary across culture [119].

CONCLUSION

There has been a tremendous recognition of the influence of sex differences in disease manifestation—effects which have had important implications, not only within health care policy and management, but also toward advancing initiatives to improve the sex balance in research [120]. Traditionally, the activational effects of sex hormones have been considered to be key determinants underlying sexual dimorphisms in the brain and target tissue. However, there is also strong evidence that mechanisms facilitating differential responsivity among men and women are likely manifold, consisting of an interactive mosaic of genetic, psychological, environmental, lifestyle/behavioral, sociocultural, and physiological factors. Several methodologies exist by which to investigate these differences. While neuroimaging techniques allow for the examination of brain organization, as well as variation in structure and function of neural processes among men and women, other strategies such as gene sequencing offer important insights into the influence of genes and sex chromosomes on differential disease expression. Further, sociocultural factors including stereotypical gender beliefs explain a portion of the variance in sex differences in health and illness. Taking these issues into consideration, it is crucial that sex differences be accounted for in the design and implementation of research in order to optimize health care among men and women, strengthen current methods of assessment, and facilitate enhanced understanding of the multiple processes that increase health-related disparities among the sexes. We offer the following considerations for adopting good practice guidelines in the measurement and study of sex differences:

1. Preclinical research has historically been hampered by the use of single-sex studies. This is particularly problematic as failure to include both sexes in the research design could impact the translational value of basic science outcomes to clinical practice. Further, it could impede a deeper understanding of the mechanisms underlying disease in men and women and hinder the development of novel therapeutics that may be more suitable or effective in one sex than the other. Unless there is a strong justification for only investigating one sex (eg, studying disease effects specific to or predominant in one sex), a more concerted effort must be made for equal representation of both males and females in both preclinical and clinical research.
2. While sex and gender are often used interchangeably in the literature, it is important that these be considered as conceptually distinct constructs that can interact with one another to shape the experience of health. Accurate usage of these terms would facilitate greater understanding and measurement of the differences that exist between men and women.
3. Though the biological consequences of sex exude a strong influence on health and disease disparities between men and women, gender also plays a prominent role in these differences. Research must not only consider the physiological presentation (eg, sex chromosomes, hormones, etc.) between men and women, but also how cultural, social, and psychological variables impact the manifestation of sex differences in health.
4. Hormone status is generally neglected in research, both in reproductive and nonreproductive women. For instance, characterization of menstrual cycle phases across studies is often inconsistent and hormone assays are seldom procured. If the variable of interest is known, or suspected, to vary as a function of sex hormones (or the menstrual cycle),

then an ideal methodology would be to conduct hormone assays at the time of testing. At minimum, ovulation should be captured through luteinizing hormone testing as an anovulatory cycle can interrupt the hormone milieu. While collection of hormone data carries with it inherent costs and complexities, it offers a unique opportunity to examine the extent to which sex hormones affect disease and health. Similarly, and when applicable, research should take into account the effect of exogenous hormone administration (ie, oral contraceptives, hormone replacement) and its impact on sex differences in health outcomes.

5. Demographic and clinical characteristics (eg, race, education, socioeconomic status) must also be acknowledged as interaction of these factors with sex and gender may pose additional health risks across various subgroups among men and women. Further, research on sex and gender should take a life span approach as age can have a prominent influence on the manifestation of disease processes among the sexes.
6. A greater collaboration between basic and clinical scientists is warranted to ensure optimal bench-to-bedside research in the study of sex differences and facilitate the development of more refined and novel therapeutics.
7. As a scientific community, we must maintain good research integrity and avoid publication bias in an effort to report both significant and null findings from studies. Circumventing this will allow for a more accurate characterization of sex differences and their impact on health-related outcomes.

References

[1] Morrow EH. The evolution of sex differences in disease. Biol Sex Differ 2015;6:5. Available from: http://dx.doi.org/10.1186/s13293-015-0023-0.

[2] Vlassoff C. Gender differences in determinants and consequences of health and illness. J Health Popul Nutr 2007;25(1):47–61.

[3] Denton M, Prus S, Walters V. Gender differences in health: a Canadian study of the psychosocial, structural and behavioural determinants of health. Soc Sci Med 2004;58(12):2585–600. Available from: http://dx.doi.org/10.1016/j.socscimed.2003.09.008.

[4] McFadden D, Bracht MS. Sex differences in the relative lengths of metacarpals and metatarsals in gorillas and chimpanzees. Horm Behav 2005;47(1):99–111. Available from: http://dx.doi.org/10.1016/j.yhbeh.2004.08.013.

[5] Bao AM, Swaab DF. Sexual differentiation of the human brain: relation to gender identity, sexual orientation and neuropsychiatric disorders. Front Neuroendocrinol 2011;32(2):214–26. Available from: http://dx.doi.org/10.1016/j.yfrne.2011.02.007.

[6] Phoenix CH. Organizing action of prenatally administered testosterone propionate on the tissues mediating mating behavior in the female guinea pig. Horm Behav 2009;55(5):566. Available from: http://dx.doi.org/10.1016/j.yhbeh.2009.01.004.

[7] Berenbaum SA, Beltz AM. Sexual differentiation of human behavior: effects of prenatal and pubertal organizational hormones. Front Neuroendocrinol 2011;32(2):183–200. Available from: http://dx.doi.org/10.1016/j.yfrne.2011.03.001.

[8] Berenbaum SA, Bryk KK, Nowak N, Quigley CA, Moffat S. Fingers as a marker of prenatal androgen exposure. Endocrinology 2009;150(11):5119–24. Available from: http://dx.doi.org/10.1210/en.2009-0774.

[9] Brown WM, Hines M, Fane BA, Breedlove SM. Masculinized finger length patterns in human males and females with congenital adrenal hyperplasia. Horm Behav 2002;42(4):380–6. Available from: http://dx.doi.org/10.1006/hbeh.2002.1830.

[10] Csatho A, Osvath A, Bicsak E, Karadi K, Manning J, Kallai J. Sex role identity related to the ratio of second to fourth digit length in women. Biol Psychol 2003;62(2):147–56. Available from: http://dx.doi.org/10.1016/S0301-0511(02)00127-8.

[11] Garn SM, Burdi AR, Babler WJ, Stinson S. Early prenatal attainment of adult metacarpal-phalangeal rankings and proportions. Am J Phys Anthropol 1975;43(3):327–32. Available from: http://dx.doi.org/10.1002/ajpa.1330430305.

[12] Trivers R, Manning J, Jacobson A. A longitudinal study of digit ratio (2D:4D) and other finger ratios in Jamaican children. Horm Behav 2006;49(2):150–6. Available from: http://dx.doi.org/10.1016/j.yhbeh.2005.05.023.

[13] Hell B, Päßler K. Are occupational interests hormonally influenced? The 2D:4D-interest nexus. Pers Individ Dif 2011;51(4):376–80. Available from: http://dx.doi.org/10.1016/j.paid.2010.05.033.

[14] Williams JH, Greenhalgh KD, Manning JT. Second to fourth finger ratio and possible precursors of developmental psychopathology in preschool children. Early Hum Dev 2003;72(1):57–65 http://dx.doi.org/10.1016/S0378-3782(03)00012-4

[15] Millet K, Dewitte S. Digit ratio (2D:4D) moderates the impact of an aggressive music video on aggression. Pers Individ Dif 2007;43(2):289–94. Available from: http://dx.doi.org/10.1016/j.paid.2006.11.024.

[16] Moskowitz DS, Sutton R, Zuroff DC, Young SN. Fetal exposure to androgens, as indicated by digit ratios (2D:4D), increases men's agreeableness with women. Pers Individ Dif 2015;75(0):97–101. Available from: http://dx.doi.org/10.1016/j.paid.2014.11.008.

[17] McFadden D, Shubel E. Relative lengths of fingers and toes in human males and females. Horm Behav 2002;42(4):492–500. Available from: http://dx.doi.org/10.1006/hbeh.2002.1833.

[18] Manning JT, Bundred PE, Newton DJ, Flanagan BF. The second to fourth digit ratio and variation in the androgen receptor gene. Evol Hum Behav 2003;24(6):399–405. Available from: http://dx.doi.org/10.1016/S1090-5138(03)00052-7

[19] van de Beek C, Thijssen JH, Cohen-Kettenis PT, van Goozen SH, Buitelaar JK. Relationships between sex hormones assessed in amniotic fluid, and maternal and umbilical cord serum: what is the best source of information to investigate the effects of fetal hormonal exposure? Horm Behav 2004;46(5):663–9. Available from: http://dx.doi.org/10.1016/j.yhbeh.2004.06.010.

[20] Simmons D. Interrelation between umbilical cord serum sex hormones, sex hormone-binding globulin, insulin-like growth factor I, and insulin in neonates from normal pregnancies and pregnancies complicated by diabetes. J Clin Endocrinol Metab 1995;80(7):2217–21. Available from: http://dx.doi.org/10.1210/jcem.80.7.7608282.

[21] Simmons D, France JT, Keelan JA, Song L, Knox BS. Sex differences in umbilical cord serum levels of inhibin, testosterone, oestradiol, dehydroepiandrosterone sulphate, and sex hormone-binding globulin in human term neonates. Biol Neonate 1994;65(5):287–94. Available from: http://dx.doi.org/10.1159/000244074.

[22] Fillingim RB, Ness TJ. The influence of menstrual cycle and sex hormones on pain responses in humans. In: Fillingim RB, editor. Sex, gender, and pain, vol. 17. Seattle, WA: IASP Press; 2000. p. 191–207.

[23] Kuba T, Quinones-Jenab V. The role of female gonadal hormones in behavioral sex differences in persistent and chronic pain: clinical versus preclinical studies. Brain Res Bull 2005;66 (3):179.

[24] Houghton LA, Lea R, Jackson N, Whorwell PJ. The menstrual cycle affects rectal sensitivity in patients with irritable bowel syndrome but not healthy volunteers. Gut 2002;50(4):471–4. Available from: http://dx.doi.org/10.1136/gut.50.4.471.

[25] LeResche L, Mancl L, Sherman JJ, Gandara BK, Dworkin SF. Changes in temporomandibular pain and other symptoms across the menstrual cycle. Pain 2003;106(3):253–61. Available from: http://dx.doi.org/10.1016/j.pain.2003.06.001.

[26] Pinkerton JV, Guico-Pabia CJ, Taylor HS. Menstrual cycle-related exacerbation of disease. Am J Obstet Gynecol 2010;202 (3):221–31. Available from: http://dx.doi.org/10.1016/j.ajog.2009.07.061.

[27] Powell-Boone T, Ness TJ, Cannon R, Lloyd LK, Weigent DA, Fillingim RB. Menstrual cycle affects bladder pain sensation in subjects with interstitial cystitis. J Urol 2005;174(5):1832–6. Available from: http://dx.doi.org/10.1097/01.ju.0000176747.40242.3d.

[28] Granger DA, Shirtcliff EA, Booth A, Kivlighan KT, Schwartz EB. The "trouble" with salivary testosterone. Psychoneuroendocrinology 2004;29(10):1229–40. Available from: http://dx.doi.org/10.1016/j.psyneuen.2004.02.005.

[29] Shirtcliff EA, Granger DA, Likos A. Gender differences in the validity of testosterone measured in saliva by immunoassay. Horm Behav 2002;42(1):62–9. Available from: http://dx.doi.org/10.1006/hbeh.2002.1798.

[30] Shirtcliff EA, Granger DA, Schwartz EB, Curran MJ, Booth A, Overman WH. Assessing estradiol in biobehavioral studies using saliva and blood spots: simple radioimmunoassay protocols, reliability, and comparative validity. Horm Behav 2000;38(2):137–47. Available from: http://dx.doi.org/10.1006/hbeh.2000.1614.

[31] Gillies GE, McArthur S. Estrogen actions in the brain and the basis for differential action in men and women: a case for sex-specific medicines. Pharmacol Rev 2010;62(2):155–98. Available from: http://dx.doi.org/10.1124/pr.109.002071.

[32] Sniekers YH, Weinans H, van Osch GJ, van Leeuwen JP. Oestrogen is important for maintenance of cartilage and subchondral bone in a murine model of knee osteoarthritis. Arthritis Res Ther 2010;12(5):R182. Available from: http://dx.doi.org/10.1186/ar3148.

[33] Brincat SD, Borg M, Camilleri G, Calleja-Agius J. The role of cytokines in postmenopausal osteoporosis. Minerva Ginecol 2014;66(4):391–407.

[34] Iwasa T, Matsuzaki T, Tungalagsuvd A, Munkhzaya M, Kawami T, Kato T, et al. Effects of ovariectomy on the inflammatory responses of female rats to the central injection of lipopolysaccharide. J Neuroimmunol 2014;277(1-2):50–6. Available from: http://dx.doi.org/10.1016/j.jneuroim.2014.09.017.

[35] Kluft C, Leuven JA, Helmerhorst FM, Krans HM. Pro-inflammatory effects of oestrogens during use of oral contraceptives and hormone replacement treatment. Vascul Pharmacol 2002;39(3):149–54 http://dx.doi.org/10.1016/S1537-1891(02)00304-X

[36] Gandara BK, Leresche L, Mancl L. Patterns of salivary estradiol and progesterone across the menstrual cycle. Ann N Y Acad Sci 2007;1098(1):446–50. Available from: http://dx.doi.org/10.1196/annals.1384.022.

[37] Gann PH, Giovanazzi S, Van Horn L, Branning A, Chatterton RT. Saliva as a medium for investigating intra-and interindividual differences in sex hormone levels in premenopausal women. Cancer Epidemiol Biomarkers Prev 2001;10(1):59–64.

[38] Sherman JJ, LeResche L, Mancl LA, Huggins K, Sage JC, Dworkin SF. Cyclic effects on experimental pain response in women with temporomandibular disorders. J Orofac Pain 2004;19(2):133–43.

[39] Bellem A, Meiyappan S, Romans S, Einstein G. Measuring estrogens and progestagens in humans: an overview of methods. Gender Med 2011;8(5):283–99. Available from: http://dx.doi.org/10.1016/j.genm.2011.07.001.

[40] Greenspan JD, Craft RM, LeResche L, Arendt-Nielsen L, Berkley KJ, Fillingim RB, et al. Studying sex and gender differences in pain and analgesia: a consensus report. Pain 2007;132: S26–45. Available from: http://dx.doi.org/10.1016/j.pain.2007.10.014.

[41] Quissell DO. Steroid hormone analysis in human saliva. Ann N Y Acad Sci 1993;694(1):143–5. Available from: http://dx.doi.org/10.1111/j.1749-6632.1993.tb18348.x.

[42] Lu Y-C, Bentley GR, Gann PH, Hodges KR, Chatterton RT. Salivary estradiol and progesterone levels in conception and nonconception cycles in women: evaluation of a new assay for salivary estradiol. Fertil Steril 1999;71(5):863–8. Available from: http://dx.doi.org/10.1016/S0015-0282(99)00093-X.

[43] Stumpf WE, Sar M. Steroid hormone target sites in the brain: the differential distribution of estrogin, progestin, androgen and glucocorticosteroid. J Steroid Biochem 1976;7(11–12):1163–70. Available from: http://dx.doi.org/10.1016/0022-4731(76)90050-9.

[44] Prewitt AK, Wilson ME. Changes in estrogen receptor-alpha mRNA in the mouse cortex during development. Brain Res 2007;1134(1):62–9. Available from: http://dx.doi.org/10.1016/j.brainres.2006.11.069.

[45] Shughrue PJ, Stumpf WE, MacLusky NJ, Zielinski JE, Hochberg RB. Developmental changes in estrogen receptors in mouse cerebral cortex between birth and postweaning: studied by autoradiography with 11 beta-methoxy-16 alpha-[125I]iodoestradiol. Endocrinology 1990;126(2):1112–24. Available from: http://dx.doi.org/10.1210/endo-126-2-1112.

[46] Fenzi F, Rizzzuto N. Estrogen receptors localization in the spinal trigeminal nucleus: an immunohistochemical study in humans. Eur J Pain 2011;15(10):1002–7. Available from: http://dx.doi.org/10.1016/j.ejpain.2011.05.003.

[47] Ravizza T, Galanopoulou AS, Veliskova J, Moshe SL. Sex differences in androgen and estrogen receptor expression in rat substantia nigra during development: an immunohistochemical study. Neuroscience 2002;115(3):685–96. Available from: http://dx.doi.org/10.1016/S0306-4522(02)00491-8.

[48] Karolczak M, Beyer C. Developmental sex differences in estrogen receptor-beta mRNA expression in the mouse hypothalamus/preoptic region. Neuroendocrinology 1998;68(4):229–34. Available from: http://dx.doi.org/10.1159/000054370.

[49] Smith S, Ni Gabhann J, McCarthy E, Coffey B, Mahony R, Byrne JC, et al. Estrogen receptor alpha regulates tripartite motif-containing protein 21 expression, contributing to dysregulated cytokine production in systemic lupus erythematosus. Arthritis Rheumatol 2014;66(1):163–72. Available from: http://dx.doi.org/10.1002/art.38187.

[50] Silkaitis K, Lemos B. Sex-biased chromatin and regulatory cross-talk between sex chromosomes, autosomes, and mitochondria. Biol Sex Differ 2014;5(1):2. Available from: http://dx.doi.org/10.1186/2042-6410-5-2.

[51] van Nas A, GuhaThakurta D, Wang SS, Yehya N, Horvath S, Zhang B, et al. Elucidating the role of gonadal hormones in sexually dimorphic gene coexpression networks. Endocrinology 2009;150(3):1235–49. Available from: http://dx.doi.org/10.1210/en.2008-0563.

[52] Weiss LA, Pan L, Abney M, Ober C. The sex-specific genetic architecture of quantitative traits in humans. Nat Genet 2006;38(2):218–22. Available from: http://dx.doi.org/10.1038/ng1726.

[53] Rinn JL, Snyder M. Sexual dimorphism in mammalian gene expression. Trends Genet 2005;21(5):298–305. Available from: http://dx.doi.org/10.1016/j.tig.2005.03.005.

[54] Grisel JE, Allen S, Nemmani KV, Fee JR, Carliss R. The influence of dextromethorphan on morphine analgesia in Swiss Webster mice is sex-specific. Pharmacol Biochem Behav 2005;81(1):131–8. Available from: http://dx.doi.org/10.1016/j.pbb.2005.03.001.

[55] Brynhildsen JO, Bjors E, Skarsgard C, Hammar ML. Is hormone replacement therapy a risk factor for low back pain among postmenopausal women? Spine (Phila Pa 1976) 1998;23(7):809–13. Available from: http://dx.doi.org/10.1097/00007632-199804010-00014.

[56] LeResche L, Saunders K, Von Korff MR, Barlow W, Dworkin SF. Use of exogenous hormones and risk of temporomandibular disorder pain. Pain 1997;69(1–2):153–60. Available from: http://dx.doi.org/10.1016/S0304-3959(96)03230-7.

[57] Ferry S, Hannaford P, Warskyj M, Lewis M, Croft P. Carpal tunnel syndrome: a nested case-control study of risk factors in women. Am J Epidemiol 2000;151(6):566–74. Available from: http://dx.doi.org/10.1093/oxfordjournals.aje.a010244.

[58] Brake T, Lambert PF. Estrogen contributes to the onset, persistence, and malignant progression of cervical cancer in a human papillomavirus-transgenic mouse model. Proc Natl Acad Sci USA 2005;102(7):2490–5. Available from: http://dx.doi.org/10.1073/pnas.0409883102.

[59] Aloisi AM, Bachiocco V, Costantino A, Stefani R, Ceccarelli I, Bertaccini A, et al. Cross-sex hormone administration changes pain in transsexual women and men. Pain 2007;132(Suppl. 1):S60–67. Available from: http://dx.doi.org/10.1016/j.pain.2007.02.006.

[60] Coffee AL, Sulak PJ, Kuehl TJ. Long-term assessment of symptomatology and satisfaction of an extended oral contraceptive regimen. Contraception 2007;75(6):444–9. Available from: http://dx.doi.org/10.1016/j.contraception.2007.01.014.

[61] Dao T, Knight K, Ton-That V. Modulation of myofascial pain by the reproductive hormones: a preliminary report. J Prosthet Dent 1998;79(6):663–70 http://dx.doi.org/10.1016/S0022-3913(98)70073-3

[62] Grady D, Rubin SM, Petitti DB, Fox CS, Black D, Ettinger B, et al. Hormone therapy to prevent disease and prolong life in postmenopausal women. Ann Intern Med 1992;117(12):1016–37 http://dx.doi.org/10.1016/0020-7292(93)90679-Q

[63] Purdie D, Green A, Bain C, Siskind V, Ward B, Hacker N, et al. Reproductive and other factors and risk of epithelial ovarian cancer: an Australian case-control study. Int J Cancer 1995;62(6):678–84. Available from: http://dx.doi.org/10.1002/ijc.2910620606.

[64] Cahill L. Why sex matters for neuroscience. Nat Rev Neurosci 2006;7(6):477–84. Available from: http://dx.doi.org/10.1038/nrn1909.

[65] Cosgrove KP, Mazure CM, Staley JK. Evolving knowledge of sex differences in brain structure, function, and chemistry. Biol Psychiatry 2007;62(8):847–55. Available from: http://dx.doi.org/10.1016/j.biopsych.2007.03.001.

[66] Goldstein JM, Seidman LJ, Horton NJ, Makris N, Kennedy DN, Caviness VS, et al. Normal sexual dimorphism of the adult human brain assessed by in vivo magnetic resonance imaging. Cereb Cortex 2001;11(6):490–7. Available from: http://dx.doi.org/10.1093/cercor/11.6.490.

[67] Gur RC, Turetsky BI, Matsui M, Yan M, Bilker W, Hughett P, et al. Sex differences in brain gray and white matter in healthy young adults: correlations with cognitive performance. J Neurosci 1999;19(10):4065–72.

[68] Luders E, Narr KL, Thompson PM, Rex DE, Woods RP, DeLuca H, et al. Gender effects on cortical thickness and the influence of scaling. Hum Brain Mapp 2006;27(4):314–24. Available from: http://dx.doi.org/10.1002/hbm.20187.

[69] Gong G, Rosa-Neto P, Carbonell F, Chen ZJ, He Y, Evans AC. Age-and gender-related differences in the cortical anatomical network. J Neurosci 2009;29(50):15684–93. Available from: http://dx.doi.org/10.1523/JNEUROSCI.2308-09.2009.

[70] Ingalhalikar M, Smith A, Parker D, Satterthwaite TD, Elliott MA, Ruparel K, et al. Sex differences in the structural connectome of the human brain. Proc Natl Acad Sci 2014;111(2):823–8. Available from: http://dx.doi.org/10.1073/pnas.1316909110.

[71] Kilpatrick LA, Zald DH, Pardo JV, Cahill LF. Sex-related differences in amygdala functional connectivity during resting conditions. Neuroimage 2006;30(2):452–61. Available from: http://dx.doi.org/10.1016/j.neuroimage.2005.09.065.

[72] Zubieta J-K, Dannals RF, Frost JJ. Gender and age influences on human brain mu-opioid receptor binding measured by PET. Am J Psychiatry 2014;156(6):842–8. Available from: http://dx.doi.org/10.1176/ajp.156.6.842.

[73] Nishizawa S, Benkelfat C, Young SN, Leyton M, Mzengeza S, de Montigny C, et al. Differences between males and females in rates of serotonin synthesis in human brain. Proc Natl Acad Sci 1997;94(10):5308–13. Available from: http://dx.doi.org/10.1073/pnas.94.10.5308.

[74] Wrase J, Klein S, Gruesser SM, Hermann D, Flor H, Mann K, et al. Gender differences in the processing of standardized emotional visual stimuli in humans: a functional magnetic resonance imaging study. Neurosci Lett 2003;348(1):41–5. Available from: http://dx.doi.org/10.1016/S0304-3940(03)00565-2.

[75] Naliboff BD, Berman S, Chang L, Derbyshire SWG, Suyenobu B, Vogt BA, et al. Sex-related differences in IBS patients: central processing of visceral stimuli. Gastroenterology 2003;124(7):1738–47. Available from: http://dx.doi.org/10.1016/S0016-5085(03)00400-1.

[76] Symms M, Jäger HR, Schmierer K, Yousry TA. A review of structural magnetic resonance neuroimaging. J Neurol Neurosurg Psychiatry 2004;75(9):1235–44. Available from: http://dx.doi.org/10.1136/jnnp.2003.032714.

[77] Whitwell JL. Voxel-based morphometry: an automated technique for assessing structural changes in the brain. J Neurosci 2009;29(31):9661–4. Available from: http://dx.doi.org/10.1523/jneurosci.2160-09.2009.

[78] Wright IC, McGuire PK, Poline JB, Travere JM, Murray RM, Frith CD, et al. A voxel-based method for the statistical analysis of gray and white matter density applied to schizophrenia. Neuroimage 1995;2(4):244–52. Available from: http://dx.doi.org/10.1006/nimg.1995.1032.

[79] Ashburner J, Friston KJ. Voxel-based morphometry—the methods. Neuroimage 2000;11(6 Pt 1):805–21. Available from: http://dx.doi.org/10.1006/nimg.2000.0582.

[80] Mechelli A, Price CJ, Friston KJ, Ashburner J. Voxel-based morphometry of the human brain: methods and applications. Curr Med Imaging Rev 2005;1(2):105–13. Available from: http://dx.doi.org/10.2174/1573405054038726.

[81] Basser PJ, Pierpaoli C. Microstructural and physiological features of tissues elucidated by quantitative-diffusion-tensor MRI. J Magn Reson B 1996;111(3):209–19. Available from: http://dx.doi.org/10.1006/jmrb.1996.0086.

[82] Alexander AL, Lee JE, Lazar M, Field AS. Diffusion tensor imaging of the brain. Neurotherapeutics 2007;4(3):316–29. Available from: http://dx.doi.org/10.1016/j.nurt.2007.05.011.

[83] Le Bihan D, Mangin JF, Poupon C, Clark CA, Pappata S, Molko N, et al. Diffusion tensor imaging: concepts and applications. J Magn Reson Imaging 2001;13(4):534–46. Available from: http://dx.doi.org/10.1002/jmri.1076.
[84] Pierpaoli C, Jezzard P, Basser PJ, Barnett A, Di Chiro G. Diffusion tensor MR imaging of the human brain. Radiology 1996;201(3):637–48. Available from: http://dx.doi.org/10.1148/radiology.201.3.8939209.
[85] Basser PJ, Mattiello J, Le Bihan D. MR diffusion tensor spectroscopy and imaging. Biophys J 1994;66(1):259–67. Available from: http://dx.doi.org/10.1016/S0006-3495(94)80775-1.
[86] Dong Q, Welsh RC, Chenevert TL, Carlos RC, Maly-Sundgren P, Gomez-Hassan DM, et al. Clinical applications of diffusion tensor imaging. J Magn Reson Imaging 2004;19(1):6–18. Available from: http://dx.doi.org/10.1002/jmri.10424.
[87] Bandettini PA. What's new in neuroimaging methods? Ann N Y Acad Sci 2009;1156(1):260–93. Available from: http://dx.doi.org/10.1111/j.1749-6632.2009.04420.x.
[88] Crosson B, Ford A, McGregor KM, Meinzer M, Cheshkov S, Li X, et al. Functional imaging and related techniques: an introduction for rehabilitation researchers. J Rehabil Res Dev 2010;47(2):vii–xxxiv. Available from: http://dx.doi.org/10.1682/JRRD.2010.02.0017.
[89] Gusnard DA, Raichle ME. Searching for a baseline: functional imaging and the resting human brain. Nat Rev Neurosci 2001;2(10):685–94. Available from: http://dx.doi.org/10.1038/35094500.
[90] Raichle ME. Behind the scenes of functional brain imaging: a historical and physiological perspective. Proc Natl Acad Sci USA 1998;95(3):765–72. Available from: http://dx.doi.org/10.1073/pnas.95.3.765.
[91] Wendel K, Väisänen O, Malmivuo J, Gencer NG, Vanrumste B, Durka P, et al. EEG/MEG source imaging: methods, challenges, and open issues. Comput Intell Neurosci 2009;2009:13. Available from: http://dx.doi.org/10.1155/2009/656092.
[92] Liu AK, Dale AM, Belliveau JW. Monte Carlo simulation studies of EEG and MEG localization accuracy. Hum Brain Mapp 2002;16(1):47–62. Available from: http://dx.doi.org/10.1002/hbm.10024.
[93] Malmivuo J, Suihko V, Eskola H. Sensitivity distributions of EEG and MEG measurements. IEEE Trans Biomed Eng 1997;44(3):196–208. Available from: http://dx.doi.org/10.1109/10.554766.
[94] Tamres LK, Janicki D, Helgeson VS. Sex differences in coping behavior: a meta-analytic review and an examination of relative coping. Pers Soc Psychol Rev 2002;6(1):2–30. Available from: http://dx.doi.org/10.1207/S15327957PSPR0601_1.
[95] Unruh AM, Ritchie J, Merskey H. Does gender affect appraisal of pain and pain coping strategies? Clin J Pain 1999;15(1):31–40. Available from: http://dx.doi.org/10.1097/00002508-199903000-00006.
[96] Brougham RR, Zail CM, Mendoza CM, Miller JR. Stress, sex differences, and coping strategies among college students. Curr Psychol 2009;28(2):85–97. Available from: http://dx.doi.org/10.1007/s12144-009-9047-0.
[97] Fillingim RB, King CD, Ribeiro-Dasilva MC, Rahim-Williams B, Riley III JL. Sex, gender, and pain: a review of recent clinical and experimental findings. J Pain 2009;10(5):447–85. Available from: http://dx.doi.org/10.1016/j.jpain.2008.12.001.
[98] Gentry LA, Chung JJ, Aung N, Keller S, Heinrich KM, Maddock JE. Gender differences in stress and coping among adults living in Hawaii. Californian J Health Promot 2007;5(2):89–102.
[99] Lindquist TL, Beilin LJ, Knuiman MW. Influence of lifestyle, coping, and job stress on blood pressure in men and women. Hypertension 1997;29(1):1–7. Available from: http://dx.doi.org/10.1161/01.HYP.29.1.1.
[100] Hines M. Gender development and the human brain. Annu Rev Neurosci 2011;34:69–88. Available from: http://dx.doi.org/10.1146/annurev-neuro-061010-113654.
[101] Bem SL. The measurement of psychological androgyny. J Couns Clin Psychol 1974;42:155–62. Available from: http://dx.doi.org/10.1037/h0036215.
[102] Spence JT, Helmreich R, Stapp J. Ratings of self and peers on sex role attributes and their relation to self-esteem and conceptions of masculinity and femininity. J Pers Soc Psychol 1975;32(1):29–39. Available from: http://dx.doi.org/10.1037/h0076857.
[103] Hoffman RM, Borders LD. Twenty-five years after the Bem Sex-Role Inventory: a reassessment and new issues regarding classification variability. Meas Eval Couns Dev 2001;34(1):39–55.
[104] Thomson NR, Zand DH. Analysis of the Children's Personal Attributes Questionnaire (Short Form) with a sample of African American adolescents. Sex Roles 2005;52(3–4):237–44. Available from: http://dx.doi.org/10.1007/s11199-005-1298-0.
[105] Carver LF, Vafaei A, Guerra R, Freire A, Phillips SP. Gender differences: examination of the 12-Item Bem Sex Role Inventory (BSRI-12) in an older Brazilian population. PLoS ONE 2013;8(10):e76356. Available from: http://dx.doi.org/10.1371/journal.pone.0076356.
[106] Eisler RM, Skidmore JR. Masculine gender role stress: scale development and component factors in the appraisal of stressful situations. Behav Modif 1987;11(2):123–36. Available from: http://dx.doi.org/10.1177/01454455870112001.
[107] Gillespie BL, Eisler RM. Development of the feminine gender role stress scale: a cognitive-behavioral measure of stress, appraisal, and coping for women. Behav Modif 1992;16(3):426–38. Available from: http://dx.doi.org/10.1177/01454455920163008.
[108] Tang CS-K, Lau BH-B. The assessment of gender role stress for Chinese. Sex Roles 1995;33(7–8):587–95. Available from: http://dx.doi.org/10.1007/BF01544682.
[109] Yarnitsky D. Conditioned pain modulation (the diffuse noxious inhibitory control-like effect): its relevance for acute and chronic pain states. Curr Opin Anaesthesiol 2010;23(5):611–15. Available from: http://dx.doi.org/10.1097/ACO.0b013e32833c348b.
[110] Mogil JS. Sex differences in pain and pain inhibition: multiple explanations of a controversial phenomenon. Nat Rev Neurosci 2012;13(12):859–66. Available from: http://dx.doi.org/10.1038/nrn3360.
[111] Bartley EJ, Fillingim RB. Sex differences in pain: a brief review of clinical and experimental findings. Br J Anaesth 2013;111(1):52–8. Available from: http://dx.doi.org/10.1093/bja/aet127.
[112] LeResche L. Gender considerations in the epidemiology of chronic pain. In: Crombie IK, editor. Epidemiology of pain. Seattle, WA: IASP Press; 1999. p. 43–52.
[113] Fowler SL, Rasinski HM, Geers AL, Helfer SG, France CR. Concept priming and pain: an experimental approach to understanding gender roles in sex-related pain differences. J Behav Med 2011;34(2):139–47. Available from: http://dx.doi.org/10.1007/s10865-010-9291-7.
[114] Robinson ME, Riley III JL, Myers CD, Papas RK, Wise EA, Waxenberg LB, et al. Gender role expectations of pain: relationship to sex differences in pain. J Pain 2001;2(5):251–7. Available from: http://dx.doi.org/10.1054/jpai.2001.24551.
[115] Robinson ME, Gagnon CM, Riley JL, Price DD. Altering gender role expectations: effects on pain tolerance, pain threshold, and pain ratings. J Pain 2003;4(5):284–8. Available from: http://dx.doi.org/10.1016/S1526-5900(03)00559-5.

[116] Wise EA, Price DD, Myers CD, Heft MW, Robinson ME. Gender role expectations of pain: relationship to experimental pain perception. Pain 2002;96:335–42. Available from: http://dx.doi.org/10.1016/S0304-3959(01)00473-0.

[117] Alabas OA, Tashani OA, Tabasam G, Johnson MI. Gender role affects experimental pain responses: a systematic review with meta-analysis. Eur J Pain 2012;16(9):1211–23. Available from: http://dx.doi.org/10.1002/j.1532-2149.2012.00121.x.

[118] Alabas OA, Tashani OA, Johnson MI. Gender role expectations of pain mediate sex differences in cold pain responses in healthy Libyans. Eur J Pain 2012;16(2):300–11. Available from: http://dx.doi.org/10.1016/j.ejpain.2011.05.012.

[119] Defrin R, Shramm L, Eli I. Gender role expectations of pain is associated with pain tolerance limit but not with pain threshold. Pain 2009;145(1–2):230–6. Available from: http://dx.doi.org/10.1016/j.pain.2009.06.028.

[120] McCullough LD, de Vries GJ, Miller VM, Becker JB, Sandberg K, McCarthy MM. NIH initiative to balance sex of animals in preclinical studies: generative questions to guide policy, implementation, and metrics. Biol Sex Differ 2014;5:15. Available from: http://dx.doi.org/10.1186/s13293-014-0015-5.

References for Strategies and Approaches for Studying Sex Differences in Physiology

[1] Plant TM, Zeleznik AJ. Knobil & Neill's physiology of reproduction. 4th ed. Academic Press; 2014.

[2] Beery AK, Zucker I. Sex bias in neuroscience and biomedical research. Neurosci Biobehav Rev 2010;35:565–72.

[3] Prendergast BJ, Onishi KG, Zucker I. Female mice liberated for inclusion in neuroscience and biomedical research. Neurosci Biobehav Rev 2014;40:1–5.

[4] Mogil JS, Chanda ML. The case for the inclusion of female subjects in basic science studies of pain. Pain 2005;117(1–2):1–5.

[5] Itoh Y, Arnold AP. Are females more variable than males in gene expression? Meta-analysis of microarray datasets. Biol Sex Differ 2015;6:18.

[6] Klein SL, Schiebinger L, Stefanick ML, et al. Opinion: sex inclusion in basic research drives discovery. Proc Natl Acad Sci USA 2015;112(17):5257–8.

[7] Mogil JS, Bailey AL. Sex and gender differences in pain and analgesia. Prog Brain Res 2010;186:141–57.

[8] Hill CA, Fitch RH. Sex differences in mechanisms and outcome of neonatal hypoxia-ischemia in rodent models: implications for sex-specific neuroprotection in clinical neonatal practice. Neurol Res Int 2012;2012:867531.

[9] Huang GZ, Woolley CS. Estradiol acutely suppresses inhibition in the hippocampus through a sex-specific endocannabinoid and mGluR-dependent mechanism. Neuron 2012;74(5):801–8.

[10] Amateau SK, McCarthy MM. Induction of PGE(2) by estradiol mediates developmental masculinization of sex behavior. Nat Neurosci 2004;7(6):643–50.

[11] Livernois AM, Graves JA, Waters PD. The origin and evolution of vertebrate sex chromosomes and dosage compensation. Heredity 2012;108(1):50–8.

[12] Arnold AP. The end of gonad-centric sex determination in mammals. Trends Genet 2012;28(2):55–61.

[13] Goodfellow PN, Lovell-Badge R. SRY and sex determination in mammals. Annu Rev Genet 1993;27:71–92.

[14] Harley V, Goodfellow P. The biochemical role of SRY in sex determination. Mol Reprod Dev 1994;39:184–93.

[15] Hiramatsu R, Matoba S, Kanai-Azuma M, et al. A critical time window of Sry action in gonadal sex determination in mice. Development 2009;136(1):129–38.

[16] Wallen K, Baum MJ. Masculinization and defeminization in altricial and precocial mammals: comparative aspects of steroid hormone action. In: Pfaff D, editor. Hormones brain and behavior. London, UK: Academic Press; 2002. p. 385–424.

[17] McCarthy MM, Arnold AP. Reframing sexual differentiation of the brain. Nat Neurosci 2011;14(6):677–83.

[18] Bianco SD. A potential mechanism for the sexual dimorphism in the onset of puberty and incidence of idiopathic central precocious puberty in children: sex-specific kisspeptin as an integrator of puberty signals. Front Endocrinol 2012;3:149.

[19] Ojeda SR, Lomniczi A, Sandau U, Matagne V. New concepts on the control of the onset of puberty. Endocr Dev 2010;17:44–51.

[20] Romeo RD, Richardson HN, Sisk CL. Puberty and the maturation of the male brain and sexual behavior: recasting a behavioral potential. Neurosci Biobehav Rev 2002;26(3):381–91.

[21] Sisk CL, Foster DL. The neural basis of puberty and adolescence. Nat Neurosci 2004;7:1040–7.

[22] Moore CL. Maternal contributions to the development of masculine sexual behavior in laboratory rats. Dev Psychobiol 1984;17(4):347–56.

[23] Edelmann MN, Demers CH, Auger AP. Maternal touch moderates sex differences in juvenile social play behavior. PloS ONE 2013;8(2):e57396.

[24] Edelmann MN, Auger AP. Epigenetic impact of simulated maternal grooming on estrogen receptor alpha within the developing amygdala. Brain Behav Immun 2011;25(7):1299–304.

[25] Bowers JM, Perez Pouchoulen M, Edwards NS, McCarthy MM. Foxp2 mediates sex differences in ultrasonic vocalization by rat pups and directs order of maternal retrieval. J Neurosci 2012; [in press]

[26] Markle JG, Frank DN, Mortin-Toth S, et al. Sex differences in the gut microbiome drive hormone-dependent regulation of autoimmunity. Science 2013;339(6123):1084–8.

[27] Maney DL. Just like a circus: the public consumption of sex differences. Curr Top Behav Neurosci 2015;19:279–96.

[28] Agrawal AK, Shapiro BH. Intrinsic signals in the sexually dimorphic circulating growth hormone profiles of the rat. Mol Cell Endocrinol 2001;173:167–81.

[29] Jansson JO, Ekberg S, Isaksson OG, Eden S. Influence of gonadal steroids on age- and sex-related secretory patterns of growth hormone in the rat. Endocrinology 1984;114(4):1287–94.

[30] Perrot-Sinal TS. Sex differences in performance in the Morris water maze and the effects of initial nonstationary hidden platform training. Behav Neurosci 1996;110:1309–20.

[31] Insel TR, Hulihan TJ. A gender-specific mechanism for pair bonding: oxytocin and partner preference formation in monogamous voles. Behav Neurosci 1995;109(4):782–9.

[32] De Vries GJ, Villalba C. Brain sexual dimorphism and sex differences in parental and other social behaviors. Ann N Y Acad Sci 1997;807:273–86.

[33] Lonstein JS, De Vries GJ. Sex differences in the parental behavior of rodents. Neurosci Biobehav Rev 2000;24(6):669–86.

[34] De Vries GJ. Minireview: sex differences in adult and developing brains: compensation, compensation, compensation. Endocrinology 2004;145(3):1063–8.

[35] McCarthy MM, Arnold AP, Ball GF, Blaustein JD, De Vries GJ. Sex differences in the brain: the not so inconvenient truth. J Neurosci 2012;32(7):2241–7.

[36] Shors TJ, Chua C, Falduto J. Sex differences and opposite effects of stress on dendritic spine density in the male versus female hippocampus. J Neurosci 2001;21(16):6292–7.

Index

Note: Page numbers followed by "*b*," "*f*," and "*t*" refer to boxes, figures, and tables, respectively.

Printed in the United States
by Baker & Taylor Publisher Services